CHARLES E. MERRILL PUBLISHING COMPANY
A BELL & HOWELL COMPANY
COLUMBUS TORONTO LONDON SYDNEY

THOMAS L. FLOYD

2ND EDITION

PRINCIPLES OF ELECTRIC CIRCUITS

In memory of my father, V. R. Floyd

Published by Charles E. Merrill Publishing Co.
A Bell & Howell Company
Columbus, Ohio 43216

This book was set in Times Roman.
Production Editor: Rex Davidson.
Cover Design Coordination: Cathy Watterson.
Cover photo by Jeff Perkell/Joan Kramer and Associates.

Library of Congress Catalog Card Number: 84–62242
International Standard Book Number: 0–675–20402–X
Printed in the United States of America
    3 4 5 6 7 8 9 10—88 87

Principles of Electric Circuits, 2nd Edition, is a book that students can understand. The writing style, organization, format, and special features were developed with the student in mind. All of the popular features of the first edition have been retained. Improvements and additions have been made based on extensive surveys and reviews.

In addition to more illustrations (over 1100), more examples, more problems and self-test exercises, and more troubleshooting coverage, the new features in this edition are:

1. Chapter objectives.

2. Optional computer analysis sections.

3. Sectionalized problem sets.

4. Advanced problem sections.

5. Improved section reviews.

6. Complex numbers throughout ac analysis.

7. Chapter on theorems in ac analysis.

8. Chapter on polyphase systems.

9. Separate dc and ac instruments chapters.

10. Larger type size.

11. Expanded and improved accompanying lab manual.

ORGANIZATION

This book has been carefully organized to provide improved coverage and flexibility in terms of level, content, and topic sequence. The dc coverage remains essentially unchanged from the first edition with the exception of a new Chapter 10 on dc measurements. The ac coverage has been significantly reorganized to provide better treatment and more flexibility. The organizational changes are:

1. The computer analysis section is placed last in each of the 15 chapters in which they appear, so that they may be omitted easily.

2. The Advanced Problems sections provide additional assignments for the stronger student. They can be omitted at the discretion of the instructor.

3. Capacitors and inductors are covered in consecutive chapters (12 and 13) and, by postponing the coverage of one section in each chapter, these topics can be introduced as part of the dc material.

4. The sequence of chapters 11 through 15 permits several variations in the presentational approach for ac analysis. These are shown in detail under Suggestions for Use.

5. Complex numbers are introduced in Chapter 11, at the beginning of the ac coverage, and then used throughout the ac analysis chapters. This approach strengthens the ac analysis and, because of the polar/rectangular conversions available on scientific calculators, actually makes the mathematical aspects of the ac analysis coverage easier for the student.

PREFACE

6. *RLC* circuits and resonance immediately follow *RC* and *RL* circuit analysis for continuity.

7. ac circuit analysis has been expanded to include the use of theorems in a new chapter (17), which is the ac equivalent of Chapter 8.

8. A new chapter on polyphase systems (20) can be used in those programs with an emphasis on electrical technology or power applications. It can be omitted if desired.

9. A strong treatment of pulse responses is provided. The coverage of pulse response, transients, and nonsinusoidal waveforms is now distributed as follows: Non-sinusoidal Wave Forms (sections 11–10 and 11–11), Capacitors in dc Circuits (Section 12–3), Inductors in dc Circuits (Section 13–4), Pulse Response of *RC* Circuits (Section 14–9), Response of *RC* Circuits to Periodic Pulse Wave Forms (Section 14–10), and Pulse Response of *RL* Circuits (Section 15–9). The sections in chapters 14 and 15 may be omitted if desired.

10. The coverage of dc and ac instruments has been separated into two chapters (10 and 21).

FORMAT

The widely praised and award winning format of the first edition has been retained with only minor adjustments.

Pre-chapter Overview:

1. Chapter objectives (new in this edition) are included in the introduction.

2. Section titles are listed on the chapter title page.

Within Chapter:

1. Larger typeface is used throughout (including problem sets).

2. Over 1100 illustrations (an average of over 50 per chapter).

3. Numerous boxed examples are clearly separated from text. This frees the student from having to read lengthy passages.

4. The necessity of turning pages to relate an illustration to the text or to an example has been minimized.

5. Simple words, short sentences, and numerous subheadings contribute to the clarity of the writing style.

6. At the end of each section are review questions and problems (many new to this edition) with answers at the end of the chapter.

End of Chapter:

1. Fifteen chapters have optional computer analysis sections with programs in BASIC to illustrate computer applications in circuit analysis and problem solving. The program listings were produced on a TRS-80, but minor substitutions allow these programs to run on Apple II, IBM PC, or other popular computers. These sections can be omitted easily without affecting any other material in the book.

2. Each chapter has a list of important formulas followed by a summary.

3. There are three end-of-chapter problem sets. The self-test problem set has fully worked-out solutions (not just answers) in back of book. The traditional problem set is now organized by section for easy assignment (with answers for odd-numbered

problems in back of book and all solutions in Solutions Manual). A set of advanced problems, new to this edition, is designed to challenge the stronger student. In general, the problem sets provide an upward progression in level of difficulty.

These pedagogical features lengthen the text, but they make it easier to follow, and the coverage becomes more effective. I think you will find that your students can cover more material in the same length of time with a greater understanding in this book than in texts that may have fewer pages but lack some of the features and the carefully planned format of this text.

APPROACH

This text uses conventional current throughout. A different version, *Electric Circuits, Electron Flow Version* (#20037-7) is available for those who prefer the electron flow approach.

SUGGESTIONS FOR USE `

This text is very flexible. First, some suggestions for variations in the sequence of presentation of the material in chapters 11 through 15:

1. *Straightforward approach:* ac Characteristics and Analysis (Chapter 11), The Capacitor (Chapter 12), The Inductor (Chapter 13), *RC* Circuit Analysis (Chapter 14), and *RL* Circuit Analysis (Chapter 15).

2. *Introduction of capacitors and inductors in the dc coverage:* End dc coverage with: The Capacitor (all of Chapter 12 except Section 12–6), The Inductor (all of Chapter 13, except Section 13–7). Begin ac coverage with: ac Characteristics and Analysis (Chapter 11), Capacitors in ac Circuits (Section 12–6), Inductors in ac Circuits (Section 13–7), *RC* Circuit Analysis (Chapter 14), and *RL* Circuit Analysis (Chapter 15).

3. *Consecutive coverage of capacitive topics followed by consecutive coverage of inductive topics:* ac Characteristics and Analysis (Chapter 11), The Capacitor (Chapter 12), *RC* Circuit Analysis (Chapter 14), The Inductor (Chapter 13), and *RL* Circuit Analysis (Chapter 15).

4. *Integrated coverage of capacitors and inductors:* ac Characteristics and Analysis (Chapter 11), alternate corresponding sections in chapters 12 and 13, *RC* Circuit Analysis (Chapter 14), and *RL* Circuit Analysis (Chapter 15).

5. *Reverse coverage:* Inductive topics before capacitive in any of formats just listed (i.e., chapters 13, 12, 15, 14 or chapters 13, 15, 12, 14).

The dc and ac instruments chapters (10 and 21) are placed at the end of the dc and ac coverages, respectively. The material in these two chapters may be introduced at any point in the coverage to support the accompanying laboratory work.

Suggestions for adapting this text to fit most program requirements are as follows:

1. Traditional two-semester or two-quarter dc/ac course:

> Option 1:
>> First term: Chapters 1 through 10.
>> Second term: Chapters 11 through 21.

> Option 2:
>> First term: Chapters 1 through 10, plus capacitance and inductance (chapters 12 and 13 with sections 12–6 and 13–7 omitted).
>> Second term: Chapter 11, Section 12–6, Section 13–7, and chapters 14 through 21.

The following topics can be omitted without affecting the flow of material. This should be done based on program emphasis and time considerations.

(a) Computer analysis sections.

(b) Troubleshooting sections.

(c) dc measurements (Chapter 10).

(d) Pulse response (sections 14–9, 14–10, and 15–9).

(e) Polyphase (Chapter 20).

(f) ac measurements (Chapter 21).

2. One-term dc/ac course:

Chapters 1–16 with selected omission of topics just listed and light treatment of other selected topics based on instructor preference.

The large number of illustrations, boxed examples, and other pedagogical features as well as the student-oriented writing style make this text easier to get through than the number of pages might suggest. With careful planning, use in a one-term course is very feasible.

3. Three-term dc/ac course:

Three terms, of course, allow thorough coverage of all topics. A suggested approach is chapters 1 through 8 in first term, chapters 9 through 15 in second term, and chapters 16 through 21 in the third term.

SUPPLEMENTS AND FOLLOW-UP MATERIALS

A laboratory manual, *Experiments in Electric Circuits,* 2/E by Brian Stanley (#20403–8) has been developed to accompany this text. It contains 25 dc experiments and 25 ac experiments which are cross-referenced to sectional readings from this text. Like the text, the lab manual has been extensively revised based on user feedback. Also, it has its own Solutions Manual.

An Instructor's Manual for this text provides completely worked-out solutions to all end-of-chapter problems. Odd and even solutions are separated so that the instructor can duplicate solutions to odd problems (which have answers in back of text) for student handouts without revealing the even solutions.

A short paperback, *Essentials of Electronic Devices* (#20062–8), using electron flow, is available as a brief introduction to semiconductor devices at the end of the dc/ac course.

Also, *Electronic Devices* (#20157–8) is available for a comprehensive follow-up text in a course on electronic devices and circuits. This text is designed to follow *Principles of Electric Circuits,* using the same format and approach and with a thorough coverage of discrete semiconductor devices and circuits through op amps and other linear ICs. It has an accompanying lab manual by Howard Berlin (#20234–5).

ACKNOWLEDGMENTS

Many people contributed to this revision of *Principles of Electric Circuits.* I would like to express my appreciation to the following reviewers: Ernest G. Robinson, Jr., Houston Community College; Ulrich Zeisler, Utah Technical College @ Salt Lake; Timothy J. McIntyre, Southwestern College (California); Brian H. Stanley, Foothill College (California); P. Erik Liimatta, Anne Arundel Community College (Maryland); Richard T. Burchell, Riverside City College (California); Robert Blodgett, DeVry Institute of Technology-Chicago; Ed Parsons, Golden West College (California); Salvatore J. Levanti, Queensborough Community College (New York); James T. Humphries, Southern Illinois University-Carbondale, School of Technical Careers.

While space does not permit acknowledgment of everyone who has called or written with suggestions for this new edition, I do wish to thank the following: Jerry Cox, Robert Beck, Ole Thompson, Richard Falley, Harvey Salz, George Hendrickson, John Burns, Harold Martin, Kenneth

R. Washburn, Glenn Jackson, Peter Zawasky, Richard Eichenlaub, Richard Brongel, Dave Lalond, Gordon F. Theisz, Dennis Dolan, Leo E. Keiser, Victor Michael, Sandra J. Lloyd, Michael Charek, Melvin Klahr, and Dee Bailey.

Also, my thanks to Roy Ray and David Leitch for their thorough job of proofreading the manuscript for accuracy and to Chris Conty, Rex Davidson, Jo Ellen Gohr, JoAnne Weaver, and the other people at Charles E. Merrill Publishing for the fine work in producing this textbook.

A well-deserved expression of appreciation goes to my wife, Sheila, for her total support during the revision of this book and for the many hours she has spent assisting me in putting this product together.

Every effort has been made to ensure a clear, accurate, and highly effective text, and I hope that it meets the needs of those who use it—educators and students alike. Your comments and suggestions are always welcome.

Thomas L. Floyd
January, 1985

**MERRILL'S INTERNATIONAL SERIES IN
ELECTRICAL AND ELECTRONICS TECHNOLOGY**

BATESON	Introduction to Control System Technology, Second Edition, 8255–2
BOYLESTAD	Introductory Circuit Analysis, Fourth Edition, 9938–2
	Student Guide to Accompany Introductory Circuit Analysis, Fourth Edition, 9856–4
BOYLESTAD/KOUSOUROU	Experiments in Circuit Analysis, Fourth Edition, 9858–0
BREY	Microprocessor/Hardware Interfacing and Applications, 20158–6
FLOYD	Digital Fundamentals, Second Edition, 9876–9
	Electronic Devices, 20157–8
	Essentials of Electronic Devices, 20062–8
	Principles of Electric Circuits, Second Edition, 20402–X
	Electric Circuits: Electron Flow Version, 20037–7
STANLEY, B. H.	Experiments in Electric Circuits, Second Edition, 20403–8
BERLIN	Experiments in Electronic Devices, 20234–5
GAONKAR	Microprocessor Architecture, Programming and Applications with the 8085/8080A, 20159–4
LAMIT/LLOYD	Drafting for Electronics, 20200–0
LAMIT/WAHLER/HIGGINS	Workbook in Drafting for Electronics, 20417–8
NASHELSKY/BOYLESTAD	BASIC Applied to Circuit Analysis, 20161–6
ROSENBLATT/FRIEDMAN	Direct and Alternating Current Machinery, Second Edition, 20160–8
SCHWARTZ	Survey of Electronics, Third Edition, 20162–4
SEIDMAN/WAINTRAUB	Electronics: Devices, Discrete and Integrated Circuits, 8494–6
STANLEY, W. D.	Operational Amplifiers with Linear Integrated Circuits, 20090–3
TOCCI	Fundamentals of Electronic Devices, Third Edition, 9887–4
	Electronic Devices, Third Edition, Conventional Flow Version, 20063–6
	Fundamentals of Pulse and Digital Circuits, Third Edition, 20033–4
	Introduction to Electric Circuit Analysis, Second Edition, 20002–4
WARD	Applied Digital Electronics, 9925–0
YOUNG	Electronic Communication Techniques, 20202–7

CONTENTS

This chapter briefly examines the history of electricity and electronics and discusses some of the many areas of application. Also, to aid you throughout the book, the basics of scientific notation and metric prefixes are covered, along with the quantitites and units commonly used in electronics.

The use of the computer as an aid in circuit analysis is introduced as an optional topic. Computer programs are inserted at the end of many chapters to illustrate examples of computer-aided analysis in specific areas.

In this chapter, you will learn:

1. The history of important developments in the fields of electricity and electronics.
2. Some of the important areas in which electronics technology is applied.
3. The electrical quantities and their units.
4. How to use scientific notation.
5. The metric prefixes and how to use them.
6. The essentials of the BASIC computer language.

1

INTRODUCTION

1—1 One of the first important discoveries about static electricity is attributed to William Gilbert (1540–1603). Gilbert was an English physician who, in a book published in 1600, described how amber differs from magnetic loadstones in its attraction of certain materials. He found that when amber was rubbed with a cloth, it attracted only lightweight objects, whereas loadstones attracted only iron. Gilbert also discovered that other substances, such as sulfur, glass, and resin, behave like amber. He used the Latin word *elektron* for amber and originated the word *electrica* for the other substances that acted similarly to amber. The word *electricity* was used for the first time by Sir Thomas Browne (1605–82), an English physician.

Another Englishman, Stephen Gray (1696–1736), discovered that some substances conduct electricity and some do not. Following Gray's lead, a Frenchman named Charles du Fay experimented with the conduction of electricity. These experiments led him to believe that there were two kinds of electricity. He called one type *vitreous electricity* and the other type *resinous electricity*. He found that objects charged with vitreous electricity repelled each other and those charged with resinous electricity attracted each other. It is known today that two types of electrical *charge* do exist. They are called *positive* and *negative*.

Benjamin Franklin (1706–90) conducted studies in electricity in the mid-1700s. He theorized that electricity consisted of a single *fluid,* and he was the first to use the terms *positive* and *negative*. In his famous kite experiment, Franklin showed that lightning is electricity.

Charles Augustin de Coulomb (1736–1806), a French physicist, in 1785 proposed the laws that govern the attraction and repulsion between electrically charged bodies. Today, the unit of electrical charge is called the *coulomb*.

Luigi Galvani (1737–98) experimented with current electricity in 1786. Galvani was a professor of anatomy at the University of Bologna in Italy. Electrical current was once known as *galvanism* in his honor.

In 1800, Alessandro Volta (1745–1827), an Italian professor of physics, discovered that the chemical action between moisture and two different metals produced electricity. Volta constructed the first battery, using copper and zinc plates separated by paper that had been moistened with a salt solution. This battery, called the *voltaic pile,* was the first source of steady electric current. Today, the unit of electrical potential energy is called the *volt* in honor of Volta.

A Danish scientist, Hans Christian Oersted (1777–1851), is credited with the discovery of electromagnetism, in 1820. He found that electrical current flowing through a wire caused the needle of a compass to move. This finding showed that a magnetic field exists around a current-carrying conductor and that the field is produced by the current.

The modern unit of electrical current is the *ampere* (also called *amp*) in honor of the French physicist André Ampère (1775–1836). In 1820, Ampère measured the magnetic effect of an electrical current. He found that two wires carrying current can attract and repel each other, just as magnets can. By 1822, Ampère had developed the fundamental laws that are basic to the study of electricity.

One of the most well known and widely used laws in electrical circuits today is *Ohm's law*. It was formulated by Georg Simon Ohm (1789–1854), a German teacher, in 1826. Ohm's law gives us the relationship among the three important electrical quantities of resistance, voltage, and current.

Although it was Oersted who discovered electromagnetism, it was Michael Faraday (1791–1867) who carried the study further. Faraday was an English physicist who believed that if electricity could produce magnetic effects, then magnetism could produce electricity. In 1831 he found that a moving magnet caused an electric current in a coil of wire placed within the field of the magnet. This effect, known today as *electromagnetic induction,* is the basic principle of electric generators and transformers.

Joseph Henry (1797–1878), an American physicist, independently discovered the same principle in 1831, and it is in his honor that the unit of inductance is called the *henry*. The unit of capacitance, the *farad,* is named in honor of Michael Faraday.

In the 1860s, James Clerk Maxwell (1831–79), a Scottish physicist, produced a set of mathematical equations that expressed the laws governing electricity and magnetism. These are known as *Maxwell's equations*. Maxwell also predicted that electromagnetic waves (radio waves) that travel at the speed of light in space could be produced.

It was left to Heinrich Rudolph Hertz (1857–1894), a German physicist, to actually produce these waves that Maxwell predicted. Hertz performed this work in the late 1880s. Today, the unit of frequency is called the *hertz*.

The Beginning of Electronics

The early experiments in electronics involved electric currents flowing in glass tubes. One of the first to conduct such experiments was a German named Heinrich Geissler (1814–79). Geissler removed most of the air from a glass tube and found that the tube glowed when an electrical potential was placed across it.

Around 1878, Sir William Crookes (1832–1919), a British scientist, experimented with tubes similar to those of Geissler. In his experiments, Crooke found that the current flowing in the tubes seemed to consist of particles.

Thomas Edison (1847–1931), experimenting with the carbon-filament light bulb that he had invented, made another important finding. He inserted a small metal plate in the bulb. When the plate was positively charged, a current flowed from the filament to the plate. This device was the first *thermionic diode*. Edison patented it but never used it.

The electron was discovered in the 1890s. The French physicist Jean Baptiste Perrin (1870–1942) demonstrated that the current in a vacuum tube consists of negatively charged particles. Some of the properties of these particles were measured by Sir Joseph Thomson (1856–1940), a British physicist, in experiments he performed between 1895 and 1897. These negatively charged particles later became known as *electrons*. The charge on the electron was accurately measured by an American physicist, Robert A. Millikan (1868–1953), in 1909. As a result of these discoveries, electrons could be controlled, and the electronic age was ushered in.

Putting the Electron to Work

A vacuum tube that allowed electrical current to flow in only one direction was constructed in 1904 by John A. Fleming, a British scientist. The tube was used to detect electromagnetic waves. Called the *Fleming valve,* it was the forerunner of the more recent vacuum diode tubes.

Major progress in electronics, however, awaited the development of a device that could boost, or *amplify,* a weak electromagnetic wave or radio signal. This device was the *audion,* patented in 1907 by Lee de Forest, an American. It was a triode vacuum tube capable of amplifying small electrical signals.

Two other Americans, Harold Arnold and Irving Langmuir, made great improvements in the triode tube between 1912 and 1914. About the same time, de Forest and Edwin Armstrong, an electrical engineer, used the triode tube in an *oscillator circuit.* In 1914, the triode was incorporated in the telephone system and made the transcontinental telephone network possible.

The tetrode tube was invented in 1916 by Walter Schottky, a German. The tetrode, along with the pentode (invented in 1926 by Tellegen, a Dutch engineer), provided great improvements over the triode. The first television picture tube, called the *kinescope,* was developed in the 1920s by Vladimir Zworykin, an American researcher.

During World War II, several types of microwave tubes were developed that made possible modern microwave radar and other communications systems. In 1939, the *magnetron* was invented in Britain by Henry Boot and John Randall. In the same year, the *klystron* microwave tube was developed by two Americans, Russell Varian and his brother Sigurd Varian. The *traveling-wave* tube was invented in 1943 by Rudolf Komphner, an Austrian-American.

The Computer

The computer probably has had more impact on modern technology than any other single type of electronic system. The first electronic digital computer was completed in 1946 at the University of Pennsylvania. It was called the Electronic Numerical Integrator and Computer (ENIAC). One of the most significant developments in computers was the *stored program* concept, developed in the 1940s by John von Neumann, an American mathematician.

Solid State Electronics

The crystal detectors used in the early radios were the forerunners of modern solid state devices. However, the era of solid state electronics began with the invention of the *transistor* in 1947 at Bell Labs. The inventors were Walter Brattain, John Bardeen, and William Shockley. Figure 1–1 shows these three men, along with the notebook entry describing the historic discovery.

In the early 1960s, the integrated circuit was developed. It incorporated many transistors and other components on a single small *chip* of semiconductor material. Integrated circuit technology continues to be developed and improved, allowing more complex circuits to be built on smaller chips. The introduction of the microprocessor in the early 1970s created another electronics revolution: the

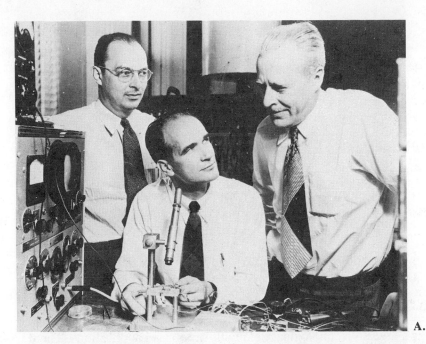

A.

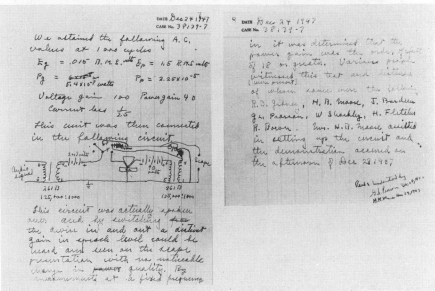

B.

FIGURE 1–1   A. *Nobel Prize winners Drs. John Bardeen, William Shockley, and Walter Brattain, shown left to right, with apparatus used in their first investigations that led to the invention of the transistor. The trio received the 1956 Nobel Physics award for their invention of the transistor, which was announced by Bell Laboratories in 1948.*   B. *The laboratory notebook entry of scientist Walter H. Brattain recorded the events of December 23, 1947, when the transistor effect was discovered at Bell Telephone Laboratories. The notebook entry describes the event and adds, "This circuit was actually spoken over and by switching the device in and out a distinct gain in speech level could be heard and seen on the scope presentation with no noticeable change in quality." (A and B, courtesy of Bell Laboratories)*

5

entire processing portion of a computer placed on a single, small, silicon chip. Continued development brought about complete computers on a single chip by the late 1970s.

Review for 1-1

1. Who developed the first battery?
2. The unit of what electrical quantity is named after André Ampère?
3. What contribution did Georg Simon Ohm make to the study of electricity?
4. In what year was the transistor invented?
5. What major development followed invention of the transistor?

APPLICATIONS OF ELECTRONICS

1-2

Electronics is a diverse technological field with very broad applications. There is hardly an area of human endeavor that is not dependent to some extent on electronics. Some of the applications are discussed here in a general way to give you an idea of the scope of the field.

Computers

One of the most important electronic systems is the digital computer; its applications are broad and diverse. For example, computers have applications in business for record keeping, accounting, payrolls, inventory control, market analysis, and statistics, to name but a few.

Scientific fields utilize the computer to process huge amounts of data and to perform complex and lengthy calculations. In industry, the computer is used for controlling and monitoring intricate manufacturing processes. Communications, navigation, medical, military, and home uses are a few of the other areas in which the computer is used extensively.

The computer's success is based on its ability to perform mathematical operations extremely fast and to process and store large amounts of information.

Computers vary in complexity and capability, ranging from very large systems with vast capabilities down to a computer on a chip with much more limited performance. Figure 1–2 shows some typical computers of varying sizes.

Communications Systems

Communications electronics encompasses a wide range of specialized fields. Included are space and satellite communications, commercial radio and television, citizens' band and amateur radio, data communications, navigation systems, radar, telephone systems, military applications, and specialized radio applications such as police, aircraft, and so on. Computers are used to a great extent in many communications systems. Figure 1–3 (page 8) shows a telephone switching system as an example of electronic communications.

A.

B.

C.

FIGURE 1–2 **A.** *TRS-80 Model II Microcomputer System. (Courtesy of Radio Shack)*
B. *PDP-11V23 Computer. (Courtesy of Digital Equipment Corporation)* **C.** *V-8600 Computer. (Courtesy of NCR Corporation)*

Automation

Electronic systems are employed extensively in the control of manufacturing processes. Computers and specialized electronic systems are used in industry for various purposes, for example, control of ingredient mixes, operation of machine tools, product inspection, and control and distribution of power.

FIGURE 1–3 *A digital long-distance telephone switching system, GTE No. 3 EAX (Electronic Automatic Exchange). (Courtesy of GTE Automatic Electric Incorporated)*

Medicine

Electronic devices and systems are finding ever-increasing applications in the medical field. The familiar *electrocardiograph* (ECG), used for the diagnosis of heart and other circulatory ailments, is a widely used medical electronic instrument. A closely related instrument is the *electromyograph,* which uses a cathode ray tube display rather than an ink trace.

The *diagnostic sounder* uses ultrasonic sound waves for various diagnostic procedures in neurology, for heart chamber measurement, and for detection of certain types of tumors. The *electroencephalograph* (EEG) is similar to the electrocardiograph. It records the electrical activity of the brain rather than heart activity. Another electronic instrument used in medical procedures is the *coagulograph*. This instrument is used in blood clot analysis.

Electronic instrumentation is also used extensively in intensive-care facilities. Heart rate, pulse, body temperature, respiration, and blood pressure can be monitored on a continuous basis. Monitoring equipment is also used a great deal in operating rooms. Some typical medical electronic equipment is pictured in Figure 1–4.

Consumer Products

Electronic products used directly by the consumers for information, entertainment, recreation, or work around the home are an important segment of the total electronics market. For example, the electronic calculator and digital watch are popular examples of consumer electronics. The small personal computer is used widely by hobbyists and is also becoming a common household appliance.

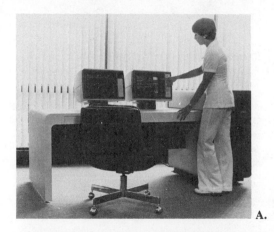

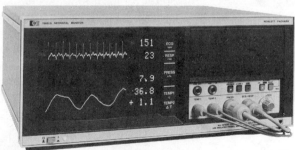

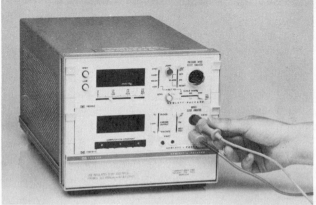

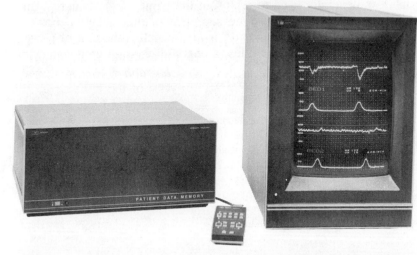

FIGURE 1–4 **A.** *Patient information center displays ECG wave forms and alphanumeric data.* **B.** *Neonatal monitor shows both wave form and computed numeric data of vital patient information.* **C.** *Cardiac output module computes cardiac output and measures continuous pulmonary artery blood pressure.* **D.** *Patient data memory system for fetal monitoring enables hospital staff to select and control information from a small, hand-held keyboard.* (**A–D,** *courtesy of Hewlett-Packard Company*)

Electronic systems are used in automobiles to control and monitor engine functions, control braking, provide entertainment, and display useful information to the driver.

Most appliances such as microwave ovens, washers, and dryers are available with electronic controls. Home entertainment, of course, is largely electronic. Examples are television, radio, stereo, and recorders. Also, many new games for adults and children incorporate electronic devices.

Review for 1–2

1. Name some of the areas in which electronics is used.

2. The computerization of manufacturing processes is an example of _____ .

ELECTRICAL UNITS

1–3

In electronics work, you must deal with measurable quantities. For example, you must be able to express how many volts are measured at a certain test point, how much current is flowing through a wire, or how much power a certain amplifier produces.

In this section you will learn the units and symbols for many of the electrical quantities that are used throughout the book. Definitions of the quantities are presented as they are needed in later chapters.

Symbols are used in electronics to represent both quantities and their units. One symbol is generally used to represent the name of the quantity, and another is used to represent the unit of measurement of that quantity. For example, P stands for *power,* and W stands for *watts,* which is the unit of power. Table 1–1 lists the most important quantities, along with their SI units and sym-

TABLE 1–1 *Electrical quantities and units with SI symbols.*

Quantity	Symbol	Unit	Symbol
Capacitance	C	farad	F
Charge	Q	coulomb	C
Conductance	G	siemen	S
Current	I	ampere	A
Energy	$\mathcal{E}$	joule	J
Frequency	f	hertz	Hz
Impedance	Z	ohm	Ω
Inductance	L	henry	H
Power	P	watt	W
Reactance	X	ohm	Ω
Resistance	R	ohm	Ω
Time	t	second	s
Voltage	V	volt	V
Wavelength	λ	meter	m

bols. The term *SI* is the French abbreviation for *International System* (*Système International* in French).

Review for 1–3

1. What does *SI* stand for?

2. Without referring to Table 1–1, list as many electrical quantities as possible, including their symbols, units, and unit symbols.

SCIENTIFIC NOTATION

1–4

In electronics work, you will encounter both very small and very large numbers. For example, it is common to have electrical current values of only a few thousandths or even a few millionths of an ampere. On the other hand, you will find resistance values of several thousand or several million ohms. This range of values is typical of many other electrical quantities also.

Powers of Ten

Scientific notation uses *powers of ten*, a method that makes it much easier to express large and small numbers and to do calculations involving such numbers.

Table 1–2 lists some powers of ten, both positive and negative. The power of ten is expressed as an exponent of the base 10 in each case. The exponent indicates the number of decimal places to the right or left of the decimal point in the expanded number. If the power is *positive*, the decimal point is moved to the *right*. For example,

$$10^4 = 1 \times 10^4 = 1.0000. = 10,000.$$

If the power is *negative*, the decimal point is moved to the *left*. For example,

$$10^{-4} = 1 \times 10^{-4} = .0001. = 0.0001$$

TABLE 1–2 *Some positive and negative powers of ten.*

$1,000,000 = 10^6$	$0.000001 = 10^{-6}$
$100,000 = 10^5$	$0.00001 = 10^{-5}$
$10,000 = 10^4$	$0.0001 = 10^{-4}$
$1,000 = 10^3$	$0.001 = 10^{-3}$
$100 = 10^2$	$0.01 = 10^{-2}$
$10 = 10^1$	$0.1 = 10^{-1}$
$1 = 10^0$	

Example 1–1

Express each number as a positive power of ten.
(a) 200   (b) 5000   (c) 85,000   (d) 3,000,000

Solution:

In each case there are many possibilities for expressing the number in powers of ten. We do not show all possibilities in this example but include the most common powers of ten used in electrical work.
(a) $200 = 0.0002 \times 10^6 = 0.2 \times 10^3$
(b) $5000 = 0.005 \times 10^6 = 5 \times 10^3$
(c) $85,000 = 0.085 \times 10^6 = 8.5 \times 10^4 = 85 \times 10^3$
(d) $3,000,000 = 3 \times 10^6 = 3000 \times 10^3$

Example 1–2

Express each number as a negative power of ten.
(a) 0.2   (b) 0.005   (c) 0.00063   (d) 0.000015

Solution:

Again, all the possible ways to express each number as a power of ten are not given. The most commonly used powers are included, however.
(a) $0.2 = 2 \times 10^{-1} = 200 \times 10^{-3} = 200,000 \times 10^{-6}$
(b) $0.005 = 5 \times 10^{-3} = 5000 \times 10^{-6}$
(c) $0.00063 = 0.63 \times 10^{-3} = 6.3 \times 10^{-4} = 630 \times 10^{-6}$
(d) $0.000015 = 0.015 \times 10^{-3} = 1.5 \times 10^{-5} = 15 \times 10^{-6}$

Example 1–3

Express each of the following powers of ten as a regular decimal number:
(a) 10^5 (b) 2×10^3 (c) 3.2×10^{-2} (d) 250×10^{-6}

Solution:

(a) $10^5 = 1 \times 10^5 = 100,000$ (b) $2 \times 10^3 = 2000$
(c) $3.2 \times 10^{-2} = 0.032$ (d) $250 \times 10^{-6} = 0.000250$

Calculating in Powers of Ten

The great convenience of scientific notation is in addition, subtraction, multiplication, and division of very small or very large numbers.

Rules for Addition: The rules for adding numbers of powers of ten are as follows:

1. Convert the numbers to be added to the *same* power of ten.
2. Add the numbers directly to get the sum.
3. Bring down the common power of ten, which is the power of ten of the sum.

Example 1–4

Add 2×10^6 and 5×10^7.

Solution:

1. Convert both numbers to the same power of ten:
$$(2 \times 10^6) + (50 \times 10^6)$$
2. Add $2 + 50 = 52$.
3. Bring down the common power of ten (10^6), and the sum is 52×10^6.

Rules for Subtraction: The rules for subtracting numbers in powers of ten are as follows:

1. Convert the numbers to be subtracted to the *same* power of ten.
2. Subtract the numbers directly to get the difference.
3. Bring down the common power of ten, which is the power of ten of the difference.

Example 1–5

Subtract 25×10^{-12} from 75×10^{-11}.

Solution:

1. Convert each number to the same power of ten:
$$(75 \times 10^{-11}) - (2.5 \times 10^{-11})$$
2. Subtract $75 - 2.5 = 72.5$.
3. Bring down the common power of ten (10^{-11}), and the difference is 72.5×10^{-11}.

Rules for Multiplication: The rules for multiplying numbers in powers of ten are as follows:

1. Multiply the numbers directly.
2. Add the powers of ten algebraically (the powers do not have to be the same).

Example 1–6

Multiply 5×10^{12} and 3×10^{-6}.

Solution:

Multiply the numbers, and algebraically add the powers:

$$(5 \times 10^{12})(3 \times 10^{-6}) = 15 \times 10^{[12+(-6)]} = 15 \times 10^{6}$$

Rules for Division: The rules for dividing numbers in powers of ten are as follows:

1. Divide the numbers directly.
2. Subtract the power of ten in the denominator from the power of ten in the numerator.

Example 1–7

Divide 50×10^{8} by 25×10^{3}.

Solution:

The division problem is written with a numerator and denominator as

$$\frac{50 \times 10^{8}}{25 \times 10^{3}}$$

Dividing the numbers and subtracting 3 from 8, we get

$$\frac{50 \times 10^{8}}{25 \times 10^{3}} = 2 \times 10^{8-3} = 2 \times 10^{5}$$

Review for 1–4

1. Scientific notation uses powers of ten (T or F).

2. Express 100 as a power of ten.

3. Do the following operations:
(a) $(1 \times 10^5) + (2 \times 10^5)$ (b) $(3 \times 10^6)(2 \times 10^4)$
(c) $(8 \times 10^3) \div (4 \times 10^2)$

METRIC PREFIXES

1–5

In electrical and electronics work, certain powers of ten are used more often than others. The most frequently used powers of ten are 10^9, 10^6, 10^3, 10^{-3}, 10^{-6}, 10^{-9}, and 10^{-12}.

It is common practice to use *metric prefixes* to represent these quantities. Table 1–3 lists the metric prefix for each of the commonly used powers of ten.

Use of Metric Prefixes

Now we use examples to illustrate use of metric prefixes. The number 2000 can be expressed in scientific notation as 2×10^3. Suppose we wish to represent 2000 watts (W) with a metric prefix. Since $2000 = 2 \times 10^3$, the metric prefix *kilo* (k) is used for 10^3. So we can express 2000 W as 2 kW (2 kilowatts).

As another example, 0.015 ampere (A) can be expressed as 15×10^{-3} A. The metric prefix *milli* (m) is used for 10^{-3}. So 0.015 becomes 15 mA (15 milliamperes).

Example 1–8

Express each quantity using a metric prefix.
(a) 50,000 V   (b) 25,000,000 Ω   (c) 0.000036 A

Solution:

(a) $50,000 \text{ V} = 50 \times 10^3 \text{ V} = 50 \text{ kV}$
(b) $25,000,000 \text{ Ω} = 25 \times 10^6 \text{ Ω} = 25 \text{ MΩ}$
(c) $0.000036 \text{ A} = 36 \times 10^{-6} \text{ A} = 36 \text{ μA}$

TABLE 1–3 *Metric prefixes and their symbols.*

Power of Ten	Value	Metric Prefix	Metric Symbol	Power of Ten	Value	Metric Prefix	Metric Symbol
10^9	one billion	giga	G	10^{-6}	one-millionth	micro	μ
10^6	one million	mega	M	10^{-9}	one-billionth	nano	n
10^3	one thousand	kilo	k	10^{-12}	one-trillionth	pico	p
10^{-3}	one-thousandth	milli	m				

Review for 1-5

1. List the metric prefix for each of the following powers of ten: 10^6, 10^3, 10^{-3}, 10^{-6}, 10^{-9}, and 10^{-12}.

2. Use an appropriate metric prefix to express 0.000001 ampere.

THE COMPUTER AS AN ANALYSIS TOOL

1-6   The computer is a very useful tool in the areas of circuit analysis and problem solving.

Many of the chapters in this book provide computer programs as examples of programming for the analysis of electrical circuits. These sections on computer analysis are optional and can be omitted without affecting the flow of material.

All of the programs are patterned after selected example problems that are presented in the text. The programs are limited to fundamental program statements so that few if any changes are necessary for use on machines other than the TRS-80, for which they were written.

The BASIC Language

A computer program is a *series of instructions* that tell the computer, step-by-step, what to do. Generally, in BASIC, each instruction must have a *line number*. Instruction words such as CLS, PRINT, LET, INPUT, GOTO, FOR/NEXT, and IF/THEN tell the computer what operation to perform at each step in the program.

Mathematical operators such as $+$, $-$, $*$, $/$, $\uparrow$, and $=$ are used for addition, subtraction, multiplication, division, exponentiation, and equating, respectively.

These instruction words and operators, of course, do not represent the extent of the BASIC language, but most of the programs found throughout this text are limited to these. Table 1-4 lists each of the BASIC language components mentioned above and gives a brief description of each. For a detailed and thorough coverage of BASIC, refer to your computer instruction manual or a BASIC language textbook.

Review for 1-6

1. List seven BASIC instruction words or sets of words.

2. List six BASIC mathematical operators.

Self-Test

1. List the units of the following electrical quantities: current, voltage, resistance, power, and energy.

TABLE 1–4 *Some BASIC instruction words and mathematical operators.*

Instruction Word	Description	Example
CLS	Clears video screen	CLS
PRINT	Causes the computer to display or print out any message that follows in quotes or the value of a designated variable or the result of a specified calculation.	PRINT "HELLO"→HELLO PRINT X→Value of X PRINT 2+3→5
LET	Assigns a value to a variable. Can be omitted in some versions of BASIC for simplicity.	LET X=8 or simply X=8 LET Y=X/2 or simply Y=X/2
INPUT	Allows data to be entered into computer, such as variable values.	INPUT X (The computer stops and waits for you to enter a value for X from the keyboard.)
GOTO	Causes the computer to branch from its place in the program to a specified line number and skip everything in between.	GOTO 50
FOR/NEXT	These two instruction words are used together to set up loops.	A value for Y is calculated for each of three values of X (1, 2, and 3). 10 FOR X=1 TO 3 20 Y=X*2 30 NEXT X
IF/THEN	These two instruction words are used together to set up conditional statements.	IF X=6 THEN GOTO 50 or IF X=6 THEN 50

Mathematical Operator	Description	Example
=	Equals	X=3 (X equals 3)
+	Addition	Y=X+5 (X plus 5)
−	Subtraction	A=10−X (10 minus X)
*	Multiplication	Z=2*X (2 times X)
/	Division	W=Y/4 (Y divided by 4)
↑ (↑)	Exponentiation	X↑2 (X squared)

2. List the symbol for each unit in Problem 1.

3. List the symbol for each quantity in Problem 1.

4. Express the following numbers in powers of ten:
   (a) 100   (b) 12,000   (c) 5,600,000   (d) 78,000,000

5. Express the following numbers in powers of ten:
   (a) 0.03   **(b)** 0.0005   **(c)** 0.00058   **(d)** 0.0000224

6. Express each of the following powers of ten as a regular decimal number:
   (a) 7×10^4 **(b)** 45×10^3 **(c)** 100×10^{-3} **(d)** 4×10^{-1}

7. Add 12×10^5 and 25×10^6.

8. Subtract 5×10^{-3} from 8×10^{-3}.

9. Multiply 33×10^3 and 20×10^{-4}.

10. Divide 4×10^2 by 2×10^{-3}.

Problems

Section 1–4

1–1. Express each of the following numbers as a power of ten:
   (a) 3000   **(b)** 75,000   **(c)** 2,000,000

1–2. Express each number as a power of ten.
   (a) 1/500   **(b)** 1/2000   **(c)** 1/5,000,000

1–3. Express each of the following numbers in three ways, using 10^3, 10^4, and 10^5:
   (a) 8400   **(b)** 99,000   **(c)** 0.2×10^6

1–4. Express each of the following numbers in three ways, using 10^{-3}, 10^{-4}, and 10^{-5}:
   (a) 0.0002   **(b)** 0.6   **(c)** 7.8×10^{-2}

1–5. Express each power of ten in regular decimal form.
   (a) 2.5×10^{-6} **(b)** 50×10^2 **(c)** 3.9×10^{-1}

1–6. Express each power of ten in regular decimal form.
   (a) 45×10^{-6} **(b)** 8×10^{-9} **(c)** 40×10^{-12}

1–7. Add the following numbers:
   (a) $(92 \times 10^6) + (3.4 \times 10^7)$ **(b)** $(5 \times 10^3) + (85 \times 10^{-2})$
   (c) $(560 \times 10^{-8}) + (460 \times 10^{-9})$

1–8. Perform the following subtractions:
   (a) $(3.2 \times 10^{12}) - (1.1 \times 10^{12})$ **(b)** $(26 \times 10^8) - (1.3 \times 10^9)$
   (c) $(150 \times 10^{-12}) - (8 \times 10^{-11})$

1–9. Perform the following multiplications:
   (a) $(5 \times 10^3)(4 \times 10^5)$ **(b)** $(12 \times 10^{12})(3 \times 10^2)$
   (c) $(2.2 \times 10^{-9})(7 \times 10^{-6})$

1–10. Divide the following:
   (a) $(10 \times 10^3) \div (2.5 \times 10^2)$ **(b)** $(250 \times 10^{-6}) \div (50 \times 10^{-8})$
   (c) $(4.2 \times 10^8) \div (2 \times 10^{-5})$

Section 1–5

1–11. Express each of the following as a quantity having a metric prefix:
   (a) 31×10^{-3} A   **(b)** 5.5×10^3 V   **(c)** 200×10^{-12} F

1–12. Express the following using metric prefixes:
  (a) 3×10^{-6} F   (b) $3.3 \times 10^6 \ \Omega$ (c) 350×10^{-9} A

Advanced Problems

1–13. Convert each of the following numbers to another number having a multiplier of 10^{-6}:
  (a) 2.37×10^{-3} (b) 0.001856×10^{-2} (c) 5743.89×10^{-12}
  (d) 100×10^3

1–14. Perform the following indicated operations:
  (a) $(2.8 \times 10^3)(3 \times 10^2)/2 \times 10^2$ (b) $(46)(10^{-3})(10^5)/10^6$
  (c) $(7.35)(0.5 \times 10^{12})/[(2)(10 \times 10^{10})]$ (d) $(30)^2 (5)^3/10^{-2}$

1–15. Perform the indicated conversions:
  (a) 5 mA to microamperes.   (b) 3200 μW to milliwatts.
  (c) 5000 kV to megavolts.   (d) 10 MW to kilowatts.

1–16. Determine the following:
  (a) The number of microamperes in 1 milliampere.
  (b) The number of millivolts in 0.05 kilovolt.
  (c) The number of megohms in 0.02 kilohm.
  (d) The number of kilowatts in 155 milliwatts.

Answers to Section Reviews

Section 1–1:
1. Volta. **2.** Current. **3.** He established the relationship among current, voltage, and resistance as expressed in Ohm's law. **4.** 1947. **5.** Integrated circuits.

Section 1–2:
1. Computers, communications, automation, medicine, and consumer products.
2. Automation.

Section 1–3:
1. The abbreviation for Système International. **2.** Refer to Table 1–1 after you have compiled your list.

Section 1–4:
1. T. **2.** 10^2. **3.** (a) 3×10^5; (b) 6×10^{10}; (c) 2×10^1.

Section 1–5:
1. Mega (M), kilo (k), milli (m), micro (μ), nano (n), and pico (p). **2.** 1 μA (one microampere).

Section 1–6:
1. CLS, PRINT, LET, INPUT, GOTO, FOR/NEXT, IF/THEN.
2. =, +, −, *, /, ↑.

Three of the most important electrical quantities are presented in this chapter: current, voltage, and resistance. No matter what type of electrical or electronic equipment you may work with, these quantities will always be of primary importance.

To aid your understanding of electrical current, the basic structure of the atom is discussed. The types of materials and the kinds of components that are basic to most electrical applications are also covered. Finally, the electric circuit is defined, and basic measurements of the electrical quantities are discussed.

Specifically, you will learn:

1. The structure of atoms.
2. The definition of atomic weight and atomic number.
3. That the electron is the basic particle of electrical charge.
4. What electrical current is.
5. The definition of voltage.
6. The meaning of resistance in an electrical circuit.
7. Various types of common resistors and how to determine their values.
8. How conductors, semiconductors, and insulators differ.
9. How wires are sized according to the American Wire Gage.
10. The basic definition of an electrical circuit.
11. How to measure current, voltage, and resistance.
12. Various types of protective devices.

2

CURRENT, VOLTAGE, AND RESISTANCE

2–1

An atom is the smallest particle of an element that still retains the characteristics of that element. Different elements have different types of atoms. In fact, every element has a unique atomic structure.

Atoms have a *planetary* type of structure, consisting of a central *nucleus* surrounded by orbiting *electrons*. The nucleus consists of positively charged particles called *protons* and uncharged particles called *neutrons*. The electrons are the basic particles of *negative charge*.

Each type of atom has a certain number of electrons and protons that distinguishes the atom from all other atoms of other elements. For example, the simplest atom is that of hydrogen. It has one proton and one electron, as pictured in Figure 2–1A. The helium atom, shown in Figure 2–1B, has two protons and two neutrons in the nucleus, which is orbited by two electrons.

Atomic Weight and Number

All elements are arranged in the *periodic table of the elements* in order according to their *atomic number,* which is the number of electrons in the orbits of the atom. The elements can also be arranged by their *atomic weight,* which is approximately the number of protons and neutrons in the nucleus. For example, hydrogen has an atomic number of *one* and an atomic weight of *one*. The atomic number of helium is *two,* and its atomic weight is *four*.

In their normal, or *neutral,* state, all atoms of a given element have the same number of electrons as protons. So the positive charges cancel the negative charges, and the atom has a net charge of zero.

The Copper Atom

Since copper is the most commonly used metal in electrical applications, let us examine its atomic structure. The copper atom has 29 electrons in orbit around the nucleus. They do not all occupy the same orbit, however. They move in orbits at varying distances from the nucleus. The orbits in which the electrons revolve are called *shells*. The number of electrons in each shell follows a predictable pattern.

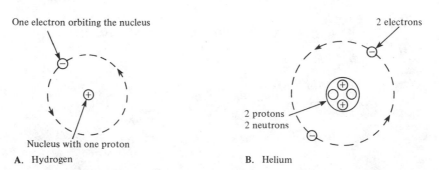

A. Hydrogen   B. Helium

FIGURE 2–1 *Hydrogen and helium atoms.*

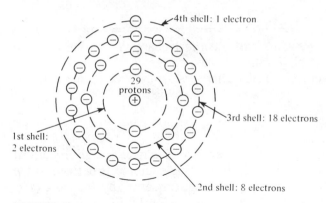

FIGURE 2–2 *The copper atom.*

The first shell of any atom can have up to two electrons. The second shell can have up to eight electrons. The third shell can have up to 18 electrons. The fourth shell can have up to 32 electrons. A copper atom is shown in Figure 2–2. Notice that the fourth or outermost shell has only one electron, called the *valence* electron.

Attraction between Positive and Negative Charges

The positively charged nucleus attracts the orbiting electrons. The electrons remain in a stable orbit because the centrifugal force counteracts the attractive force. The closer an electron is to the nucleus, the greater is the attractive force and the more tightly bound the electron is to the atom. The single electron in the outer shell of the copper atom is loosely bound to the atom because of its distance from the nucleus.

Free Electrons

When the 29th electron in the outer shell of the copper atom gains sufficient energy from the surrounding media, it can break away from the parent atom and become what is called a *free electron*. The free electrons in the copper material are capable of moving from one atom to another in the material. In other words, they drift randomly from atom to atom within the copper material. As you will see, the free electrons make electrical *current* possible.

Review for 2–1

1. What is the basic particle of negative charge?
2. Define *atom*.
3. What does a typical atom consist of?
4. Do all elements have the same types of atoms?
5. What is a free electron?

CURRENT

2–2

Electrical current is the net movement of electrical charge from one point to another in a conductive material. In other words, current is the *rate of flow of electrical charge* in a conductor. Electrical charge is symbolized by the letter Q, current by the letter I.

Unit of Charge

The electron is the smallest unit of negative electrical charge. Electrical charge is measured in *coulombs,* abbreviated C. One coulomb is defined as the charge carried by 6.25×10^{18} electrons. The charge possessed by one electron is 1.6×10^{-19} C.

Formula for Current

Since current is the rate of charge flow, it can be stated as follows:

$$I = \frac{Q}{t}$$

(2–1)

where I is current, Q is charge, and t is time.

The Ampere

Current is measured in a unit called the *ampere,* or *amp* for short, abbreviated by the letter A. *One ampere is the amount of current flowing in a conductor when one coulomb of charge moves past a given point in one second.*

Example 2–1

Ten coulombs of charge flow past a given point in a wire in 2 seconds. How many amperes of current are flowing?

Solution:

$$I = \frac{Q}{t} = \frac{10 \text{ C}}{2 \text{ s}} = 5 \text{ A}$$

Review for 2–2

1. Define *current.*
2. How many electrons make up one coulomb of charge?
3. If 20 coulombs flow past a point in a wire in 4 seconds, what is the current?

2–3

Normally, in a conductive material such as copper wire, the free electrons are in random motion and have no net direction. In order to produce current, the free electrons *must* move in the same general direction. To produce motion in a given direction, energy must be imparted to the electrons. This energy comes from a voltage source connected to a conductor.

Voltage (symbolized V) represents the energy required to move a certain amount of charge from one point to another. Voltage is also known as *electromotive force* (emf) or *potential difference*.

Formula for Voltage

Voltage is the amount of energy, $\mathscr{E}$, used to move an amount of charge, symbolized by Q:

$$V = \frac{\mathscr{E}}{Q} \qquad\qquad (2\text{–}2)$$

Energy is expressed in joules (J), and charge in coulombs (C). A joule is the energy required to move an object a distance of one meter against an opposing force of one newton (0.225 lb).

The Volt

Voltage is measured in a unit called the *volt,* abbreviated V. *One volt is the amount of potential difference between two points when one joule of energy is used to move one coulomb of charge from one point to the other.*

Example 2–2

If 50 joules of energy are used to move 10 coulombs of charge through a conductor, what is the voltage across the conductor?

Solution:

$$V = \frac{\mathscr{E}}{Q} = \frac{50 \text{ J}}{10 \text{ C}} = 5 \text{ V}$$

The Voltage Source

The battery is a typical energy source that provides voltage. Figure 2–3 shows a battery connected to a conductor to light a bulb. The voltage source has a positive terminal and a negative terminal as indicated. Energy is imparted to the electrons at the negative terminal; then the electrons move through the conductor and the bulb to the positive terminal. As a result, current flows through the

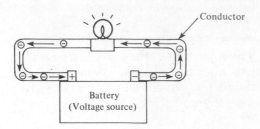

FIGURE 2–3 *Electron flow in a conductor is produced by a voltage source.*

conductor and the filament of the bulb. In other words, the negative terminal repels electrons away from it, and the positive terminal attracts electrons toward it. Keep in mind that *you must have voltage in order to have current.*

Review for 2–3

1. Define *voltage.*

2. Voltage is required to produce current (T or F).

3. Current is required to produce voltage (T or F).

4. One hundred joules of energy are used to move 20 coulombs of charge from the negative terminal of a battery through a wire to the positive terminal. What is the battery voltage?

RESISTANCE

2–4 *Resistance is the opposition to current.* It is used in electrical circuits to *limit* or control the amount of current that flows.

When current flows through resistance, heat is produced. In certain applications, the main purpose of resistance is to produce heat (an electric heater is an example). In many other applications, the heat produced represents an unwanted loss of energy.

The Ohm

Resistance, *R*, is measured in the unit of *ohms,* symbolized by the Greek letter omega, Ω. Figure 2–4 shows the schematic symbol for a 1-Ω resistance. *The resistance is one ohm when one ampere of current flows with one volt applied.*

Conductance

Later, in circuit analysis problems, you will find conductance to be useful. Conductance, symbolized by *G*, is simply the *reciprocal* of resistance, and is a measure of the ease with which current is able to flow. The formula is

$$G = \frac{1}{R} \qquad\qquad (2–3)$$

R

$1\ \Omega$

FIGURE 2–4 *Resistance symbol.*

The unit of conductance is the siemen, abbreviated S.

Review for 2–4

1. Define *resistance*.

2. What is the unit of resistance?

3. What two things does resistance do in an electrical circuit?

An electrical component having the property of resistance is called a *resistor*. There are many types of resistors in common use, but generally they can be placed in two main categories: *fixed* and *variable*.

Fixed Resistors

Fixed resistors have ohmic values set by the manufacturer and cannot be changed easily. Some common types of fixed resistors are shown in Figure 2–5. Various sizes and construction methods are used to control the heat-dissipating capabilities, the resistance value, and the precision.

Types of Fixed Resistors

Fixed resistors are constructed using various methods and materials. One common resistor is the *carbon-composition* type, which is made with a mixture of finely ground carbon, insulating filler, and a resin binder. The ratio of carbon to insulating filler sets the resistance value. The mixture is formed into rods, and lead connections are made. The entire resistor is then encapsulated in an insulated coating for protection. Figure 2–6 shows the construction of a typical carbon-composition resistor.

Other types of fixed resistors include carbon film, metal oxide, metal film, metal glaze, and wire-wound. Film resistors, for example, are made by depositing a resistive material evenly onto a high-grade ceramic rod. The resistive film may be carbon (carbon film), nickel chromium (metal film), a mixture of metals and glass (metal glaze), or metal and insulating oxide (metal oxide). In these types of resistors, the desired resistance value is obtained by removing part of the resistive material in a helical pattern along the rod using a *spiraling* technique. Very close tolerance can be achieved with this method.

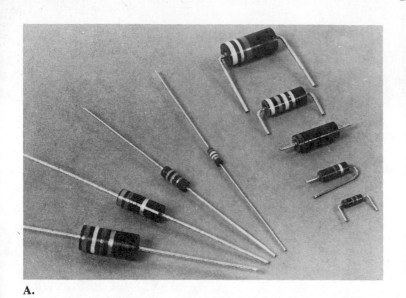

B.

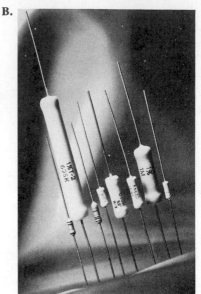

A.

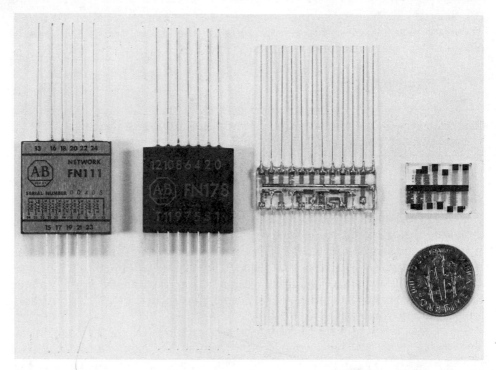

C.

FIGURE 2–5 *Typical fixed resistors.* **A.** *Carbon-composition resistors. (Courtesy of Stackpole Carbon Company)* **B.** *Metal film resistors. (Courtesy of Stackpole Carbon Company)* **C.** *Resistor networks. (Courtesy of Allen-Bradley Company)*

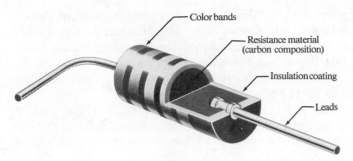

FIGURE 2–6 *Cutaway view of carbon-composition resistor. (Courtesy of Allen-Bradley Company)*

Wire-wound resistors are constructed with resistance wire wound around an insulating rod and then sealed. Normally, wire-wound resistors are used because of their relatively high power ratings.

Resistor Color Codes

Fixed resistors with value tolerances of 5%, 10%, or 20% are color coded with four bands to indicate the resistance value and the tolerance. This color-code band system is shown in Figure 2–7, and the color code is listed in Table 2–1.

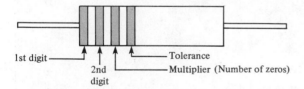

FIGURE 2–7 *Color-code bands on a resistor.*

TABLE 2–1 *Resistor color code.*

	Digit	Color
	0	Black
	1	Brown
	2	Red
	3	Orange
Resistance value, first three bands	4	Yellow
	5	Green
	6	Blue
	7	Violet
	8	Gray
	9	White
	5%	Gold
Tolerance, fourth band	10%	Silver
	20%	No band

Beginning at the banded end, the first band is the first digit of the resistance value. The second band is the second digit. The third band is the number of zeros, or the *multiplier*. The fourth band indicates the tolerance. For example, a 5% tolerance means that the *actual* resistance value is within ±5% of the color-coded value. So, a 100-Ω resistor with a tolerance of ±5% can have acceptable values as low as 95 Ω and as high as 105 Ω.

For resistance values less than 10 Ω, the third band is either gold or silver. Gold represents a multiplier of 0.1, and silver represents 0.01. For example, a color code of red, violet, gold, and silver represents 2.7 Ω with a tolerance of ±10%.

Example 2–3

Find the resistance value in ohms and the tolerance in percent for the resistor pictured in Figure 2–8.

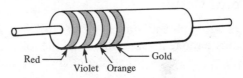

Red ——⟋   ⟍—— Gold
           Violet   Orange

FIGURE 2–8

Solution:

First band is red: 2
Second band is violet: 7
Third band is orange: 000
Fourth band is gold: 5% tolerance

The resistance value is 27,000 Ω, ±5%.

Certain precision resistors with tolerances of 1% or 2% are color coded with five bands. Beginning at the banded end, the first band is the first digit of the resistance value, the second band is the second digit, the third band is the third digit, the fourth band is the number of zeros, and the fifth band indicates the tolerance. Table 2–1 applies except that gold indicates 1% and silver indicates 2%.

Numerical labels are also commonly used on certain types of resistors where the resistance value and tolerance are stamped on the body of the resistor. For example, a common system uses R to designate the decimal point and uses letters to indicate tolerance as follows:

F = ±1%,   G = ±2%,   J = ±5%,   K = ±10%,   M = ±20%

For values above 100 Ω, three digits are used to indicate resistance value followed by a fourth digit that specifies the number of zeros. For values less than 100 Ω, R indicates the decimal point.

Some examples are as follows: 6R8M is a 6.8-Ω ±20% resistor; 3301F is a 3300-Ω ±1% resistor; and 2202J is a 22000-Ω ±5% resistor.

Variable Resistors

Variable resistors are designed so that their resistance values can be changed easily with a manual or an automatic adjustment.

Two basic types of manually adjustable resistors are the *potentiometer* and the *rheostat*. Schematic symbols for these types are shown in Figure 2–9. The potentiometer is a *three-terminal device,* as indicated in Part A. Terminals 1 and 2 have a fixed resistance between them, which is the total resistance. Terminal 3 is connected to a *moving contact.* We can vary the resistance between 3 and 1 or between 3 and 2 by moving the contact up or down.

Figure 2–9B shows the rheostat as a *two-terminal* variable resistor. Part C shows how we can use a potentiometer as a rheostat by connecting terminal 3 to either terminal 1 or terminal 2. Some typical potentiometers are pictured in Figure 2–10.

Potentiometers and rheostats can be classified as *linear* or *tapered*. In a linear potentiometer, the resistance between either terminal and the moving contact varies linearly with the position of the moving contact. For example, one-half of a turn results in one-half the total resistance. Three-quarters of a turn results in three-quarters of the total resistance between the moving contact and one terminal, or one-quarter of the total resistance between the other terminal and the moving contact.

In the tapered potentiometer, the resistance varies nonlinearly with the position of the moving contact, so that one-half of a turn does not necessarily result in one-half the total resistance. This concept is illustrated in Figure 2–11, where a potentiometer with a total resistance of 100 Ω is used as an example. The nonlinear values are arbitrary.

Thermistors and Photoconductive Cells

A thermistor is a type of variable resistor that is temperature-sensitive. Its resistance changes inversely with temperature. That is, it has a *negative temperature coefficient*. When temperature increases, the resistance decreases, and vice versa.

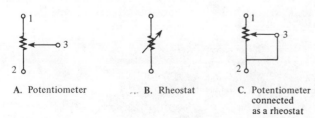

A. Potentiometer   B. Rheostat   C. Potentiometer
                                                   connected
                                                   as a rheostat

FIGURE 2–9 *Potentiometer and rheostat symbols.*

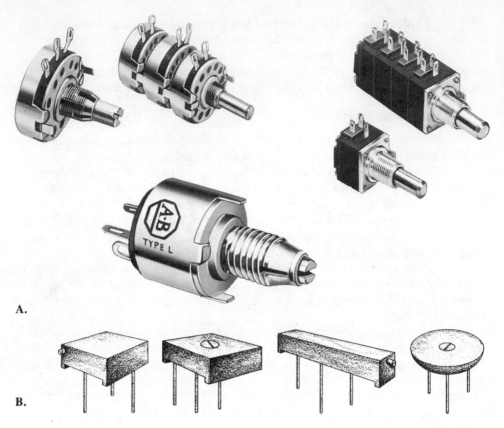

A.

B.

FIGURE 2–10 **A.** *Typical potentiometers. (Courtesy of Allen-Bradley Company)* **B.** *Trimmer potentiometers.*

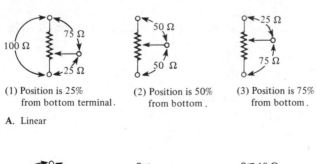

(1) Position is 25% from bottom terminal.

(2) Position is 50% from bottom.

(3) Position is 75% from bottom.

A. Linear

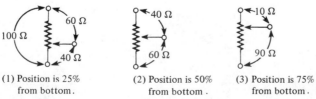

(1) Position is 25% from bottom.

(2) Position is 50% from bottom.

(3) Position is 75% from bottom.

B. Tapered

FIGURE 2–11 *Examples of linear and tapered potentiometers.*

32

Photovoltaic cell acts like battery (sattelite)

A. Thermistor

B. Photoconductive cell

FIGURE 2–12 *Resistive devices with sensitivities to temperature and light.*

The resistance of a photoconductive cell changes with a change in light intensity. This cell also has a negative temperature coefficient. Symbols for both of these devices are shown in Figure 2–12.

Review for 2–5

1. Name the two main categories of resistors, and briefly explain the difference between them.
2. Some resistors have the resistance stamped on their bodies, and others are color coded (T or F).
3. On a carbon-composition resistor, what does each of the four colored bands stand for?
4. What is the main difference between a rheostat and a potentiometer?
5. What is a thermistor?
6. What does *negative temperature coefficient* mean?

CONDUCTORS, SEMICONDUCTORS, AND INSULATORS

There are three categories of materials used in electronics: conductors, semiconductors, and insulators.

2–6

Conductors

Conductors are materials that allow current to flow easily. They have a large number of free electrons in their structure. Most metals are good conductors. Silver is the best conductor, and copper is next. Copper is the most widely used conductive material because it is less expensive than silver.

Semiconductors

These materials are classed below the conductors in their ability to carry current. However, because of their unique characteristics, certain semiconductor materials are the basis for modern electronic devices such as the diode,

TABLE 2–2 *Breakdown strengths of insulating materials*

Material	Average Breakdown Strength (kV/cm)
Mica	2000
Glass	900
Teflon®	600
Paper (paraffin)	500
Rubber	270
Bakelite®	150
Oil	140
Porcelain	70
Air	30

transistor, and integrated circuit. Silicon and germanium are common semiconductor materials.

Insulators

Insulating materials are poor conductors of electric current. In fact, they are used to *prevent* current where it is not wanted. Compared to conductive materials, insulators have very few free electrons.

Although insulators do not normally carry current, they will break down and conduct if a sufficiently large voltage is applied. The *breakdown strength* of an insulator specifies how much voltage across a certain thickness of the material is required to break the insulating ability down. Table 2–2 lists some common insulating materials and their average breakdown strengths in thousands of volts (kV) per centimeter of thickness.

Review for 2–6

1. What is a conductor?
2. What are semiconductor materials used for?
3. What is an insulator?
4. What happens when an insulator breaks down?

WIRE SIZE

2–7   Wires are the most common form of conductive material used in electrical applications. They vary in diameter size and are arranged according to standard *gage numbers,* called *American Wire Gage* (AWG) sizes. The larger the gage number is, the smaller the wire diameter is. The AWG sizes are listed in Table 2–3.

As Table 2–3 shows, the size of a wire is also specified in terms of its *cross-sectional area,* as illustrated also in Figure 2–13. The unit of cross-sectional area is the *circular mil,* abbreviated CM. One circular mil is the area of a wire with a diameter of 0.001 inch (1 mil). The cross-sectional area is

TABLE 2–3 *American Wire Gage (AWG) sizes for solid round copper.*

AWG #	Area (CM)	Ω/1000 ft at 20C°	AWG #	Area (CM)	Ω/1000 ft at 20C°
0000	211,600	0.0490	19	1,288.1	8.051
000	167,810	0.0618	20	1,021.5	10.15
00	133,080	0.0780	21	810.10	12.80
0	105,530	0.0983	22	642.40	16.14
1	83,694	0.1240	23	509.45	20.36
2	66,373	0.1563	24	404.01	25.67
3	52,634	0.1970	25	320.40	32.37
4	41,742	0.2485	26	254.10	40.81
5	33,102	0.3133	27	201.50	51.47
6	26,250	0.3951	28	159.79	64.90
7	20,816	0.4982	29	126.72	81.83
8	16,509	0.6282	30	100.50	103.2
9	13,094	0.7921	31	79.70	130.1
10	10,381	0.9989	32	63.21	164.1
11	8,234.0	1.260	33	50.13	206.9
12	6,529.0	1.588	34	39.75	260.9
13	5,178.4	2.003	35	31.52	329.0
14	4,106.8	2.525	36	25.00	414.8
15	3,256.7	3.184	37	19.83	523.1
16	2,582.9	4.016	38	15.72	659.6
17	2,048.2	5.064	39	12.47	831.8
18	1,624.3	6.385	40	9.89	1049.0

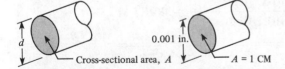

FIGURE 2–13 *Cross-sectional area of a wire.*

found by expressing the diameter in thousandths of an inch (mils) and squaring it, as follows:

$$A = d^2 \qquad (2\text{--}4)$$

where A is the cross-sectional area in circular mils and d is the diameter in mils.

Example 2–4

What is the cross-sectional area of a wire with a diameter of 0.005 inch?

Solution:

$$d = 0.005 \text{ in.} = 5 \text{ mils}$$
$$A = d^2 = 5^2 = 25 \text{ CM}$$

Wire Resistance

Although copper wire conducts electricity extremely well, it still has some resistance, as do all conductors. The resistance of a wire depends on four factors: (1) type of material; (2) length of wire; (3) cross-sectional area; and (4) temperature.

Each type of conductive material has a characteristic called its *resistivity*, ρ. For each material, ρ is a constant value at a given temperature. The formula for the resistance of a wire of length l and cross-sectional area A is

$$R = \frac{\rho l}{A} \qquad\qquad (2\text{–}5)$$

This formula tells us that resistance increases with resistivity and length, and decreases with cross-sectional area. For resistance to be calculated in ohms, the length must be in feet, the cross-sectional area in circular mils, and the resistivity in CM-Ω/ft.

Example 2–5

Find the resistance of a 100-ft length of copper wire with a cross-sectional area of 810.1 CM. The resistivity of copper is 10.4 CM-Ω/ft.

Solution:

$$R = \frac{\rho l}{A} = \frac{(10.4 \text{ CM-}\Omega/\text{ft})\,(100 \text{ ft})}{810.1 \text{ CM}} = 1.284 \; \Omega$$

Table 2–3 lists the resistance of the various standard wire sizes in ohms per 1000 feet at 20°C. For example, a 1000-ft length of 14-gage copper wire has a resistance of 2.525 Ω. A 1000-ft length of 22-gage wire has a resistance of 16.14 Ω. For a given length, the smaller wire has more resistance. Thus, for a given voltage, larger wires can carry more current than smaller ones.

Review for 2–7

1. The larger the wire gage number is, the smaller the wire is (T or F).
2. What does "AWG" stand for?
3. Name the four factors that determine the resistance of a wire.
4. What is the resistance of 500 feet of 18-gage copper wire?

THE ELECTRIC CIRCUIT

2–8 In its simplest form, an electric circuit consists basically of a source, a load, and a current path. The *source* can be a battery or any other type of energy source that produces voltage. The *load* can be a simple resistor or any other type of electrical device or more complex circuit. The *current path* is the conductors connecting the source to the load. A simple circuit is shown in Figure 2–14.

Closed Circuit and Open Circuit

A *closed* circuit is one in which the current has a complete path, as indicated in Figure 2–15A. An *open* circuit is one in which the current path is *broken* and the current cannot flow, as illustrated in Figure 2–15B. An open circuit represents an infinitely large resistance. A *switch,* symbolized at the top of the diagram, is the device commonly used to open or close a circuit. An open circuit sometimes is a result of the failure of a component in a circuit, such as a burned-out resistor or lamp bulb, as illustrated in Figure 2–15C.

Short Circuit

A short circuit is a near-zero resistance path and occurs when two points accidentally become connected, and current flows through the shorted contact. A short across a component such as a resistor will cause all of the current to flow through the short, bypassing the resistor, as illustrated in Figure 2–16.

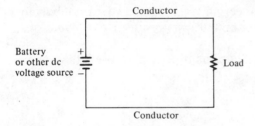

FIGURE 2–14 *A simple electrical circuit.*

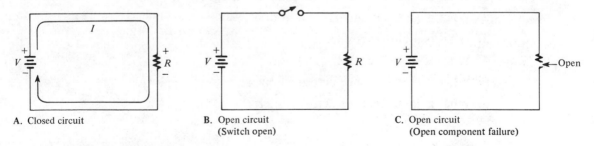

A. Closed circuit

B. Open circuit
(Switch open)

C. Open circuit
(Open component failure)

FIGURE 2–15 *Closed and open circuits.*

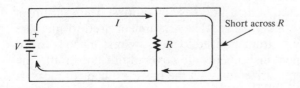

FIGURE 2–16 *Short circuit.*

Direction of Current

Early in the history of man's knowledge of electricity, it was assumed that all current consisted of positive moving charges. Later, of course, the electron was identified as the charge carrier in current that flows in conductive materials.

Today, there are two accepted conventions for the direction of electrical current. *Electron flow direction,* preferred by many in the fields of electrical and electronics technology, assumes current flowing out of the negative terminal of a voltage source, through the circuit, and into the positive terminal of the source. *Conventional current direction* assumes current flowing out of the positive terminal of a voltage source, through the circuit, and into the negative terminal of the source. By following the direction of conventional current flow, there is a *rise* in voltage across a source (negative to positive) and a *drop* in voltage across a resistor (positive to negative), as indicated in Figure 2–15A.

It actually makes little difference which direction of current is assumed as long as we are *consistent.* The outcome of electrical circuit analysis is not affected by the direction of current that is assumed for analytical purposes. The direction assumed is largely a matter of preference.

Conventional current direction is used widely in electronics technology and is used almost exclusively at the engineering level. The standard symbols for most electronic devices such as diodes and transistors utilize conventional current direction indicators.

For these reasons, conventional current direction is used throughout this text.

Review for 2–8

1. What are the basic elements of an electric circuit?

2. What is an open circuit?

3. What is a closed circuit?

4. What is a short circuit?

5. What is the resistance across an open?

6. What is the resistance across a short?

MEASUREMENTS

2–9 Current, voltage, and resistance measurements are common in electrical work. Special types of instruments are used to measure these quantities. The instrument used to measure voltage is called a *voltmeter.* The instrument used to measure current is called an *ammeter.* The instrument used to measure resistance is called an *ohmmeter.* Test and measurement instruments are covered in Chapter 10. The purpose of this discussion is to familiarize you with the basic methods of measuring current, voltage, and resistance in a circuit.

Meter Symbols

Throughout the book, we will use certain symbols to represent different meters. You will encounter either of two types of both voltmeter and ammeter symbols, depending on which is more useful in a given diagram for conveying the information required. These symbols are shown in Figure 2–17.

Measuring Current

Current is measured with an ammeter connected *in the current path* by breaking the circuit and inserting the meter, as shown in Figure 2–18. As you will learn later, such a connection is called a *series* connection. The positive side of the meter is connected toward the positive terminal of the voltage source. Either placement of the meter, as indicated in Figure 2–18, will measure the same current.

Measuring Voltage

Voltage is measured with a voltmeter connected *across the component,* as shown in Figure 2–19. As you will learn later, such a connection is called a *parallel* connection. The positive side of the meter must be connected toward the positive terminal of the voltage source. Either placement of the meter in Figure 2–19 will measure the same voltage.

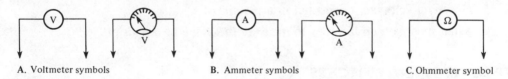

A. Voltmeter symbols   **B.** Ammeter symbols   **C.** Ohmmeter symbol

FIGURE 2–17   *Meter symbols.*

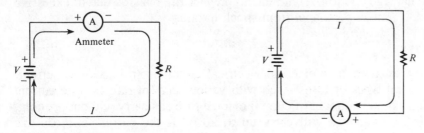

FIGURE 2–18   *Current measurement with an ammeter.*

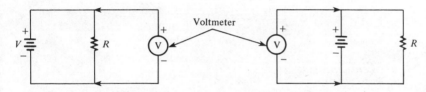

FIGURE 2–19   *Voltage measurement with a voltmeter.*

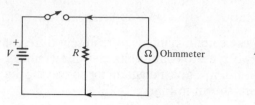

A. *R* disconnected from circuit with open switch   B. *R* removed from circuit

FIGURE 2–20   *Resistance measurement with an ohmmeter.*

Measuring Resistance

Resistance is measured with an ohmmeter connected across the resistor, as shown in Figure 2–20. The resistor *must* be removed from the circuit or disconnected from the voltage source in some way, as indicated in the figure. Failure to disconnect the voltage source will result in damage to the ohmmeter.

Review for 2–9

1. Name the meters for measurement of current, voltage, and resistance.
2. How is an ammeter connected in a circuit?
3. How is a voltmeter connected in a circuit?
4. What must you do in order to measure the value of a resistor?

PROTECTIVE DEVICES AND SWITCHES

2–10   Protective devices are used in electrical and electronic circuits to protect the circuit from damage due to overcurrent, to prevent fire hazards due to excessive current, and to protect personnel from shock hazards.

Fuses

Fuses are used to protect an electrical circuit from excessive current. There are several types of fuses, each with various current ratings. The current rating is the maximum amount of current that the fuse can carry without opening. For example, a 20-A fuse will carry up to 20 A. If the current exceeds this

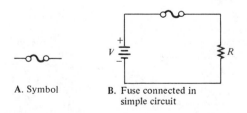

A. Symbol   B. Fuse connected in
                      simple circuit

FIGURE 2–21   *Fuse.*

amount, the fuse will blow and cause an open that stops the current. The schematic symbol for a fuse is shown in Figure 2–21A, and a fuse connected in a simple circuit is shown in Figure 2–21B.

Two types of fuses are found in power applications such as residential wiring: the *plug* type and the *cartridge* type, shown in Figure 2–22.

Fuses are also commonly used to protect electronic instruments. The type shown in Figure 2–23A is normally used; it has lower ampere ratings than a power fuse. Common types of fuse holders are shown in Figure 2–23B and C.

Circuit Breakers

In power applications such as commercial, industrial, and residential wiring, circuit breakers are replacing fuses in new installations. A circuit breaker can be reset and reused repeatedly — an advantage over fuses, which must be replaced when they go out. Circuit breakers are also commonly used in electronic equipment. The schematic symbol for circuit breakers is shown in Figure 2–24.

There are two basic types of circuit breakers: magnetic and thermal. Some typical circuit breakers are pictured in Figure 2–25.

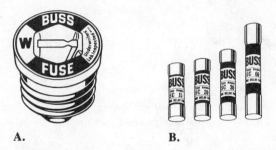

A.   **B.**

FIGURE 2–22 *Power fuses.* **A.** *Plug type.* **B.** *Cartridge type. (Courtesy of Bussmann Manufacturing, a division of McGraw-Edison Co.)*

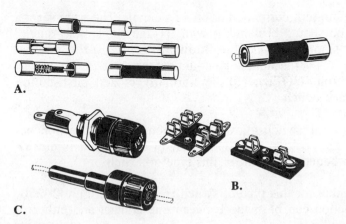

A.

B.

C.

FIGURE 2–23 **A.** *Fuses* **B.,C.** *Fuse holders. (Courtesy of Bussmann Manufacturing, a division of McGraw-Edison Co.)*

FIGURE 2–24 *Circuit breaker symbol.*

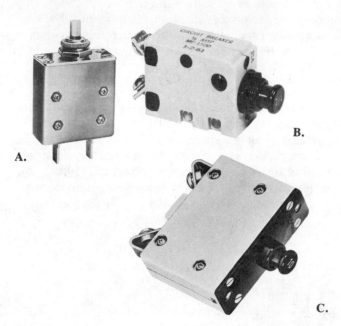

FIGURE 2–25 *Typical circuit breakers.* **A.** *Used in commercial applications or electrical equipment, for example, vending machines.* **B.,C.** *Used in aircraft.* (**A–C,** *courtesy of Mechanical Products)*

Switches

Switches are used to turn current on or off in a circuit. There are several types of switches that you should be familiar with: (1) the single-pole–single-throw (SPST); (2) the single-pole–double-throw (SPDT); (3) the double-pole–single-throw (DPST); (4) the double-pole–double-throw (DPDT); (5) the normally open push-button (NOPB); (6) the normally closed push-button (NCPB); and (7) the rotary switch.

SPST Switch: This type of switch allows connection or disconnection between two contacts. In one position it is open, and in the other it is closed. Figure 2–26A shows a schematic symbol for this type of switch.

SPDT Switch: The symbol for this type of switch is shown in Figure 2–26B. With this switch, connection can be made between one contact and either of two others.

DPST Switch: This switch allows simultaneous connection or disconnection of two sets of contacts. The symbol is shown in Figure 2–26C. The dashed line

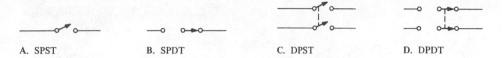

A. SPST   B. SPDT   C. DPST   D. DPDT

FIGURE 2–26 *Switch symbols.*

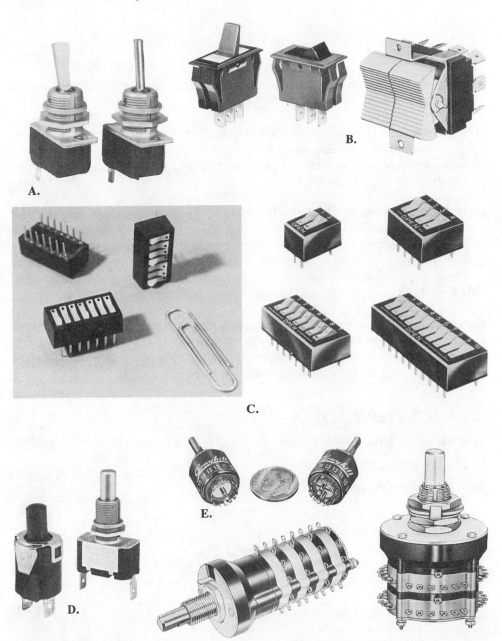

FIGURE 2–27 *Switches.* **A.** *Typical toggle-lever switches. (Courtesy of Eaton Corporation, Specialty Products Operations)* **B.** *Rocker switches. (Courtesy of Eaton Corporation, Specialty Products Operations)* **C.** *Rocker DIP switches. (Courtesy of Amp, Inc., and Grayhill, Inc., La Grange, Ill.)* **D.** *Push-button switches. (Courtesy of Eaton Corporation, Specialty Products Operations)* **E.** *Rotary-position switches. (Courtesy of Grayhill, Inc., La Grange, Ill.)*

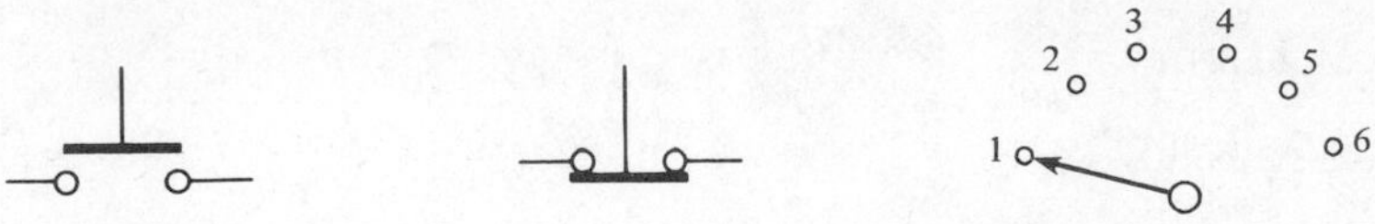

A. NOPB   **B.** NCPB   **C.** Rotary (6-position)

FIGURE 2–28 *Push-button and rotary switch symbols.*

indicates that the contact arms are grouped together so that both move with a single switch action.

DPDT Switch: This switch provides connection from one set of contacts to either of two other sets. The symbol is shown in Figure 2–26D. These types of switches are normally found in toggle (Figure 2–27A), slide, or rocker configurations (Figure 2–27B and C).

Push-Button Switches: In the NOPB switch, connection is made between two contacts when the button is depressed, and the connection is broken when the button is released. In the NCPB switch, connection between the two contacts is broken when the button is depressed and is made again when the button is released. Typical push-button switches are pictured in Figure 2–27D. The symbols for NOPB and NCPB switches are shown in Figure 2–28A and B.

Rotary Switch: In a rotary switch, a knob is turned to make connection between one contact and any of several others, as symbolized in Figure 2–28C. The one shown is a six-position switch; however, switches with many more contacts are available. Typical rotary switches are shown in Figure 2–27E.

Review for 2–10

1. What is the difference between a fuse and a circuit breaker?

2. List the main types of switches.

3. What is the resistance of a blown fuse?

Formulas

$$I = \frac{Q}{t} \tag{2-1}$$

$$V = \frac{\mathscr{E}}{Q} \tag{2-2}$$

$$G = \frac{1}{R} \tag{2-3}$$

$$A = d^2 \tag{2-4}$$

$$R = \frac{\rho l}{A} \tag{2-5}$$

Summary

1. An atom is the smallest particle of an element that retains the characteristics of that element.

2. An atom consists of a positively charged nucleus surrounded by orbiting electrons.

3. The electron is the basic particle of negative charge.

4. The proton is the particle of positive charge.

5. The neutron is an uncharged particle.

6. The nucleus of an atom contains protons and neutrons.

7. The atomic number is the number of electrons in an atom.

8. The atomic weight is the number of protons and neutrons in an atom.

9. Free electrons are loosely bound to the atom and provide for current in a material.

10. Current is the rate of flow of electrons.

11. The unit of charge is the coulomb (C).

12. The current is one ampere (A) when one coulomb passes a point in one second.

13. Voltage is the amount of energy used to move an amount of charge between two points in a circuit.

14. One volt (V) is the amount of potential difference when one joule (J) is used to move one coulomb.

15. Resistance is the opposition to current.

16. One ohm (Ω) is the resistance when one ampere flows with one volt applied.

17. Conductance is the reciprocal of resistance.

18. The unit of conductance is the siemen (S).

19. Larger wires have smaller gage numbers.

20. The resistance of a wire depends on its material, size, length, and temperature.

21. An open circuit prevents current. It is a break in the current path.

22. A closed circuit allows current. It completes a current path.

23. A short is an unintentional (usually), near-zero resistance path that bypasses the normal current path between two points.

24. An ammeter measures current, a voltmeter measures voltage, and an ohmmeter measures resistance.

Self-Test

1. How many electrons are in an atom with an atomic number of three?

2. How many protons and neutrons are in an atom with an atomic weight of six?

3. Fifty coulombs of charge flow past a point in a circuit in 5 seconds. What is the current?

4. If 2 A of current exist in a circuit, how many coulombs pass a given point in 10 seconds?

5. Five hundred joules of energy are used to move 100 C of charge through a resistor. What is the voltage across the resistor?

6. What is the conductance of a 10-Ω resistor?

7. Figure 2–29 shows a color-coded resistor. What are the resistance and the tolerance values?

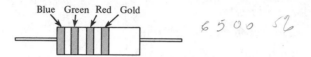 6 5 0 0 5%

FIGURE 2–29

8. The adjustable contact of a linear potentiometer is set at the center of its adjustment. If the total resistance is 1000 Ω, what is the resistance between the end terminal and the contact?

9. What is the maximum voltage that a 10-cm thickness of mica can withstand?

10. What is the resistance of a 10-ft length of copper wire with a diameter of 0.003 in. ($\rho = 10.4$)?

Problems

Section 2–2

2–1. Six-tenths coulomb flows past a point in 3 seconds (3 s). What is the current?

2–2. How long does it take 10 C to move past a point if the current is 5 A?

2–3. How many coulombs pass a point in 0.1 s if the current is 1.5 A?

2–4. Determine the current in each of the following cases:
(a) 75 C in 1 s   (b) 10 C in 0.5 s   (c) 5 C in 2 s   (d) 12 C in 12 s

Section 2–3

2–5. What is the voltage of a battery that uses 800 J of energy to move 40 C of charge through a resistor?

2–6. How much energy does a 12-V battery use to move 2.5 C through a circuit?

2–7. Determine the voltage in each of the following cases:
(a) 10 J/C   (b) 5 J/2 C   (c) 100 J/25 C   (d) 1 J/5 C

Section 2–4

2–8. Find the conductance for each of the following resistance values:
(a) 5 Ω (b) 25 Ω (c) 100 Ω (d) 1000 Ω

2–9. Find the resistance corresponding to the following conductances:
(a) 0.1 S   (b) 0.5 S   (c) 0.02 S   (d) 3 S

Section 2–5

2–10. Find the resistance and the tolerance for each color code:
    (a) First band is orange;   **(b)** First band is yellow;
        second band is orange;   second band is violet;
        third band is black;   third band is orange;
        fourth band is gold.   fourth band is silver.

2–11. Find the minimum and the maximum resistance for the two resistors in Problem 2–10.

Section 2–6

2–12. A piece of rubber is 50 cm thick. How many volts will it withstand before it breaks down?

2–13. Two conductive plates have a separation of 75 cm. What is the breakdown voltage of the air between them?

Section 2–7

2–14. What is the cross-sectional area of a wire having a diameter of 0.008 in.?

2–15. Determine the resistance of a 150-ft length of copper wire with a diameter of 10 mils.

2–16. Find the resistance of 50 ft of the following gage copper wire at 20°C:
    (a) 00   **(b)** 6   **(c)** 10   **(d)** 12   **(e)** 14
    (f) 18   **(g)** 22   **(h)** 28   **(i)** 40

Section 2–9

2–17. Show the placement of an ammeter and a voltmeter to measure the current and the source voltage in Figure 2–30.

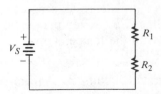

FIGURE 2–30

2–18. Show how you would measure the resistance of R_2 in Figure 2–30.

Advanced Problems

2–19. 574×10^{15} electrons flow through a wire in 250 ms. What is the current in amperes?

2–20. If a resistor with a current of 2 A through it converts 1000 joules of electrical energy into heat energy in 15 s, what is the voltage across the resistor?

2–21. A 120-V source is to be connected to a 1500-Ω load by a wire cable. The voltage source is to be located 50 ft from the load. Determine the gage number of the smallest sized wire that can be used if the *total* resistance across the wire cable is not to exceed 6 Ω.

2–22. Determine the diameter in meters, centimeters, and inches for each of the following wire gages:

(a) 0000   (b) 5   (c) 18   (d) 22

2–23. Devise a switch arrangement whereby two voltage sources (V_{S1} and V_{S2}) can be connected simultaneously to either of two loads (R_1 and R_2) as follows: V_{S1} connected to R_1 and V_{S2} to R_2, or V_{S1} connected to R_2 and V_{S2} connected to R_1.

Answers to Section Reviews

Section 2–1:
1. Electron. **2.** The smallest particle of an element that retains the unique characteristics of the element. **3.** A positively charged nucleus surrounded by orbiting electrons. **4.** No. **5.** An outer-shell electron that has drifted away from the parent atom.

Section 2–2:
1. Rate of flow of charge. **2.** 6.25×10^{18}. **3.** 5 A.

Section 2–3:
1. The amount of energy per unit charge. **2.** T. **3.** F. **4.** 5 V.

Section 2–4:
1. Opposition to current. **2.** Ohm (Ω). **3.** It limits current and dissipates heat.

Section 2–5:
1. Fixed, value cannot be changed. Variable, value easily changed. **2.** T. **3.** First band, first digit; second band, second digit; third band, multiplier; fourth band, tolerance. **4.** A rheostat is a two-terminal device that controls current; a potentiometer is a three-terminal device that controls voltage. **5.** A temperature-sensitive resistor. **6.** The resistance changes inversely with the temperature.

Section 2–6:
1. A good carrier of electrons. **2.** Transistors and integrated circuits. **3.** A material that does not conduct electric current. **4.** It conducts.

Section 2–7:
1. T. **2.** American Wire Gage. **3.** Material, cross-sectional area, length, and temperature. **4.** 3.1925 Ω.

Section 2–8:
1. Energy source, load, and current path. **2.** A break in the current path. **3.** A complete current path. **4.** A low-resistance connection that bypasses the normal current path. **5.** Infinitely high. **6.** Zero ohms.

Section 2–9:
1. Ammeter, voltmeter, and ohmmeter, respectively. **2.** In series. **3.** In parallel. **4.** Remove the resistor from the circuit.

Section 2–10:
1. A circuit breaker is resettable; a fuse is not. **2.** SPST, DPST, SPDT, DPDT, NOPB, NCPB, and rotary. **3.** Infinitely high.

Georg Simon Ohm found that voltage, current, and resistance are related in a specific way. This basic relationship, known as **Ohm's law,** is one of the most important laws in electrical and electronics work. In this chapter, you will learn Ohm's law and how to use it in solving circuit problems. Also, a computer program is introduced that can be used for solving problems related to Ohm's law.

In this chapter, you will learn:

1. The definition of Ohm's law.
2. The formula for Ohm's law.
3. How to use the Ohm's law formula to determine current, voltage, or resistance.
4. How to deal with various units of current, voltage, and resistance.
5. The meaning of the linear relationship between current and voltage.
6. How to create a flow chart for a problem and write a computer program in BASIC.

3

OHM'S LAW

3–1

Ohm's law tells us how current, voltage, and resistance are related. Georg Ohm determined experimentally that if the voltage across a resistor is increased, the current will also increase, and, likewise, if the voltage is decreased, the current will decrease. For example, if the voltage is doubled, the current will double. If the voltage is halved, the current will also be halved. This relationship is illustrated in Figure 3–1, with meter indications of voltage and current.

Ohm's law also states that if the voltage is kept constant, *less* resistance results in *more* current, and, also, *more* resistance results in *less* current. For example, if the resistance is halved, the current doubles. If the resistance is doubled, the current is halved. This concept is illustrated by the meter indications in Figure 3–2, where the resistance is varied and the voltage is constant.

Formula for Current

Ohm's law can be stated as follows:

$$I = \frac{V}{R} \tag{3–1}$$

This formula describes what was indicated by the circuits of Figures 3–1 and 3–2. For a constant value of R, if the value of V is increased, the value of I increases; if V is decreased, I decreases. Also notice in Equation (3–1) that if V is constant and R is increased, I decreases. Similarly, if V is constant and R is decreased, I increases.

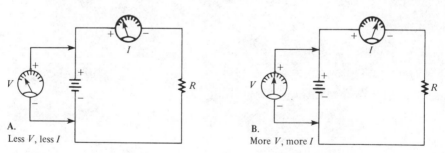

FIGURE 3–1 *Effect of changing the voltage with the same resistance in both circuits.*

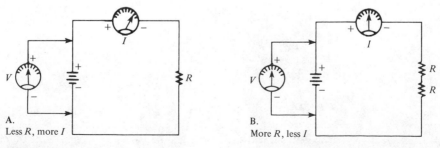

FIGURE 3–2 *Effect of changing the resistance with the same voltage in both circuits.*

Using Equation (3–1), we can calculate the *current* if the values of voltage and resistance are known.

Formula for Voltage

Ohm's law can also be stated another way. By multiplying both sides of Equation (3–1) by R, we obtain an *equivalent* form of Ohm's law, as follows:

$$V = IR \qquad (3\text{--}2)$$

With this equation, we can calculate *voltage* if the current and resistance are known.

Formula for Resistance

There is a third *equivalent* way to state Ohm's law. By dividing both sides of Equation (3–2) by I, we obtain

$$R = \frac{V}{I} \qquad (3\text{--}3)$$

This form of Ohm's law is used to determine *resistance* if voltage and current values are known.

Remember, *the three formulas you have learned in this section are all equivalent.* They are simply three different ways of expressing Ohm's law.

Review for 3–1

1. Ohm's law defines how three basic quantities are related. What are these quantities?
2. What is the Ohm's law formula for current?
3. What is the Ohm's law formula for voltage?
4. What is the Ohm's law formula for resistance?
5. If the voltage across a fixed value resistor is tripled, does the current increase or decrease, and by how much?
6. If the voltage across a fixed resistor is cut in half, how much will the current change?
7. There is a fixed voltage across a resistor, and you measure a current of 1 A. If you replace the resistor with one that has twice the resistance value, how much current will you measure?
8. In a circuit the voltage is doubled and the resistance is cut in half. Would you observe any change in the current value?

CALCULATING CURRENT

3–2

In this section you will learn to determine current values when you know the values of voltage and resistance. In these problems, the formula $I = V/R$ is used. In order to get current in *amperes,* you must express the value of *V* in *volts* and the value of *R* in *ohms.*

Example 3–1

How many amperes of current are flowing in the circuit of Figure 3–3?

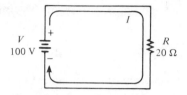

FIGURE 3–3

Solution:

Substitute into the formula $I = V/R$, 100 V for *V*, and 20 Ω for *R*. Divide 20 Ω into 100 V as follows:

$$I = \frac{V}{R} = \frac{100 \text{ V}}{20 \text{ } \Omega} = 5 \text{ A}$$

There are 5 A of current in this circuit.

Example 3–2

If the resistance in Figure 3–3 is changed to 50 Ω, what is the new value of current?

Solution:

We still have 100 V. Substituting 50 Ω into the formula for *I* gives 2 A as follows:

$$I = \frac{V}{R} = \frac{100 \text{ V}}{50 \text{ } \Omega} = 2 \text{ A}$$

Larger Units of Resistance

In electronics work, resistance values of thousands of ohms or even millions of ohms are common. As you learned in Chapter 1, large values of resistance are indicated by the metric system prefixes *kilo* (k) and *mega* (M).

Thus, thousands of ohms are expressed in kilohms (kΩ), and millions of ohms in megohms (MΩ). The following examples illustrate use of kilohms and megohms when using Ohm's law to calculate current.

Example 3–3

Calculate the current in Figure 3–4.

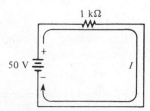

FIGURE 3–4

Solution:

Remember that 1 kΩ is the same as 1×10^3 Ω. Substituting 50 V for V and 1×10^3 Ω for R gives the current in amperes as follows:

$$I = \frac{V}{R} = \frac{50 \text{ V}}{1 \times 10^3 \ \Omega} = 50 \times 10^{-3} \text{ A}$$

$$= 0.05 \text{ A}$$

In Example 3–3, 50×10^{-3} A can be expressed as 50 milliamperes (50 mA). This fact can be used to advantage when we divide *volts by kilohms.* The current will be in *milliamperes,* as Example 3–4 illustrates.

Example 3–4

How many milliamperes flow in the circuit of Figure 3–5?

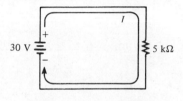

FIGURE 3–5

Example 3–4 (continued)

Solution:

When we divide *volts* by *kilohms,* we get *milliamperes.* In this case, 30 V divided by 5 kΩ gives 6 mA as follows:

$$I = \frac{V}{R} = \frac{30 \text{ V}}{5 \text{ k}\Omega} = 6 \text{ mA}$$

If *volts* are applied when resistance values are in *megohms,* the current is in *microamperes* (μA), as Example 3–5 shows.

Example 3–5

Determine the amount of current in the circuit of Figure 3–6.

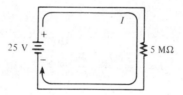

FIGURE 3–6

Solution:

Recall that 5 MΩ equals 5×10^6 Ω. Substituting 25 V for *V* and 5×10^6 Ω for *R* gives the following result:

$$I = \frac{V}{R} = \frac{25 \text{ V}}{5 \times 10^6 \text{ }\Omega} = 5 \times 10^{-6} \text{ A}$$

Notice that 5×10^{-6} A equals 5 microamperes (5 μA).

Example 3–6

Change the value of *R* in Figure 3–6 to 2 MΩ. What is the new value of current?

Solution:

When we divide *volts* by *megohms,* we get *microamperes.* In this case, 25 V divided by 2 MΩ gives 12.5 μA as follows:

$$I = \frac{V}{R} = \frac{25 \text{ V}}{2 \text{ M}\Omega} = 12.5 \ \mu\text{A}$$

Larger Units of Voltage

Small voltages, usually less than 50 volts, are common in semiconductor circuits. Occasionally, however, large voltages are encountered. For example, the high-voltage supply in a television receiver is around 20,000 volts (20 kilovolts, or 20 kV), and transmission voltages generated by the power companies may be as high as 345,000 V (345 kV). The following two examples illustrate use of voltage values in the kilovolt range when calculating current.

Example 3–7

How much current is produced by a voltage of 24 kV across a 12-kΩ resistance?

Solution:

Since we are dividing kV by kΩ, the prefixes cancel, and we get amperes:

$$I = \frac{V}{R} = \frac{24 \text{ kV}}{12 \text{ k}\Omega}$$

$$= \frac{24 \times 10^3 \text{ V}}{12 \times 10^3 \text{ V}} = 2 \text{ A}$$

Example 3–8

How much current will flow through 100 MΩ when 50 kV are applied?

Solution:

In this case, we divide 50 kV by 100 MΩ to get the current. Using 50×10^3 V for 50 kV and 100×10^6 Ω for 100 MΩ, we obtain the current as follows:

$$I = \frac{V}{R} = \frac{50 \text{ kV}}{100 \text{ M}\Omega} = \frac{50 \times 10^3 \text{ V}}{100 \times 10^6 \ \Omega}$$

$$= 0.5 \times 10^{-3} \text{ A} = 0.5 \text{ mA}$$

Remember that the power of ten in the denominator is subtracted from the power of ten in the numerator. So 50 was divided by 100, giving 0.5, and 6 was subtracted from 3, giving 10^{-3}.

Review for 3–2

In Problems 1–4, calculate I when

1. $V = 10$ V and $R = 5\ \Omega$.
2. $V = 100$ V and $R = 500\ \Omega$.
3. $V = 5$ V and $R = 2.5$ kΩ.
4. $V = 15$ V and $R = 5$ MΩ.
5. If a 5-MΩ resistor has 20 kV across it, how much current flows?
6. How much current will 10 kV across 2 kΩ produce?

CALCULATING VOLTAGE

3–3

In this section you will learn to determine voltage values when the current and resistance are known. In these problems, the formula $V = IR$ is used. To obtain voltage in *volts*, you must express the value of I in *amperes* and the value of R in *ohms*.

Example 3–9

In the circuit of Figure 3–7, how much voltage is needed to produce 5 A of current?

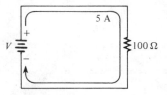

FIGURE 3–7

Solution:

Substitute 5 A for I and 100 Ω for R into the formula $V = IR$ as follows:

$$V = IR = (5\ \text{A})(100\ \Omega) = 500\ \text{V}$$

Thus, 500 V are required to produce 5 A of current through a 100-Ω resistor.

Smaller Units of Current

The following two examples illustrate use of current values in the milliampere (mA) and microampere (μA) ranges when calculating voltage.

Example 3–10

How much voltage will be measured across the resistor in Figure 3–8?

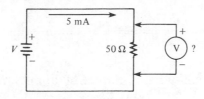

FIGURE 3–8

Solution:

Note that 5 mA equals 5×10^{-3} A. Substituting the values for I and R into the formula $V = IR$, we get the following result:

$$V = IR = (5 \text{ mA})(50 \text{ } \Omega)$$
$$= (5 \times 10^{-3} \text{ A})(50 \text{ } \Omega) = 250 \times 10^{-3} \text{ V}$$

Since 250×10^{-3} V equals 250 mV, when *milliamperes* are multiplied by *ohms*, we get *millivolts*.

Example 3–11

Suppose that 8 μA are flowing through a 10-Ω resistor. How much voltage is across the resistor?

Solution:

Note that 8 μA equals 8×10^{-6} A. Substituting the values for I and R into the formula $V = IR$, we get the voltage as follows:

$$V = IR = (8 \text{ } \mu\text{A})(10 \text{ } \Omega)$$
$$= (8 \times 10^{-6} \text{ A})(10 \text{ } \Omega) = 80 \times 10^{-6} \text{ V}$$

Since 80×10^{-6} V equals 80 μV, when *microamperes* are multiplied by *ohms*, we get *microvolts*.

These examples have demonstrated that when we multiply *milliamperes* and *ohms*, we get *millivolts*. When we multiply *microamperes* and *ohms*, we get *microvolts*.

Larger Units of Resistance

The following two examples illustrate use of resistance values in the kilohm (kΩ) and megohm (MΩ) ranges when calculating voltage.

Example 3–12

The circuit in Figure 3–9 has a current of 10 mA. What is the voltage?

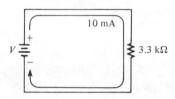

FIGURE 3–9

Solution:

Note that 10 mA equals 10×10^{-3} A and that 3.3 kΩ equals 3.3×10^3 Ω. Substituting these values into the formula $V = IR$, we get

$$V = IR = (10 \text{ mA})(3.3 \text{ k}\Omega)$$
$$= (10 \times 10^{-3} \text{ A})(3.3 \times 10^3 \text{ }\Omega) = 33 \text{ V}$$

Since 10^{-3} and 10^3 cancel, *milliamperes* cancel *kilohms* when multiplied, and the result is *volts*.

Example 3–13

If 50 μA are flowing through a 5-MΩ resistor, what is the voltage?

Solution:

Note that 50 μA equals 50×10^{-6} A and that 5 MΩ is 5×10^6 Ω. Substituting these values into $V = IR$, we get

$$V = IR = (50 \text{ μA})(5 \text{ M}\Omega)$$
$$= (50 \times 10^{-6} \text{ A})(5 \times 10^6 \text{ }\Omega) = 250 \text{ V}$$

Since 10^{-6} and 10^6 cancel, *microamperes* cancel *megohms* when multiplied, and the result is *volts*.

Review for 3–3

In Problems 1–7, calculate V when

1. $I = 1$ A and $R = 10$ Ω.
2. $I = 8$ A and $R = 470$ Ω.
3. $I = 3$ mA and $R = 100$ Ω.
4. $I = 25$ μA and $R = 50$ Ω.
5. $I = 2$ mA and $R = 1.8$ kΩ.
6. $I = 5$ mA and $R = 100$ MΩ.
7. $I = 10$ μA and $R = 2$ MΩ.
8. How much voltage is required to produce 100 mA through 4.7 kΩ?
9. What voltage do you need to cause 3 mA of current in a 3-kΩ resistance?
10. A battery produces 2 A of current into a 6-Ω resistive load. What is the battery voltage?

CALCULATING RESISTANCE

3–4

In this section you will learn to determine resistance values when the current and voltage are known. In these problems, the formula $R = V/I$ is used. To get resistance in *ohms*, you must express the value of I in *amperes* and the value of V in *volts*.

Example 3–14

In the circuit of Figure 3–10, how much resistance is needed to draw 3 A of current from the battery?

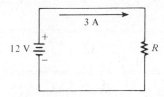

FIGURE 3–10

Solution:

Substitute 12 V for V and 3 A for I into the formula $R = V/I$:

$$R = \frac{V}{I} = \frac{12 \text{ V}}{3 \text{ A}} = 4 \text{ }\Omega$$

Smaller Units of Current

The following two examples illustrate use of current values in the milli-ampere (mA) and microampere (μA) ranges when calculating resistance.

Example 3–15

Suppose that the ammeter in Figure 3–11 indicates 5 mA of current and the voltmeter reads 150 V. What is the value of R?

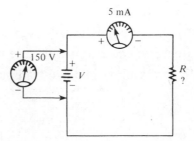

FIGURE 3–11

Solution:

Note that 5 mA equals 5×10^{-3} A. Substituting the voltage and current values into the formula $R = V/I$, we get

$$R = \frac{V}{I} = \frac{150 \text{ V}}{5 \text{ mA}} = \frac{150 \text{ V}}{5 \times 10^{-3} \text{ A}}$$

$$= 30 \times 10^{3} \ \Omega = 30 \text{ k}\Omega$$

Thus, if *volts* are divided by *milliamperes*, the resistance will be in *kilohms*.

Example 3–16

Suppose that the value of the resistor in Figure 3–11 is changed. If the battery voltage is still 150 V and the ammeter reads 75 μA, what is the new resistor value?

Solution:

Note that 75 μA equals 75×10^{-6} A. Substituting V and I values into the equation for R, we get

$$R = \frac{V}{I} = \frac{150 \text{ V}}{75 \text{ } \mu\text{A}} = \frac{150 \text{ V}}{75 \times 10^{-6} \text{ A}}$$

$$= 2 \times 10^6 \text{ } \Omega = 2 \text{ M}\Omega$$

Thus, if *volts* are divided by *microamperes,* the resistance has units of *megohms*.

Review for 3–4

In Problems 1–5, calculate R when

1. $V = 10$ V and $I = 2$ A.

2. $V = 250$ V and $I = 10$ A.

3. $V = 20$ kV and $I = 5$ A.

4. $V = 15$ V and $I = 3$ mA.

5. $V = 5$ V and $I = 2$ μA.

6. You have a resistor across which you measure 25 V, and your ammeter indicates 50 mA of current. What is the resistor's value in kilohms? In ohms?

THE LINEAR RELATIONSHIP OF CURRENT AND VOLTAGE

3–5

Ohm's law brings out a very important relationship between current and voltage: They are *linearly proportional*. You may have already recognized this relationship from our discussions in the previous sections.

When we say that the current and voltage are linearly proportional, we mean that if one is increased or decreased by a certain percentage, the other will increase or decrease by the same percentage, assuming that the resistance is constant in value. For example, if the voltage across a resistor is tripled, the current will triple.

Example 3–17

Show that if the voltage in the circuit of Figure 3–12 is increased to three times its present value, the current will triple in value.

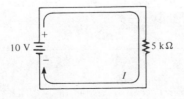

FIGURE 3–12

64

Example 3–17 (continued)

Solution:

With 10 V, the current is

$$I = \frac{V}{R} = \frac{10 \text{ V}}{5 \text{ k}\Omega} = 2 \text{ mA}$$

If the voltage is increased to 30 V, the current will be

$$I = \frac{V}{R} = \frac{30 \text{ V}}{5 \text{ k}\Omega} = 6 \text{ mA}$$

The current went from 2 mA to 6 mA (tripled) when the voltage was tripled to 30 V.

A Graph of Current versus Voltage

Let us take a constant value of resistance, for example, 10 Ω, and calculate the current for several values of voltage ranging from 10 V to 100 V. The values obtained are shown in Figure 3–13A. The graph of the *I* values versus the *V* values is shown in Figure 3–13B. Note that it is a straight line graph. This graph tells us that a change in voltage results in a linearly proportional change in current.

No matter what value *R* is, assuming that *R* is constant, the graph of *I* versus *V* will always be a straight line. Example 3–18 illustrates a use for the linear relationship between voltage and current in a resistive circuit.

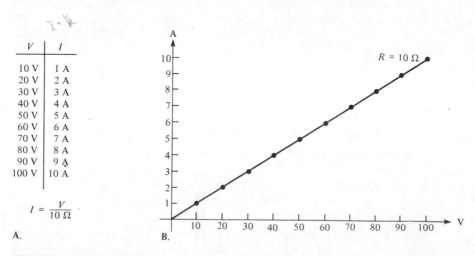

FIGURE 3–13 *Graph of current versus voltage for R = 10 Ω.*

Example 3–18

Assume that you are measuring the current in a circuit that is operating with 25 V. The ammeter reads 50 mA. Later, you notice that the current has dropped to 40 mA. Assuming that the resistance did not change, you must conclude that the voltage has changed. How much has the voltage changed, and what is its new value?

Solution:

The current has dropped from 50 mA to 40 mA, which is a decrease of 20%. Since the voltage is linearly proportional to the current, the voltage has decreased by the same percentage that the current did. Taking 20% of 25 V, we get

$$\text{Change in voltage} = (0.2)(25 \text{ V}) = 5 \text{ V}$$

Subtracting this change from the original voltage, we get the new voltage as follows:

$$\text{New voltage} = 25 \text{ V} - 5 \text{ V} = 20 \text{ V}$$

Notice that we did not need the resistance value in order to find the new voltage.

Review for 3–5

1. What does *linearly proportional* mean?
2. In a circuit, $V = 2$ V and $I = 10$ mA. If V is changed to 1 V, what will I equal?
3. If $I = 3$ A at a certain voltage, what will it be if the voltage is doubled?
4. By how many volts must you increase a 12-V source in order to increase the current in a circuit by 50%?

COMPUTER ANALYSIS

3–6

As mentioned in Chapter 1, several examples of computer-aided analysis are presented throughout the book. These examples illustrate how computer programs can be written in BASIC to solve typical circuit problems. The program in this chapter is the first of these examples.

This program is written in level II BASIC for the TRS-80, but it can be run with little or no change on other machines.

This program is based on Ohm's law and computes the unknown value when two known values are provided. For example, if voltage and current values are provided, the resistance is calculated. If voltage and resistance values are

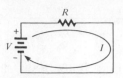

FIGURE 3–14

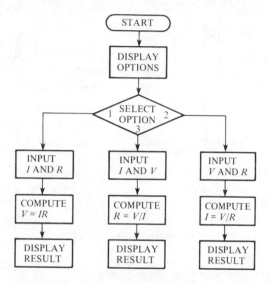

FIGURE 3–15

given, the current is calculated. If current and resistance values are given, the voltage is calculated. The generalized circuit for which solutions are computed is shown in Figure 3–14.

The *flow chart* in Figure 3–15 shows the sequence of operations that the computer follows when the program is executed.

The listing of the program itself appears below. Note that it consists of 34 lines. Lines 10 through 80 display the program options and allow the selection of a desired option. Line 100 is a branching statement that causes the computer to go to the segment of the program required for the selected option. Lines 110 through 180 are for option 1; lines 190 through 260 are for option 2; and lines 270 through 340 are for option 3.

The program must be entered into the computer via the keyboard exactly as it appears and then executed (RUN).

```
10 CLS
20 PRINT "THIS PROGRAM PROVIDES THREE OPTIONS:"
30 PRINT
40 PRINT "(1) COMPUTE VOLTAGE IF CURRENT AND RESISTANCE ARE KNOWN"
50 PRINT "(2) COMPUTE CURRENT IF VOLTAGE AND RESISTANCE ARE KNOWN"
60 PRINT "(3) COMPUTE RESISTANCE IF VOLTAGE AND CURRENT ARE KNOWN"
```

```
70 PRINT:PRINT
80 INPUT "SELECT OPTION 1, 2, OR 3";X
90 CLS
100 ON X GOTO 110, 190, 270
110 INPUT "CURRENT IN AMPS";I
120 INPUT "RESISTANCE IN OHMS";R
130 CLS
140 V=I*R
150 PRINT "V=";V;"VOLTS"
160 PRINT:PRINT:PRINT
170 INPUT "FOR NEXT COMPUTATION PRESS 'ENTER'";Y
180 GOTO 10
190 INPUT "VOLTAGE IN VOLTS";V
200 INPUT "RESISTANCE IN OHMS";R
210 CLS
220 I=V/R
230 PRINT "I=";I;"AMPS"
240 PRINT:PRINT:PRINT
250 INPUT "FOR NEXT COMPUTATION PRESS 'ENTER'";Y
260 GOTO 10
270 INPUT "VOLTAGE IN VOLTS";V
280 INPUT "CURRENT IN AMPS";I
290 CLS
300 R=V/I
310 PRINT "R=";R;"OHMS"
320 PRINT:PRINT:PRINT
330 INPUT "FOR NEXT COMPUTATION PRESS 'ENTER'";Y
340 GOTO 10
```

Review for 3–6

1. What function do the INPUT statements perform?

2. What is the purpose of line 100?

Formulas

$$I = \frac{V}{R} \qquad\qquad\qquad\qquad (3\text{--}1)$$

$$V = IR \qquad\qquad\qquad\qquad (3\text{--}2)$$

$$R = \frac{V}{I} \qquad\qquad\qquad\qquad (3\text{--}3)$$

Summary

1. There are three forms of Ohm's law, all of which are equivalent.

2. Use $I = V/R$ when calculating the current.

3. Use $V = IR$ when calculating the voltage.

4. Use $R = V/I$ when calculating the resistance.

5. Voltage and current are linearly proportional.

Self-Test

1. Write the three forms of Ohm's law from memory.
2. Which of the three formulas is used for finding current? Voltage? Resistance?
3. Calculate I for $V = 10$ V and $R = 5\ \Omega$.
4. Calculate I for $V = 75$ V and $R = 10$ kΩ.
5. Calculate V for $I = 2.5$ A and $R = 20\ \Omega$.
6. Calculate V for $I = 30$ mA and $R = 2.2$ kΩ.
7. Calculate R for $V = 12$ V and $I = 6$ A.
8. Calculate R for $V = 9$ V and $I = 100\ \mu$A.
9. Determine the current if $V = 50$ kV and $R = 2.5$ kΩ.
10. Determine the current if $V = 15$ mV and $R = 10$ kΩ.
11. Determine the voltage if $I = 4\ \mu$A and $R = 100$ kΩ.
12. Determine the voltage if $I = 100$ mA and $R = 1$ MΩ.
13. Determine the resistance if $V = 5$ V and $I = 2$ mA.
14. If a 1.5-V battery is connected across a 5-kΩ resistor, what is the current?
15. Determine the current in each circuit of Figure 3–16.

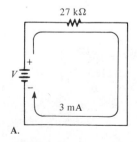
A.

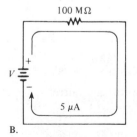
B.

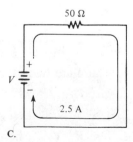
C.

FIGURE 3–16

16. Assign a voltage value to each source in the circuits of Figure 3–17 to obtain the indicated amounts of current.

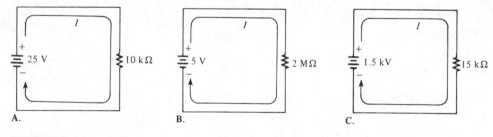
A.   B.   C.

FIGURE 3–17

17. Choose the correct value of resistance to get the current values indicated in each circuit of Figure 3–18.

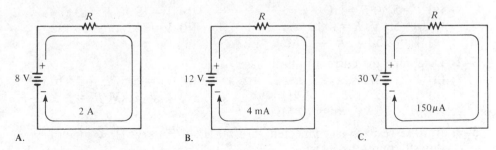

FIGURE 3–18

18. You are measuring the current in a circuit that is operated on a 10-V battery. The ammeter reads 50 mA. Later, you notice that the current has dropped to 30 mA. Eliminating the possibility of a resistance change, you must conclude that the voltage has changed. How much has the voltage of the battery changed, and what is its new value?

19. If you wish to increase the amount of current in a resistor from 100 mA to 150 mA by changing the 20-V source, by how many volts should you change the source? To what new value should you set it?

20. By varying the rheostat (variable resistor) in the circuit of Figure 3–19, you can change the amount of current. The setting of the rheostat is such that the current is 750 mA. What is the ohmic value of this setting? To adjust the current to 1 A, to what ohmic value must you set the rheostat?

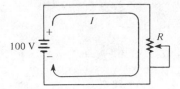

FIGURE 3–19

21. Use the computer program to compute the current in each circuit of Figure 3–16.

22. Use the computer program to compute the voltage in each circuit of Figure 3–17.

23. Use the computer program to compute the resistance in each circuit of Figure 3–18.

70

Problems

Section 3–2

3–1. Determine the current in each case.
   (a) $V = 5$ V, $R = 1$ Ω   **(b)** $V = 15$ V, $R = 10$ Ω
   (c) $V = 50$ V, $R = 100$ Ω   **(d)** $V = 30$ V, $R = 15$ kΩ
   (e) $V = 250$ V, $R = 5$ MΩ

3–2. Determine the current in each case.
   (a) $V = 9$ V, $R = 2.7$ kΩ   **(b)** $V = 5.5$ V, $R = 10$ kΩ
   (c) $V = 40$ V, $R = 68$ kΩ   **(d)** $V = 1$ kV, $R = 2$ kΩ
   (e) $V = 66$ kV, $R = 10$ MΩ

3–3. A 10-Ω resistor is connected across a 12-V battery. How much current flows through the resistor?

Section 3–3

3–4. Calculate the voltage for each value of I and R.
   (a) $I = 2$ A, $R = 18$ Ω   **(b)** $I = 5$ A, $R = 50$ Ω
   (c) $I = 2.5$ A, $R = 600$ Ω   **(d)** $I = 0.6$ A, $R = 47$ Ω
   (e) $I = 0.1$ A, $R = 500$ Ω

3–5. Calculate the voltage for each value of I and R.
   (a) $I = 1$ mA, $R = 10$ Ω   **(b)** $I = 50$ mA, $R = 33$ Ω
   (c) $I = 3$ A, $R = 5$ kΩ   **(d)** $I = 1.6$ mA, $R = 2.2$ kΩ
   (e) $I = 250$ μA, $R = 1$ kΩ   **(f)** $I = 500$ mA, $R = 1.5$ MΩ
   (g) $I = 850$ μA, $R = 10$ MΩ   **(h)** $I = 75$ μA, $R = 50$ Ω

3–6. Three amperes of current are measured through a 27-Ω resistor connected across a voltage source. How much voltage does the source produce?

Section 3–4

3–7. Calculate the resistance for each value of V and I.
   (a) $V = 10$ V, $I = 2$ A   **(b)** $V = 90$ V, $I = 45$ A
   (c) $V = 50$ V, $I = 5$ A   **(d)** $V = 5.5$ V, $I = 10$ A
   (e) $V = 150$ V, $I = 0.5$ A

3–8. Calculate R for each set of V and I values.
   (a) $V = 10$ kV, $I = 5$ A   **(b)** $V = 7$ V, $I = 2$ mA
   (c) $V = 500$ V, $I = 250$ mA   **(d)** $V = 50$ V, $I = 500$ μA
   (e) $V = 1$ kV, $I = 1$ mA

3–9. Six volts are applied across a resistor. A current of 2 mA is measured. What is the value of the resistor?

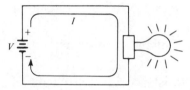

A.

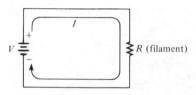

B.

FIGURE 3–20

3–10. The filament of a light bulb in the circuit of Figure 3–20A has a certain amount of resistance, represented by an equivalent resistance in Figure 3–20B. If the bulb operates with 120 V and 0.8 A of current, what is the resistance of its filament?

3–11. A certain electrical device has an unknown resistance. You have available a 12-V battery and an ammeter. How would you determine the value of the unknown resistance? Draw the necessary circuit connections.

Section 3–5

3–12. A variable voltage source is connected to the circuit of Figure 3–21. Start at 0 V and increase the voltage in 10-V steps up to 100 V. Determine the current at each voltage point, and plot a graph of V versus I. Is the graph a straight line? What does the graph indicate?

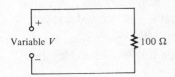

FIGURE 3–21

3–13. In a certain circuit, $V = 1$ V and $I = 5$ mA. Determine the current for each of the following voltages in the same circuit:
(a) $V = 1.5$ V   (b) $V = 2$ V   (c) $V = 3$ V
(d) $V = 4$ V   (e) $V = 10$ V

3–14. Figure 3–22 is a graph of voltage versus current for three resistance values. Determine R_1, R_2, and R_3.

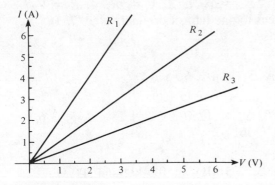

FIGURE 3–22

3–15. Which circuit in Figure 3–23 (page 72) has the most current? The least current?

Section 3–6

3–16. Modify the program in Section 3–6 so that the two known values are displayed along with the final result for each option.

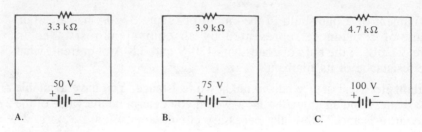

A.   B.   C.

FIGURE 3–23

3–17. Write a program to compute values of current for a specified range of voltage values and for a single resistor value.

Advanced Problems

3–18. A certain resistor has the following color code: orange, orange, red, gold. Determine the maximum and minimum currents you should expect to measure when a 12-V source is connected across the resistor.

3–19. A 6-V source is connected to a 100-Ω resistor by two 12-ft lengths of 18-gage copper wire. Determine the following:
(a) Current
(b) Resistor voltage
(c) Voltage drop across each length of wire

3–20. Plot a graph of current versus voltage for voltage values ranging from 10 V to 100 V in a 10-V steps for each of the following resistance values:
(a) 1 Ω (b) 5 Ω (c) 20 Ω (d) 100 Ω

Answers to Section Reviews

Section 3–1:
1. Current, voltage, and resistance. **2.** $I = V/R$. **3.** $V = IR$. **4.** $R = V/I$.
5. Increases by three times. **6.** Reduces to one-half of original value. **7.** 0.5 A.
8. Yes, it would increase by four times.

Section 3–2:
1. 2 A. **2.** 0.2 A. **3.** 2 mA. **4.** 3 μA. **5.** 4 mA. **6.** 5 A.

Section 3–3:
1. 10 V. **2.** 3760 V. **3.** 300 mV or 0.3 V. **4.** 1250 μV or 1.25 mV. **5.** 3.6 V.
6. 500 kV. **7.** 20 V. **8.** 470 V. **9.** 9 V. **10.** 12 V.

Section 3–4:
1. 5 Ω. **2.** 25 Ω. **3.** 4 kΩ. **4.** 5 kΩ. **5.** 2.5 MΩ. **6.** 0.5 kΩ and 500 Ω.

Section 3–5:
1. The same percentage change occurs in two quantities. **2.** 5 mA. **3.** 6 A.
4. 6 V.

Section 3–6:
1. To allow data to be input to the computer. **2.** To branch the computer to the selected option.

Energy is the ability to do work, and power is the rate at which energy is used. Current carries electrical energy through a circuit. As the electrons pass through the resistance of the circuit, they give up their energy when they collide with atoms in the resistive material. The electrical energy given up by the electrons is converted into heat energy. The rate at which the electrical energy is lost is the power in the circuit.

In this chapter, you will learn:

1. The definition of power and its unit, the watt (W).
2. How to determine the amount of power in a resistive circuit.
3. How to choose resistors to handle the power.
4. The definition of the kilowatthour (kWh).
5. How energy loss and voltage drop are related.
6. The characteristics of power supplies and batteries.
7. What the ampere-hour rating means.
8. The definition of power supply efficiency.
9. The use of a simple computer program in analysis work.

4

ENERGY AND POWER

4–1

As mentioned, *power is the rate at which energy is used*. In other words, power is a certain amount of energy used in a certain length of time, expressed as follows:

$$\text{Power} = \frac{\text{energy}}{\text{time}} \qquad (4\text{–}1)$$

The symbol for energy is $\mathscr{E}$, the symbol for time is t, and the symbol for power is P. Using these symbols, we can rewrite Equation (4–1) more concisely:

$$P = \frac{\mathscr{E}}{t} \qquad (4\text{–}2)$$

Units

Energy is measured in *joules* (J), time is measured in *seconds* (s), and power is measured in *watts* (W).

Energy in *joules* divided by time in *seconds* gives power in *watts*. For example, if 50 J of energy are used in 2 s, the power is 50 J/2 s = 25 W. Equation (4–3) expresses power in terms of units:

$$\text{Watts} = \frac{\text{joules}}{\text{seconds}} \qquad (4\text{–}3)$$

One Watt

By definition, *one watt* is the amount of power when *one joule* of energy is consumed in *one second*. Thus, the number of joules consumed in one second is always equal to the number of watts. For example, if 75 J are used in 1 s, the power is 75 W. The following two examples illustrate use of these units when calculating power.

Example 4–1

An amount of energy equal to 100 J is used in 5 s. What is the power in watts?

Solution:

$$P = \frac{\text{energy}}{\text{time}} = \frac{100 \text{ J}}{5 \text{ s}} = 20 \text{ W}$$

Example 4–2

If 1000 J are used in 1 s, what is the power?

Solution:

$$P = \frac{1000 \text{ J}}{1 \text{ s}} = 1000 \text{ W}$$

Smaller and Larger Units

Amounts of power much less than one watt are common in certain areas of electronics. As with small current and voltage values, metric prefixes are used to designate small amounts of power. Thus, *milliwatts* (mW) and *microwatts* (μW) are commonly found in some applications.

In the electrical utilities field, *kilowatts* (kW) and *megawatts* (MW) are common units. Radio and television stations also use large amounts of power to transmit signals.

Example 4–3

Convert the following powers from watts to the appropriate metric units:
(a) 0.045 W   (b) 0.000012 W   (c) 3500 W   (d) 10,000,000 W

Solution:

(a) 0.045 W = 45 mW   (b) 0.000012 W = 12 μW
(c) 3500 W = 3.5 kW   (d) 10,000,000 W = 10 MW

Review for 4–1

1. Define *power*.
2. Write the formula for power in terms of energy and time.
3. Define *watt*.
4. If 25 J of energy are used in 5 s, what is the rate of energy consumption in watts?
5. If 100 J are used in 10 ms, what is the power?
6. Express each of the following values of power in the most appropriate units:
   (a) 68,000 W   (b) 0.005 W   (c) 0.000025 W

POWER FORMULAS

4–2

When current flows through a resistance, energy is dissipated. In this section you will learn three formulas for calculating power in a resistance. First, recall from Chapter 2 that voltage can be expressed in terms of energy and charge as

$V = \mathscr{E}/Q$. Also recall that current can be expressed in terms of charge and time as $I = Q/t$. When V and I are multiplied, the result is

$$VI = \left(\frac{\mathscr{E}}{Q}\right)\left(\frac{Q}{t}\right)$$

$$= \frac{\mathscr{E}}{t} = \text{power}$$

The power in a resistor is the product of the voltage across the resistor and the current through it:

$$P = VI \qquad\qquad (4\text{–}4)$$

Ohm's law states that $V = IR$. If IR is substituted for V in the formula $P = VI$, we get a second form of the power equation:

$$P = VI$$

$$= (IR)I$$

$$P = I^2R \qquad\qquad (4\text{–}5)$$

Ohm's law also states that $I = V/R$. If V/R is substituted for I in the formula $P = VI$, we get a third form of the power equation:

$$P = VI$$

$$= V\left(\frac{V}{R}\right)$$

$$P = \frac{V^2}{R} \qquad\qquad (4\text{–}6)$$

Equations (4–4), (4–5), and (4–6) are all equivalent.

Calculation of Power

To calculate the power in a resistance, you can use any one of the three power formulas, depending on what information you have. For example, in Figure 4–1, assume that you know the values of current and voltage. In this case you calculate the power with the formula $P = VI$. If you know I and R, use the formula $P = I^2R$. If you know V and R, use the formula $P = V^2/R$.

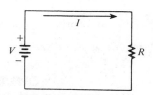

FIGURE 4–1

Example 4–4

Calculate the power in each of the three circuits of Figure 4–2.

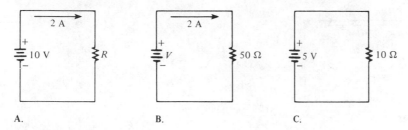

A.   B.   C.

FIGURE 4–2

Solution:

In circuit A, V and I are known. The power is determined as follows:

$$P = VI = (10 \text{ V})(2 \text{ A}) = 20 \text{ W}$$

In circuit B, I and R are known. The power is determined as follows:

$$P = I^2R = (2 \text{ A})^2(50 \text{ } \Omega) = 200 \text{ W}$$

In circuit C, V and R are known. The power is determined as follows:

$$P = \frac{V^2}{R} = \frac{(5 \text{ V})^2}{10 \text{ } \Omega} = 2.5 \text{ W}$$

Example 4–5

A 100-W light bulb operates on 120 V. How much current does it require?

Solution:

Use the formula $P = VI$ and solve for I as follows:

$$I = \frac{P}{V} = \frac{100 \text{ W}}{120 \text{ V}} = 0.833 \text{ A}$$

Review for 4–2

1. Write the three power formulas from memory.

2. If there are 10 V across a resistor and a current of 3 A flowing through it, what is the power?

3. How much power does the source in Figure 4–3 generate? What is the power in the resistor? Are the two values the same? Why?

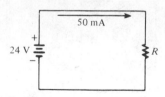

FIGURE 4–3

4. If a current of 5 A is flowing through a 50-Ω resistor, what is the power?

5. How much power is produced by 20 mA through a 5-kΩ resistor?

6. Five volts are applied to a 10-Ω resistor. What is the power?

7. How much power does a 2-kΩ resistor with 8 V across it produce?

8. What is the resistance of a 75-W bulb that takes 0.5 A?

RESISTOR POWER RATINGS

4–3

As you have learned, a resistor dissipates energy, and, as a result, it heats up. A resistor must be able to dissipate a sufficient amount of heat, depending on how much power it is required to handle. The amount of power that a resistor can handle is determined by the physical size and shape of the resistor. *The larger the surface area of a resistor, the more power it can handle.* Figure 4–4 shows carbon-composition resistors with standard available power ratings.

When a resistor is used in a circuit, its power rating must be greater than the power that it actually handles. For example, if a carbon-composition resistor such as those shown in Figure 4–4 is to handle 0.75 W in a circuit, its rating should be 1 watt or greater. Figure 4–5 shows some resistors with power ratings

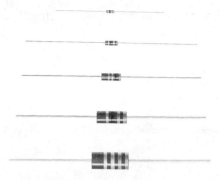

FIGURE 4–4 *Carbon-composition resistors with standard power ratings of 1/8 W, 1/4 W, 1/2 W, 1 W, and 2 W. (Courtesy of Allen-Bradley Company)*

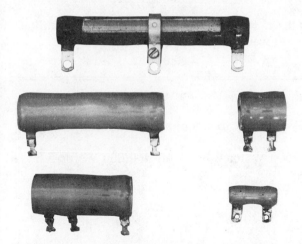

FIGURE 4–5 *Typical power resistors.*

higher than those available in standard carbon-composition resistors. These are usually wire-wound resistors.

Review for 4–3

1. Name two important values associated with a resistor.

2. How does the physical size of a resistor determine the amount of power that it can handle?

3. List the standard power ratings of carbon-composition resistors.

4. A resistor must handle 0.3 W. What size carbon resistor should be used to dissipate the energy properly?

ENERGY

4–4

Since power is the rate of energy usage, power utilized over a period of time represents energy consumption. If we multiply *power* and *time,* we have *energy:*

$$\text{Energy} = (\text{power})(\text{time})$$

$$\mathscr{E} = Pt \qquad\qquad (4\text{–}7)$$

Unit of Energy

Earlier, the joule was defined as a unit of energy. However, there is another way of expressing energy. Since power is expressed in *watts* and time in *seconds,* we can use units of energy called the *wattsecond* (Ws), *watthour* (Wh), and *kilowatthour* (kWh).

When you pay your electric bill, you are charged on the basis of the amount of *energy* you use. Because power companies deal in huge amounts of energy, the most practical unit is the *kilowatthour*. You have used a kilowatthour of energy when you have used 1000 watts of power for one hour.

Example 4–6

Determine the number of kilowatthours for each of the following energy consumptions:

(a) 1400 W for 1 h   (b) 2500 W for 2 h   (c) 100,000 W for 5 h

Solution:

(a) 1400 W = 1.4 kW
Energy = (1.4 kW)(1 h) = 1.4 kWh
(b) 2500 W = 2.5 kW
Energy = (2.5 kW)(2 h) = 5 kWh
(c) 100,000 W = 100 kW
Energy = (100 kW)(5 h) = 500 kWh

Review for 4–4

1. Distinguish between energy and power.

2. Write the formula for energy in terms of power and time.

3. What is the most practical unit of energy?

4. If you use 100 W of power for 10 h, how much energy (in kilowatthours) have you consumed?

5. Convert 2000 Wh to kilowatthours.

6. Convert 360,000 Ws to kilowatthours.

ENERGY LOSS AND VOLTAGE DROP

4–5   When current flows through a resistance, energy is dissipated in the form of heat. This heat loss is caused by collisions of the free electrons within the atomic structure of the resistive material. When a collision occurs, heat is given off, and the electron loses some of its acquired energy.

In Figure 4–6, electrons are flowing out of the negative terminal of the battery. They have acquired energy from the battery and are at their highest energy level at the negative side of the circuit. As the electrons move through the resistor, they *lose* energy. The electrons emerging from the upper end of the resistor are at a lower energy level than those entering the lower end. The drop in energy level through the resistor creates a potential difference, or *voltage drop,* across the resistor having the polarity shown in Figure 4–6.

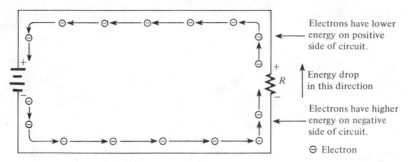

FIGURE 4–6 *Illustration of electron flow.*

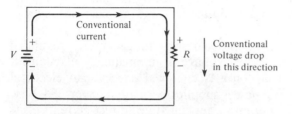

FIGURE 4–7 *Voltage drop with conventional current.*

The upper end of the resistor in Figure 4–6 is *less negative* (more positive) than the lower end. When we follow the flow of electrons, this change corresponds to a voltage drop from a more negative potential to a less negative potential. When we follow *conventional current,* as is the practice in this book, we think of voltage drop as being from a more positive potential to a less positive (negative) potential. This concept is illustrated in Figure 4–7.

Review for 4–5

1. What is the basic reason for energy loss in a resistor?

2. What is a voltage drop?

3. What is the polarity of a voltage drop when electron flow is used?

4. What is the polarity of a voltage drop when conventional current is used?

POWER SUPPLIES

4–6

A *power supply* is a device that provides power to a load. A *load* is any electrical device or circuit that is connected to the output of the power supply and draws current from the supply.

Figure 4–8 shows a block diagram of a power supply with a loading device connected to it. The load can be anything from a light bulb to a computer. The power supply produces a voltage across its two output terminals and provides current through the load, as indicated in the figure. The product $V_{OUT}I$ is the

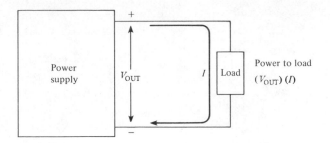

FIGURE 4–8 *Block diagram of power supply and load.*

amount of power produced by the supply and consumed by the load. For a given output voltage (V_{OUT}), more current drawn by the load means more power from the supply.

Power supplies range from simple batteries to accurately regulated electronic circuits where an accurate output voltage is automatically maintained. A battery is a dc power supply that converts chemical energy into electrical energy. (See Appendix B.) Electronic power supplies normally convert 115 V ac (alternating current) from a wall outlet into a regulated dc (direct current) or ac voltage. Figure 4–9A shows some typical batteries, and Figure 4–9B and C illustrate typical regulated power supplies used in the laboratory or shop.

Ampere-Hour Ratings of Batteries

Batteries convert chemical energy into electrical energy. Because of their limited source of chemical energy, batteries have a certain *capacity* which limits the amount of time over which they can produce a given power level. This

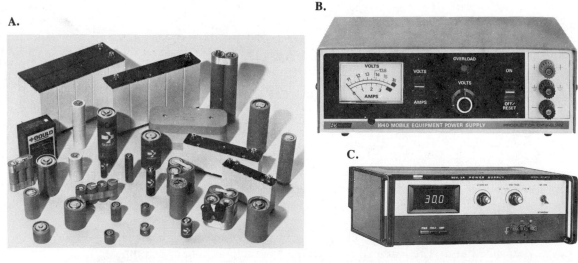

FIGURE 4–9 **A.** *Sealed Nicad® nickel-cadmium cells and batteries for consumer applications. (Courtesy of Gould, Inc., Portable Battery Division, St. Paul, Minn.)* **B.,C.** *Typical regulated power supplies. (**B.,** courtesy of B&K-Precision Test Instruments, Dynascan Corp.* **C.,** *courtesy of Health/Schlumberger Instruments)*

capacity is measured in *ampere-hours (Ah). The ampere-hour rating determines the length of time that a battery can deliver a certain amount of current to a load at the rated voltage.*

   A rating of one ampere-hour means that a battery can deliver *one* ampere of current to a load for *one* hour at the rated voltage output. This same battery can deliver two amperes for one-half hour. The more current the battery is required to deliver, the shorter is the life of the battery. In practice, a battery usually is rated for a specified current level and output voltage. For example, a 12-V automobile battery may be rated for 70 Ah at 3.5 A. This means that it can produce 3.5 A for 20 hours.

Example 4–7

For how many hours can a battery deliver 2 A if it is rated at 70 Ah?

Solution:

The ampere-hour rating is the current times the hours:

$$70 \text{ Ah} = (2 \text{ A})(x \text{ h})$$

Solving for x hours, we get

$$x = \frac{70 \text{ Ah}}{2 \text{ A}} = 35 \text{ h}$$

Power Supply Efficiency

   An important characteristic of electronic power supplies is efficiency. *Efficiency is the ratio of the output power to the input power:*

$$\text{Efficiency} = \frac{\text{output power}}{\text{input power}}$$

$$\text{Efficiency} = \frac{P_{\text{OUT}}.}{P_{\text{IN}}} \qquad (4\text{–}8)$$

Efficiency is usually expressed as a percentage. For example, if the input power is 100 W and the output power is 50 W, the efficiency is (50 W/100 W) × (100) = 50%.

   All power supplies require that power be put into them. For example, an electronic power supply might use the ac power from a wall outlet as its input. Its output may be regulated dc or ac. The output power is *always* less than the input power because some of the total power must be used internally to operate the power supply circuitry. This amount is normally called the *power loss*. The output power is the input power minus the amount of internal power loss:

$$P_{\text{OUT}} = P_{\text{IN}} - P_{\text{LOSS}} \qquad (4\text{–}9)$$

High efficiency means that little power is lost and there is a higher proportion of output power for a given input power.

Example 4–8

A certain power supply unit requires 25 W of input power. It can produce an output power of 20 W. What is its efficiency, and what is the power loss?

Solution:

$$\text{Efficiency} = \left(\frac{P_{OUT}}{P_{IN}}\right)100\% = \left(\frac{20\text{ W}}{25\text{ W}}\right)100\%$$

$$= 80\%$$

$$P_{LOSS} = P_{IN} - P_{OUT} = 25\text{ W} - 20\text{ W}$$

$$= 5\text{ W}$$

Review for 4–6

1. When a loading device draws an increased amount of current from a power supply, does this change represent a greater or a smaller load on the supply?

2. A power supply produces an output voltage of 10 V. If the supply provides 0.5 A to a load, what is the power output?

3. If a battery has an ampere-hour rating of 100 Ah, how long can it provide 5 A to a load?

4. If the battery in Problem 3 is a 12-V device, what is its power output for the specified value of current?

5. An electronic power supply used in the lab operates with an input power of 1 W. It can provide an output power of 750 mW. What is its efficiency?

COMPUTER ANALYSIS

4–7   The following program is a simple example of how the computer can be used to quickly perform many calculations and display the results in a convenient format.

The program computes the power for a specified resistance and for each of a specified number of current values within a specified range and displays the results in tabular form.

```
10 CLS
20 PRINT "THIS PROGRAM COMPUTES THE POWER FOR EACH OF A SPECIFIED"
30 PRINT "NUMBER OF CURRENT VALUES AND A SPECIFIED RESISTANCE."
40 PRINT "THE RESULTS ARE DISPLAYED IN TABULAR FORM."
50 PRINT:PRINT:PRINT
60 INPUT "TO CONTINUE PRESS 'ENTER'";X
70 CLS
80 INPUT "RESISTANCE VALUE IN OHMS";R
```

```
90 INPUT "MINIMUM CURRENT IN AMPS";I1
100 INPUT "MAXIMUM CURRENT IN AMPS";I2
110 INPUT "CURRENT INCREMENTS IN AMPS";II
120 CLS
130 PRINT TAB(15),"CURRENT (AMPS)"; TAB(35),"POWER (WATTS)"
140 FOR I=I1 TO I2 STEP II
150 P=I*I*R
160 PRINT TAB(15),I; TAB(35), P
170 NEXT
```

Review for 4–7

1. State the purpose of lines 80 through 110.

2. In which line is the power computation performed?

Formulas

$$P = \frac{\mathscr{E}}{t} \qquad\qquad (4\text{–}2)$$

$$P = VI \qquad\qquad (4\text{–}4)$$

$$P = I^2R \qquad\qquad (4\text{–}5)$$

$$P = \frac{V^2}{R} \qquad\qquad (4\text{–}6)$$

$$\mathscr{E} = Pt \qquad\qquad (4\text{–}7)$$

$$\text{Efficiency} = \frac{P_{\text{OUT}}}{P_{\text{IN}}} \qquad\qquad (4\text{–}8)$$

$$P_{\text{OUT}} = P_{\text{IN}} - P_{\text{LOSS}} \qquad\qquad (4\text{–}9)$$

Summary

1. Power is the rate at which energy is used.

2. One watt equals one joule per second.

3. Watt is the unit of power, joule is a unit of energy, and second is a unit of time.

4. Use $P = VI$ to calculate power when voltage and current are known.

5. Use $P = I^2R$ to calculate power when current and resistance are known.

6. Use $P = V^2/R$ to calculate power when voltage and resistance are known.

7. The power rating of a resistor determines the maximum power that it can handle safely.

8. Resistors with a larger physical size can dissipate more power in the form of heat than smaller ones.

9. A resistor should have a power rating higher than the maximum power that it is expected to handle in the circuit.

10. Energy is equal to power multiplied by time.

11. The kilowatthour is a unit of energy.

12. One kilowatthour = 1000 watts used for one hour.

13. When conventional current direction is used, the polarity of a voltage drop is from plus (+) to minus (−).

14. A power supply is an energy source used to operate electrical and electronic devices.

15. A battery is one type of power supply that converts chemical energy into electrical energy.

16. An electronic power supply converts commercial energy (ac from the power company) to regulated dc or ac at various voltage levels.

17. The output power of a supply is the output voltage times the load current.

18. A load is a device that draws current from the power supply.

19. The capacity of a battery is measured in ampere-hours (Ah).

20. One ampere-hour equals one ampere used for one hour, or any other combination of amperes and hours that has a product of one.

21. Efficiency of a power supply is equal to the output power divided by the input power. Multiply by 100 to get percentage of efficiency.

22. A power supply with a high efficiency wastes less power than one with a lower efficiency.

Self-Test

1. Two hundred joules of energy are consumed in 10 s. What is the power?

2. If it takes 300 ms to use 10,000 J of energy, what is the power?

3. How many watts are there in 50 kW?

4. How many milliwatts are there in 0.045 W?

5. If a resistive load draws 350 mA from a 10-V battery, how much power is handled by the resistor? How much power is delivered by the battery?

6. If a 1000-Ω resistor is connected across a 50-V source, what is the power?

7. A 5-kΩ resistor draws 500 mA from a voltage source. How much power is handled by the resistor?

8. What is the power of a light bulb that operates on 115 V and 2 A of current?

9. If 15 W of power are handled by a resistor, how many joules of energy are used in one minute?

10. If you have used 500 W of power for 24 h, how many kilowatthours have you used?

11. How many watthours represent 75 W used for 10 h?

12. If your average daily power usage is 750 W, how many kilowatthours will be read on your meter at the end of April? The reading at the end of March was 1000 kWh.

13. Convert 1,500,000 Ws to kWh.

14. For each circuit in Figure 4–10, assign the proper polarity for the voltage drop across the resistor.

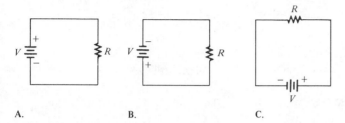

A.   B.   C.

FIGURE 4–10

15. Two 25-V power supplies are sitting on the lab bench. Power supply 1 is providing 100 mA to a load. Power supply 2 is supplying 0.5 A to a load. Which supply has the greater load? Which resistive load has the smaller ohmic value?

16. What is the power output of each power supply in Problem 15?

17. A 12-V battery is connected to a 600-Ω load. Under these conditions, it is rated at 50 Ah. How long can it supply current to the load?

18. A given power supply is capable of providing 8 A for 2.5 h. What is its ampere-hour rating?

19. A laboratory power supply requires 250 mW internally. If it takes 2 W of input power, what is the output power?

20. A power supply produces a 0.5-W output with an input of 0.6 W. What is its percentage of efficiency?

21. Use the computer program to compute powers in a circuit with a resistance of 1500 Ω and with current values ranging from 0.01 A to 0.1 A in 0.01-A steps.

Problems

Section 4–1

4–1. What is the power when energy is consumed at the rate of 350 J/s?

4–2. How many watts are used when 7500 J of energy are consumed in 5 h?

4–3. How many watts does 1000 J in 50 ms equal?

4–4. Convert the following to kilowatts:
   (a) 1000 W   (b) 3750 W   (c) 160 W   (d) 50,000 W

4–5. Convert the following to megawatts:
   (a) 1,000,000 W   **(b)** 3×10^6 W   **(c)** 15×10^7 W
   (d) 8700 kW

4–6. Convert the following to milliwatts:
   (a) 1 W   **(b)** 0.4 W   **(c)** 0.002 W   **(d)** 0.0125 W

4–7. Convert the following to microwatts:
   (a) 2 W   **(b)** 0.0005 W   **(c)** 0.25 mW   **(d)** 0.00667 mW

4–8. Convert the following to watts:
   (a) 1.5 kW   **(b)** 0.5 MW   **(c)** 350 mW   **(d)** 9000 μW

Section 4–2

4–9. If a 75-V source is supplying 2 A to a load, what is the ohmic value of the load?

4–10. If a resistor has 5.5 V across it and 3 mA flowing through it, what is the power?

4–11. An electric heater works on 115 V and draws 3 A of current. How much power does it use?

4–12. How much power is produced by 500 mA of current through a 4.7-kΩ resistor?

4–13. Calculate the power handled by a 10-kΩ resistor carrying 100 μA.

4–14. If there are 60 V across a 600-Ω resistor, what is the power?

4–15. A 50-Ω resistor is connected across the terminals of a 1.5-V battery. What is the power dissipation in the resistor?

4–16. If a resistor is to carry 2 A of current and handle 100 W of power, how many ohms must it be? Assume that the voltage can be adjusted to any required value.

Section 4–3

4–17. A 6.8-kΩ resistor has burned out in a circuit. You must replace it with another resistor with the same ohmic value. If the resistor carries 10 mA, what should its power rating be? Assume that you have available carbon-composition resistors in all the standard power ratings.

4–18. A certain type of power resistor comes in the following ratings: 3 W, 5 W, 8 W, 12 W, 20 W. Your particular application requires a resistor that can handle approximately 8 W. Which rating would you use? Why?

Section 4–4

4–19. A particular electronic device uses 100 mW of power. If it runs for 24 h, how many joules of energy does it consume?

4–20. How many watthours does 50 W used for 12 h equal? How many kilowatthours?

4–21. A certain appliance uses 300 W. If it is allowed to run continuously for 30 days, how many kilowatthours of energy does it consume?

4–22. At the end of a 31-day period, your utility bill shows that you have used 1500 kWh. What is your average daily power?

4–23. Convert 5×10^6 wattminutes to kWh.

4–24. Convert 6700 wattseconds to kWh.

Section 4–6

4–25. If a power supply has an output voltage of 50 V and provides 2 A to a load, what is its output power?

4–26. A 50-Ω load consumes 1 W of power. What is the output voltage of the power supply?

4–27. A battery can provide 1.5 A of current for 24 h. What is its ampere-hour rating?

4–28. How much continuous current can be drawn from an 80-Ah battery for 10 h?

4–29. If a battery is rated at 650 mAh, how much current will it provide for 48 h?

4–30. If the input power is 500 mW and the output power is 400 mW, how much power is lost? What is the efficiency of this power supply?

4–31. To operate at 85% efficiency, how much output power must a source produce if the input power is 5 W?

4–32. A certain power supply provides a continuous 2 W to a loading device. It is operating at 60% efficiency. In a 24-h period, how many watthours does it consume?

Section 4–7

4–33. Modify the program in Section 4–7 to compute the powers for a range of voltage values rather than for current values.

4–34. Modify the program in Section 4–7 to compute the resistor voltage for each current value, and display these values in a third column.

Advanced Problems

4–35. For how many seconds must a 5-A current flow through a 50-Ω resistor in order to consume 25 J?

4–36. A 12-V source is connected across a 10-Ω resistor.
  (a) How much energy is used in 2 minutes?
  (b) If the resistor is disconnected after 1 minute, does the power increase or decrease?

4–37. Devise a computer program to compute the energy consumption (kWh) for a home during any specified period of time, given the wattage rating of each appliance (for any number of appliances), and the number of hours per day that each appliance is used. Also, the program should compute the cost of energy during the specified period given the cost per kWh.

Answers to Section Reviews

Section 4–1:

1. Power is the rate at which energy is used. **2.** Power = energy/time. **3.** One watt is one joule of energy consumed in one second. **4.** 5 W. **5.** 10 kW. **6.** (a) 68 kW; (b) 5 mW; (c) 25 μW.

Section 4–2:
1. $P = VI$, $P = I^2R$, $P = V^2/R$. **2.** 30 W. **3.** 1.2 W, 1.2 W. Yes, all energy produced by source is dissipated by resistance. **4.** 1250 W. **5.** 2 W. **6.** 2.5 W. **7.** 32 mW. **8.** 300 Ω.

Section 4–3:
1. Ohmic value, power rating. **2.** A larger surface area dissipates more energy. **3.** 0.125 W, 0.25 W, 0.5 W, 1 W, 2 W. **4.** 0.5 W or greater.

Section 4–4:
1. Energy is power times time. **2.** Energy = (power)(time). **3.** Kilowatthour. **4.** 1 kWh. **5.** 2 kWh. **6.** 0.1 kWh.

Section 4–5:
1. Collisions of the electrons within the atomic structure. **2.** Potential difference between two points due to energy loss. **3.** Minus to plus. **4.** Plus to minus.

Section 4–6:
1. Greater. **2.** 5 W. **3.** 20 h. **4.** 60 W. **5.** 75%.

Section 4–7:
1. For data input. **2.** 150.

Resistors can be connected in a circuit in two basic ways: in series or parallel. This chapter discusses circuits with series resistance. Specifically, you will learn:

1. How to identify a series circuit.
2. How to determine current in a series circuit.
3. How to determine total resistance in a series circuit.
4. How to apply Ohm's law to series circuits.
5. How to connect voltage sources in series.
6. How to use Kirchhoff's voltage law.
7. How to assign polarities to voltage drops.
8. How to analyze and apply voltage dividers.
9. How to use the potentiometer as a voltage divider.
10. How to determine power in a series circuit.
11. How to identify open and closed circuits.
12. How to troubleshoot series circuits.
13. How to use a simple computer program for circuit analysis.

5
SERIES RESISTIVE CIRCUITS

RESISTORS IN SERIES

5–1

Resistors in series are connected end-to-end or in a "string," as shown in Figure 5–1. Figure 5–1A shows two resistors connected in series between point *A* and point *B*. Part B of the figure shows three in series, and Part C shows four in series. Of course, there can be any number of resistors in a series connection.

The only way for electrons to get from point *A* to point *B* in any of the connections of Figure 5–1 is to go through *each* of the resistors. The following is an important way to identify a series connection: *A series connection provides only one path for current between two points in a circuit so that the same current flows through each series resistor.*

Identifying Series Connections

In an actual circuit diagram, a series connection may not always be as easy to identify as those in Figure 5–1. For example, Figure 5–2 shows series resistors drawn in other ways. Remember, *if there is only one current path between two points, the resistors between those two points are in series,* no matter how they appear in a diagram.

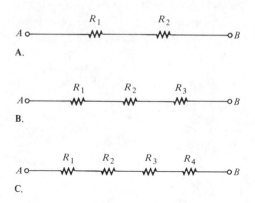

FIGURE 5–1 *Resistors in series.*

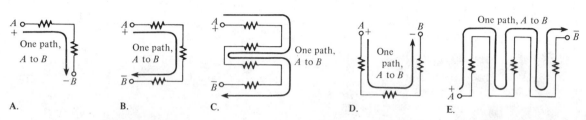

FIGURE 5–2 *Some examples of series connections. Notice that the current must be the same at all points.*

96

Example 5–1

Suppose that there are five resistors positioned on a circuit board as shown in Figure 5–3. Wire them together in series so that, starting from the positive (+) terminal, R_1 is first, R_2 is second, R_3 is third, and so on. Draw a schematic diagram showing this connection.

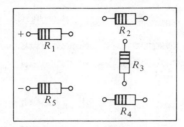

FIGURE 5–3

Solution:

The wires are connected as shown in Figure 5–4A, which is the *assembly diagram*. The *schematic diagram* is shown in Figure 5–4B. Note that the schematic diagram does not necessarily show the actual physical arrangement of the resistors as does the assembly diagram. The purpose of the *schematic* is to show how components are connected *electrically*. The purpose of the *assembly* diagram is to show how components are arranged *physically*.

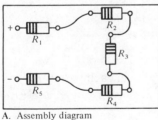
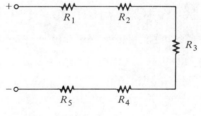

A. Assembly diagram   B. Schematic diagram

FIGURE 5–4

Example 5–2

Describe how the resistors on the printed circuit (PC) board in Figure 5–5 are related electrically.

Example 5–2 (continued)

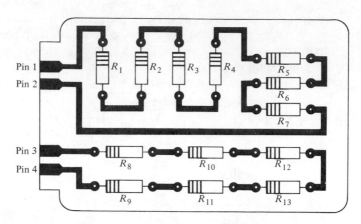

FIGURE 5–5

Solution:

Resistors R_1 through R_7 are in series with each other. This series combination is connected between pins 1 and 2 on the PC board.

Resistors R_8 through R_{13} are in series with each other. This series combination is connected between pins 3 and 4 on the PC board.

Review for 5–1

1. How are the resistors connected in a series circuit?

2. How can you identify a series connection?

3. Complete the schematic diagrams for the circuits in each part of Figure 5–6 by connecting the resistors in series in numerical order from A to B.

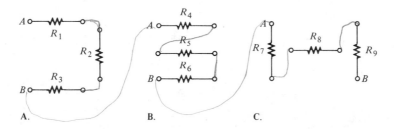

FIGURE 5–6

4. Now connect each *group* of series resistors in Figure 5–6 in series.

The same amount of current flows through all points in a series circuit. Figure 5–7 shows three series resistors connected to a voltage source. *At any point in this circuit, the current entering that point must equal the current leaving that point,* as illustrated by the current directional arrows at points *A*, *B*, *C*, and *D*.

5–2

    Notice also that the current flowing out of each of the resistors must equal the current flowing in, because there is no place where part of the current can branch off and go somewhere else. Therefore, the current in each section of the circuit is the same as the current in all other sections. It has only one path in which to flow going from the positive (+) side of the source to the negative (−) side.

    Let us assume that the battery in Figure 5–7 supplies one ampere of current to the series resistance. One ampere is flowing out of the positive terminal. If we connect ammeters at several points in the circuit as shown in Figure 5–8, *each* meter will read one ampere.

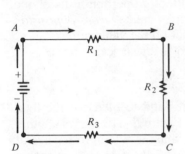

FIGURE 5–7   *Current in a series circuit.*

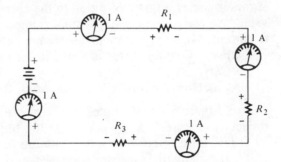

FIGURE 5–8   *Current is the same at all points in a series circuit.*

Review for 5–2

1. In a series circuit with a 10-Ω and a 5-Ω resistor in series, 1 A flows through the 10-Ω resistor. How much current flows through the 5-Ω resistor?

2. A milliammeter is connected between points *A* and *B* in Figure 5–9. It measures 50 mA. If you move the meter and connect it between points *C* and *D*, how much current will it indicate? Between *E* and *F*?

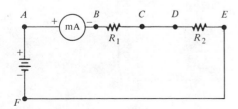

FIGURE 5–9

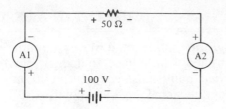

FIGURE 5–10

3. In Figure 5–10, how much current does ammeter 1 indicate? How much current does ammeter 2 indicate?

4. What statement can you make about the amount of current in a series circuit?

TOTAL SERIES RESISTANCE

5–3

The total resistance of a series connection is equal to the sum of the resistances of each individual resistor. This fact is understandable because each of the resistors in series offers opposition to the current in direct proportion to its ohmic value. A greater number of resistors connected in series creates *more opposition* to current. More opposition to current implies a higher ohmic value of resistance. Thus, every time a resistor is added in series, the total resistance increases.

Series Resistor Values Add

Figure 5–11 illustrates how series resistances add to *increase* the total resistance. Figure 5–11A has a single 10-Ω resistor. Figure 5–11B shows another 10-Ω resistor connected in series with the first one, making a total resistance of 20 Ω. If a third 10-Ω resistor is connected in series with the first two, as shown in Figure 5–11C, the total resistance becomes 30 Ω.

Series Resistance Formula

For *any number* of individual resistors connected in series, the total resistance is the sum of each of the individual values:

$$R_T = R_1 + R_2 + R_3 + \cdots + R_n \qquad \text{(5–1)}$$

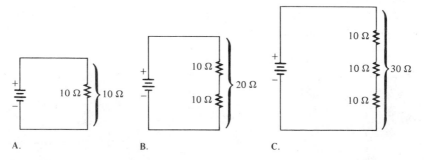

FIGURE 5–11   *Total resistance increases with each additional series resistor.*

where R_T is the total resistance and R_n is the last resistor in the series string (n can be any positive integer equal to the number of resistors in series). For example, if we have four resistors in series ($n = 4$), the total resistance formula is

$$R_T = R_1 + R_2 + R_3 + R_4$$

If we have six resistors in series ($n = 6$), the total resistance formula is

$$R_T = R_1 + R_2 + R_3 + R_4 + R_5 + R_6$$

To illustrate the calculation of total series resistance, let us take the circuit of Figure 5–12 and determine its R_T. (V_S is the source voltage.)

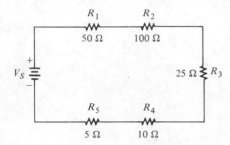

FIGURE 5–12 *Example of five resistors in series.*

The circuit of Figure 5–12 has five resistors in series. To get the total resistance, we simply add the values as follows:

$$R_T = 50\ \Omega + 100\ \Omega + 25\ \Omega + 10\ \Omega + 5\ \Omega = 190\ \Omega$$

This equation illustrates an important point: In Figure 5–12, the order in which the resistances are added does not matter; we still get the same total. Also, we can physically change the positions of the resistors in the circuit without affecting the total resistance.

Example 5–3

What is the total resistance (R_T) in the circuit of Figure 5–13?

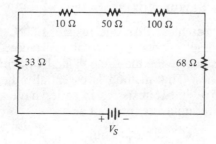

FIGURE 5–13

Example 5–3 (continued)

Solution:

Sum all the values as follows:

$$R_T = 33\ \Omega + 10\ \Omega + 50\ \Omega + 100\ \Omega + 68\ \Omega$$
$$= 261\ \Omega$$

Example 5–4

Calculate R_T for the circuit of Figure 5–14.

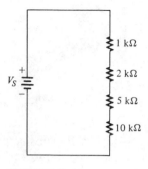

FIGURE 5–14

Solution:

Sum the resistor values:

$$R_T = 1\ k\Omega + 2\ k\Omega + 5\ k\Omega + 10\ k\Omega$$
$$= 18\ k\Omega$$

Notice that we can change the positions of the resistors without changing the total resistance.

Equal-Value Series Resistors

When a circuit has more than one resistor of the *same* value in series, there is a shortcut method to obtain the total resistance: Simply multiply the *ohmic value* of the resistors having the same value by the *number* of equal-value resistors that are in series. This method is essentially the same as adding the values. For example, five 100-Ω resistors in series have an R_T of 5(100 Ω) = 500 Ω. In general, the formula is expressed as

$$R_T = nR \qquad (5\text{–}2)$$

where n is the number of equal-value resistors and R is the value.

Example 5–5

Find the R_T of eight 20-Ω resistors in series.

Solution:

We find R_T by adding the values as follows:

$R_T = 20\ \Omega + 20\ \Omega + 20\ \Omega + 20\ \Omega + 20\ \Omega + 20\ \Omega + 20\ \Omega + 20\ \Omega$
     $= 160\ \Omega$

However, it is much easier to multiply:

$$R_T = 8(20\ \Omega) = 160\ \Omega$$

Review for 5–3

1. The following resistors (one each) are in series: 1-Ω, 2-Ω, 3-Ω, and 4-Ω. What is the total resistance?

2. Calculate R_T for each circuit in Figure 5–15.

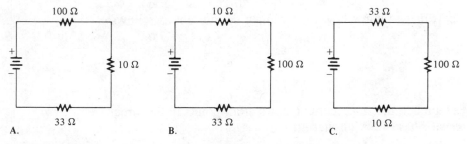

100 Ω	10 Ω	33 Ω
10 Ω	100 Ω	100 Ω
33 Ω	33 Ω	10 Ω
A.	B.	C.

FIGURE 5–15

3. The following resistors are in series: one 100-Ω, two 50-Ω, four 12-Ω, and one 330-Ω. What is the total resistance?

4. Suppose that you have one resistor each of the following values: 1-kΩ, 2.5-kΩ, 5-kΩ, and 500-Ω. To get a total resistance of 10-kΩ, you need one more resistor. What should its value be?

5. What is the R_T for twelve 50-Ω resistors in series?

6. What is the R_T for twenty 5-Ω resistors and thirty 8-Ω resistors in series?

APPLYING OHM'S LAW

5–4

In this section we solve several circuit problems to see how Ohm's law can be applied to series circuit analysis.

Example 5–6

Find the current in the circuit of Figure 5–16.

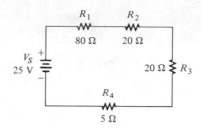

FIGURE 5–16

Solution:

The current is determined by the voltage and the *total resistance*. First, we calculate the total resistance as follows:

$$R_T = R_1 + R_2 + R_3 + R_4$$
$$= 80 \ \Omega + 20 \ \Omega + 20 \ \Omega + 5 \ \Omega$$
$$= 125 \ \Omega$$

Next, using Ohm's law, we calculate the current as follows:

$$I = \frac{V_S}{R_T} = \frac{25 \text{ V}}{125 \ \Omega}$$
$$= 0.2 \text{ A}$$

Remember, the *same* current flows at all points in the circuit. Thus, *each* resistor has 0.2 A through it.

Example 5–7

In the circuit of Figure 5–17, 1 mA of current flows. For this amount of current to flow, what must the source voltage V_S be?

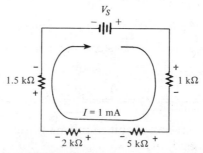

FIGURE 5–17

Solution:

In order to calculate V_S, we must determine R_T as follows:

$$R_T = 1 \text{ k}\Omega + 5 \text{ k}\Omega + 2 \text{ k}\Omega + 1.5 \text{ k}\Omega$$
$$= 9.5 \text{ k}\Omega$$

Now use Ohm's law to get V_S:

$$V_S = IR_T = (1 \text{ mA})(9.5 \text{ k}\Omega)$$
$$= 9.5 \text{ V}$$

Example 5–8

Calculate the voltage across each resistor in Figure 5–18, and find the value of V_S.

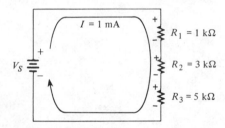

FIGURE 5–18

Solution:

By Ohm's law, the voltage across each resistor is equal to its resistance multiplied by the current through it. Using the Ohm's law formula $V = IR$, we determine the voltage across each of the resistors. Keep in mind that the same current flows through each series resistor.

Voltage across R_1:

$$V_1 = IR_1 = (1 \text{ mA})(1 \text{ k}\Omega) = 1 \text{ V}$$

Voltage across R_2:

$$V_2 = IR_2 = (1 \text{ mA})(3 \text{ k}\Omega) = 3 \text{ V}$$

Voltage across R_3:

$$V_3 = IR_3 = (1 \text{ mA})(5 \text{ k}\Omega) = 5 \text{ V}$$

The source voltage V_S is equal to the current times the *total resistance:*

$$R_T = 1 \text{ k}\Omega + 3 \text{ k}\Omega + 5 \text{ k}\Omega$$
$$= 9 \text{ k}\Omega$$

Example 5–8 (continued)

$$V_S = (1 \text{ mA})(9 \text{ k}\Omega)$$
$$= 9 \text{ V}$$

Notice that if you add the voltage drops of the resistors, they total 9 V, which is the same as the source voltage.

Review for 5–4

1. A 10-V battery is connected across three 100-Ω resistors in series. What is the current through each resistor?

2. How much voltage is required to produce 5 A through the circuit of Figure 5–19?

3. How much voltage is dropped across each resistor in Figure 5–19?

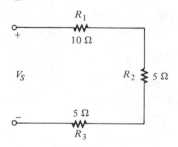

R_1
$10 \text{ }\Omega$

V_S

$R_2 \lessgtr 5 \text{ }\Omega$

$5 \text{ }\Omega$
R_3

FIGURE 5–19

4. There are four equal-value resistors connected in series with a 5-V source. Five milliamperes of current are measured. What is the value of each resistor?

VOLTAGE SOURCES IN SERIES

5–5

A voltage source is an energy source that provides a constant voltage to a load. Batteries and dc power supplies are practical examples.

When two or more voltage sources are in series, the total voltage is equal to the algebraic sum of the individual source voltages. The *algebraic sum* means that the polarities of the sources must be included when the sources are combined in series. Sources with opposite polarities have voltages with opposite signs.

$$V_T = V_{S1} + V_{S2} + \cdots + V_{Sn}$$

When the sources are all in the same direction in terms of their polarities, as in Figure 5–20A, all of the voltages have the same sign when added, and we get a total of 4.5 V with terminal A more positive than terminal B:

$$V_{AB} = 1.5 \text{ V} + 1.5 \text{ V} + 1.5 \text{ V} = +4.5 \text{ V}$$

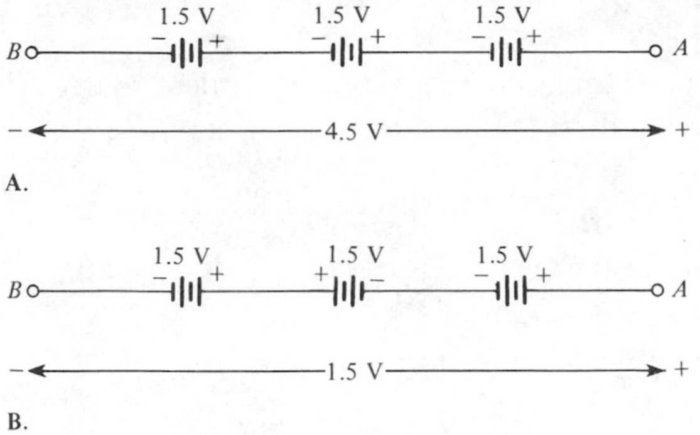

1.5 V   1.5 V   1.5 V

B ○—————|||———|||———|||———○ A

◄————————— 4.5 V —————————►

A.

1.5 V   1.5 V   1.5 V

B ○—————|||———|||———|||———○ A

◄————————— 1.5 V —————————►

B.

FIGURE 5–20 *Voltage sources in series add algebraically.*

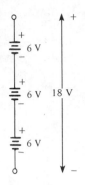

FIGURE 5–21 *Connection of three 6-V batteries to get 18 V.*

In Figure 5–20B, the middle source is opposite to the other two; so its voltage has an opposite sign when added to the others. For this case the total voltage is

$$V_{AB} = +1.5 \text{ V} - 1.5 \text{ V} + 1.5 \text{ V} = +1.5 \text{ V}$$

Terminal A is 1.5 V more positive than terminal B.

    A familiar example of sources in series is the flashlight. When you put two 1.5-V batteries in your flashlight, they are connected in *series,* giving a total of 3 V. When connecting batteries or other voltage sources in series to increase the total voltage, always connect from the positive (+) terminal of one to the negative (−) of another. Such a connection is illustrated in Figure 5–21.

    The following two examples illustrate calculation of total source voltage.

Example 5–9

What is the total source voltage (V_{ST}) in Figure 5–22?

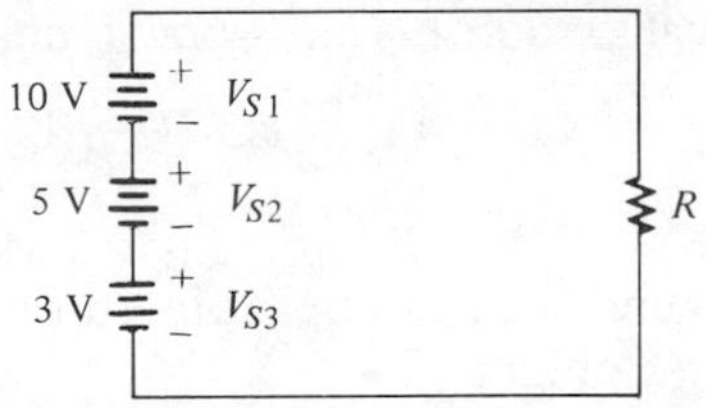

10 V V_{S1}
5 V V_{S2} R
3 V V_{S3}

FIGURE 5–22

Solution:

The polarity of each source is the same (the sources are connected in the same direction in the circuit). So we sum the three voltages to get the total:

$$V_{ST} = V_{S1} + V_{S2} + V_{S3} = 10 \text{ V} + 5 \text{ V} + 3 \text{ V}$$
$$= 18 \text{ V}$$

Example 5–9 (continued)

The three individual sources can be replaced by a single equivalent source of 18 V with its polarity as shown in Figure 5–23.

FIGURE 5–23

Example 5–10

Determine V_{ST} in Figure 5–24.

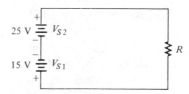

FIGURE 5–24

Solution:

These sources are connected in *opposing* directions. If you go clockwise around the circuit, you go from plus to minus through V_{S1}, and minus to plus through V_{S2}. The total voltage is the *difference* of the two source voltages (algebraic sum of oppositely signed values). The total voltage has the same polarity as the larger-value source. Here we will choose V_{S2} to be positive:

$$V_{ST} = V_{S2} - V_{S1} = 25 \text{ V} - 15 \text{ V}$$
$$= 10 \text{ V}$$

The two sources in Figure 5–24 can be replaced by a 10-V equivalent one with polarity as shown in Figure 5–25.

FIGURE 5–25

Review for 5–5

1. Four 1.5-V flashlight batteries are connected in series plus to minus. What is the total voltage of all four cells?

2. How many 12-V batteries must be connected in series to produce 60 V? Sketch a schematic that shows the battery connections.

3. The resistive circuit in Figure 5–26 is used to bias a transistor amplifier. Show how to connect two 15-V power supplies in order to get 30 V across the two resistors.

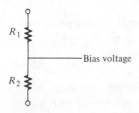

FIGURE 5–26

4. Determine the total source voltage in each circuit of Figure 5–27.

5. Sketch the *equivalent single source* circuit for each circuit of Figure 5–27.

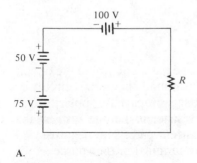

A.

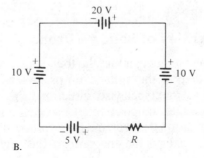

B.

FIGURE 5–27

KIRCHHOFF'S VOLTAGE LAW

5–6

Kirchhoff's voltage law states that *the sum of all the voltages around a closed path is zero*. In other words, *the sum of the voltage drops equals the total source voltage*.

For example, in the circuit of Figure 5–28, there are three voltage drops and one voltage source. If we sum all of the voltages around the circuit, we get

$$V_S - V_1 - V_2 - V_3 = 0 \qquad (5\text{–}3)$$

Notice that the source voltage has a sign opposite to that of the voltage drops.

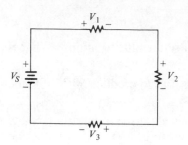

FIGURE 5–28 *The sum of the voltage drops equals the source voltage, V_S.*

Thus, the algebraic sum equals zero. Equation (5–3) can be written another way by transposing of the voltage drop terms to the right side of the equation:

$$V_S = V_1 + V_2 + V_3 \qquad\qquad (5\text{–}4)$$

In words, Equation (5–4) says that the source voltage equals the sum of the voltage drops. Both Equations (5–3) and (5–4) are equivalent ways of expressing Kirchhoff's voltage law for the circuit in Figure 5–28.

The previous illustration of Kirchhoff's voltage law was for the special case of three voltage drops and one voltage source. In general, Kirchhoff's voltage law applies to any number of series voltage drops and any number of series voltage sources, as illustrated in Figure 5–29.

The *general form* of Kirchhoff's voltage law is

$$V_{S1} + V_{S2} + \cdots + V_{Sn} = V_1 + V_2 + \cdots + V_n \qquad\qquad (5\text{–}5)$$

Polarities of Voltage Drops

Always remember that the total source voltage has a polarity in the circuit *opposite* to that of the voltage drops. As we go around the series circuit of Figure 5–30 in the clockwise direction, the voltage drops have polarities as shown. Notice that the positive (+) side of each resistor is the one *nearest* the positive terminal of the source as we follow the current path. The negative (−) side of each resistor is the one *nearest* the negative terminal of the source as we follow the current path.

Let us examine polarities in a series loop from the conventional current standpoint. Refer to Figure 5–30. *The current flows out of the positive side of the*

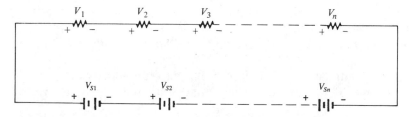

FIGURE 5–29 *Sum of the voltage drops equals the total source voltage.*

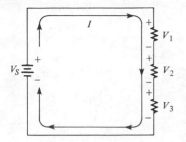

FIGURE 5–30 *The polarity of the voltage drops is opposite to that of the source voltage.*

source and through the resistors as shown. *The current flows into the positive side of each resistor and out of the negative side.* We naturally tend to think of a *drop* as going from a higher to a lower value. If we go from a more positive voltage to a more negative voltage, this seems to fit our concept of a *drop* in voltage, as was discussed in Chapter 4. The direction of conventional current through the circuit supports this concept, because it is considered to be from the positive side to the more negative side of a resistor. As mentioned earlier, for *analysis purposes* we will think in terms of the conventional direction of current rather than the actual electron flow direction.

The following three examples use Kirchhoff's voltage law to solve circuit problems.

Example 5–11

Determine the applied voltage V_S in Figure 5–31 where the two voltage drops are given.

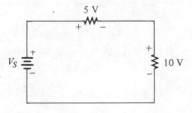

FIGURE 5–31

Solution:

By Kirchhoff's voltage law, the source voltage (applied voltage) must equal the sum of the voltage drops. Adding the voltage drops gives us the value of the source voltage:

$$V_S = 5 \text{ V} + 10 \text{ V} = 15 \text{ V}$$

Example 5–12

Determine the unknown voltage drop, V_3, in Figure 5–32.

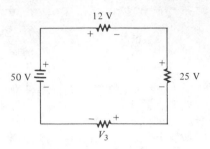

FIGURE 5–32

Solution:

By Kirchhoff's voltage law, the algebraic sum of all the voltages around the circuit is zero:

$$V_S - V_1 - V_2 - V_3 = 0$$

The value of each voltage drop except V_3 is known. Substitute these values into the equation as follows:

$$50\ V - 12\ V - 25\ V - V_3 = 0$$

Next combine the known values:

$$13\ V - V_3 = 0$$

Transpose 13 V to the right side of the equation, and cancel the minus signs:

$$-V_3 = -13\ V$$

$$V_3 = 13\ V$$

The voltage drop across R_3 is 13 V, and its polarity is as shown in Figure 5–32.

Example 5–13

Find the value of R_4 in Figure 5–33.

Solution:

In this problem we must use both Ohm's law *and* Kirchhoff's voltage law. Follow this procedure carefully.

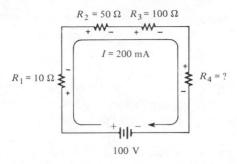

FIGURE 5–33

First find the voltage drop across each of the *known* resistors. Use Ohm's law:

$$V_1 = IR_1 = (200 \text{ mA})(10 \text{ } \Omega) = 2 \text{ V}$$

$$V_2 = IR_2 = (200 \text{ mA})(50 \text{ } \Omega) = 10 \text{ V}$$

$$V_3 = IR_3 = (200 \text{ mA})(100 \text{ } \Omega) = 20 \text{ V}$$

Next, use Kirchhoff's voltage law to find V_4, the voltage drop across the *unknown* resistor:

$$V_S - V_1 - V_2 - V_3 - V_4 = 0$$

$$100 \text{ V} - 2 \text{ V} - 10 \text{ V} - 20 \text{ V} - V_4 = 0$$

$$68 \text{ V} - V_4 = 0$$

$$V_4 = 68 \text{ V}$$

Now that we know V_4, we can use Ohm's law to calculate R_4 as follows:

$$R_4 = \frac{V_4}{I} = \frac{68 \text{ V}}{200 \text{ mA}} = 340 \text{ } \Omega$$

Review for 5–6

1. State Kirchhoff's voltage law in two ways.

2. A 50-V source is connected to a series resistive circuit. What is the total of the voltage drops in this circuit?

3. Two equal-value resistors are connected in series across a 10-V battery. What is the voltage drop across each resistor?

4. In a series circuit with a 25-V source, there are three resistors. One voltage drop is 5 V, and the other is 10 V. What is the value of the third voltage drop?

5. The individual voltage drops in a series string are as follows: 1 V, 3 V, 5 V, 8 V, and 7 V. What is the total voltage applied across the series string?

VOLTAGE DIVIDERS

5–7

A series circuit acts as a *voltage divider*. In this section you will see what this term means and why voltage dividers are an important application of series circuits.

To illustrate how a series string of resistors acts as a voltage divider, we will examine Figure 5–34 where there are two resistors in series. As you already know, there are two voltage drops: one across R_1 and one across R_2. We call these voltage drops V_1 and V_2, respectively, as indicated in the diagram.

Since the same current flows through each resistor, the voltage drops are proportional to the ohmic values of the resistors. For example, if the value of R_2 is twice that of R_1, then the value of V_2 is twice that of V_1. In other words, the total voltage drop *divides* among the series resistors in amounts directly proportional to the resistance values.

For example, in Figure 5–34, if V_S is 10 V, R_1 is 50 Ω, and R_2 is 100 Ω, then V_1 is one-third the total voltage, or 3.33 V, because R_1 is one-third the *total* resistance. Likewise, V_2 is two-thirds V_S, or 6.67 V.

Voltage Divider Formula

With a few calculations, a formula for determining how the voltages divide among series resistors can be developed. Let us assume that we have several resistors in series as shown in Figure 5–35. This figure shows five resistors, but there can be any number.

Let us call the voltage drop across any one of the resistors V_x, where x represents the number of a particular resistor (1, 2, 3, and so on). By Ohm's law, the voltage drop across any of the resistors in Figure 5–35 can be written as follows:

$$V_x = IR_x$$

where x = 1, 2, 3, 4, or 5 and R_x is any one of the series resistors.

The current is equal to the source voltage divided by the total resistance. For our example circuit of Figure 5–35, the total resistance is $R_1 + R_2 + R_3 + R_4 + R_5$, and

$$I = \frac{V_S}{R_T}$$

Substituting V_S/R_T for I in the expression for V_x, we get

$$V_x = \left(\frac{V_S}{R_T}\right) R_x$$

By rearranging, we get

$$V_x = \left(\frac{R_x}{R_T}\right) V_S \qquad \textbf{(5–6)}$$

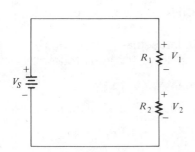

FIGURE 5–34 *Two-resistor voltage divider.*

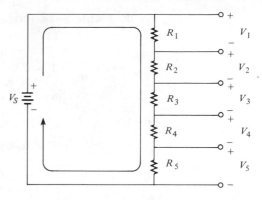

FIGURE 5–35 *Five-resistor voltage divider.*

Equation (5–6) is the *general voltage divider formula*. It tells us the following: *The voltage drop across any resistor or combination of resistors in a series circuit is equal to the ratio of that resistance value to the total resistance, multiplied by the source voltage.*

The following three examples illustrate use of the voltage divider formula.

Example 5–14

Determine the voltage across R_1 and the voltage across R_2 in the voltage divider in Figure 5–36.

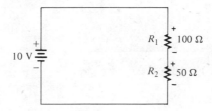

FIGURE 5–36

Solution:

Use the voltage divider formula, $V_x = (R_x/R_T)V_S$. In this problem we are looking for V_1; so $V_x = V_1$ and $R_x = R_1$. The total resistance is

$$R_T = R_1 + R_2 = 100\,\Omega + 50\,\Omega = 150\,\Omega$$

R_1 is 100 Ω and V_S is 10 V. Substituting these values into the voltage divider formula, we have

$$V_1 = \left(\frac{R_1}{R_T}\right)V_S = \left(\frac{100\,\Omega}{150\,\Omega}\right)10\text{ V}$$
$$= 6.67\text{ V}$$

Example 5–14 (continued)

There are two ways to find the value of V_2 in this problem: Kirchhoff's voltage law or the voltage divider formula.

First, using Kirchhoff's voltage law, we know that $V_S = V_1 + V_2$. By substituting the values for V_S and V_1, we can solve for V_2 as follows:

$$V_2 = 10 \text{ V} - 6.67 \text{ V} = 3.33 \text{ V}$$

A second way is to use the voltage divider formula to find V_2 as follows:

$$V_2 = \left(\frac{R_2}{R_T}\right)V_S = \left(\frac{50 \text{ }\Omega}{150 \text{ }\Omega}\right)10 \text{ V}$$

$$= 3.33 \text{ V}$$

We get the same result either way.

Example 5–15

Calculate the voltage drop across each resistor in the voltage divider of Figure 5–37.

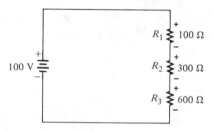

FIGURE 5–37

Solution:

Look at the circuit for a moment and consider the following: The total resistance is 1000 Ω. We can examine the circuit and determine that 10% of the total voltage is across R_1 because it is 10% of the total resistance (100 Ω is 10% of 1000 Ω). Likewise, we see that 30% of the total voltage is dropped across R_2 because it is 30% of the total resistance (300 Ω is 30% of 1000 Ω). Finally, R_3 drops 60% of the total voltage because 600 Ω is 60% of 1000 Ω.

Because of the convenient values in this problem, it is easy to figure the voltages mentally. Such is not always the case, but sometimes a little thinking will produce a result more efficiently and eliminate some calculating.

Although we have already reasoned through this problem, the calculations will verify our results:

$$V_1 = \left(\frac{R_1}{R_T}\right)V_S = \left(\frac{100 \ \Omega}{1000 \ \Omega}\right)100 \ \text{V}$$

$$= 10 \ \text{V}$$

$$V_2 = \left(\frac{R_2}{R_T}\right)V_S = \left(\frac{300 \ \Omega}{1000 \ \Omega}\right)100 \ \text{V}$$

$$= 30 \ \text{V}$$

$$V_3 = \left(\frac{R_3}{R_T}\right)V_S = \left(\frac{600 \ \Omega}{1000 \ \Omega}\right)100 \ \text{V}$$

$$= 60 \ \text{V}$$

Notice that the sum of the voltage drops is equal to the source voltage, in accordance with Kirchhoff's voltage law. This check is a good way to verify your results.

Example 5–16

Determine the voltages between the following points in the voltage divider of Figure 5–38:
(a) A to B (b) A to C (c) B to C (d) B to D (e) C to D

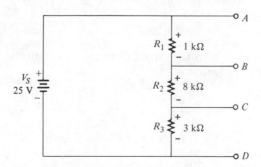

FIGURE 5–38

Solution:

First determine R_T:

$$R_T = 1 \ \text{k}\Omega + 8 \ \text{k}\Omega + 3 \ \text{k}\Omega = 12 \ \text{k}\Omega$$

Now apply the voltage divider formula to obtain each required voltage.

Example 5–16 (continued)

(a) The voltage A to B is the voltage drop across R_1. The calculation is as follows:

$$V_{AB} = \left(\frac{R_1}{R_T}\right)V_S = \left(\frac{1 \text{ k}\Omega}{12 \text{ k}\Omega}\right)25 \text{ V}$$

$$= 2.08 \text{ V}$$

(b) The voltage from A to C is the combined voltage drop across both R_1 and R_2. In this case, R_x in the general formula [Equation (5–6)] is $R_1 + R_2$. The calculation is as follows:

$$V_{AC} = \left(\frac{R_1 + R_2}{R_T}\right)V_S = \left(\frac{9 \text{ k}\Omega}{12 \text{ k}\Omega}\right)25 \text{ V}$$

$$= 18.75 \text{ V}$$

(c) The voltage from B to C is the voltage drop across R_2. The calculation is as follows:

$$V_{BC} = \left(\frac{R_2}{R_T}\right)V_S = \left(\frac{8 \text{ k}\Omega}{12 \text{ k}\Omega}\right)25 \text{ V}$$

$$= 16.67 \text{ V}$$

(d) The voltage from B to D is the combined voltage drop across both R_2 and R_3. In this case, R_x in the general formula is $R_2 + R_3$. The calculation is as follows:

$$V_{BD} = \left(\frac{R_2 + R_3}{R_T}\right)V_S = \left(\frac{11 \text{ k}\Omega}{12 \text{ k}\Omega}\right)25 \text{ V}$$

$$= 22.92 \text{ V}$$

(e) Finally, the voltage from C to D is the voltage drop across R_3. The calculation is as follows:

$$V_{CD} = \left(\frac{R_3}{R_T}\right)V_S = \left(\frac{3 \text{ k}\Omega}{12 \text{ k}\Omega}\right)25 \text{ V}$$

$$= 6.25 \text{ V}$$

If you connect this voltage divider in the lab, you can verify each of the calculated voltages by connecting a voltmeter between the appropriate points in each case.

The Potentiometer as an Adjustable Voltage Divider

Recall from Chapter 2 that a potentiometer is a variable resistor with three terminals. A potentiometer connected to a voltage source is shown in

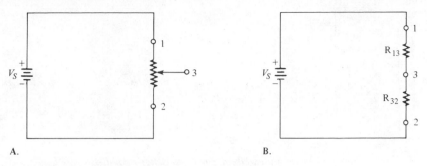

FIGURE 5–39 *The potentiometer as a voltage divider.*

Figure 5–39A. Notice that the two end terminals are labeled 1 and 2. The adjustable terminal or wiper is labeled 3. The potentiometer acts as a voltage divider. We can illustrate this concept better by separating the total resistance into two parts, as shown in Figure 5–39B. The resistance between terminal 1 and terminal 3 (R_{13}) is one part, and the resistance between terminal 3 and terminal 2 (R_{32}) is the other part. So this potentiometer actually is a two-resistor voltage divider that can be manually adjusted.

Figure 5–40 shows what happens when the wiper terminal (3) is moved. In Part A of Figure 5–40, the wiper is exactly centered, making the two resistances equal. If we measure the voltage across terminals 3 to 2 as indicated by the voltmeter symbol, we have one-half of the total source voltage. When the wiper is moved up, as in Figure 5–40B, the resistance between terminals 3 and 2 increases, and the voltage across it increases proportionally. When the wiper is moved down, as in Figure 5–40C, the resistance between terminals 3 and 2 decreases, and the voltage decreases proportionally.

Voltage Divider Applications

The volume control of radio or TV receivers is a common application of a potentiometer used as a voltage divider. Since the loudness of the sound is dependent on the amount of voltage associated with the audio signal, you can increase or decrease the volume by adjusting the potentiometer, that is, by turning the knob of the volume control on the set.

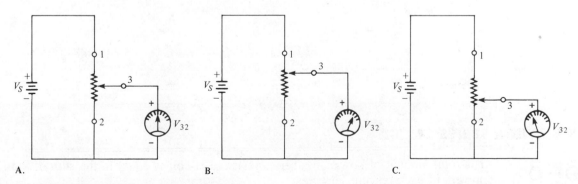

FIGURE 5–40 *Adjusting the voltage divider.*

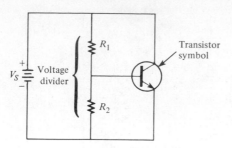

FIGURE 5–41 *The voltage divider as a bias circuit for a transistor amplifier.*

Another application for voltage dividers is in setting the dc operating voltage (*bias*) in transistor amplifiers. Figure 5–41 shows a voltage divider used for this purpose. You will study transistor amplifiers and biasing in a later course; so it is important that you understand the basics of voltage dividers at this point.

These examples are only two out of an almost endless number of applications of voltage dividers.

Review for 5–7

1. What is a voltage divider?
2. How many resistors can there be in a series voltage divider circuit?
3. Write the general formula for voltage dividers.
4. If two series resistors of equal value are connected across a 10-V source, how much voltage is there across each resistor?
5. A 50-Ω resistor and a 75-Ω resistor are connected as a voltage divider. The source voltage is 100 V. Sketch the circuit, and determine the voltage across each of the resistors.
6. The circuit of Figure 5–42 is an adjustable voltage divider. If the potentiometer is linear, where would you set the wiper in order to get 5 V from *A* to *B* and 5 V from *B* to *C*?

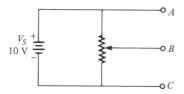

FIGURE 5–42

POWER IN A SERIES CIRCUIT

5–8   The total amount of power in a series resistive circuit is equal to the sum of the powers in each resistor in series:

$$P_T = P_1 + P_2 + P_3 + \cdots + P_n \qquad (5\text{--}7)$$

where n is the number of resistors in series, P_T is the total power, and P_n is the power in the last resistor in series. In other words, the powers are additive.

The power formulas that you learned in Chapter 4 are, of course, directly applicable to series circuits. Since the same current flows through each resistor in series, the following formulas are used to calculate the total power:

$$P_T = V_S I$$

$$P_T = I^2 R_T$$

$$P_T = \frac{V_S^2}{R_T}$$

where V_S is the total source voltage across the series connection and R_T is the total resistance. Example 5–17 illustrates how to calculate total power in a series circuit.

Example 5–17

Determine the total amount of power in the series circuit in Figure 5–43.

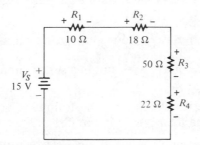

FIGURE 5–43

Solution:

We know that the source voltage is 15 V. The total resistance is

$$R_T = 10 \ \Omega + 18 \ \Omega + 50 \ \Omega + 22 \ \Omega = 100 \ \Omega$$

The easiest formula to use is $P_T = V_S^2/R_T$ since we know both V_S and R_T:

$$P_T = \frac{V_S^2}{R_T} = \frac{(15 \ \text{V})^2}{100 \ \Omega} = \frac{225 \ \text{V}^2}{100 \ \Omega}$$

$$= 2.25 \ \text{W}$$

If the power of each resistor is determined separately and all of these powers are added, the same result is obtained. We will work through another calculation to illustrate this formula.

Example 5–17 (continued)

First, find the current as follows:

$$I = \frac{V_S}{R_T} = \frac{15 \text{ V}}{100 \text{ }\Omega}$$

$$= 0.15 \text{ A}$$

Next, calculate the power for each resistor using $P = I^2R$:

$$P_1 = (0.15 \text{ A})^2 (10 \text{ }\Omega) = 0.225 \text{ W}$$

$$P_2 = (0.15 \text{ A})^2 (18 \text{ }\Omega) = 0.405 \text{ W}$$

$$P_3 = (0.15 \text{ A})^2 (50 \text{ }\Omega) = 1.125 \text{ W}$$

$$P_4 = (0.15 \text{ A})^2 (22 \text{ }\Omega) = 0.495 \text{ W}$$

Now, add these powers to get the total power:

$$P_T = 0.225 \text{ W} + 0.405 \text{ W} + 1.125 \text{ W} + 0.495 \text{ W}$$

$$= 2.25 \text{ W}$$

This result shows that the sum of the individual powers is equal to the total power as determined by one of the power formulas.

Review for 5–8

1. If you know the power in each resistor in a series circuit, how can you find the total power?

2. The resistors in a series circuit have the following powers: 2 W, 5 W, 1 W, and 8 W. What is the total power in the circuit?

3. A circuit has a 100-Ω, a 300-Ω, and a 600-Ω resistor in series. A current of 1 A flows through the circuit. What is the total power?

OPEN AND CLOSED CIRCUITS

5–9

An *open* circuit is one in which the current path is interrupted and no current flows. In an actual circuit, an open can be caused by a switch connected in series, as shown in Figure 5–44. Here, the switch symbol represents a simple *single-pole–single-throw* switch (see Chapter 2). In one position the switch disconnects and causes an *open* circuit. In the *open* position, no current flows and the light is off, as illustrated in Figure 5–44A. When the switch is moved to the *closed* position, contact is made, allowing the current to flow and the light to turn on, as shown in Figure 5–44B.

Sometimes a component failure will cause a circuit to open. There are many ways such a failure can occur, but we will use a familiar example to

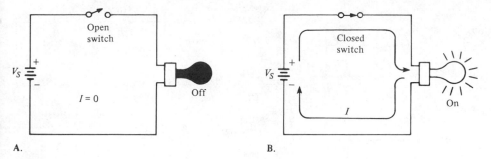

FIGURE 5–44 *Opening and closing a circuit with a switch.*

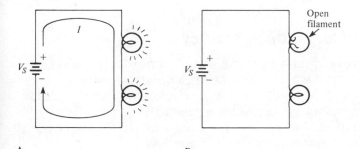

FIGURE 5–45 *Example of open circuit caused by a component failure.*

illustrate. In Figure 5–45 there are two light bulbs in series. As long as the bulbs are good, there is a *closed* circuit and current flows, keeping the lights on, as in Part A of the figure. A common failure in a light bulb is filament burnout. When burnout occurs, an *open* results and current ceases to flow, as in Part B. Since no current flows, both bulbs are off.

Review for 5–9

1. Define *closed circuit*.
2. Define *open circuit*.
3. What happens when a series circuit opens?
4. Name two general ways in which an open circuit can occur in practice.
5. What is the main purpose of a switch?

TROUBLESHOOTING SERIES CIRCUITS

The most common failure of a resistor is an *open*. A resistor will overheat when the power is greater than the power rating of the resistor. When the excess power is sufficient, the resistor will be damaged by the excessive heat, and an open resistor can result.

5–10

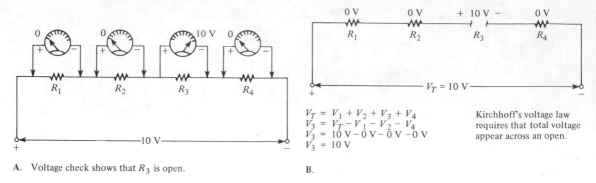

$$V_T = V_1 + V_2 + V_3 + V_4$$
$$V_3 = V_T - V_1 - V_2 - V_4$$
$$V_3 = 10\,V - 0\,V - 0\,V - 0\,V$$
$$V_3 = 10\,V$$

Kirchhoff's voltage law
requires that total voltage
appear across an open.

A. Voltage check shows that R_3 is open.   B.

FIGURE 5–46 *Troubleshooting a series circuit.*

How to Check for an Open Resistor

Sometimes a visual check will reveal a charred resistor. However, it is possible for a resistor to open without showing visible signs of damage. In this situation, a *voltage check* of the series circuit is required. The general procedure is as follows: Measure the voltage across each resistor in series. *The voltage across all of the good resistors will be zero. The voltage across the open resistor will equal the total voltage across the series combination.*

The above condition occurs because an open resistor will prevent current through the series circuit. With no current, there can be no voltage drop across any of the good resistors. Since $IR = 0$, in accordance with *Ohm's law,* the voltage on each side of a good resistor is the same. The total voltage must then appear across the open resistor in accordance with *Kirchhoff's voltage law,* as illustrated in Figure 5–46.

Review for 5–10

1. When a resistor fails, it will normally open (T or F).

2. The total voltage across a string of series resistors is 24 V. If one of the resistors is open, how much voltage is there across it? How much is there across each of the good resistors?

COMPUTER ANALYSIS

5–11   The program that follows can be used to compute the voltage across each resistor in a voltage divider with any specified number of resistors and a specified source voltage. The program computes and tabulates the results.

```
10 CLS
20 PRINT "THIS PROGRAM COMPUTES AND TABULATES THE VOLTAGES"
30 PRINT "ACROSS THE RESISTORS IN A SPECIFIED VOLTAGE"
40 PRINT "DIVIDER CIRCUIT."
50 PRINT:PRINT:PRINT
60 PRINT "YOU ARE REQUIRED TO ENTER THE NUMBER OF RESISTORS,"
70 PRINT "THE RESISTOR VALUES, AND THE SOURCE VOLTAGE."
80 PRINT:PRINT:PRINT
```

```
90 INPUT "TO CONTINUE PRESS 'ENTER'.";X
100 CLS
110 INPUT "HOW MANY RESISTORS ARE IN THE VOLTAGE DIVIDER";N
120 CLS
130 FOR X=1 TO N
140 PRINT "ENTER THE VALUE OF R";X;"IN OHMS"
150 INPUT R(X)
160 NEXT X
170 CLS
180 INPUT "THE SOURCE VOLTAGE";VS
190 CLS
200 FOR X=1 TO N
210 RT=RT+R(X)
220 NEXT X
230 FOR X=1 TO N
240 V(X)=(R(X)/RT)*VS
250 PRINT "R";X;"=";R(X);"OHMS","V";X;"=";V(X);"VOLTS"
260 NEXT X
```

Review for Section 5–11

1. There are three FOR/NEXT loops in the program. Identify each by line numbers, and explain their purpose.

2. What variable stands for the number of resistors in the divider circuit?

Formulas

$$R_T = R_1 + R_2 + R_3 + \cdots + R_n \qquad \text{(5–1)}$$

$$R_T = nR \qquad \text{(5–2)}$$

$$V_S - V_1 - V_2 - V_3 = 0 \qquad \text{(5–3)}$$

$$V_S = V_1 + V_2 + V_3 \qquad \text{(5–4)}$$

$$V_{S1} + V_{S2} + \cdots + V_{Sn} = V_1 + V_2 + V_3 + \cdots + V_n \qquad \text{(5–5)}$$

$$V_x = \left(\frac{R_x}{R_T}\right)V_S \qquad \text{(5–6)}$$

$$P_T = P_1 + P_2 + P_3 + \cdots + P_n \qquad \text{(5–7)}$$

Summary

1. Resistors in series are connected end-to-end.

2. A series connection has only *one* path for current.

3. The same amount of current flows at all points in a series circuit.

4. The total series resistance is the sum of all resistors in the series circuit.

5. The total resistance between any two points in a series circuit is equal to the sum of all resistors connected in series between those two points.

6. If all of the resistors in a series connection are of equal ohmic value, the total resistance is the number of resistors multiplied by the ohmic value.

7. Voltage sources in series add algebraically.

8. First statement of Kirchhoff's voltage law: The sum of all the voltages around a closed path is zero.

9. Second statement of Kirchhoff's voltage law: The sum of the voltage drops equals the total source voltage. (Both 8 and 9 say the same thing.)

10. The voltage drops in a circuit are always opposite in polarity to the total source voltage.

11. Current flows out of the positive side of a source in the conventional direction and into the negative side.

12. Conventional current flows into the positive side of each resistor and out of the more negative (less positive) side.

13. A voltage drop is considered to be from a more positive voltage to a more negative voltage.

14. A voltage divider is a series arrangement of resistors.

15. A voltage divider is so named because the voltage drop across any resistor in the series circuit is divided down from the total voltage by an amount proportional to that resistance value in relation to the total resistance.

16. A potentiometer can be used as an adjustable voltage divider.

17. The total power in a resistive circuit is the sum of all the individual powers of the resistors making up the series circuit.

18. An open circuit is one in which the current is interrupted.

19. A closed circuit is one having a complete current path.

Self-Test

1. Sketch a series circuit having four resistors in series with a voltage source.

2. Five amperes of current flow into a series string of ten resistors. How much current flows out of the sixth resistor? The tenth resistor?

3. Connect each set of resistors in Figure 5–47 in series between points A and B.

4. If you measured the current between each of the points marked in Figure 5–48, how much current would you read in each case? The total current out of the source is 3 A.

A.

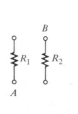

B.

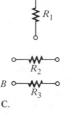

C.

FIGURE 5–47

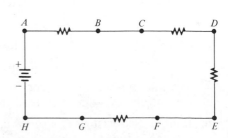

FIGURE 5–48

5. If a 75-Ω resistor and a 470-Ω resistor are connected in series, what is the total resistance?

6. Eight 56-Ω resistors are in series. Determine the total resistance.

7. Suppose that you need a total resistance of 20 kΩ. The resistors that you have available are two 1-kΩ, one 5-kΩ, and one 3-kΩ. What other single resistor do you need to make the required total?

8. Three 1-kΩ resistors are connected in series, and 5 V are applied across the series circuit. How much current is there through each resistor?

9. Which circuit in Figure 5–49 has more current?

10. Six resistors of equal value are in series with a 12-V source. The current through the circuit is 2 mA. What is the total resistance? Determine the value of each resistor.

11. Determine the total source voltage in the circuit in Figure 5–50.

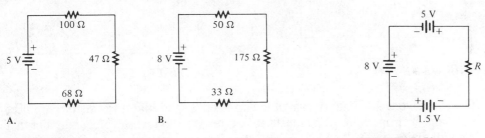

A.

FIGURE 5–49

B.

FIGURE 5–50

12. Two 9-V batteries are connected in opposite directions in a series circuit. What is the total source voltage?

13. Determine the value and the polarity of the total source voltage in each circuit of Figure 5–51. Sketch the equivalent single-source circuit.

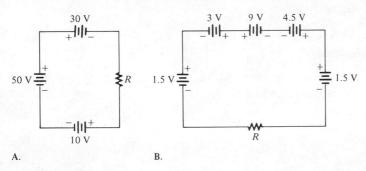

A.

B.

FIGURE 5–51

14. Five equal-value resistors are in series with a voltage source. The voltage drop across the first resistor is 2 V. What is the value of the source voltage?

15. A circuit with a 50-V source has three resistors in series. The voltage drop of the first resistor is 10 V. The voltage drop of the second resistor is 15 V. What is the voltage drop of the third resistor?

16. Determine the unknown resistance (R_3) in the circuit in Figure 5–52.

17. The following resistors (one each) are connected in series: 10-Ω, 50-Ω, 100-Ω, and 40-Ω. The total voltage is 20 V. Using the voltage divider formula, calculate the voltage drop across the 50-Ω resistor.

18. Calculate the voltage across each resistor in the voltage divider of Figure 5–53.

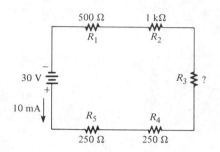

FIGURE 5–52

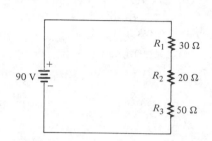

FIGURE 5–53

19. You have a 9-V battery available and need 3 V to bias a transistor amplifier for proper operation. You must use two resistors to form a voltage divider that will divide the 9 V down to 3 V. Determine the ideal values needed to achieve this if their total must be 100 kΩ.

20. In a series circuit with three resistors, one handles 2.5 W, one handles 5 W, and one handles 1.2 W. What is the total power in the circuit?

21. Run the program in Section 5–11 for a voltage divider with four resistors having the following values: 10 kΩ, 15 kΩ, 22 kΩ, and 33 kΩ. The source voltage is 12 V.

Problems

Section 5–1

5–1. Connect the resistors in Figure 5–54 in series, with R_1 closest to the positive terminal, R_2 next, and so forth.

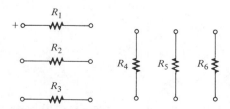

FIGURE 5–54

Section 5–2

5–2. What is the current through each resistor of a series circuit if the total voltage is 12 V and the total resistance is 120 Ω?

5–3. The current flowing out of the source in Figure 5–55 is 5 mA. How much current does each milliammeter in the circuit indicate?

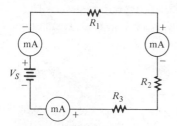

FIGURE 5–55

Section 5–3

5–4. The following resistors (one each) are connected in a series circuit: 1-Ω, 2-Ω, 5-Ω, 12-Ω, and 22-Ω. Determine the total resistance.

5–5. Find the total resistance of each of the following groups of series resistors:
   (a) 560-Ω and 1000-Ω **(b)** 47-Ω and 56-Ω
   (c) 1.5-kΩ, 2.2-kΩ, and 10-kΩ **(d)** 1-MΩ, 470-kΩ, 1-kΩ, 2.5-MΩ

5–6. Calculate R_T for each circuit of Figure 5–56.

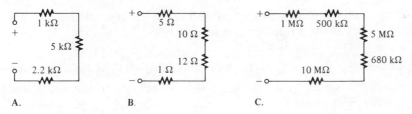

A.   B.   C.

FIGURE 5–56

5–7. If the total resistance in Figure 5–57 is 18 kΩ, what is the value of R_5?

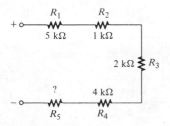

FIGURE 5–57

5–8. What is the total resistance of twelve 5.6-kΩ resistors in series?

5–9. Six 50-Ω resistors, eight 100-Ω resistors, and two 22-Ω resistors are all connected in series. What is the total resistance?

5–10. You have the following resistor values available to you in the lab in unlimited quantities: 10 Ω, 100 Ω, 470 Ω, 560 Ω, 680 Ω, 1 kΩ, 2.2 kΩ, and 5.6 kΩ. All of the other standard values are out of stock. A project that you are working on requires an 18-kΩ resistance. What combinations of the available values would you use in series to achieve this total resistance?

Section 5–4
5–11. What is the current in each circuit of Figure 5–58?

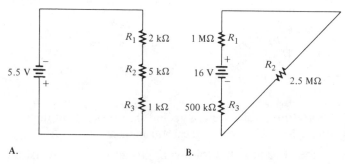

A.

B.

FIGURE 5–58

5–12. Three 470-Ω resistors are connected in series with a 500-V source. How much current is flowing in the circuit?

5–13. Four equal-value resistors are in series with a 5-V battery, and 2.5 mA are measured. What is the value of each resistor?

Section 5–5
5–14. *Series aiding* is a term sometimes used to describe voltage sources of the same polarity in series. If a 5-V and a 9-V source are connected in this manner, what is the total voltage?

5–15. The term *series opposing* means that sources are in series with opposite polarities. If a 12-V and a 3-V battery are series opposing, what is the total voltage?

5–16. Determine the total source voltage in each circuit of Figure 5–59.

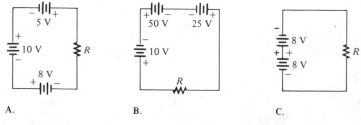

A.

B.

C.

FIGURE 5–59

Section 5–6

5–17. The following voltage drops are measured across three resistors in series: 5.5 V, 8.2 V, and 12.3 V. What is the value of the source voltage to which these resistors are connected?

5–18. Five resistors are in series with a 20-V source. The voltage drops across four of the resistors are 1.5 V, 5.5 V, 3 V, and 6 V. How much voltage is dropped across the fifth resistor?

5–19. In the circuit of Figure 5–60, determine the resistance of R_4.

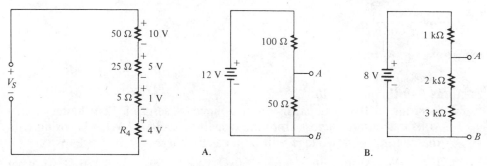

FIGURE 5–60   **FIGURE 5–61**

Section 5–7

5–20. The total resistance of a circuit is 500 Ω. What percentage of the total voltage appears across a 25-Ω resistor that makes up part of the total series resistance?

5–21. Determine the voltage between points A and B in each voltage divider of Figure 5–61.

5–22. What is the voltage across each resistor in Figure 5–62? R is the lowest-value resistor, and all others are multiples of that value as indicated.

5–23. Determine the voltage at each point in Figure 5–63 with respect to the negative side of the battery.

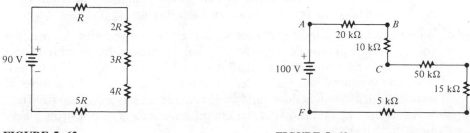

FIGURE 5–62   **FIGURE 5–63**

5–24. If there are 10 V across R_1 in Figure 5–64, what is the voltage across each of the other resistors?

Section 5–8

5–25. Five series resistors each handle 50 mW. What is the total power?

5–26. What is the total power in the circuit in Figure 5–64? Use the results of Problem 5–24.

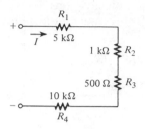

FIGURE 5–64

Sections 5–9 through 5–10

5–27. A string of five series resistors is connected across a 12-V battery. Zero volts is measured across all of the resistors except R_2. What is wrong with the circuit? What voltage will be measured across R_2?

5–28. Four $\frac{1}{2}$-W resistors are in series: 50-Ω, 68-Ω, 100-Ω, and 120-Ω. To what maximum value can the current be raised before the wattage rating of one of the resistors is exceeded? Which resistor will burn out first if the current is increased above its maximum value?

Section 5–11

5–29. Modify the program in Section 5–11 to compute and tabulate the power dissipation in each resistor in addition to the voltages.

5–30. Modify the program in Section 5–11 to compute and tabulate the voltages across any specified combination of resistors in a series voltage divider rather than just the voltages across the individual resistors.

Advanced Problems

5–31. A certain series circuit consists of a $\frac{1}{8}$-W resistor, a $\frac{1}{4}$-W resistor, and a $\frac{1}{2}$-W resistor. The total resistance is 2400 Ω. If each of the resistors is operating in the circuit at its maximum power dissipation, determine the following:
(a) I (b) V_T (c) The value of each resistor

5–32. With the table of standard resistor values given in Appendix A, design a voltage divider to provide the following approximate voltages with respect to ground using a 30-V source: 8.18 V, 14.73 V, and 24.55 V. The current drain on the source must be limited to no more than 1 mA. The number of resistors, their ohmic values, and their wattage ratings must be specified. A schematic diagram showing the circuit arrangement and resistor placement must be provided.

5–33. Using 1.5-V batteries, a switch, and three lamps, devise a circuit to apply 4.5 V across either one lamp, two lamps in series, or three lamps in series with a single-control switch selection. Draw the schematic diagram.

5–34. Design a variable voltage divider to provide output voltages ranging from a minimum of 10 V to a maximum of 100 V using a 120-V source. The maximum voltage must occur at the maximum resistance setting of the potentiometer, and the minimum voltage must occur at the minimum resistance (zero) setting. The maximum current is to be 10 mA.

Answers to Section Reviews

Section 5–1:
1. End-to-end. **2.** There is a single current path. **3.** See Figure 5–65. **4.** See Figure 5–66.

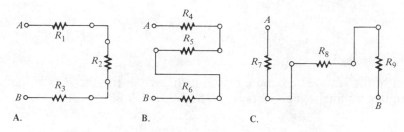

FIGURE 5–65

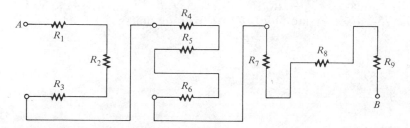

FIGURE 5–66

Section 5–2:
1. 1 A. **2.** 50 mA between C and D; 50 mA between E and F. **3.** 2 A, 2 A. **4.** Current is the same at all points.

Section 5–3:
1. 10 Ω. **2.** A, 143 Ω; B, 143 Ω; C, 143 Ω. **3.** 578 Ω. **4.** 1 kΩ. **5.** 600 Ω. **6.** 340 Ω.

Section 5–4:
1. 0.033 A or 33 mA. **2.** 100 V. **3.** $V_1 = 50$ V, $V_2 = 25$ V, $V_3 = 25$ V. **4.** 250 Ω.

Section 5–5:
1. 6 V. **2.** Five; see Figure 5–67 (next page). **3.** See Figure 5–68 (next page). **4.** A, 75 V; B, 15 V. **5.** See Figure 5–69 (next page).

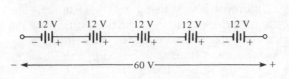

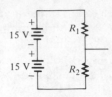

FIGURE 5–67   **FIGURE 5–68**

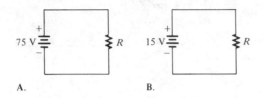

A.   B.

FIGURE 5–69

Section 5–6:
1. (a) The sum of the voltages around a closed path is zero; **(b)** the sum of the voltage drops equals the total source voltage. **2.** 50 V. **3.** 5 V. **4.** 10 V. **5.** 24 V.

Section 5–7:
1. Two or more resistors in a series connection in which the voltage taken across any resistor or combination of resistors is proportional to the value of that resistance. **2.** Any number. **3.** $V_x = (R_x/R_T)V_S$. **4.** 5 V. **5.** 40 V across the 50-Ω; 60 V across the 75-Ω; see Figure 5–70. **6.** At the midpoint.

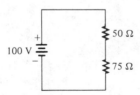

FIGURE 5–70

Section 5–8:
1. Add the power in each resistor. **2.** 16 W. **3.** 1000 W or 1 kW.

Section 5–9:
1. A circuit with a complete current path. **2.** A circuit with an interrupted current path. **3.** Current ceases to flow. **4.** Switch or other component failure. **5.** To turn current on or off in a circuit.

Section 5–10:
1. T. **2.** 24 V, 0 V.

Section 5–11:
1. Lines 130–160: Inputs resistor values. Lines 200–220: Calculates total resistance. Lines 230–260: Computes each voltage and prints out results. **2.** N.

In Chapter 5, you learned about series circuits. In this chapter, you will study circuits with parallel resistors. Specifically, you will learn:

1. How to identify a parallel circuit.
2. How to determine total resistance in a parallel circuit.
3. How to apply Ohm's law to parallel circuits.
4. How to use Kirchhoff's current law.
5. How to analyze and apply current dividers.
6. How to determine power in a parallel circuit.
7. How to recognize the effects of open paths.
8. How to troubleshoot parallel circuits.
9. How to use a simple computer program for circuit analysis.

6

PARALLEL RESISTIVE CIRCUITS

6–1

When two or more components are connected across the same voltage source, they are in parallel. A parallel circuit provides more than one path for current. Each parallel path is called a branch. *Two resistors connected in parallel are shown in Figure 6–1A.*

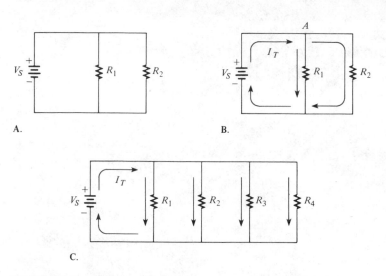

A.

B.

C.

FIGURE 6–1   *Resistors in parallel.*

In Figure 6–1B, the current flowing out of the source divides when it gets to point A. Part of it goes through R_1 and part through R_2. If additional resistors are connected in parallel with the first two, more current paths are provided, as shown in Figure 6–1C.

Identifying Parallel Connections

In Figure 6–1, the resistors obviously are connected in parallel. Often, in actual circuit diagrams, the parallel relationship is not so clear. It is important that you learn to recognize parallel connections regardless of how they may be drawn.

A rule for identifying parallel circuits is as follows: *If there is more than one current path (branch) between two points, and if the voltage between those two points appears across each of the branches, then there is a parallel circuit between those two points.* Figure 6–2 shows parallel resistors drawn in different ways between two points labeled A and B. Notice that in each case, the current "travels" two paths going from A to B, and the voltage across each branch is the same. Although these figures show only two parallel paths, there can be any number of resistors in parallel.

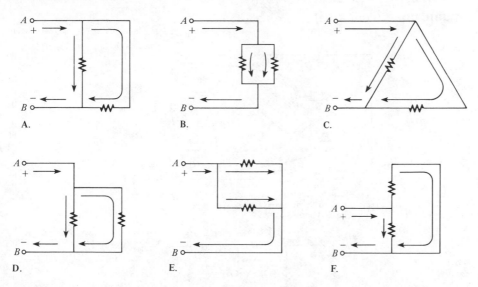

A. B. C.

D. E. F.

FIGURE 6–2 *Examples of circuits with two parallel paths.*

Example 6–1

Suppose that there are five resistors positioned on a circuit board as shown in Figure 6–3. Wire them together in parallel between the positive (+) and the negative (−) terminals. Draw a schematic diagram showing this connection.

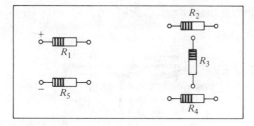

FIGURE 6–3

Solution:

Wires are connected as shown in the assembly diagram of Figure 6–4A. The schematic diagram is shown in Figure 6–4B. Again, note that the schematic does not necessarily have to show the actual physical arrangement of the resistors. The purpose of the schematic is to show how components are connected electrically.

Example 6–1 (continued)

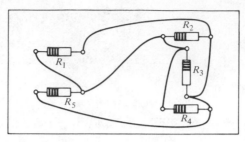

A. Assembly diagram. Compare this to the
same resistor arrangement connected in
series in Figure 5–4A.

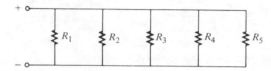

B. Schematic diagram.

FIGURE 6–4

Example 6–2

Describe how the resistors on the PC board in Figure 6–5 are related
electrically.

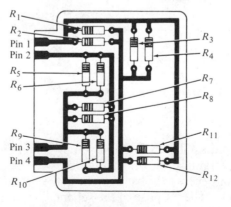

FIGURE 6–5

Solution:

Resistors R_1 through R_4 and R_{11} and R_{12} are all in parallel. This parallel combination is connected to pins 1 and 4.

Resistors R_5 through R_{10} are all in parallel. This combination is connected to pins 2 and 3.

Review for 6–1

1. How are the resistors connected in a parallel circuit?
2. How do you identify a parallel connection?
3. Complete the schematic diagrams for the circuits in each part of Figure 6–6 by connecting the resistors in parallel between points *A* and *B*.
4. Now connect each *group* of parallel resistors in Figure 6–6 in parallel with each other.

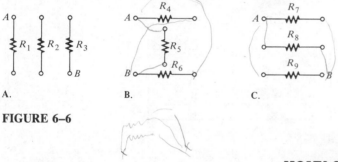

A.   B.   C.

FIGURE 6–6

VOLTAGE DROP IN A PARALLEL CIRCUIT

6–2

As mentioned, each path in a parallel circuit is sometimes called a *branch. The voltage across any branch of a parallel combination is equal to the voltage across each of the other branches in parallel.*

To illustrate voltage drop in a parallel circuit, let us examine Figure 6–7A. Points *A*, *B*, *C*, and *D* along the top of the parallel circuit are *electrically the same point* because the voltage is the same along this line. You can think of all of these points as being connected by a single wire to the positive terminal of the battery. The points *E*, *F*, *G*, and *H* along the bottom of the circuit are all at a potential equal to the negative terminal of the source. Thus, each voltage across each parallel resistor is the same, and each is equal to the source voltage.

Figure 6–7B is the same circuit as in Part A, drawn in a slightly different way. Here the tops of the resistors are connected to a single point, which is the positive battery terminal. The bottoms of the resistors are all connected to the

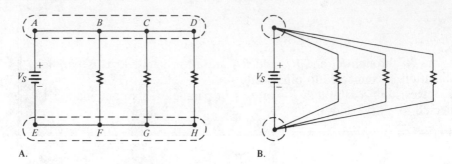

A.   B.

FIGURE 6–7   *Voltage across parallel branches is the same.*

same point, which is the negative battery terminal. The resistors are still all in parallel across the source.

Example 6–3

Determine the voltage across each resistor in Figure 6–8.

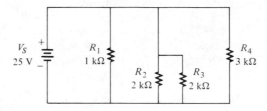

FIGURE 6–8

Solution:

The four resistors are in parallel; so the voltage drop across each one is equal to the applied source voltage:

$$V_S = V_1 = V_2 = V_3 = V_4$$
$$= 25 \text{ V}$$

Review for 6–2

1. A 10-Ω and a 20-Ω resistor are connected in parallel with a 5-V source. What is the voltage across each of the resistors?

2. A voltmeter is connected across R_1 in Figure 6–9. It measures 118 V. If you move the meter and connect it across R_2, how much voltage will it indicate? What is the source voltage?

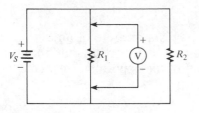

FIGURE 6–9

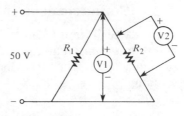

FIGURE 6–10

3. In Figure 6–10, how much voltage does voltmeter 1 indicate? Voltmeter 2?
4. How are voltages across each branch of a parallel circuit related?

TOTAL PARALLEL RESISTANCE

6–3

When resistors are connected in parallel, the total resistance of the circuit decreases. The total resistance of a parallel combination is always less than the value of the smallest resistor. For example, if a 10-Ω resistor and a 100-Ω resistor are connected in parallel, the total resistance is less than 10 Ω. The exact value must be calculated, and you will learn how to do so later in this section.

More Current Paths Mean Less Opposition

As you know, when resistors are connected in parallel, the current has more than one path. The number of current paths is equal to the number of parallel branches.

For example, in Figure 6–11A, there is only one current path since it is a series circuit. A certain amount of current, I_1, flows through R_1. If resistor R_2 is connected in parallel with R_1, as shown in Figure 6–11B, an additional amount of current, I_2, flows through R_2. The total current coming from the source has increased with the addition of the parallel branch. An increase in the *total current* from the source means that the total resistance has decreased, in accordance with Ohm's law. Additional resistors connected in parallel will further reduce the resistance and increase the total current.

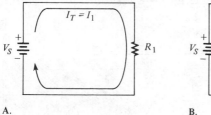

A.

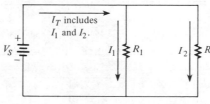

B.

FIGURE 6–11 *Resistors in parallel reduce total resistance and increase total current.*

Conductances in Parallel

Recall from Chapter 2 that conductance (G) is a measure of a resistor's ability to *conduct* current and is measured in *siemens* (S), which is the SI unit of conductance. Thus, it is the inverse (reciprocal) of the opposition (resistance) to current:

$$G = \frac{1}{R} \qquad \qquad \textbf{(6–1)}$$

Every resistor has a certain value of conductance, just as it has a certain value of resistance. Starting with the concept of total parallel conductance, you will see how the formula for calculating total parallel resistance is developed.

The key idea to keep in mind here is that an *increase* in conductance makes current flow more easily, and therefore an increase in G corresponds to a *decrease* in R. When resistors are connected in parallel, current flows more easily because a parallel circuit provides multiple current paths. Thus, conductance *increases* as parallel paths are added, as illustrated in Figure 6–12.

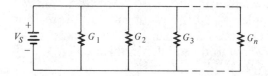

FIGURE 6–12   *Conductances add in parallel because more total current will flow as more paths are added.*

Each time a new resistor is connected in parallel in a parallel circuit, the conductance of the circuit increases by the value of the conductance of that new resistor. Therefore, *the total conductance of a parallel circuit is equal to the sum of the conductances of the individual resistors:*

$$G_T = G_1 + G_2 + G_3 + \cdots + G_n \qquad \qquad \textbf{(6–2)}$$

where G_T is the total conductance, G_n is the conductance of the last resistor in parallel, and n is the number of parallel branches.

Example 6–4 illustrates how to add parallel conductances to obtain the total conductance.

Example 6–4

Determine the total conductance of the circuit of Figure 6–13.

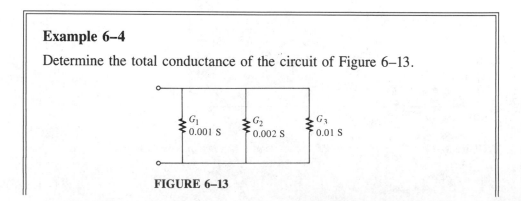

FIGURE 6–13

Solution:

The total conductance G_T is the sum of each individual conductance. In this circuit there are three conductances in parallel, and G_T is calculated as follows:

$$G_T = 0.001 \text{ S} + 0.002 \text{ S} + 0.01 \text{ S} = 0.013 \text{ S}$$

where "S" stands for siemens.

Formula for Total Parallel Resistance

Now, using the relationship between resistance and conductance, $G = 1/R$, we can rewrite Equation (6–2) as follows:

$$\frac{1}{R_T} = \frac{1}{R_1} + \frac{1}{R_2} + \frac{1}{R_3} + \cdots + \frac{1}{R_n} \qquad \textbf{(6–3)}$$

If we take the reciprocal of both sides of Equation (6–3), we get the general formula for total parallel resistance:

$$R_T = \frac{1}{\left(\dfrac{1}{R_1}\right) + \left(\dfrac{1}{R_2}\right) + \left(\dfrac{1}{R_3}\right) + \cdots + \left(\dfrac{1}{R_n}\right)} \qquad \textbf{(6–4)}$$

Equation (6–4) states that the total parallel resistance is found by adding the conductances $(1/R)$ of all of the resistors in parallel and then taking the reciprocal of this sum. Example 6–5 shows how to use this formula in a specific case.

Example 6–5

Calculate the total parallel resistance between points A and B of the circuit in Figure 6–14.

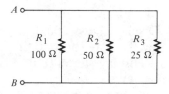

FIGURE 6–14

Solution:

First, find the conductance of each of the three resistors as follows:

$$G_1 = \frac{1}{R_1} = \frac{1}{100 \ \Omega} = 0.01 \text{ S}$$

Example 6–5 (continued)

$$G_2 = \frac{1}{R_2} = \frac{1}{50 \ \Omega} = 0.02 \ S$$

$$G_3 = \frac{1}{R_3} = \frac{1}{25 \ \Omega} = 0.04 \ S$$

Next, calculate R_T by adding G_1, G_2, and G_3 and taking the reciprocal of the sum as follows:

$$R_T = \frac{1}{0.01 \ S + 0.02 \ S + 0.04 \ S} = \frac{1}{0.07 \ S}$$

$$= 14.29 \ \Omega$$

Notice that the value of R_T (14.29 Ω) is smaller than the smallest value in parallel, which is R_3 (25 Ω).

Calculator Solution

The parallel resistance formula is easily solved on an electronic calculator. The general procedure is to enter the value of R_1 and then take its reciprocal by pressing the $1/x$ key. Next, press the + key; then enter the value of R_2 and take its reciprocal. Repeat this procedure until all of the resistor values have been entered and the reciprocal of each has been added. The final step is to press the $1/x$ key to convert $1/R_T$ to R_T. The total parallel resistance is now on the display. This calculator procedure is illustrated in Example 6–6.

Example 6–6

Show the steps required for a calculator solution of Example 6–5.

Solution:

Step 1: Enter 100. Display shows 100.
Step 2: Press $1/x$ key. Display shows 0.01.
Step 3: Press + key. Display shows 0.01.
Step 4: Enter 50. Display shows 50.
Step 5: Press $1/x$ key. Display shows 0.02.
Step 6: Press + key. Display shows 0.03.
Step 7: Enter 25. Display shows 25.
Step 8: Press $1/x$ key. Display shows 0.04.
Step 9: Press = key. Display shows 0.07.
Step 10: Press $1/x$ key. Display shows 14.2857 . . . .

The number displayed in Step 9 is the total *conductance* in siemens. The number displayed in Step 10 is the total *resistance* in ohms.

Two Resistors in Parallel

Equation (6–4) is a general formula for finding the total resistance for any number of resistors in parallel. It is often useful to consider only two resistors in parallel because this setup occurs commonly in practice. Also, any number of resistors in parallel can be broken down into *pairs* as an alternate way to find the R_T. Based on Equation (6–4), the formula for two resistors in parallel is

$$R_T = \frac{1}{\left(\dfrac{1}{R_1}\right) + \left(\dfrac{1}{R_2}\right)}$$

Combining the terms in the denominator, we get

$$R_T = \frac{1}{\dfrac{R_1 + R_2}{R_1 R_2}}$$

This equation can be rewritten as follows:

$$R_T = \frac{R_1 R_2}{R_1 + R_2} \tag{6–5}$$

Equation (6–5) states that *the total resistance for two resistors in parallel is equal to the product of the two resistors divided by the sum of the two resistors.* This equation is sometimes referred to as the "product over the sum" formula. Example 6–7 illustrates how to use it.

Example 6–7

Calculate the total resistance between the positive and negative terminals of the source of the circuit in Figure 6–15.

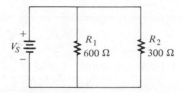

FIGURE 6–15

Solution:

Use Equation (6–5) as follows:

Example 6–7 (continued)

$$R_T = \frac{R_1 R_2}{R_1 + R_2} = \frac{(600 \ \Omega)(300 \ \Omega)}{600 \ \Omega + 300 \ \Omega}$$

$$= \frac{180,000 \ \Omega^2}{900 \ \Omega} = 200 \ \Omega$$

Resistors of Equal Value in Parallel

Another special case of parallel circuits is the parallel connection of several resistors having the same ohmic value. There is a shortcut method of calculating R_T when this case occurs.

If several resistors in parallel have the same resistance, they can be assigned the same symbol R. For example, $R_1 = R_2 = R_3 = \cdots = R_n = R$. Starting with Equation (6–4), we can develop a special formula for finding R_T:

$$R_T = \frac{1}{\left(\dfrac{1}{R}\right) + \left(\dfrac{1}{R}\right) + \left(\dfrac{1}{R}\right) + \cdots + \left(\dfrac{1}{R}\right)}$$

Notice that in the denominator, the same term, $1/R$, is added n times (n is the number of equal resistors in parallel). Therefore, the formula can be written as

$$R_T = \frac{1}{n/R}$$

Rewriting, we obtain

$$R_T = \frac{R}{n} \tag{6–6}$$

Equation (6–6) says that when any number of resistors (n), all having the same resistance (R), are connected in parallel, R_T is equal to the resistance divided by the number of resistors in parallel. Example 6–8 shows how to use this formula.

Example 6–8

Find the total resistance between points A and B in Figure 6–16.

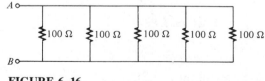

FIGURE 6–16

Solution:

There are five 100-Ω resistors in parallel. Use Equation (6–6) as follows:

$$R_T = \frac{R}{n} = \frac{100\ \Omega}{5} = 20\ \Omega$$

Determining an Unknown Parallel Resistor

Sometimes it is necessary to determine the values of resistors that are to be combined to produce a desired total resistance. For example, consider the case where *two* parallel resistors are used to obtain a desired total resistance. One resistor value is arbitrarily chosen, and then the second resistor value is calculated using Equation (6–7). This equation is derived from the formula for two parallel resistors as follows:

$$R_T = \frac{R_x R_A}{R_x + R_A}$$

$$R_T(R_x + R_A) = R_x R_A$$

$$R_T R_x + R_T R_A = R_x R_A$$

$$R_A R_x - R_T R_x = R_T R_A$$

$$R_x(R_A - R_T) = R_T R_A$$

$$R_x = \frac{R_A R_T}{R_A - R_T} \tag{6–7}$$

where R_x is the unknown resistor and R_A is the selected value. Example 6–9 illustrates use of this formula.

Example 6–9

Suppose that you wished to obtain a resistance of 200 Ω by combining two resistors in parallel. There is a 300-Ω resistor available. What other value is needed?

Solution:

$$R_T = 200\ \Omega \quad \text{and} \quad R_A = 300\ \Omega$$

$$R_x = \frac{R_A R_T}{R_A - R_T} = \frac{(300\ \Omega)(200\ \Omega)}{300\ \Omega - 200\ \Omega}$$

$$= 600\ \Omega$$

Notation for Parallel Resistors

Sometimes, for convenience, parallel resistors are designated by two parallel vertical marks. For example, R_1 in parallel with R_2 can be written as $R_1 \| R_2$. Also, when several resistors are in parallel with each other, this notation can be used. For example, $R_1 \| R_2 \| R_3 \| R_4 \| R_5$ indicates that R_1 through R_5 are all in parallel.

This notation is also used with resistance values. For example,

$$10 \text{ k}\Omega \| 5 \text{ k}\Omega$$

means that 10 kΩ are in parallel with 5 kΩ.

Review for 6-3

1. Does the total resistance increase or decrease as more resistors are connected in parallel?

2. The total parallel resistance is always less than _____ .

3. From memory, write the general formula for R_T with any number of resistors in parallel.

4. Write the special formula for two resistors in parallel.

5. Write the special formula for any number of equal-value resistors in parallel.

6. Calculate R_T for Figure 6–17.

7. Determine R_T for Figure 6–18.

8. Find R_T for Figure 6–19.

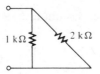

FIGURE 6–17

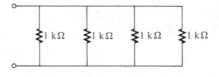

FIGURE 6–18

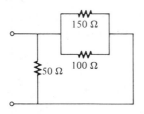

FIGURE 6–19

APPLYING OHM'S LAW

6-4   In this section you will see how Ohm's law can be applied to parallel circuit problems. First you will learn how to find the total current in a parallel circuit, and then you will learn how to determine branch currents using Ohm's law. Let us start with Example 6–10. Then, in Example 6–11, we use Ohm's law to find branch currents.

Example 6–10

Find the total current produced by the battery in Figure 6–20.

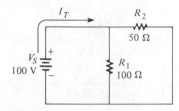

FIGURE 6–20

Solution:

The battery "sees" a total parallel resistance which determines the amount of current that it generates. First, we calculate R_T:

$$R_T = \frac{R_1 R_2}{R_1 + R_2} = \frac{(100 \ \Omega)(50 \ \Omega)}{100 \ \Omega + 50 \ \Omega}$$

$$= \frac{5000 \ \Omega^2}{150 \ \Omega} = 33.33 \ \Omega$$

The battery voltage is 100 V. Use Ohm's law to find I_T:

$$I_T = \frac{100 \ \text{V}}{33.33 \ \Omega} = 3 \ \text{A}$$

Example 6–11

Determine the current through each resistor in the parallel circuit of Figure 6–21.

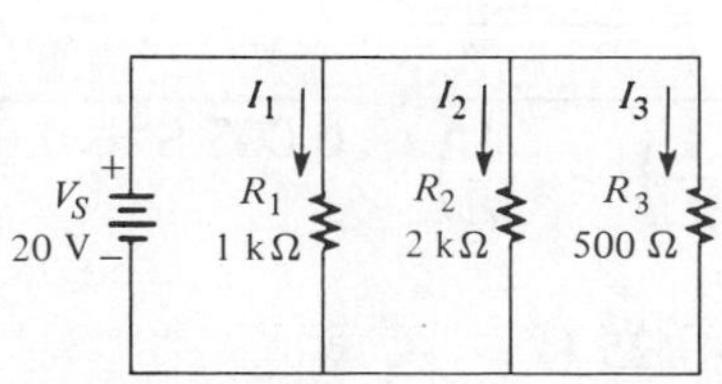

FIGURE 6–21

Example 6–11 (continued)

Solution:

The voltage across each resistor (branch) is equal to the source voltage. That is, the voltage across R_1 is 20 V, the voltage across R_2 is 20 V, and the voltage across R_3 is 20 V. The current through each resistor is determined as follows:

$$I_1 = \frac{V_S}{R_1} = \frac{20 \text{ V}}{1 \text{ k}\Omega} = 20 \text{ mA}$$

$$I_2 = \frac{V_S}{R_2} = \frac{20 \text{ V}}{2 \text{ k}\Omega} = 10 \text{ mA}$$

$$I_3 = \frac{V_S}{R_3} = \frac{20 \text{ V}}{500 \text{ }\Omega} = 40 \text{ mA}$$

In Example 6–12, we use Ohm's law to determine the unknown voltage across a parallel circuit.

Example 6–12

Find the voltage across the parallel circuit in Figure 6–22.

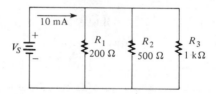

FIGURE 6–22

Solution:

We know the total current into the parallel circuit. We need to know the total resistance, and then we can apply Ohm's law to get the voltage. The total resistance is

$$R_T = \frac{1}{\left(\dfrac{1}{R_1}\right) + \left(\dfrac{1}{R_2}\right) + \left(\dfrac{1}{R_3}\right)} = \frac{1}{0.005 \text{ S} + 0.002 \text{ S} + 0.001 \text{ S}}$$

$$= \frac{1}{0.008 \text{ S}} = 125 \text{ }\Omega$$

$$V_S = I_T R_T = (10 \text{ mA}) (125 \ \Omega)$$
$$= 1.25 \text{ V}$$

Review for 6–4

1. A 10-V battery is connected across three 60-Ω resistors that are in parallel. What is the total current from the battery?

2. How much voltage is required to produce 2 A of current through the circuit of Figure 6–23?

3. How much current is there through each resistor of Figure 6–23?

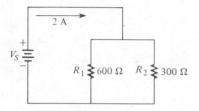

FIGURE 6–23

4. There are four equal-value resistors in parallel with a 12-V source, and 6 mA of current from the source. What is the value of each resistor?

5. A 1-kΩ and a 2-kΩ resistor are connected in parallel. A total of 100 mA flows through the parallel combination. How much voltage is dropped across the resistors?

CURRENT SOURCES IN PARALLEL

6–5

A current source is an energy source that provides a constant value of current to a load even when that load changes in resistance value. A transistor can be used as a current source, and thus current sources are important in circuit analysis. At this point, you are not prepared to study transistor circuits in detail, but you do need to understand how current sources act in circuits. The only practical case that we need to consider is that of current sources connected in parallel.

The general rule to remember is that the total current produced by current sources in parallel is equal to the algebraic sum of the individual current sources. The *algebraic sum* means that you must consider the direction of current when combining the sources in parallel. For example, in Figure 6–24A (next page), the three current sources in parallel provide current in the same direction (into point A). So the total current into point A is $I_T = 1 \text{ A} + 2 \text{ A} + 2 \text{ A} = 5 \text{ A}$.

In Figure 6–24B, the 1-A source provides current in a direction opposite to the other two. The total current into point A in this case is $I_T = 2 \text{ A} + 2 \text{ A} - 1 \text{ A} = 3 \text{ A}$.

154

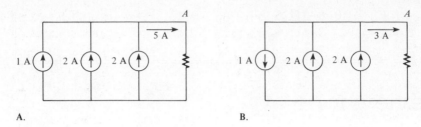

A.   B.

FIGURE 6–24

Example 6–13

Determine the current through R_L in Figure 6–25.

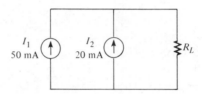

FIGURE 6–25

Solution:

The two current sources are in the same direction; so the current through R_L is

$$I_L = I_1 + I_2 = 50 \text{ mA} + 20 \text{ mA} = 70 \text{ mA}$$

Review for 6–5

1. Four 0.5-A current sources are connected in parallel in the same direction. What current will be produced through a load resistor?

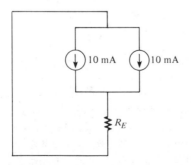

FIGURE 6–26

2. How many 1-A current sources must be connected in parallel to produce a total current output of 3 A? Sketch a schematic showing the sources connected.

3. In a transistor amplifier circuit, the transistor can be represented by a 10-mA current source, as shown in Figure 6–26. The transistors act in parallel, as in a differential amplifier. How much current is flowing through the resistor R_E?

KIRCHHOFF'S CURRENT LAW

6–6

Kirchhoff's current law states that *the sum of the currents into a junction is equal to the sum of the currents out of that junction*. A *junction* is any point in a circuit where two or more circuit paths come together. In a parallel circuit, a junction is where the parallel branches connect together. Another way to state Kirchhoff's current law is to say that *the total current into a junction is equal to the total current out of that junction*.

For example, in the circuit of Figure 6–27, point A is one junction and point B is another. Let us start at the positive terminal of the source and follow the current. The total current I_T flows from the source and *into* the junction at point A. At this point, the current splits up among the three branches as indicated. Each of the three branch currents (I_1, I_2, and I_3) flows *out of* junction A. Kirchhoff's current law says that the total current into junction A is equal to the total current out of junction A; that is,

$$I_T = I_1 + I_2 + I_3$$

Now, following the currents in Figure 6–27 through the three branches, you see that they come back together at point B. Currents I_1, I_2, and I_3 flow into junction B, and I_T flows out. Kirchhoff's current law formula at this junction is therefore the same as at junction A:

$$I_T = I_1 + I_2 + I_3$$

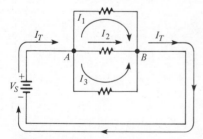

FIGURE 6–27 *Total current into junction A equals sum of currents out of junction A. Sum of currents into junction B equals total current out of junction B.*

General Formula for Kirchhoff's Current Law

The previous discussion was a specific case to illustrate Kirchhoff's current law. Now let us look at the general case. Figure 6–28 shows a *generalized* circuit junction where a number of branches are connected to a point in the circuit. Currents $I_{IN(1)}$ through $I_{IN(n)}$ flow into the junction (n can be any number).

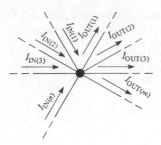

FIGURE 6–28 *Generalized circuit junction for illustrating Kirchhoff's current law.*

Currents $I_{OUT(1)}$ through $I_{OUT(m)}$ flow out of the junction (m can be any number, but not necessarily equal to n). By Kirchhoff's current law, the sum of the currents into a junction must equal the sum of the currents out of the junction. With reference to Figure 6–28, the general formula for Kirchhoff's current law is

$$I_{IN(1)} + I_{IN(2)} + \cdots + I_{IN(n)} = I_{OUT(1)} + I_{OUT(2)} + \cdots + I_{OUT(m)} \quad \textbf{(6–8)}$$

If all of the terms on the right side of Equation (6–8) are brought over to the left side, their signs change to negative, and a zero is left on the right side. Kirchhoff's current law is sometimes stated in this way: *The algebraic sum of all the currents entering and leaving a junction is equal to zero.* This statement is just another, equivalent way of stating what we have just discussed. The following three examples illustrate use of Kirchhoff's current law.

Example 6–14

The branch currents in the circuit of Figure 6–29 are known. Determine the total current entering junction A and the total current leaving junction B.

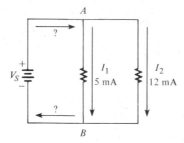

FIGURE 6–29

Solution:

The total current flowing out of junction A is the sum of the two branch currents. So the total current flowing into A is

$$I_T = I_1 + I_2 = 5 \text{ mA} + 12 \text{ mA} = 17 \text{ mA}$$

The total current entering point B is the sum of the two branch currents. So the total current flowing out of B is

$$I_T = I_1 + I_2 = 5 \text{ mA} + 12 \text{ mA} = 17 \text{ mA}$$

Example 6–15

Determine the current through R_2 in Figure 6–30.

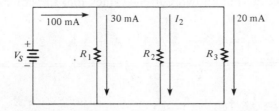

FIGURE 6–30

Solution:

The total current flowing into the junction of the three branches is known. Two of the branch currents are known. The current equation at this junction is

$$I_T = I_1 + I_2 + I_3$$

Solving for I_2, we get

$$I_2 = I_T - I_1 - I_3 = 100 \text{ mA} - 30 \text{ mA} - 20 \text{ mA}$$
$$= 50 \text{ mA}$$

Example 6–16

Find I_1 in Figure 6–31.

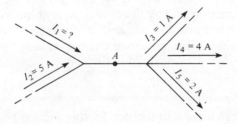

FIGURE 6–31

Example 6–16 (continued)

Solution:

Currents I_1 and I_2 are flowing into junction A, and I_3, I_4, and I_5 are flowing out. Kirchhoff's current equation at this junction is

$$I_1 + I_2 = I_3 + I_4 + I_5$$

Solving for I_1, we have

$$I_1 = I_3 + I_4 + I_5 - I_2$$

Now we substitute the known current values into the equation and get I_1 as follows:

$$I_1 = 1\text{ A} + 4\text{ A} + 2\text{ A} - 5\text{ A} = 2\text{ A}$$

Review for 6–6

1. State Kirchhoff's current law in two ways.

2. A total current of 2.5 A flows into the junction of three parallel branches. What is the sum of all three branch currents?

3. In Figure 6–32, 100 mA and 300 mA flow into the junction. What is the amount of current flowing out of the junction?

4. Determine I_1 in the circuit of Figure 6–33.

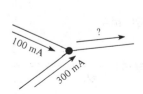

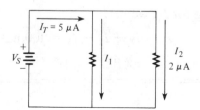

FIGURE 6–32   **FIGURE 6–33**

5. Two branch currents enter a junction, and two branch currents leave the same junction. One of the currents entering the junction is 1 A, and one of the currents leaving the junction is 3 A. The total current entering and leaving the junction is 8 A. Determine the value of the unknown current entering the junction and the value of the unknown current leaving the junction.

CURRENT DIVIDERS

6–7   A parallel circuit acts as a *current divider.* In this section you will see what this term means and why current dividers are an important application of parallel circuits.

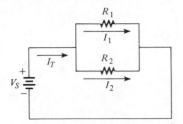

FIGURE 6–34 *Total current divides between two branches.*

To understand how a parallel connection of resistors acts as a current divider, look at Figure 6–34, where there are two resistors in parallel. As you already know, there is a current through R_1 and a current through R_2. We call these branch currents I_1 and I_2, respectively, as indicated in the diagram.

Since the same voltage is across each of the resistors in parallel, the branch currents are inversely proportional to the ohmic values of the resistors. For example, if the value of R_2 is twice that of R_1, then the value of I_2 is one-half that of I_1. In other words, *the total current divides among parallel resistors in a manner inversely proportional to the resistance values*. The branches with higher resistance have less current, and the branches with lower resistance have more current, in accordance with Ohm's law.

General Current Divider Formula

With a few steps, a formula for determining how currents divide among parallel resistors can be developed. Let us assume that we have several resistors in parallel, as shown in Figure 6–35. This figure shows five resistors, but there can be any number.

Let us call the current through any one of the parallel resistors I_x, where x represents the number of a particular resistor (1, 2, 3, and so on). By Ohm's law, the current through any one of the resistors in Figure 6–35 can be written as follows:

$$I_x = \frac{V_S}{R_x}$$

where $x = 1, 2, 3, 4,$ or 5. The source voltage V_S appears across each of the parallel resistors, and R_x represents any one of the parallel resistors. The total source voltage V_S is equal to the total current times the total parallel resistance:

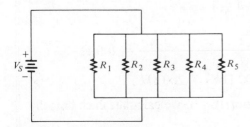

FIGURE 6–35 *Current divider.*

160

$$V_S = I_T R_T$$

Substituting $I_T R_T$ for V_S in the expression for I_x, we get

$$I_x = \frac{I_T R_T}{R_x}$$

By rearranging, we get

$$I_x = \left(\frac{R_T}{R_x}\right) I_T \qquad \textbf{(6–9)}$$

Equation (6–9) is the general current divider formula. It tells us that the current (I_x) through any branch equals the total parallel resistance (R_T), divided by the resistance (R_x) of that branch and multiplied by the total current (I_T) flowing into the parallel junction. This formula applies to a parallel circuit with any number of branches.

Example 6–17 illustrates application of Equation (6–9).

Example 6–17

Determine the current through each resistor in the circuit of Figure 6–36.

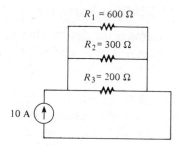

FIGURE 6–36

Solution:

First calculate the total parallel resistance:

$$R_T = \frac{1}{\left(\frac{1}{R_1}\right) + \left(\frac{1}{R_2}\right) + \left(\frac{1}{R_3}\right)}$$

$$= \frac{1}{\left(\frac{1}{600\ \Omega}\right) + \left(\frac{1}{300\ \Omega}\right) + \left(\frac{1}{200\ \Omega}\right)} = 100\ \Omega$$

The total current is 10 A. Using Equation (6–9), we calculate each branch current as follows:

$$I_1 = \left(\frac{R_T}{R_1}\right)I_T = \left(\frac{100\ \Omega}{600\ \Omega}\right)10\text{ A}$$

$$= 1.67\text{ A}$$

$$I_2 = \left(\frac{R_T}{R_2}\right)I_T = \left(\frac{100\ \Omega}{300\ \Omega}\right)10\text{ A}$$

$$= 3.33\text{ A}$$

$$I_3 = \left(\frac{R_T}{R_3}\right)I_T = \left(\frac{100\ \Omega}{200\ \Omega}\right)10\text{ A}$$

$$= 5.00\text{ A}$$

Current Divider Formula for Two Branches

For the special case of two parallel branches, Equation (6–9) can be modified. The reason for doing so is that two parallel resistors are often found in practical circuits. We start by restating Equation (6–5), the formula for the total resistance of two parallel branches:

$$R_T = \frac{R_1 R_2}{R_1 + R_2}$$

When we have two parallel resistors, as in Figure 6–37, we may want to find the current through either or both of the branches. To do so, we need two special formulas.

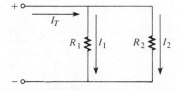

FIGURE 6–37

Using Equation (6–9), we write the formulas for I_1 and I_2 as follows:

$$I_1 = \left(\frac{R_T}{R_1}\right)I_T$$

$$I_2 = \left(\frac{R_T}{R_2}\right)I_T$$

Substituting $R_1 R_2/(R_1 + R_2)$ for R_T, we get

$$I_1 = \left(\frac{R_1 R_2/(R_1 + R_2)}{R_1}\right)I_T$$

$$I_2 = \left(\frac{R_1 R_2/(R_1 + R_2)}{R_2}\right)I_T$$

Canceling as shown, we get the final formulas:

$$I_1 = \left(\frac{R_2}{R_1 + R_2}\right)I_T \tag{6-10}$$

$$I_2 = \left(\frac{R_1}{R_1 + R_2}\right)I_T \tag{6-11}$$

When there are only *two* resistors in parallel, these equations are a little easier to use than Equation (6–9), because it is not necessary to know R_T.

Note that in Equations (6–10) and (6–11), the current in one of the branches is equal to the *opposite* branch resistance over the *sum* of the two resistors, all times the total current. In all applications of the current divider equations, you must know the total current going into the parallel branches. Example 6–18 illustrates application of these special current divider formulas.

Example 6–18

Find I_1 and I_2 in Figure 6–38.

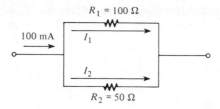

FIGURE 6–38

Solution:

Using Equation (6–10), we determine I_1 as follows:

$$I_1 = \left(\frac{R_2}{R_1 + R_2}\right)I_T = \left(\frac{50\ \Omega}{150\ \Omega}\right)100\ mA$$

$$= 33.33\ mA$$

Using Equation (6–11), we determine I_2 as follows:

$$I_2 = \left(\frac{R_1}{R_1 + R_2}\right)I_T = \left(\frac{100\ \Omega}{150\ \Omega}\right)100\ mA$$

$$= 66.67\ mA$$

Review for 6–7

1. Write the general current divider formula.

2. Write the two special formulas for calculating each branch current for a two-branch circuit.

3. A parallel circuit has the following resistors in parallel: 200-Ω, 100-Ω, 75-Ω, 50-Ω, and 22-Ω. Which resistor has the most current through it? The least current?

4. Determine the current through R_3 in Figure 6–39.

5. Find I_1 and I_2 in the circuit of Figure 6–40.

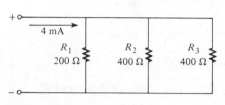

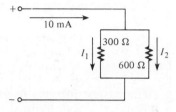

FIGURE 6–39   **FIGURE 6–40**

POWER IN A PARALLEL CIRCUIT

6–8

The total amount of power in a parallel resistive circuit is equal to the sum of the powers in each resistor in parallel. Equation (6–12) states this in a concise way for any number of resistors in parallel:

$$P_T = P_1 + P_2 + P_3 + \cdots + P_n \qquad (6\text{–}12)$$

where P_T is the total power and P_n is the power in the last resistor in parallel. As you can see, the power losses are additive, just as in the series circuit.

The power formulas in Chapter 4 are directly applicable to parallel circuits. The following formulas are used to calculate the total power P_T:

$$P_T = VI_T$$

$$P_T = I_T^2 R_T$$

$$P_T = \frac{V^2}{R_T}$$

where V is the voltage across the parallel circuit, I_T is the total current flowing into the parallel circuit, and R_T is the total resistance of the parallel circuit. Example 6–19 shows how total power can be calculated in a parallel circuit.

Example 6–19

Determine the total amount of power in the parallel circuit in Figure 6–41.

Example 6–19 (continued)

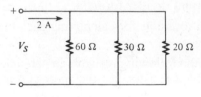

FIGURE 6–41

Solution:

We know that the total current is 2 A. The total resistance is

$$R_T = \cfrac{1}{\left(\cfrac{1}{60\ \Omega}\right) + \left(\cfrac{1}{30\ \Omega}\right) + \left(\cfrac{1}{20\ \Omega}\right)} = 10\ \Omega$$

The easiest formula to use is $P_T = I_T^2 R_T$ since we know both I_T and R_T. Thus,

$$P_T = I_T^2 R_T = (2\ \text{A})^2\,(10\ \Omega) = 40\ \text{W}$$

To demonstrate that if the power in each resistor is determined and if all of these values are added together, you get the same result, we will work through another calculation. First, find the voltage across each branch of the circuit:

$$V_S = I_T R_T = (2\ \text{A})\,(10\ \Omega) = 20\ \text{V}$$

Remember that the voltage across all branches is the same.

Next, calculate the power for each resistor using $P = V^2/R$:

$$P_1 = \frac{(20\ \text{V})^2}{60\ \Omega} = 6.67\ \text{W}$$

$$P_2 = \frac{(20\ \text{V})^2}{30\ \Omega} = 13.33\ \text{W}$$

$$P_3 = \frac{(20\ \text{V})^2}{20\ \Omega} = 20\ \text{W}$$

Now, add these powers to get the total power:

$$P_T = 6.67\ \text{W} + 13.33\ \text{W} + 20\ \text{W}$$
$$= 40\ \text{W}$$

This calculation shows that the sum of the individual powers is equal to the total power as determined by one of the power formulas.

Review for 6–8

1. If you know the power in each resistor in a parallel circuit, how can you find the total power?

2. The resistors in a parallel circuit have the following powers: 2 W, 5 W, 1 W, and 8 W. What is the total power in the circuit?

3. A circuit has a 1-kΩ, a 2-kΩ, and a 4-kΩ resistor in parallel. A total current of 1 A flows into the parallel circuit. What is the total power?

EFFECTS OF OPEN PATHS

6–9

Recall that an open circuit is one in which the current path is interrupted and no current flows. In this section we examine what happens when a branch of a parallel circuit opens.

If a switch is connected in a branch of a parallel circuit, as shown in Figure 6–42, an open or a closed path can be made by the switch. When the switch is closed, as in Figure 6–42A, R_1 and R_2 are in parallel. The total resistance is 50 Ω (two 100-Ω resistors in parallel). Current flows through both resistors. If the switch is opened, as in Figure 6–42B, R_1 is effectively removed from the circuit, and the total resistance is 100 Ω. Current now flows only through R_2.

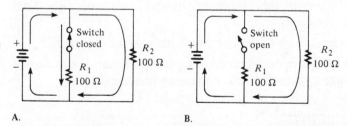

A.   B.

FIGURE 6–42 *When switch opens, total current decreases and current through R_2 remains unchanged.*

In general, *when an open circuit occurs in a parallel branch, the total resistance increases, the total current decreases, and current continues to flow through the remaining parallel paths.* The decrease in total current equals the amount of current that was previously flowing in the open branch. The other branch currents remain the same.

Consider the lamp circuit in Figure 6–43. There are four bulbs in parallel with a 120-V source. In Part A, current is flowing through each bulb. Now suppose that one of the bulbs burns out, creating an open path as shown in Figure 6–43B. This light will go out because no current is flowing through the open path. Notice, however, that current continues to flow through all the other parallel bulbs, and they continue to glow. The open branch does not change the voltage across the parallel branches; it remains at 120 V.

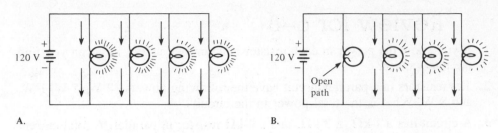

A.   B.

FIGURE 6–43 *When lamp opens, total current decreases and other branch currents remain unchanged.*

You can see that a parallel circuit has an advantage over a series connection in lighting systems because if one or more of the parallel bulbs burn out, the others will stay on. In a *series* circuit, when one bulb goes out, all of the others go out also because the current path is *completely* interrupted.

Review for 6–9

1. In a parallel circuit, what happens when one of the branches opens?

2. If several light bulbs are connected in parallel and one of the bulbs opens (burns out), will the others continue to glow?

3. In a parallel circuit, if one of the branch resistances opens, does the total resistance increase or decrease?

4. If a parallel branch opens, does the total current increase or decrease?

5. One ampere of current flows in each branch of a parallel circuit. If one branch opens, what is the current in each of the remaining branches?

TROUBLESHOOTING PARALLEL CIRCUITS

6–10   When a resistor in a parallel circuit opens, the open resistor cannot be located by measurement of the voltage across the branches, because the same voltage exists across all the branches. Thus, there is no way to tell which resistor is open by simply measuring voltage. The good resistors will always have the same voltage as the open one, as illustrated in Figure 6–44 (note that the middle resistor is open).

If a visual inspection does not reveal the open resistor, it must be located by current measurements. In practice, measuring current is more difficult than

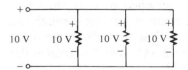

FIGURE 6–44 *Parallel branches all have the same voltage.*

measuring voltage because you must insert the ammeter in *series* to measure the current. Thus, a wire or a PC connection must be cut or disconnected, or one end of a component must be lifted off the circuit board, in order to connect the ammeter in series. This procedure, of course, is not required when voltage measurements are made, because the meter leads are simply connected *across* a component.

Where to Measure the Current

In a parallel circuit, the total current should be measured. When a parallel resistor opens, I_T is less than its normal value. Once I_T and the voltage across the branches are known, a few calculations will determine the open resistor when all the resistors are of different ohmic values.

Consider the two-branch circuit in Figure 6–45A. If one of the resistors opens, the total current will equal the current in the good resistor. Ohm's law quickly tells us what the current in each resistor should be:

$$I_1 = \frac{50 \text{ V}}{500 \ \Omega} = 0.1 \text{ A}$$

$$I_2 = \frac{50 \text{ V}}{100 \ \Omega} = 0.5 \text{ A}$$

If R_2 is open, the total current is 0.1 A, as indicated in Figure 6–45B. If R_1 is open, the total current is 0.5 A, as indicated in Figure 6–45C.

This procedure can be extended to any number of branches having unequal resistances. If the parallel resistances are all equal, the current in each branch must be checked until a branch is found with no current. This is the open resistor.

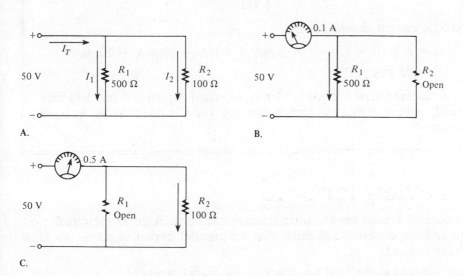

A.

B.

C.

FIGURE 6–45 *Finding an open path by current measurement.*

Example 6-20

In Figure 6-46, there is a total current of 32 mA, and the voltage across the parallel branches is 20 V. Is there an open resistor, and, if so, which one is it?

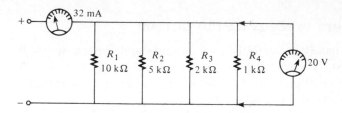

FIGURE 6-46

Solution:

Calculate the current in each branch:

$$I_1 = \frac{V}{R_1} = \frac{20 \text{ V}}{10 \text{ k}\Omega} = 2 \text{ mA}$$

$$I_2 = \frac{V}{R_2} = \frac{20 \text{ V}}{5 \text{ k}\Omega} = 4 \text{ mA}$$

$$I_3 = \frac{V}{R_3} = \frac{20 \text{ V}}{2 \text{ k}\Omega} = 10 \text{ mA}$$

$$I_4 = \frac{V}{R_4} = \frac{20 \text{ V}}{1 \text{ k}\Omega} = 20 \text{ mA}$$

The total current *should* be

$$I_T = I_1 + I_2 + I_3 + I_4 = 2 \text{ mA} + 4 \text{ mA} + 10 \text{ mA} + 20 \text{ mA}$$
$$= 36 \text{ mA}$$

The actual *measured* current is 32 mA, as stated, which is 4 mA less than normal, indicating that the branch carrying 4 mA is open. Thus, R_2 must be open.

Review for 6-10

1. If a parallel branch opens, what changes can be detected in the circuit voltage and the currents, assuming that the parallel circuit is across an ideal voltage source?

2. What happens to the total resistance if one branch opens?

The program listed in this section can be used to compute the currents in each branch of any specified parallel circuit. Also, the total resistance and total current are provided in the display. The inputs required are the number of resistors, the value of the resistors, and the value of the source voltage.

6–11

```
10 CLS
20 PRINT "THIS PROGRAM COMPUTES AND TABULATES THE TOTAL RESISTANCE,"
30 PRINT "TOTAL CURRENT, AND THE CURRENTS THROUGH EACH OF THE"
40 PRINT "INDIVIDUAL RESISTORS IN A SPECIFIED PARALLEL CIRCUIT.
50 PRINT:PRINT:PRINT
60 PRINT "YOU ARE REQUIRED TO ENTER THE NUMBER OF RESISTORS,"
70 PRINT "THE RESISTOR VALUES, AND THE SOURCE VOLTAGE."
80 PRINT:PRINT:PRINT
90 INPUT "TO CONTINUE PRESS 'ENTER'.";X:CLS
100 INPUT "HOW MANY RESISTORS ARE IN THE PARALLEL CIRCUIT";N:CLS
110 FOR X=1 TO N
120 PRINT "ENTER THE VALUE OF R";X;"IN OHMS"
130 INPUT R(X)
140 NEXT X:CLS
150 INPUT "THE SOURCE VOLTAGE IN VOLTS";VS:CLS
160 RT=R(1)
170 FOR X=2 TO N
180 RT=1/(1/RT+1/R(X))
190 NEXT X
200 IT=VS/RT
210 PRINT "TOTAL RESISTANCE = ";RT;"OHMS"
220 PRINT:PRINT "TOTAL CURRENT = ";IT;"AMPS":PRINT
230 FOR X=1 TO N
240 I(X)=(RT/R(X))*IT
250 PRINT "R";X;"=";R(X);"OHMS","I";X;"=";I(X);"AMPS"
260 NEXT X
```

Review for Section 6–11

1. Identify by line numbers the series of instructions that permit entering the resistor values.

2. Explain the purpose of line 185.

Formulas

$$G = \frac{1}{R} \tag{6-1}$$

$$G_T = G_1 + G_2 + G_3 + \cdots + G_n \tag{6-2}$$

$$R_T = \frac{1}{\left(\dfrac{1}{R_1}\right) + \left(\dfrac{1}{R_2}\right) + \left(\dfrac{1}{R_3}\right) + \cdots + \left(\dfrac{1}{R_n}\right)} \tag{6-4}$$

$$R_T = \frac{R_1 R_2}{R_1 + R_2} \tag{6-5}$$

$$R_T = \frac{R}{n} \tag{6-6}$$

$$R_x = \frac{R_A R_T}{R_A - R_T} \tag{6-7}$$

$$I_{\text{IN}(1)} + I_{\text{IN}(2)} + \cdots + I_{\text{IN}(n)} = I_{\text{OUT}(1)} + I_{\text{OUT}(2)} + \cdots + I_{\text{OUT}(m)} \tag{6-8}$$

$$I_x = \left(\frac{R_T}{R_x}\right) I_T \tag{6-9}$$

$$I_1 = \left(\frac{R_2}{R_1 + R_2}\right) I_T \tag{6-10}$$

$$I_2 = \left(\frac{R_1}{R_1 + R_2}\right) I_T \tag{6-11}$$

$$P_T = P_1 + P_2 + P_3 + \cdots + P_n \tag{6-12}$$

Summary

1. Resistors in parallel are connected across the same points.
2. A parallel combination has more than one path for current.
3. The number of current paths equals the number of resistors in parallel.
4. The total parallel resistance is less than the lowest-value resistor.
5. The voltages across all branches of a parallel circuit are the same.
6. Current sources in parallel add algebraically.
7. One way to state Kirchhoff's current law: The algebraic sum of all the currents at a junction is zero.
8. Another way to state Kirchhoff's current law: The sum of the currents into a junction (total current in) equals the sum of the currents out of the junction (total current out).
9. A parallel circuit is a current divider, so called because the total current entering the parallel junction divides up into each of the branches.
10. If all of the branches of a parallel circuit have equal resistance, the currents through all of the branches are equal.
11. The total power in a parallel resistive circuit is the sum of all of the individual powers of the resistors making up the parallel circuit.
12. The total power for a parallel circuit can be calculated with the power formulas using values of total current, total resistance, or total voltage.
13. If one of the branches of a parallel circuit opens, the total resistance increases, and therefore the total current decreases.

14. If a branch of a parallel circuit opens, current still flows through the remaining branches.

Self-Test

1. Sketch a parallel circuit having four resistors and a voltage source.

2. Five amperes of current flow into a parallel circuit having two branches of equal resistance. How much current flows through each of the branches?

3. Connect each set of resistors in Figure 6–47 in parallel between points A and B.

4. If you measured the voltage across each of the points marked in Figure 6–48, how much voltage would you read in each case?

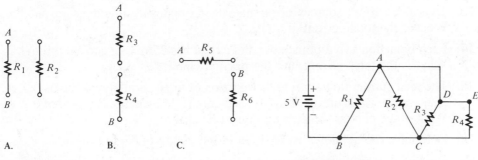

A.   B.   C.

FIGURE 6–47   **FIGURE 6–48**

5. If an 80-Ω resistor and a 150-Ω resistor are connected in parallel, what is the total resistance?

6. The following resistors are in parallel: 1000-Ω, 800-Ω, 500-Ω, 200-Ω, and 100-Ω. What is the total resistance?

7. Eight 56-Ω resistors are connected in parallel. Determine the total resistance.

8. Suppose that you need a total resistance of 100 Ω. The only resistors that are immediately available are one 200-Ω and several 400-Ω resistors. How would you connect these resistors to get a total of 100 Ω?

9. Three 600-Ω resistors are connected in parallel, and 5 V are applied across the parallel circuit. How much current is flowing out of the source?

10. Which circuit of Figure 6–49 has more total current?

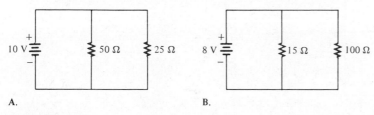

A.   B.

FIGURE 6–49

11. Six resistors of equal value are in parallel, with 12 V across them. The total current is 3 mA. Determine the value of each resistor. What is the total resistance?

12. Determine the current through R_L in the multiple-current-source circuit of Figure 6–50.

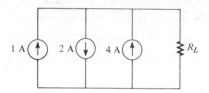

FIGURE 6–50

13. Two 1-A current sources are connected in opposite directions in parallel. What is the total current produced?

14. Each branch in a five-branch parallel circuit has 25 mA of current through it. What is the total current into the parallel circuit?

15. The total current flowing into the junction of three parallel resistors is 0.5 A. One branch has a current of 0.1 A, and another branch has a current of 0.2 A. What is the current through the third branch?

16. Determine the unknown resistances in the circuit of Figure 6–51.

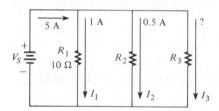

FIGURE 6–51

17. The following resistors are connected in parallel: 2-kΩ, 6-kΩ, 3-kΩ, and 1-kΩ. The total current is 1 A. Using the general current divider formula, calculate the current through each of the resistors.

18. Determine the current through each resistor in Figure 6–52.

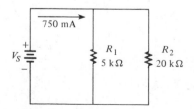

FIGURE 6–52

19. Determine the total power in the circuit of Figure 6–52.

20. There are two resistors in parallel across a 10-V source. One is 200 Ω, and the other is 500 Ω. You measure a total current of 20 mA. What is wrong with the circuit?

21. What resistance value in parallel with 100 Ω produces a total resistance of 40 Ω?

22. Run the program in Section 6–11 for a parallel circuit with five resistors having the following values: 1-kΩ, 1.5-kΩ, 2.2-kΩ, 3.3-kΩ, and 4.7-kΩ. Use a source voltage of 12 V.

Problems

Section 6–1

6–1. Connect the resistors in Figure 6–53A in parallel across the battery.

6–2. Determine whether or not all the resistors in Figure 6–53B are connected in parallel on the printed circuit (PC) board.

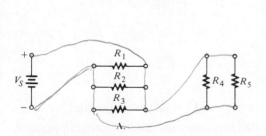
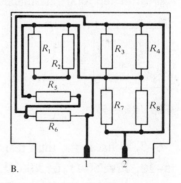

FIGURE 6–53

Section 6–2

6–3. What is the voltage across and the current through each parallel resistor if the total voltage is 12 V and the total resistance is 600 Ω? There are four resistors, all of equal value.

6–4. The source voltage in Figure 6–54 is 100 V. How much voltage does each of the meters read?

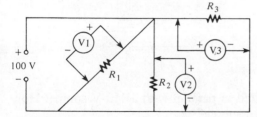

FIGURE 6–54

Section 6–3

6–5. The following resistors are connected in parallel: 1-MΩ, 2-MΩ, 5-MΩ, 12-MΩ, and 20-MΩ. Determine the total resistance.

6–6. Find the total resistance for each following group of parallel resistors:
(a) 560-Ω and 1000-Ω.   (b) 47-Ω and 56-Ω.
(c) 1.5-kΩ, 2.2-kΩ, 10-kΩ.   (d) 1-MΩ, 470-kΩ, 1-kΩ, 2.5-MΩ.

6–7. Calculate R_T for each circuit in Figure 6–55.

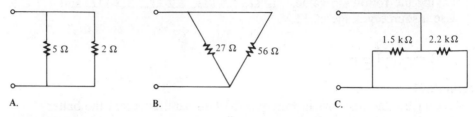

A.   B.   C.

FIGURE 6–55

6–8. If the total resistance in Figure 6–56 is 200 Ω, what is the value of R_2?

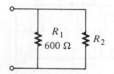

FIGURE 6–56

6–9. What is the total resistance of twelve 6-kΩ resistors in parallel?

6–10. Five 50-Ω, ten 100-Ω, and two 10-Ω resistors are all connected in parallel. What is the total resistance?

Section 6–4

6–11. What is the total current in each circuit of Figure 6–57?

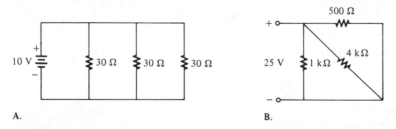

A.   B.

FIGURE 6–57

6–12. Three 33-Ω resistors are connected in parallel with a 110-V source. How much current is flowing from the source?

6–13. Four equal-value resistors are connected in parallel. Five volts are applied across the parallel circuit, and 2.5 mA are measured from the source. What is the value of each resistor?

6–14. A 10-mA and a 20-mA current source are connected in parallel in the same direction. What is the total current that can be provided to a load?

Section 6–5
6–15. Determine the current through R_L in each circuit in Figure 6–58.

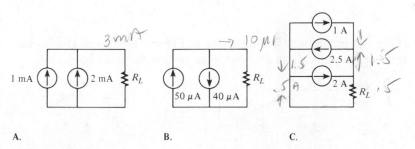

A.   B.   C.

FIGURE 6–58

Section 6–6
6–16. The following currents are measured in the same direction in a three-branch parallel circuit: 250-mA, 300-mA, and 800-mA. What is the value of the current into the junction of these three branches?

6–17. Five hundred milliamperes (500 mA) flow into five parallel resistors. The currents through four of the resistors are 50 mA, 150 mA, 25 mA, and 100 mA. How much current flows through the fifth resistor?

6–18. In the circuit of Figure 6–59, determine the resistances R_2, R_3, and R_4.

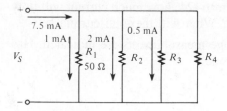

FIGURE 6–59

6–19. The total resistance of a parallel circuit is 25 Ω. How much current flows through a 200-Ω resistor that makes up part of the parallel circuit if the total current is 100 mA?

Section 6–7
6–20. Determine the current in each branch of the current dividers of Figure 6–60.

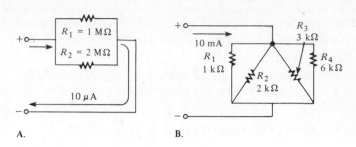

A.   B.

FIGURE 6–60

6–21. What is the current through each resistor in Figure 6–61? R is the lowest-value resistor, and all others are multiples of that value as indicated.

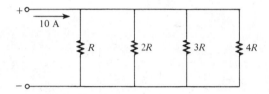

FIGURE 6–61

Section 6–8
6–22. Five parallel resistors each handle 40 mW. What is the total power?

6–23. Determine the total power in each circuit of Figure 6–60.

Sections 6–9 through 6–10
6–24. Six light bulbs are connected in parallel across 110 V. Each bulb is rated at 75 W. How much current flows through each bulb, and what is the total current?

6–25. If one of the bulbs burns out in Problem 6–24, how much current will flow through each of the remaining bulbs? What will the total current be?

6–26. In Figure 6–62, the current and voltage measurements are indicated. Has a resistor opened, and, if so, which one?

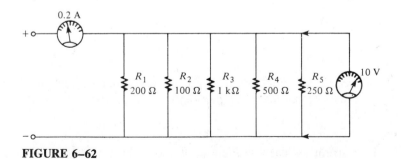

FIGURE 6–62

6–27. What is wrong with the circuit in Figure 6–63?

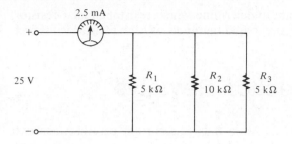

FIGURE 6-63

Section 6-11

6-28. Modify the program in Section 6-11 to include computation and display of the power dissipation in each branch and the total power.

6-29. Write a program to compute the value of an unknown resistor in an n-resistor parallel network when the desired total resistance and the values of the other $n - 1$ resistors are specified.

Advanced Problems

6-30. A certain parallel network consists of only $\frac{1}{2}$-W resistors. The total resistance is 1 kΩ, and the total current is 50 mA. If each resistor is operating at one-half its maximum power level, determine the following:
   (a) The number of resistors.   **(b)** The value of each resistor.
   (c) The current in each branch.   **(d)** The applied voltage.

6-31. Find the values of the unspecified labeled quantities in each circuit of Figure 6-64.

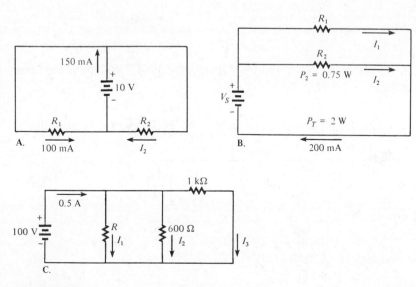

FIGURE 6-64

6–32. Develop a computer program to determine which resistor in an *n*-resistor parallel network is open (if any) based on inputs of each resistor value and the total measured current.

Answers to Section Reviews

Section 6–1:

1. Between the same two points. **2.** More than one current path between two given points. **3.** See Figure 6–65. **4.** See Figure 6–66.

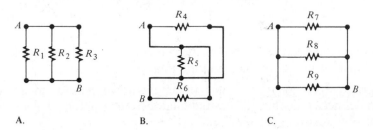

FIGURE 6–65

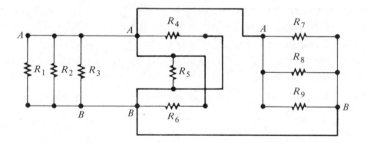

FIGURE 6–66

Section 6–2:

1. 5 V. **2.** 118 V, 118 V. **3.** 50 V and 50 V, respectively. **4.** Voltage is the same across all branches.

Section 6–3:

1. Decrease. **2.** The smallest resistance value. **3.** The equation is as follows:

$$R_T = \cfrac{1}{\left(\dfrac{1}{R_1}\right) + \left(\dfrac{1}{R_2}\right) + \cdots + \left(\dfrac{1}{R_n}\right)}$$

4. $R_T = \dfrac{R_1 R_2}{R_1 + R_2}$.

5. $R_T = R/n$. **6.** 667 Ω. **7.** 250 Ω. **8.** 27.27 Ω.

Section 6–4:
1. 0.5 A. **2.** 400 V. **3.** 0.667 A through the 600-Ω; 1.33 A through the 300-Ω.
4. 8 kΩ. **5.** 66.67 V.

Section 6–5:
1. 2 A. **2.** Three. See Figure 6–67. **3.** 20 mA.

FIGURE 6–67

Section 6–6:
1. **(a)** The algebraic sum of all the currents at a junction is zero; **(b)** the sum of
the currents entering a junction equals the sum of the currents leaving that junction. **2.** 2.5 A. **3.** 400 mA. **4.** 3 μA. **5.** 7 A entering, 5 A leaving.

Section 6–7:
1. The equation is as follows:

$$I_x = \left(\frac{R_T}{R_x}\right)I_T$$

2. The equations are as follows:

$$I_1 = \left(\frac{R_2}{R_1 + R_2}\right)I_T$$

$$I_2 = \left(\frac{R_1}{R_1 + R_2}\right)I_T$$

3. The 22-Ω has most current; the 200-Ω has least current. **4.** 1 mA.
5. I_1 = 6.67 mA, I_2 = 3.33 mA.

Section 6–8:
1. Add the power of each resistor. **2.** 16 W. **3.** 571.43 W.

Section 6–9:
1. Current continues to flow in the other branches. **2.** Yes. **3.** Increase.
4. Decrease. **5.** 1 A.

Section 6–10:
1. There is no change in voltage; the total current decreases. **2.** It increases.

Section 6–11:
1. Lines 130–150. **2.** Line 185 initializes RT to a value equal to R1.

Circuits having some combination of *both* series and parallel resistors are often found in practice. In this chapter, various combinations of series-parallel circuits are studied.

In this chapter, you will learn:

1. How to apply the circuit laws and principles in the analysis of series-parallel circuits.
2. How to identify specific series and parallel portions of a more complex circuit.
3. How to determine total resistance in series-parallel circuits.
4. How to determine the currents and voltages in series-parallel circuits.
5. How to utilize reference points in voltage measurements.
6. How to troubleshoot series-parallel circuits.
7. How to identify shorts and opens in a circuit.
8. How to determine the effects of loading on the operation of a voltage divider circuit.
9. How to analyze ladder networks.
10. How to analyze and use the Wheatstone bridge in resistance measurements.
11. How to use a computer program for the analysis of a ladder network.

7

SERIES-PARALLEL COMBINATIONS

DEFINITION OF A SERIES-PARALLEL CIRCUIT

7–1

A *series-parallel circuit* consists of a combination of both series and parallel current paths. Figure 7–1A illustrates a simple series-parallel combination of resistors. Observe that R_2 and R_3 are in parallel and that this parallel combination is in series with R_1.

If the circuit of Figure 7–1A is connected to a voltage source, the *total* current flows through the resistor R_1 and divides at point B. Part of it flows through each of the parallel resistors R_2 and R_3. The currents are shown in Figure 7–1B. The voltage across R_2 is the same as the voltage across R_3.

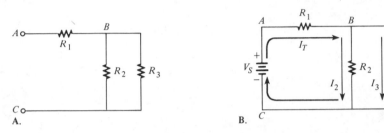

A.   B.

FIGURE 7–1 *A simple series-parallel circuit.*

Review for 7–1

1. Define *series-parallel resistive circuit*.

2. A certain series-parallel circuit is described as follows: R_1 and R_2 are in parallel. This parallel combination is in series with another parallel combination of R_3 and R_4. Sketch the circuit.

3. In the circuit of Figure 7–2, describe the series-parallel relationships of the resistors.

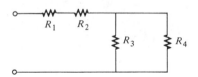

FIGURE 7–2

CIRCUIT IDENTIFICATION

7–2

In this section you will learn to distinguish the series and the parallel relationships of various circuit arrangements. Keep the following rules in mind as you analyze various circuits:

1. When the total current between two points flows through a resistor or a combination of resistors, then that resistor or combination of resistors is in series with any other resistive combination that also appears between those two points.

2. When the total current between two points divides and flows through more than one branch, then those branches are in parallel with each other, if the same voltage appears across each one.

Now let us examine several examples of series-parallel circuits in order to learn to recognize the component relationships.

Example 7–1

Identify the series-parallel relationships in Figure 7–3.

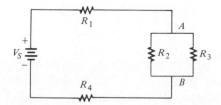

FIGURE 7–3

Solution:

Starting at the positive terminal of the source, follow the current paths. All of the current produced by the source must go through R_1. Therefore, R_1 is in series with the rest of the circuit.

The total current takes two paths when it gets to point A. Part of it flows through R_2, and part of it through R_3; the same voltage (not V_S) is across both. Therefore, R_2 and R_3 are in parallel with each other. This parallel combination is in series with R_1.

At point B, the currents through R_2 and R_3 come together again. Thus, the total current flows through R_4, and so R_4 is in series with R_1 and the parallel combination of R_2 and R_3. The currents are shown in Figure 7–4, where I_T is the total current.

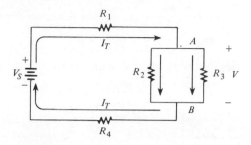

FIGURE 7–4

In summary, R_1 and R_4 are in series with the parallel combination of R_2 and R_3.

Example 7–2

Identify the series-parallel relationships in Figure 7–5.

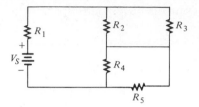

FIGURE 7–5

Solution:

Sometimes it is easier to see a particular circuit arrangement if it is drawn in a different way. In this case, the circuit schematic is redrawn in Figure 7–6, which better illustrates the series-parallel relationships. Now you can see that R_2 and R_3 are in parallel with each other and also that R_4 and R_5 are in parallel with each other. Both parallel combinations are in series with each other and with R_1.

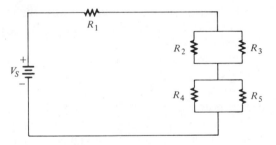

FIGURE 7–6

Example 7–3

Describe the series-parallel combination between points A and D in Figure 7–7.

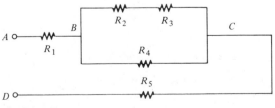

FIGURE 7–7

Solution:

Between points B and C, there are two parallel paths. The lower path consists of R_4, and the upper path consists of a *series* combination of R_2 and R_3. This parallel combination is in series with both R_1 and R_5.

In summary, R_1 and R_5 are in series with the parallel combination of R_4 and $R_2 + R_3$.

Example 7–4

Describe the total resistance between each pair of points in Figure 7–8.

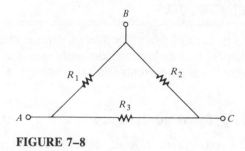

FIGURE 7–8

Solution:

From point A to B: R_1 is in *parallel* with the *series* combination of R_2 and R_3.

From point A to C: R_3 is in *parallel* with the *series* combination of R_1 and R_2.

From point B to C: R_2 is in *parallel* with the *series* combination of R_1 and R_3.

Review for 7–2

1. Describe the resistor combination in Figure 7–9.

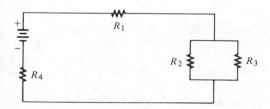

FIGURE 7–9

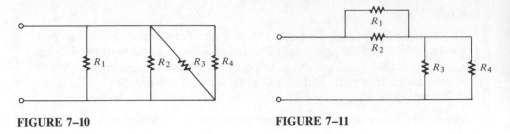

FIGURE 7–10   **FIGURE 7–11**

2. Which resistors are in parallel in Figure 7–10?

3. Describe the parallel arrangements in Figure 7–11.

4. Are the parallel combinations in Figure 7–11 in series?

ANALYSIS OF SERIES-PARALLEL CIRCUITS

7–3 Several quantities are important when you have a circuit that is a series-parallel configuration of resistors. In this section you will learn how to determine total resistance, total current, branch currents, and the voltage across any portion of a circuit.

Determining Total Resistance

In Chapter 5, you learned how to determine total series resistance. In Chapter 6, you learned how to determine total parallel resistance.

To find the total resistance R_T of a series-parallel combination, simply define the series and parallel relationships, and then perform the calculations that you have previously learned. The following two examples illustrate this general approach.

Example 7–5

Determine R_T of the circuit in Figure 7–12 between points A and B.

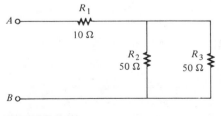

FIGURE 7–12

Solution:

First calculate the equivalent parallel resistance of R_2 and R_3. Since R_2 and R_3 are equal in value, we can use Equation (6–6):

$$R_{EQ} = \frac{R}{n} = \frac{50 \ \Omega}{2} = 25 \ \Omega$$

Notice that we used the term R_{EQ} here to designate the total resistance of a *portion* of a circuit in order to distinguish it from the total resistance R_T of the *complete* circuit.

Now, since R_1 is in series with R_{EQ}, their values are added as follows:

$$R_T = R_1 + R_{EQ} = 10 \ \Omega + 25 \ \Omega = 35 \ \Omega$$

Example 7–6

Find the total resistance between the positive and negative terminals of the battery in Figure 7–13.

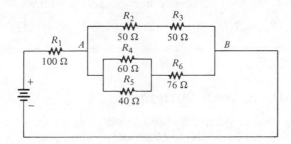

FIGURE 7–13

Solution:

In the *upper branch*, R_2 is in series with R_3. We will call the series combination R_{EQ1}. It is equal to $R_2 + R_3$:

$$R_{EQ1} = R_2 + R_3 = 50 \ \Omega + 50 \ \Omega = 100 \ \Omega$$

In the *lower branch*, R_4 and R_5 are in parallel with each other. We will call this parallel combination R_{EQ2}. It is calculated as follows:

$$R_{EQ2} = \frac{R_4 R_5}{R_4 + R_5} = \frac{(60 \ \Omega)(40 \ \Omega)}{60 \ \Omega + 40 \ \Omega} = 24 \ \Omega$$

Also in the *lower branch*, the parallel combination of R_4 and R_5 is in series with R_6. This series-parallel combination is designated R_{EQ3} and is calculated as follows:

$$R_{EQ3} = R_6 + R_{EQ2} = 76 \ \Omega + 24 \ \Omega = 100 \ \Omega$$

Figure 7–14 shows the original circuit in a simplified *equivalent* form.

Now we can find the equivalent resistance between points A and B. It is R_{EQ1} in parallel with R_{EQ3}. Since these resistances are equal, the equivalent resistance is calculated as follows:

Example 7–6 (continued)

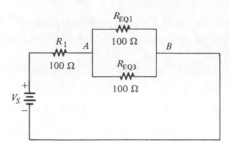

FIGURE 7–14

$$R_{AB} = \frac{100 \; \Omega}{2} = 50 \; \Omega$$

Finally, the total resistance is R_1 in series with R_{AB}:

$$R_T = R_1 + R_{AB} = 100 \; \Omega + 50 \; \Omega = 150 \; \Omega$$

Determining Total Current

Once the total resistance and the source voltage are known, we can find total current in a circuit by applying Ohm's law. Total current is the total source voltage divided by the total resistance:

$$I_T = \frac{V_S}{R_T}$$

For example, let us find the total current in the circuit of Example 7–6 (Figure 7–13). Assume that the source voltage is 30 V. The calculation is as follows:

$$I_T = \frac{V_S}{R_T} = \frac{30 \; V}{150 \; \Omega} = 0.2 \; A$$

In this case the source "sees" 150 Ω and therefore produces a current of 0.2 A.

Determining Branch Currents

Using the *current divider formula,* or *Kirchhoff's current law,* or *Ohm's law,* or combinations of these, we can find the current in any branch of a series-parallel circuit. In some cases it may take repeated application of the formula to find a given current. Working through some examples aids in understanding the procedure.

Example 7–7

Find the current through R_2 and the current through R_3 in Figure 7–15.

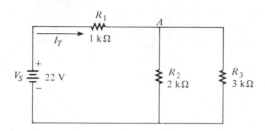

FIGURE 7–15

Solution:

First we need to know how much current is flowing into the junction (point A) of the parallel branches. This is the total circuit current. To find I_T, we need to know R_T:

$$R_T = R_1 + \frac{R_2 R_3}{R_2 + R_3} = 1 \text{ k}\Omega + \frac{(2 \text{ k}\Omega)(3 \text{ k}\Omega)}{2 \text{ k}\Omega + 3 \text{ k}\Omega}$$

$$= 1 \text{ k}\Omega + 1.2 \text{ k}\Omega = 2.2 \text{ k}\Omega$$

$$I_T = \frac{V_S}{R_T} = \frac{22 \text{ V}}{2.2 \text{ k}\Omega}$$

$$= 10 \text{ mA}$$

Using the current divider rule for two branches as given in Chapter 6, we find the current through R_2 as follows:

$$I_2 = \left(\frac{R_3}{R_2 + R_3}\right) I_T = \left(\frac{3 \text{ k}\Omega}{5 \text{ k}\Omega}\right) 10 \text{ mA}$$

$$= 6 \text{ mA}$$

Now we can use Kirchhoff's current law to find the current through R_3 as follows:

$$I_T = I_2 + I_3$$

$$I_3 = I_T - I_2 = 10 \text{ mA} - 6 \text{ mA}$$

$$= 4 \text{ mA}$$

Example 7–8

Determine the current through R_4 in Figure 7–16 if $V_S = 50$ V.

Example 7–8 (continued)

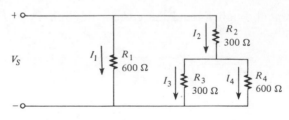

FIGURE 7–16

Solution:

First the current (I_2) into the junction of R_3 and R_4 must be found. Once we know this current, we can use the current divider formula to find I_4.

Notice that there are two main branches in the circuit. The left-most branch consists of only R_1. The right-most branch has R_2 in series with the parallel combination of R_3 and R_4. The voltage across both of these main branches is the same and equal to 50 V. We can find the current (I_2) into the junction of R_3 and R_4 by calculating the equivalent resistance (R_{EQ}) of the right-most main branch and then applying Ohm's law, because this current is the total current through this main branch. Thus,

$$R_{EQ} = R_2 + \frac{R_3 R_4}{R_3 + R_4}$$

$$= 300 \ \Omega + \frac{(600 \ \Omega)(300 \ \Omega)}{900 \ \Omega} = 500 \ \Omega$$

$$I_2 = \frac{V_S}{R_{EQ}} = \frac{50 \text{ V}}{500 \ \Omega}$$

$$= 0.1 \text{ A}$$

Using the current divider formula, we calculate I_4 as follows:

$$I_4 = \left(\frac{R_3}{R_3 + R_4}\right)I_2 = \left(\frac{300 \ \Omega}{900 \ \Omega}\right)0.1 \text{ A}$$

$$= 0.033 \text{ A}$$

Determining Voltage Drops

It is often necessary to find the voltages across certain parts of a series-parallel circuit. We can find these voltages by using the voltage divider formula given in Chapter 5, or Kirchhoff's voltage law, or Ohm's law, or combinations of each. The following three examples illustrate use of the formulas.

Example 7–9

Determine the voltage drop from A to B in Figure 7–17, and then find the voltage across R_1 (V_1).

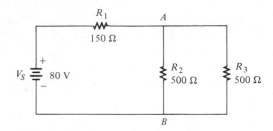

FIGURE 7–17

Solution:

Note that R_2 and R_3 are in parallel in this circuit. Since they are equal in value, their equivalent resistance is

$$R_{EQ} = \frac{500 \ \Omega}{2} = 250 \ \Omega$$

As shown in the equivalent circuit (Figure 7–18), R_1 is in series with R_{EQ}. The total circuit resistance as seen from the source is

$$R_T = R_1 + R_{EQ} = 150 \ \Omega + 250 \ \Omega = 400 \ \Omega$$

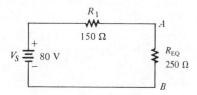

FIGURE 7–18

Now we can use the voltage divider formula to find the voltage across the parallel combination of Figure 7–17 (between points A and B). Let us call it V_{AB}:

$$V_{AB} = \left(\frac{R_{EQ}}{R_T}\right)V_S = \left(\frac{250 \ \Omega}{400 \ \Omega}\right)80 \ V$$

$$= 50 \ V$$

We can now use Kirchhoff's voltage law to find V_1 as follows:

$$V_S = V_1 + V_{AB}$$

192

Example 7–9 (continued)

$$V_1 = V_S - V_{AB} = 80 \text{ V} - 50 \text{ V}$$
$$= 30 \text{ V}$$

Example 7–10

Determine the voltages across each resistor in the circuit of Figure 7–19.

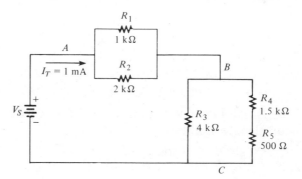

FIGURE 7–19

Solution:

The source voltage is not given, but the total current is known. Since R_1 and R_2 are in parallel, they each have the same voltage. The current through R_1 is

$$I_1 = \left(\frac{R_2}{R_1 + R_2}\right)I_T = \left(\frac{2 \text{ k}\Omega}{3 \text{ k}\Omega}\right)1 \text{ mA}$$
$$= 0.667 \text{ mA}$$

The voltages are

$$V_1 = I_1 R_1 = (0.667 \text{ mA})(1 \text{ k}\Omega)$$
$$= 0.667 \text{ V}$$
$$V_2 = V_1 = 0.667 \text{ V}$$

The current through R_3 is

$$I_3 = \left(\frac{R_4 + R_5}{R_3 + R_4 + R_5}\right)I_T = \left(\frac{2 \text{ k}\Omega}{6 \text{ k}\Omega}\right)1 \text{ mA}$$
$$= 0.333 \text{ mA}$$

The voltage across R_3 is

$$V_3 = I_3 R_3 = (0.333 \text{ mA})(4 \text{ k}\Omega)$$
$$= 1.332 \text{ V}$$

The currents through R_4 and R_5 are the same because these resistors are in series:

$$I_4 = I_5 = I_T - I_3$$
$$= 1 \text{ mA} - 0.333 \text{ mA} = 0.667 \text{ mA}$$

The voltages across R_4 and R_5 are

$$V_4 = I_4 R_4 = (0.667 \text{ mA})(1.5 \text{ k}\Omega)$$
$$= 1.001 \text{ V}$$

$$V_5 = I_5 R_5 = (0.667 \text{ mA})(500 \text{ }\Omega)$$
$$= 0.334 \text{ V}$$

The small difference in V_3 and the sum of V_4 and V_5 is a result of rounding to three places.

Example 7–11

Determine the voltage drop across each resistor in Figure 7–20.

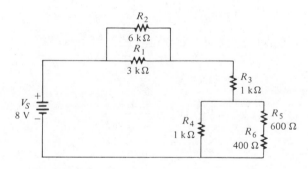

FIGURE 7–20

Solution:

Because we know the total voltage, we can solve this problem using the voltage divider formula. First reduce each parallel combination to an equivalent resistance. Since R_1 and R_2 are in parallel, we combine their values:

$$R_{EQ1} = \frac{R_1 R_2}{R_1 + R_2} = \frac{(3 \text{ k}\Omega)(6 \text{ k}\Omega)}{9 \text{ k}\Omega}$$
$$= 2 \text{ k}\Omega$$

Example 7–11 (continued)

Since R_4 is in parallel with the series combination of R_5 and R_6, we combine these values to obtain

$$R_{EQ2} = \frac{(R_4)(R_5 + R_6)}{R_4 + R_5 + R_6} = \frac{(1\ k\Omega)(1\ k\Omega)}{2\ k\Omega}$$

$$= 500\ \Omega$$

The equivalent circuit is drawn in Figure 7–21.

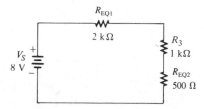

FIGURE 7–21

Applying the voltage divider formula to solve for the voltages, we get the following results:

$$V_{EQ1} = \left(\frac{R_{EQ1}}{R_T}\right)V_S = \left(\frac{2\ k\Omega}{3.5\ k\Omega}\right)8\ V$$

$$= 4.57\ V$$

$$V_{EQ2} = \left(\frac{R_{EQ2}}{R_T}\right)V_S = \left(\frac{500\ \Omega}{3.5\ k\Omega}\right)8\ V$$

$$= 1.14\ V$$

$$V_3 = \left(\frac{R_3}{R_T}\right)V_S = \left(\frac{1\ k\Omega}{3.5\ k\Omega}\right)8\ V$$

$$= 2.29\ V$$

V_{EQ1} equals the voltage across both R_1 and R_2:

$$V_1 = V_2 = V_{EQ1} = 4.57\ V$$

V_{EQ2} is the voltage across R_4 and across the series combination of R_5 and R_6:

$$V_4 = V_{EQ2} = 1.14\ V$$

Now apply the voltage divider formula to the series combination of R_5 and R_6 to get V_5 and V_6:

$$V_5 = \left(\frac{R_5}{R_5 + R_6}\right)V_{EQ2} = \left(\frac{600\ \Omega}{1000\ \Omega}\right)1.14\ V$$

$$= 0.684\ V$$

$$V_6 = \left(\frac{R_6}{R_5 + R_6}\right)V_{EQ2} = \left(\frac{400 \ \Omega}{1000 \ \Omega}\right)1.14 \ \text{V}$$

$$= 0.456 \ \text{V}$$

As you have seen in this section, the analysis of series-parallel circuits can be approached in many ways, depending on what information you need and what circuit values you know. The examples in this section do not represent an exhaustive coverage. They are meant only to give you an idea of how to approach series-parallel circuit analysis.

If you know Ohm's law, Kirchhoff's laws, the voltage divider formula, and the current divider formula, and if you know how to apply these laws, you can solve most resistive circuit analysis problems. The ability to recognize series and parallel combinations is, of course, essential.

Review for 7–3

1. List the circuit laws and formulas that may be necessary in the analysis of a series-parallel circuit.

2. Find the total resistance between A and B in the circuit of Figure 7–22.

3. Find I_3 in Figure 7–22.

4. Find V_2 in Figure 7–22.

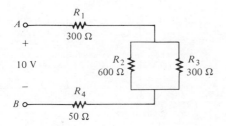

FIGURE 7–22

5. Determine R_T and I_T in Figure 7–23 as "seen" by the source.

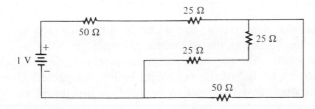

FIGURE 7–23

CIRCUIT GROUND

7-4

Voltage is relative. That is, the voltage at one point in a circuit is always measured relative to another point. For example, if we say that there are +100 V at a certain point in a circuit, we mean that the point is 100 V more positive than some *reference point* in the circuit. This *reference point* in a circuit is usually called *ground*.

The term *ground* derives from the method used in ac power lines, in which one side of the line is neutralized by connecting it to a water pipe or a metal rod driven into the ground. This method of grounding is called *earth ground*.

In most electronic equipment, the metal chassis that houses the assembly or a large conductive area on a printed circuit board is used as the *common* or *reference point,* called the *chassis ground* or *circuit ground*. This ground provides a convenient way of connecting all common points within the circuit back to one side of the battery or other energy source. The chassis or circuit ground does not necessarily have to be connected to earth ground. However, in many cases it is earth grounded in order to prevent a shock hazard due to a potential difference between chassis and earth ground.

In summary, ground is the reference point in electronic circuits. It has a potential of zero volts (0 V) *with respect to all other points in the circuit that are referenced to it,* as illustrated in Figure 7-24. In Part A, the negative side of the source is grounded, and all voltages indicated are positive with respect to ground. In Part B, the positive side of the source is ground. The voltages at all other points are therefore negative with respect to ground.

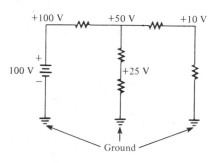
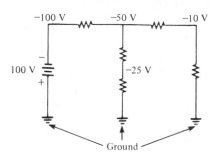

A. Negative ground   B. Positive ground

FIGURE 7-24   *Example of negative and positive grounds.*

Measurement of Voltages with Respect to Ground

When voltages are measured in a circuit, one meter lead is connected to the circuit ground, and the other to the point at which the voltage is to be measured. In a negative ground circuit, the negative meter terminal is connected to the circuit ground. The positive terminal of the voltmeter is then connected to the positive voltage point. Measurement of positive voltage is illustrated in Figure 7-25, where the meter reads the voltage at point *A* with respect to ground.

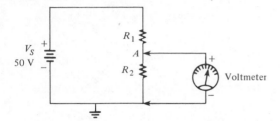

FIGURE 7–25 *Measuring a voltage with respect to negative ground.*

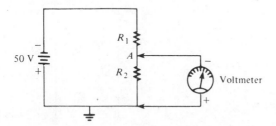

FIGURE 7–26 *Measuring a voltage with respect to positive ground.*

For a circuit with a positive ground, the positive voltmeter lead is connected to ground, and the negative lead is connected to the negative voltage point, as indicated in Figure 7–26. Here the meter reads the voltage at point A with respect to ground.

When voltages must be measured at several points in a circuit, the ground lead can be clipped to ground at one point in the circuit and left there. The other lead is then moved from point to point as the voltages are measured. This method is illustrated in Figure 7–27.

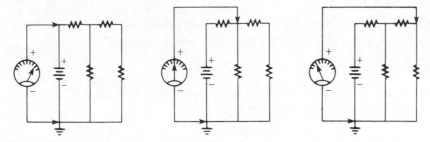

FIGURE 7–27 *Measuring voltages at several points in a circuit.*

Measurement of Voltage across an Ungrounded Resistor

Voltage can normally be measured across a resistor, as shown in Figure 7–28, even though neither side of the resistor is connected to circuit ground.

In some cases, when the meter is not isolated from power line ground, the negative lead of the meter will ground one side of the resistor and alter the

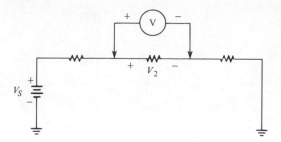

FIGURE 7–28 *Measuring voltage across a resistor.*

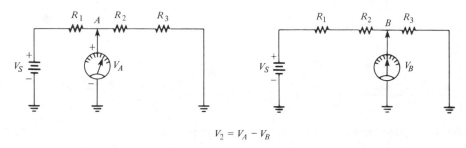

$$V_2 = V_A - V_B$$

FIGURE 7–29 *Measuring voltage across a resistor with two separate measurements to ground.*

operation of the circuit. In this situation, another method must be used, as illustrated in Figure 7–29. The voltages on each side of the resistor are measured *with respect to ground*. The *difference* of these two measurements is the voltage drop across the resistor.

Example 7–12

Determine the voltages of each of the indicated points in each circuit of Figure 7–30. Assume that 25 V are dropped across each resistor.

Solution:

In Circuit A, the voltage polarities are as shown. Point E is ground. The voltages with respect to ground are as follows:

$$V_E = 0 \text{ V}$$

$$V_D = +25 \text{ V}$$

$$V_C = +50 \text{ V}$$

$$V_B = +75 \text{ V}$$

$$V_A = +100 \text{ V}$$

In Circuit B, the voltage polarities are as shown. Point D is ground. The voltages with respect to ground are as follows:

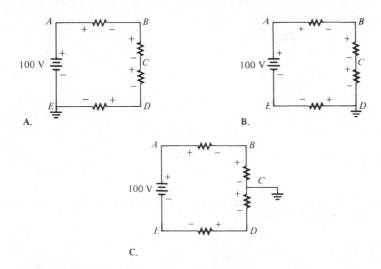

FIGURE 7–30

$$V_E = -25 \text{ V}$$
$$V_D = 0 \text{ V}$$
$$V_C = +25 \text{ V}$$
$$V_B = +50 \text{ V}$$
$$V_A = +75 \text{ V}$$

In Circuit C, the voltage polarities are as shown. Point C is ground. The voltages with respect to ground are as follows:

$$V_E = -50 \text{ V}$$
$$V_D = -25 \text{ V}$$
$$V_C = 0 \text{ V}$$
$$V_B = +25 \text{ V}$$
$$V_A = +50 \text{ V}$$

Review for 7–4

1. The common point in a circuit is called _____ .
2. Most voltages in a circuit are referenced to ground (T or F).
3. The housing or chassis is often used as circuit ground (T or F).
4. What is the symbol for ground?
5. What does *earth ground* mean?

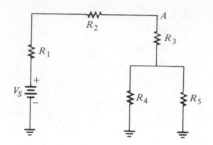

FIGURE 7–31

6. In Figure 7–31, how would you connect a voltmeter to measure the voltage at point A with respect to ground?

TROUBLESHOOTING

7–5

Troubleshooting is the process of identifying and locating a failure or problem in a circuit. Some troubleshooting techniques have already been discussed in relation to both series circuits and parallel circuits. Now we will extend these methods to the series-parallel networks.

Opens and *shorts* are typical problems that occur in electric circuits. As mentioned before, if a resistor burns out, it will normally produce an *open circuit*. Bad solder connections, broken wires, and poor contacts can also be causes of open paths. Pieces of foreign material, such as solder splashes, broken insulation on wires, and so on, can often lead to shorts in a circuit. A *short* is a zero resistance path between two points. The following three examples illustrate troubleshooting in series-parallel resistive circuits.

Example 7–13

From the indicated voltmeter reading, determine if there is a fault in Figure 7–32. If there is a fault, identify it as either a short or an open.

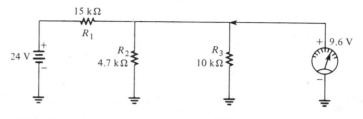

FIGURE 7–32

Solution:

First determine what the voltmeter *should* be indicating. Since R_2 and R_3 are in parallel, their equivalent resistance is

$$R_{EQ} = \frac{R_2 R_3}{R_2 + R_3} = \frac{(4.7 \text{ k}\Omega)(10 \text{ k}\Omega)}{14.7 \text{ k}\Omega}$$

$$= 3.2 \text{ k}\Omega$$

The voltage across the equivalent parallel combination is determined by the voltage divider formula as follows:

$$V_{R_{EQ}} = \left(\frac{R_{EQ}}{R_1 + R_{EQ}}\right)V_S = \left(\frac{3.2 \text{ k}\Omega}{18.2 \text{ k}\Omega}\right)24 \text{ V}$$

$$= 4.22 \text{ V}$$

Thus, 4.22 V is the voltage reading that you *should* get on the meter. But the meter reads 9.6 V instead. This value is incorrect, and, because it is higher than it should be, R_2 or R_3 is probably open. Why? Because if either of these two resistors is open, the resistance across which the meter is connected is larger than expected. A higher resistance will drop a higher voltage in this circuit, which is, in effect, a voltage divider.

Let us start by assuming that R_2 is open. If it is, the voltage across R_3 is as follows:

$$V_{R3} = \left(\frac{R_3}{R_1 + R_3}\right)V_S = \left(\frac{10 \text{ k}\Omega}{25 \text{ k}\Omega}\right)24 \text{ V}$$

$$= 9.6 \text{ V}$$

This calculation shows that R_2 is open. Replace R_2 with a new resistor.

Example 7–14

Suppose that you measure 24 V with the voltmeter in Figure 7–33. Determine if there is a fault, and, if there is, isolate it.

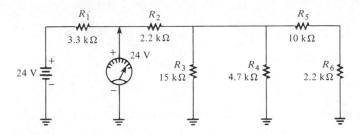

FIGURE 7–33

Solution:

There is no voltage drop across R_1 because both sides of the resistor are at +24 V. Either no current is flowing through R_1 from the source, which tells us that R_2 is open in the circuit, or R_1 is shorted.

Example 7–14 (continued)

If R_1 were open, the meter would not read 24 V. The most logical failure is in R_2. If it is open, then no current will flow from the source. To verify this, measure across R_2 with the voltmeter as shown in Figure 7–34. If R_2 is open, the meter will indicate 24 V. The right side of R_2 will be at zero volts because no current is flowing through any of the other resistors to cause a voltage drop.

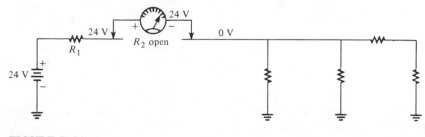

FIGURE 7–34

Example 7–15

The two voltmeters in Figure 7–35 indicate the voltages shown. Determine if there are any opens or shorts in the circuit and, if so, where they are located.

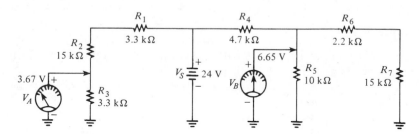

FIGURE 7–35

Solution:

First let us see if the voltmeter readings are correct. R_1, R_2, and R_3 act as a voltage divider on the left side of the source. The voltage across R_3 is calculated as follows:

$$V_{R3} = \left(\frac{R_3}{R_1 + R_2 + R_3} \right) V_S$$

$$= \left(\frac{3.3 \text{ k}\Omega}{21.6 \text{ k}\Omega} \right) 24 \text{ V} = 3.67 \text{ V}$$

The voltmeter A reading (V_A) is correct.

Now let us see if the voltmeter B reading (V_B) is correct. The part of the circuit to the right of the source also acts as a voltage divider. The series-parallel combination of R_5, R_6, and R_7 is in series with R_4. The equivalent resistance of the R_5, R_6, and R_7 combination is calculated as follows:

$$R_{EQ} = \frac{(R_6 + R_7)R_5}{R_5 + R_6 + R_7}$$

$$= \frac{(17.2 \text{ k}\Omega)(10 \text{ k}\Omega)}{27.2 \text{ k}\Omega} = 6.3 \text{ k}\Omega$$

where R_5 is in parallel with R_6 and R_7 in series. R_{EQ} and R_4 form a voltage divider. Voltmeter B is measuring the voltage across R_{EQ}. Is it correct? We check as follows:

$$V_{R_{EQ}} = \left(\frac{R_{EQ}}{R_4 + R_{EQ}}\right)V_S$$

$$= \left(\frac{6.3 \text{ k}\Omega}{11 \text{ k}\Omega}\right)24 \text{ V} = 13.75 \text{ V}$$

Thus, the actual measured voltage at this point is incorrect. Some further thought will help to isolate the problem.

We know that R_4 is not open, because if it were, the meter would read 0 V. If there were a short across it, the meter would read 24 V. Since the actual voltage is much less than it should be, R_{EQ} must be less than the calculated value. The most likely problem is a short across R_7. If there is a short from the top of R_7 to ground, R_6 is effectively in parallel with R_5. In this case, R_{EQ} is

$$R_{EQ} = \frac{R_5 R_6}{R_5 + R_6}$$

$$= \frac{(2.2 \text{ k}\Omega)(10 \text{ k}\Omega)}{12.2 \text{ k}\Omega} = 1.8 \text{ k}\Omega$$

Then V_{EQ} is

$$V_{EQ} = \left(\frac{1.8 \text{ k}\Omega}{6.5 \text{ k}\Omega}\right)24 \text{ V} = 6.65 \text{ V}$$

This value for V_{EQ} agrees with the voltmeter B reading. So there is a short across R_7. If this were an actual circuit, you would try to find the physical cause of the short.

Review for 7–5

1. Name two types of common circuit faults.

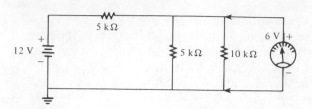

FIGURE 7–36

2. In Figure 7–36, one of the resistors in the circuit is open. Based on the meter reading, determine which is the bad resistor.

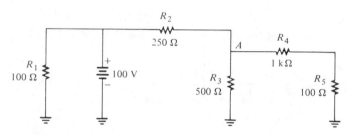

FIGURE 7–37

3. For the following faults in Figure 7–37, what voltage would be measured at point A?
   (a) No faults   **(b)** R_1 open   **(c)** Short across R_5 **(d)** R_3 and R_4 open

VOLTAGE DIVIDERS WITH RESISTIVE LOADS

7–6   Voltage dividers were discussed in Chapter 5. In this section we will discuss the effect of resistive loads on voltage dividers.

   The simple voltage divider in Figure 7–38A produces an output voltage of 5 V because the two resistors are of equal value. This voltage is the *unloaded output voltage*. If a load resistor R_L is connected across the output, as shown in Figure 7–38B, the output voltage will be reduced by an amount that depends on the value R_L. The larger R_L is compared to R_2, the less the output voltage is

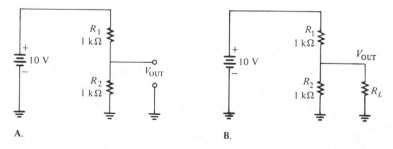

FIGURE 7–38   *A voltage divider with both unloaded and loaded output.*

reduced. Actually, as you can see in Figure 7–38B, a loaded voltage divider is a series-parallel circuit. The load resistor is connected in parallel with the resistance across which the output voltage is taken.

Example 7–16

Determine both the unloaded and the loaded output voltages of the voltage divider in Figure 7–39A for the following two values of load resistance: 10 kΩ and 100 kΩ.

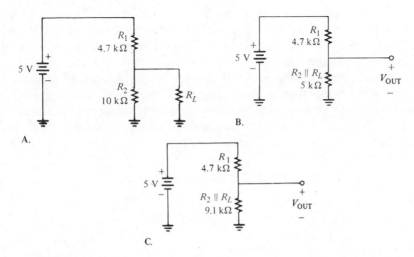

FIGURE 7–39

Solution:

The *unloaded* output voltage is

$$V_{OUT} = \left(\frac{10 \text{ k}\Omega}{14.7 \text{ k}\Omega}\right) 5 \text{ V} = 3.4 \text{ V}$$

With the 10-kΩ load resistor connected, R_L is in parallel with R_2, which gives 5 kΩ, as shown by the equivalent circuit in Figure 7–39B.
     The *loaded* output voltage is

$$V_{OUT} = \left(\frac{5 \text{ k}\Omega}{9.7 \text{ k}\Omega}\right) 5 \text{ V} = 2.6 \text{ V}$$

With the 100-kΩ load, the resistance from output to ground is

$$\frac{R_2 R_L}{R_2 + R_L} = \frac{(10 \text{ k}\Omega)(100 \text{ k}\Omega)}{110 \text{ k}\Omega} = 9.1 \text{ k}\Omega$$

as shown in Figure 7–39C.

Example 7–16 (continued)

The *loaded* output voltage is

$$V_{OUT} = \left(\frac{9.1 \text{ k}\Omega}{13.8 \text{ k}\Omega}\right) 5 \text{ V} = 3.3 \text{ V}$$

Notice that with the larger value of R_L, the output is reduced from its unloaded value by much less than it is with the smaller R_L. This problem illustrates the loading effect of R_L on the voltage divider.

Voltage dividers are sometimes useful in obtaining various voltages from a power supply. For example, suppose that we wish to derive 12 V and 6 V from a 24-V supply. To do so requires a voltage divider with two *taps*, as shown in Figure 7–40. In this example, R_1 must equal $R_2 + R_3$, and R_2 must equal R_3. The actual values of the resistors are set by the amount of current that is to be drawn from the source under unloaded conditions. This current, called the *bleeder current*, represents a continuous drain on the source. With these ideas in mind, in Example 7–17 we design a voltage divider to meet certain specified requirements.

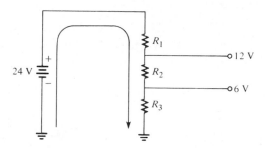

FIGURE 7–40   *Voltage divider with two output taps.*

Example 7–17

A power supply requires 12 V and 6 V from a 24-V battery. The unloaded current drain on this battery is not to exceed 1 mA. Determine the values of the resistors. Also determine the output voltage at the 12-V tap when both outputs are loaded with 100 kΩ each.

Solution:

A circuit as shown in Figure 7–40 is required. In order to have an unloaded current of 1 mA, the total resistance must be as follows:

$$R_T = \frac{V_S}{I} = \frac{24 \text{ V}}{1 \text{ mA}} = 24 \text{ k}\Omega$$

To get 12 V, R_1 must equal $R_2 + R_3 = 12 \text{ k}\Omega$. To get 6 V, R_2 must equal $R_3 = 6 \text{ k}\Omega$.

Now if the 100-kΩ loads are connected across the outputs as shown in Figure 7–41, the loaded output voltages are as determined below.

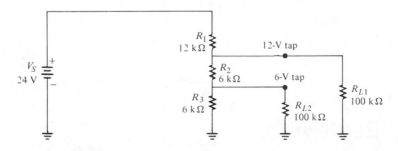

FIGURE 7–41

The equivalent resistance from the 12-V tap to ground is the 100-kΩ load resistor R_{L1} in parallel with the combination of R_2 in series with the parallel combination of R_3 and R_{L2}. This solution is as follows.
R_3 in parallel with R_{L2}:

$$R_{EQ1} = \frac{(6 \text{ k}\Omega)(100 \text{ k}\Omega)}{106 \text{ k}\Omega} = 5.66 \text{ k}\Omega$$

R_2 in series with R_{EQ1}:

$$R_{EQ2} = 6 \text{ k}\Omega + 5.66 \text{ k}\Omega = 11.66 \text{ k}\Omega$$

R_{L1} in parallel with R_{EQ2}:

$$R_{EQ3} = \frac{(100 \text{ k}\Omega)(11.66 \text{ k}\Omega)}{111.66 \text{ k}\Omega} = 10.44 \text{ k}\Omega$$

R_{EQ3} is the equivalent resistance from the 12-V tap to ground. The equivalent circuit from the 12-V tap to ground is shown in Figure 7–42. Using this equivalent circuit, we calculate the *loaded* voltage at the 12-V tap by using the voltage divider formula as follows:

$$V_{12} = \left(\frac{R_{EQ3}}{R_T}\right)V_S = \left(\frac{10.44 \text{ k}\Omega}{22.44 \text{ k}\Omega}\right)24 \text{ V} = 11.17 \text{ V}$$

As you can see, the output voltage at the 12-V tap decreases slightly from its unloaded value when the 100-kΩ loads are connected. Smaller values of load resistance would result in a greater decrease in the output voltage.

Example 7–17 (continued)

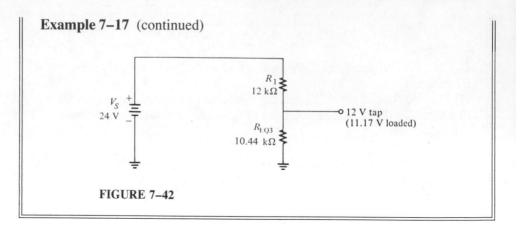

FIGURE 7–42

Review for 7–6

1. A load resistor is connected to an output tap on a voltage divider. What effect does the load resistor have on the output voltage at this tap?

2. A larger-value load resistor will cause the output voltage to change less than a smaller-value one will (T or F).

3. For the voltage divider in Figure 7–43, determine the unloaded output voltage. Also determine the output voltage with a 10-kΩ load resistor connected across the output.

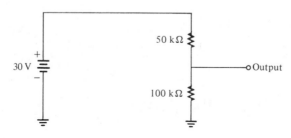

FIGURE 7–43

LADDER NETWORKS

7–7   A ladder network is a special type of series-parallel circuit. One form is commonly used to scale down voltages to certain weighted values for digital-to-analog conversion. You will study this process in later courses. In this section, we examine a basic resistive ladder of limited complexity, as shown in Figure 7–44.

One approach to the analysis of a ladder network is to simplify it one step at a time, *starting at the side farthest from the source.* In this way the current in any branch or the voltage at any point can be determined, as illustrated in Example 7–18.

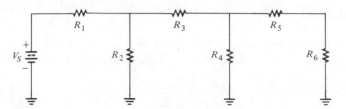

FIGURE 7–44 *Basic 3-step ladder circuit.*

Example 7–18

Determine each branch current and the voltage at each point in the ladder circuit of Figure 7–45.

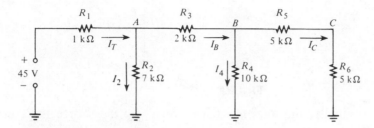

FIGURE 7–45

Solution:

To find the branch currents, we must know the total current from the source (I_T). To obtain I_T, we must find the total resistance "seen" by the source.

We determine R_T in a step-by-step process, starting at the right of the circuit diagram. First notice that R_5 and R_6 are in series across R_4. So the resistance from point B to ground is as follows:

$$R_B = \frac{R_4(R_5 + R_6)}{R_4 + (R_5 + R_6)} = \frac{(10 \text{ k}\Omega)(10 \text{ k}\Omega)}{20 \text{ k}\Omega}$$

$$= 5 \text{ k}\Omega$$

Using R_B (the resistance from point B to ground), the equivalent circuit is shown in Figure 7–46.

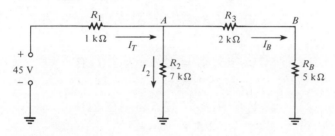

FIGURE 7–46

Example 7–18 (continued)

Next, the resistance from point A to ground (R_A) is R_2 in parallel with the series combination of R_3 and R_B. It is calculated as follows:

$$R_A = \frac{R_2(R_3 + R_B)}{R_2 + (R_3 + R_B)} = \frac{(7 \text{ k}\Omega)(7 \text{ k}\Omega)}{14 \text{ k}\Omega}$$

$$= 3.5 \text{ k}\Omega$$

Using R_A, the equivalent circuit of Figure 7–46 is further simplified to the circuit in Figure 7–47.

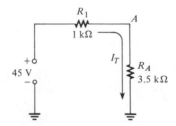

FIGURE 7–47

Finally, the total resistance "seen" by the source is R_1 in series with R_A:

$$R_T = R_1 + R_A = 1 \text{ k}\Omega + 3.5 \text{ k}\Omega = 4.5 \text{ k}\Omega$$

The total circuit current is

$$I_T = \frac{V_S}{R_T} = \frac{45 \text{ V}}{4.5 \text{ k}\Omega} = 10 \text{ mA}$$

As indicated in Figure 7–46, I_T flows into point A and divides between R_2 (I_2) and the branch containing $R_3 + R_B$ (I_B). Since the branch resistances are equal in this particular example, half the total current is through R_2 and half into point B. So I_2 is 5 mA and I_B is 5 mA.

If the branch resistances are not equal, the current divider formula is used. As indicated in Figure 7–45, I_B flows into point B and is divided equally between R_4 and the branch containing $R_5 + R_6$ because the branch resistances are equal. So I_4, I_5, and I_6 are all equal to 2.5 mA.

To determine V_A, V_B, and V_C, we apply Ohm's law as follows:

$$V_A = I_2 R_2 = (5 \text{ mA})(7 \text{ k}\Omega)$$
$$= 35 \text{ V}$$

$$V_B = I_4 R_4 = (2.5 \text{ mA})(10 \text{ k}\Omega)$$
$$= 25 \text{ V}$$

$$V_C = I_6 R_6 = (2.5 \text{ mA})(5 \text{ k}\Omega)$$
$$= 12.5 \text{ V}$$

> As you can see, the circuit values in this example have been chosen so that the computations are easily done. However, the same approach applies for more cumbersome values and more complex ladder circuits.

Review for 7–7

1. Sketch a basic four-step ladder network.
2. Determine the total circuit resistance presented to the source by the ladder network of Figure 7–48.

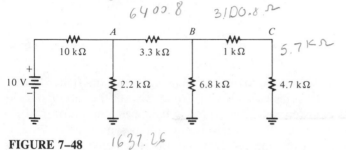

6400.8   3100.8 Ω

	A		B		C
10 kΩ		3.3 kΩ		1 kΩ	5.7 kΩ
10 V	2.2 kΩ		6.8 kΩ		4.7 kΩ

FIGURE 7–48   1637.26

3. What is the total current in Figure 7–48?   11.637 kΩ
4. What is the current through the 2.2-kΩ resistor in Figure 7–48?
5. What is the voltage at point A with respect to ground in Figure 7–48?

7–8

The *bridge circuit* is widely used in measurement devices and other applications that you will learn later. For now, we will consider the *balanced bridge,* which can be used to measure unknown resistance values. This circuit, shown in Figure 7–49A, is known as a *Wheatstone bridge.* Figure 7–49B is the same circuit electrically, but it is drawn in a different way.

A bridge is said to be *balanced* when the voltage across the output terminals C and D is *zero;* that is, $V_{AC} = V_{AD}$. If V_{AC} equals V_{AD}, then

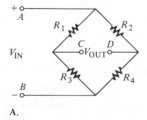

A.

B.

FIGURE 7–49 *Wheatstone bridge.*

$I_1R_1 = I_2R_2$, since one side of both R_1 and R_2 is connected to point A. Also, $I_1R_3 = I_2R_4$, since one side of both R_3 and R_4 connects to point B. Because of these equalities, we can write the ratios of the voltages as follows:

$$\frac{I_1R_1}{I_1R_3} = \frac{I_2R_2}{I_2R_4}$$

The currents cancel to give

$$\frac{R_1}{R_3} = \frac{R_2}{R_4}$$

Solving for R_1, we get

$$R_1 = R_3\left(\frac{R_2}{R_4}\right) \tag{7-1}$$

How can this formula be used to determine an unknown resistance? First, let us make R_3 a *variable* resistor and call it R_V. Also, we set the ratio R_2/R_4 to a known value. If R_V is adjusted until the bridge is balanced, the product of R_V and the ratio R_2/R_4 is equal to R_1, which is our unknown resistor (R_{UNK}). Equation (7-1) is restated in Equation (7-2), using the new subscripts:

$$R_{UNK} = R_V\left(\frac{R_2}{R_4}\right) \tag{7-2}$$

The bridge is balanced when the voltage across the output terminals equals zero ($V_{AC} = V_{AD}$). A *galvanometer* (a meter that measures small currents in either direction and is zero at center scale) is connected between the output terminals. Then R_V is adjusted until the galvanometer shows zero current ($V_{AC} = V_{AD}$), indicating a balanced condition. The setting of R_V multiplied by the ratio R_2/R_4 gives the value of R_{UNK}. Figure 7-50 shows this arrangement. For example, if $R_2/R_4 = \frac{1}{10}$ and $R_V = 680\ \Omega$, then $R_{UNK} = (680\ \Omega)(\frac{1}{10}) = 68\ \Omega$.

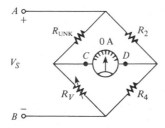

FIGURE 7-50   *Balanced Wheatstone bridge.*

Example 7-19

What is R_{UNK} under the balanced bridge conditions shown in Figure 7-51?

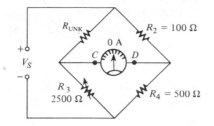

FIGURE 7–51

Solution:

$$R_{UNK} = R_V\left(\frac{R_2}{R_4}\right) = 2500\ \Omega\left(\frac{100\ \Omega}{500\ \Omega}\right)$$

$$= 500\ \Omega$$

Review for 7–8

1. Sketch a basic Wheatstone bridge circuit.

2. Under what condition is the bridge balanced?

3. What formula is used to determine the value of the unknown resistance when the bridge is balanced?

4. What is the unknown resistance for the values shown in Figure 7–52?

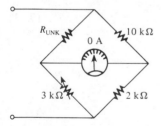

FIGURE 7–52

COMPUTER ANALYSIS

The program listed in this section provides for the analysis of a 3-step ladder network similar to that in Example 7–18. All unknown voltages and branch currents are computed and displayed when the program is run.

7–9

```
10 CLS
20 PRINT "THIS PROGRAM ANALYZES A BASIC 3-STEP LADDER NETWORK"
30 PRINT "SUCH AS THE ONE SHOWN IN FIGURE 7-45."
40 PRINT:PRINT "YOU PROVIDE THE INPUT VOLTAGE AND RESISTOR VALUES AND"
50 PRINT "THE COMPUTER WILL CALCULATE THE VOLTAGE AT EACH POINT"
60 PRINT "AND THE CURRENT IN EACH BRANCH."
70 FOR T=1 TO 5000:NEXT:CLS
```

```
80 INPUT "WHAT IS THE INPUT VOLTAGE IN VOLTS";VIN
90 PRINT:PRINT
100 FOR X=1 TO 6
110 PRINT "VALUE OF R";X;"IN OHMS"
120 INPUT R(X)
130 NEXT
140 CLS
150 RB=(R(4)*(R(5)+R(6)))/(R(4)+R(5)+R(6))
160 RA=(R(2)*(R(3)+RB))/(R(2)+R(3)+RB)
170 RT=RA+R(1)
180 IT=VIN/RT
190 VA=IT*RA
200 I(1)=IT
210 I(2)=VA/R(2)
220 I(3)=IT-I(2)
230 VB=I(3)*RB
240 I(4)=VB/R(4)
250 I(5)=I(3)-I(4)
260 I(6)=I(5)
270 VC=I(6)*R(6)
280 PRINT "VA=";VA;"VOLTS","VB=";VB;"VOLTS","VC=";VC;"VOLTS"
290 PRINT
300 FOR Y=1 TO 6
310 PRINT "I";Y;"=";I(Y);"AMPS"
320 NEXT
```

Review for 7–9

1. What is the purpose of line 70?

2. Which sequence of lines displays the results of the computations?

Formulas

$$R_1 = R_3\left(\frac{R_2}{R_4}\right) \tag{7–1}$$

$$R_{\text{UNK}} = R_V\left(\frac{R_2}{R_4}\right) \tag{7–2}$$

Summary

1. A series-parallel circuit is a combination of both series paths and parallel paths.

2. To determine total resistance in a series-parallel circuit, identify the series and parallel relationships, and then apply the formulas for series resistance and parallel resistance from Chapters 5 and 6.

3. To find the total current, divide the total voltage by the total resistance.

4. To determine branch currents, apply the current divider formula, or Kirchhoff's current law, or Ohm's law. Consider each circuit problem individually to determine the most appropriate method.

5. To determine voltage drops across any portion of a series-parallel circuit, use the voltage divider formula, or Kirchhoff's voltage law, or Ohm's law.

Consider each circuit problem individually to determine the most appropriate method.

6. *Ground* is the common or reference point in a circuit.

7. All voltages in a circuit are referenced to ground unless otherwise specified.

8. Ground is zero volts with respect to all points referenced to it in the circuit.

9. *Negative ground* is the term used when the negative side of the source is grounded.

10. *Positive ground* is the term used when the positive side of the source is grounded.

11. *Troubleshooting* is the process of identifying and locating a fault in a circuit.

12. *Open circuits* and *short circuits* are typical circuit faults.

13. Resistors normally open when they burn out.

14. When a load resistor is connected across a voltage divider output, the output voltage decreases.

15. The load resistor should be large compared to the resistance across which it is connected, in order that the loading effect may be minimized. A *10-times* value is sometimes used as a rule of thumb, but the value depends on the accuracy required for the output voltage.

16. To find total resistance of a ladder network, start at the point farthest from the source and reduce the resistance in steps.

17. A Wheatstone bridge can be used to measure an unknown resistance.

18. A bridge is *balanced* when the output voltage is *zero*. The balanced condition produces zero current through a load connected across the output terminals of the bridge.

Self-Test

1. Identify the series-parallel relationships in Figure 7–53.

2. For the circuit of Figure 7–53, determine the following:
   (a) Total resistance as "seen" by the source
   (b) Total current drawn from the source
   (c) Current through R_3
   (d) Voltage across R_4

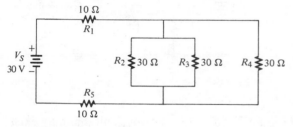

FIGURE 7–53

216

3. For the circuit in Figure 7–54, determine the following:
   (a) Total resistance   (b) Total current
   (c) Current through R_1 (d) Voltage across R_6

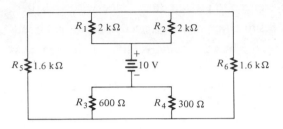

FIGURE 7–54

4. In Figure 7–55, find the following:
   (a) Total resistance between terminals A and B
   (b) Total current drawn from a 6-V source connected from A to B
   (c) Current through R_5
   (d) Voltage across R_2

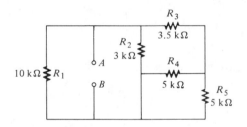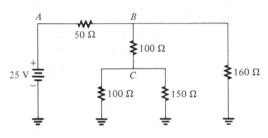

FIGURE 7–55   **FIGURE 7–56**

5. Determine the voltages with respect to ground in Figure 7–56.

6. If R_2 in Figure 7–57 opens, what voltages will be read at points A, B, and C?

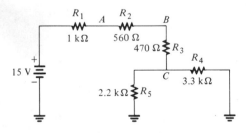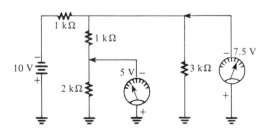

FIGURE 7–57   **FIGURE 7–58**

7. Check the meter readings in Figure 7–58 and locate any fault that may exist.

8. Determine the unloaded output voltage in Figure 7–59. If a 200-kΩ load is connected, what is the loaded output voltage?

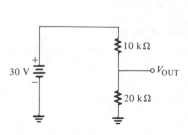

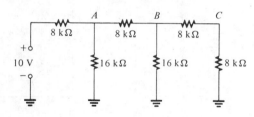

FIGURE 7–59   **FIGURE 7–60**

9. In Figure 7–60, determine the voltage at point A when the switch is open. Also determine the voltage at point A when the switch is closed.

10. For the ladder network in Figure 7–61, determine the following:
(a) Total resistance   (b) Total current   (c) Current through R_3
(d) Current through R_4 (e) Voltage at point A (f) Voltage at point B

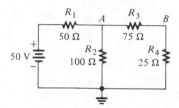

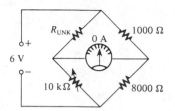

FIGURE 7–61   **FIGURE 7–62**

11. Calculate V_A, V_B, and V_C for the ladder in Figure 7–62.

12. In the bridge circuit of Figure 7–63, what is the value of the unknown resistance when the other values are as shown?

FIGURE 7-63 image goes here

FIGURE 7–63

13. Run the program in Section 7–9 for the following circuit values: $V_{IN} = 24$ V, $R_1 = 1.5$ kΩ, $R_2 = 1.5$ kΩ, $R_3 = 3.3$ kΩ, $R_4 = 10$ kΩ, $R_5 = 6.8$ kΩ, and $R_6 = 5$ kΩ.

Problems

Sections 7–1 and 7–2

7–1. Visualize and sketch the following series-parallel combinations:

(a) R_1 in series with the parallel combination of R_2 and R_3

(b) R_1 in parallel with the series combination of R_2 and R_3

(c) R_1 in parallel with a branch containing R_2 in series with a parallel combination of four other resistors

7–2. Visualize and sketch the following series-parallel circuits:

(a) A parallel combination of three branches, each containing two series resistors

(b) A series combination of three parallel circuits, each containing two resistors

7–3. In each circuit of Figure 7–64, identify the series and parallel relationships of the resistors viewed from the source.

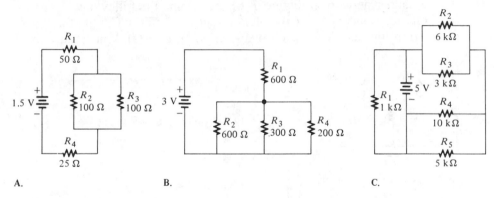

A.   B.   C.

FIGURE 7–64

7–4. For each circuit in Figure 7–65, identify the series and parallel relationships of the resistors viewed from the source.

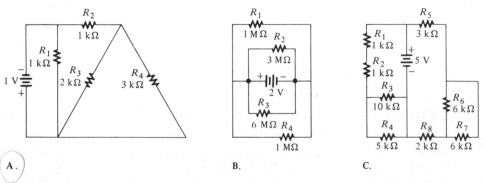

A.   B.   C.

FIGURE 7–65

Section 7–3

7–5. A certain circuit is composed of two parallel resistors. The total resistance is 667 Ω. One of the resistors is 1 kΩ. What is the other resistor?

7–6. For each circuit in Figure 7–64, determine the total resistance presented to the source.

7–7. Repeat Problem 7–6 for each circuit in Figure 7–65.

7–8. Determine the current through each resistor in Figure 7–64; then calculate each voltage drop.

7–9. Determine the current through each resistor in Figure 7–65; then calculate each voltage drop.

Section 7–4

7–10. Determine the voltage at each point with respect to ground in Figure 7–66.

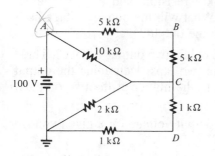

FIGURE 7–66

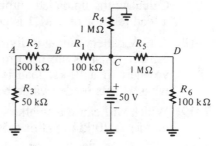

FIGURE 7–67

7–11. Determine the voltage at each point with respect to ground in Figure 7–67.

7–12. In Figure 7–67, how would you determine the voltage across R_2 by measuring without connecting a meter directly across the resistor?

Section 7–5

7–13. Is the voltmeter reading in Figure 7–68 correct?

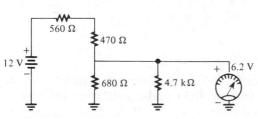

FIGURE 7–68

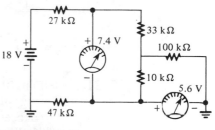

FIGURE 7–69

7–14. Are the meter readings in Figure 7–69 correct?

7–15. There is one fault in Figure 7–70. Based on the meter indications, determine what the fault is.

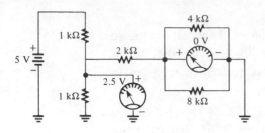

FIGURE 7–70

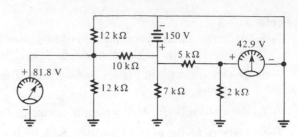

FIGURE 7–71

7–16. Look at the meters in Figure 7–71 and determine if there is a fault in the circuit. If there is a fault, identify it.

Section 7–6

7–17. A voltage divider consists of two 50-kΩ resistors and a 15-V source. Calculate the unloaded output voltage. What will the output voltage be if a load resistor of 1 MΩ is connected to the output?

7–18. A 12-V battery output is divided down to obtain two output voltages. Three 3.3-kΩ resistors are used to provide the two taps. Determine the output voltages. If a 10-kΩ load is connected to the higher of the two outputs, what will its loaded value be?

7–19. Which will cause a smaller decrease in output voltage for a given voltage divider, a 10-kΩ load or a 50-kΩ load?

7–20. In Figure 7–72, determine the continuous current drain on the battery with no load across the output terminals. With a 10-kΩ load, what is the battery current?

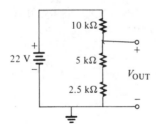

FIGURE 7–72

7–21. Determine the resistance values for a voltage divider that must meet the following specifications: The current drain under unloaded condition is not to exceed 5 mA. The source voltage is to be 10 V. A 5-V output and a 2.5-V output are required. Sketch the circuit. Determine the effect on the output voltages if a 1-kΩ load is connected to each tap.

Section 7–7

7–22. For the circuit shown in Figure 7–73, calculate the following:
(a) Total resistance across the source
(b) Total current from the source

   (c) Current through the 900-Ω resistor

   (d) Voltage from point A to point B

7–23. Determine the total resistance and the voltage at points A, B, and C in the ladder network of Figure 7–74.

7–24. Determine the total resistance between terminals A and B of the ladder network in Figure 7–75. Also calculate the current in each branch with 10 V between A and B.

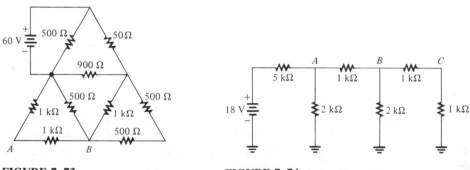

FIGURE 7–73   **FIGURE 7–74**

7–25. What is the voltage across each resistor in Figure 7–75?

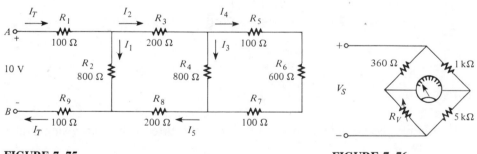

FIGURE 7–75   **FIGURE 7–76**

Section 7–8

7–26. A resistor of unknown value is connected to a Wheatstone bridge circuit. The bridge parameters are set as follows: $R_V = 18$ kΩ and $R_2/R_4 = 0.02$. What is R_{UNK}?

7–27. A bridge network is shown in Figure 7–76. To what value must R_V be set in order to balance the bridge?

Section 7–9

7–28. Modify the program in Section 7–9 to compute and display the power dissipation in each resistor.

7–29. Modify the program in Section 7–9 to analyze a basic 4-step ladder circuit rather than the 3-step circuit.

Advanced Problems

7–30. Determine the voltage, V_{AB}, in Figure 7–77.

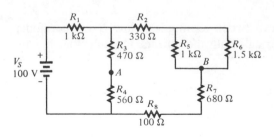 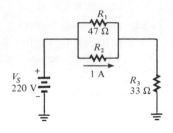

FIGURE 7–77   **FIGURE 7–78**

7–31. Find the value of R_2 in Figure 7–78.

7–32. Design a voltage divider to provide a 6-V output with no load and a minimum of 5.5 V across a 1-kΩ load. The source voltage is 24 V, and the unloaded current drain is not to exceed 100 mA.

7–33. Find I_T and V_{OUT} in Figure 7–79.

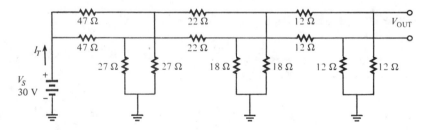

FIGURE 7–79

7–34. Modify the computer program in Section 7–9 to analyze an *n*-step ladder network.

Answers to Section Reviews

Section 7–1:
1. A circuit consisting of both series and parallel connections. **2.** See Figure 7–80. **3.** R_1 and R_2 are in series with the parallel combination of R_3 and R_4.

FIGURE 7–80

Section 7–2:
1. R_1 and R_4 are in series with each other and with the parallel combination of R_2 and R_3. **2.** All resistors. **3.** R_1 and R_2 are in parallel; R_3 and R_4 are in parallel. **4.** Yes.

Section 7–3:
1. Voltage and current divider formulas, Kirchhoff's laws, and Ohm's law. **2.** 550 Ω. **3.** 0.012 A. **4.** 3.6 V. **5.** 100 Ω, 0.01 A.

Section 7–4:
1. Ground. **2.** T. **3.** T. **4.** See Figure 7–81. **5.** A connection to earth through a metal rod or a water pipe. **6.** Positive terminal to A, negative terminal to ground.

FIGURE 7–81

Section 7–5:
1. Opens and shorts. **2.** The 10-kΩ resistor. **3.** **(a)** 57.9 V; **(b)** 57.9 V; **(c)** 57.1 V; **(d)** 100 V.

Section 7–6:
1. It decreases the output voltage. **2.** T. **3.** 20 V, 4.62 V.

Section 7–7:
1. See Figure 7–82. **2.** 11.64 kΩ. **3.** 0.859 mA. **4.** 0.639 mA. **5.** 1.41 V.

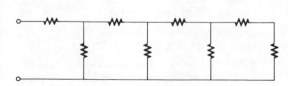

FIGURE 7–82

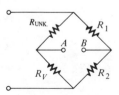

FIGURE 7–83

Section 7–8:
1. See Figure 7–83. **2.** $V_A = V_B$. **3.** $R_V(R_1/R_2)$. **4.** 15 kΩ.

Section 7–9:
1. It provides a delay to keep the initial message on the screen for a fixed time and then clears the screen. **2.** Lines 280–320.

There are many ways to simplify circuits in order to ease analysis or, in some cases, to make the analysis a manageable task that would otherwise be almost impossible.

The purpose of the theorems and conversions presented in this chapter is not to change the nature of a given dc circuit but to change the circuit to an *equivalent* form that is easier to analyze.

In this chapter, you will learn:

1. The characteristics of voltage sources.
2. The characteristics of current sources.
3. How to convert voltage sources to current sources and vice versa.
4. The superposition theorem and how to apply it to circuit analysis.
5. Thevenin's theorem and how to use it to reduce any dc circuit to a simple equivalent form in order to simplify the analysis.
6. How to determine the values for the Thevenin circuit by measurement.
7. Norton's theorem and its application to circuit analysis.
8. How Millman's theorem is used to reduce parallel voltage sources to a single equivalent voltage source.
9. How to determine when maximum power is being transferred from a source to a load.
10. How to make conversion between delta-wye and wye-delta networks.
11. How to use a computer program for the analysis of delta-wye networks.

8

CIRCUIT THEOREMS AND CONVERSIONS

8–1

Figure 8–1A is the familiar symbol for an ideal dc voltage source. The voltage across its terminals A and B remains fixed regardless of the value of load resistance that may be connected across its output. Figure 8–1B shows a load resistor R_L connected. All of the source voltage, V_S, is dropped across R_L. R_L can be changed to any value except zero, and the voltage will remain fixed. The ideal voltage source has an internal resistance of *zero*.

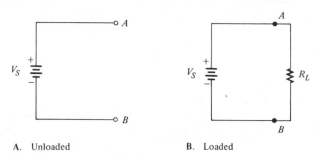

A. Unloaded   B. Loaded

FIGURE 8–1   *Ideal dc voltage source.*

In reality, no voltage source is ideal. That is, all have some inherent internal resistance as a result of their physical and/or chemical makeup, which can be represented by a resistor in series with an ideal source, as shown in Figure 8–2A. R_S is the internal source resistance and V_S is the source voltage. With no load, the output voltage (voltage from A to B) is V_S. This voltage is sometimes called the *open circuit voltage*.

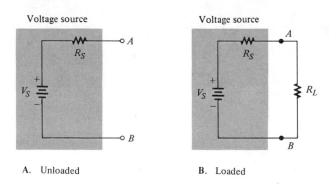

A. Unloaded   B. Loaded

FIGURE 8–2   *Practical voltage source.*

Loading of the Voltage Source

When a load resistor is connected across the output terminals, as shown in Figure 8–2B, all of the source voltage does not appear across R_L. Some of the voltage is dropped across R_S because of the current flowing through R_S to the load, R_L.

If R_S is very small compared to R_L, the source approaches ideal, because almost all of the source voltage, V_S, appears across the larger resistance R_L. Very little voltage is dropped across the internal resistance, R_S. If R_L changes, most of the source voltage remains across the output as long as R_L is much larger than R_S. As a result, very little change occurs in the output voltage. The larger R_L is compared to R_S, the less change there is in the output voltage. As a rule, R_L should be at least ten times R_S ($R_L \geq 10R_S$).

Example 8–1 illustrates the effect of changes in R_L on the output voltage when R_L is much greater than R_S. Example 8–2 shows the effect of smaller load resistances.

Example 8–1

Calculate the voltage output of the source in Figure 8–3 for the following values of R_L: 100 Ω, 500 Ω, and 1 kΩ.

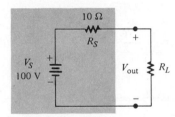

FIGURE 8–3

Solution:

For $R_L = 100$ Ω:

$$V_{OUT} = \left(\frac{R_L}{R_S + R_L}\right)V_S = \left(\frac{100\ \Omega}{110\ \Omega}\right)100\ \text{V}$$

$$= 91\ \text{V}$$

For $R_L = 500$ Ω:

$$V_{OUT} = \left(\frac{500\ \Omega}{510\ \Omega}\right)100\ \text{V}$$

$$= 98\ \text{V}$$

For $R_L = 1$ kΩ:

$$V_{OUT} = \left(\frac{1000\ \Omega}{1010\ \Omega}\right)100\ \text{V}$$

$$= 99\ \text{V}$$

Example 8–1 (continued)

Notice that the output voltage is within 10% of the source voltage, V_S, for all three values of R_L, because R_L is at least ten times R_S.

Example 8–2

Determine V_{OUT} for $R_L = 10\ \Omega$ and $R_L = 1\ \Omega$ in Figure 8–3.

Solution:

For $R_L = 10\ \Omega$:

$$V_{OUT} = \left(\frac{R_L}{R_S + R_L}\right)V_S = \left(\frac{10\ \Omega}{20\ \Omega}\right)100\ V$$

$$= 50\ V$$

For $R_L = 1\ \Omega$:

$$V_{OUT} = \left(\frac{R_L}{R_S + R_L}\right)V_S = \left(\frac{1\ \Omega}{11\ \Omega}\right)100\ V$$

$$= 9.1\ V$$

Notice in Example 8–2 that the output voltage decreases significantly as R_L is made smaller compared to R_S. This example illustrates the requirement that R_L must be much larger than R_S in order to maintain the output voltage near its open circuit value.

Review for 8–1

1. What is the symbol for the ideal voltage source?

2. Sketch a practical voltage source.

3. What is the internal resistance of the ideal voltage source?

4. What effect does the load have on the output voltage of the practical voltage source?

THE CURRENT SOURCE

8–2   Figure 8–4A shows a symbol for the ideal current source. The arrow indicates the direction of current, and I_S is the value of the source current. An ideal current source produces a *fixed* or constant value of current through a load, regardless of

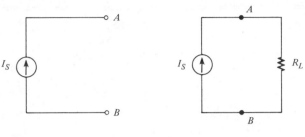

A. Unloaded   B. Loaded

FIGURE 8–4 *Ideal current source.*

the value of the load. This concept is illustrated in Figure 8–4B, where a load resistor is connected to the current source between terminals A and B. The ideal current source has an infinitely large internal resistance.

Transistors act basically as current sources, and for this reason knowledge of the current source concept is important. You will find that the equivalent model of a transistor does contain a current source.

Although the ideal current source can be used in most analysis work, no actual device is ideal. A practical current source representation is shown in Figure 8–5. Here the internal resistance appears in parallel with the ideal current source.

If the internal source resistance, R_S, is much *larger* than a load resistor, the practical source approaches ideal. The reason is illustrated in the practical current source shown in Figure 8–5. Part of the current I_S flows through R_S, and part through R_L. R_S and R_L act as a current divider. If R_S is much larger than R_L, most of the current will flow through R_L, and very little through R_S. As long as R_L remains much smaller than R_S, the current through it will stay almost constant, no matter how much R_L changes.

If we have a constant current source, we normally assume that R_S is so much larger than the load that it can be neglected. This simplifies the source to ideal, making the analysis easier.

Example 8–3 illustrates the effect of changes in R_L on the load current when R_L is much smaller than R_S. Generally, R_L should be at least ten times smaller ($10R_L \leq R_S$).

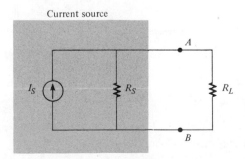

Current source

FIGURE 8–5 *Practical current source with load.*

Example 8–3

Calculate the load current in Figure 8–6 for the following values of R_L: 100 Ω, 500 Ω, and 1 kΩ.

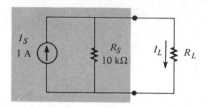

FIGURE 8–6

Solution:

For $R_L = 100$ Ω:

$$I_L = \left(\frac{R_S}{R_S + R_L}\right)I_S = \left(\frac{10 \text{ k}\Omega}{10.1 \text{ k}\Omega}\right)1 \text{ A}$$

$$= 0.99 \text{ A}$$

For $R_L = 500$ Ω:

$$I_L = \left(\frac{10 \text{ k}\Omega}{10.5 \text{ k}\Omega}\right)1 \text{ A}$$

$$= 0.95 \text{ A}$$

For $R_L = 1$ kΩ:

$$I_L = \left(\frac{10 \text{ k}\Omega}{11 \text{ k}\Omega}\right)1 \text{ A}$$

$$= 0.91 \text{ A}$$

Notice that the load current I_L is within 10% of the source current for each value of R_L.

Review for 8–2

1. What is the symbol for an ideal current source?
2. Sketch the practical current source.
3. What is the internal resistance of the ideal current source?
4. What effect does the load have on the load current of the practical current source?

SOURCE CONVERSIONS

In circuit analysis, it is sometimes useful to convert a voltage source to an *equivalent* current source, or vice versa.

8–3

Converting a Voltage Source into a Current Source

The source voltage, V_S, divided by the source resistance, R_S, gives the value of the equivalent source current:

$$I_S = \frac{V_S}{R_S} \tag{8–1}$$

The value of R_S is the same for both sources. As illustrated in Figure 8–7, the directional arrow for the current points from minus to plus. The equivalent current source is the source in parallel with R_S.

Equivalency of two sources means that for any given load resistance connected to the two sources, the same load voltage and current are produced by both sources. This concept is called *terminal equivalency.*

We can show that the voltage source and the current source in Figure 8–7 are equivalent by connecting a load resistor to each, as shown in Figure 8–8, and then calculating the load current as follows: For the voltage source,

$$I_L = \frac{V_S}{R_S + R_L}$$

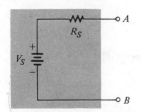

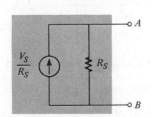

A. Voltage source   B. Current source

FIGURE 8–7 *Conversion of voltage source to equivalent current source.*

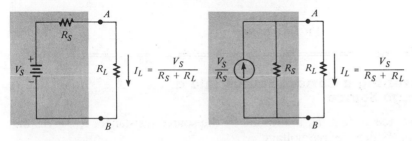

A. Loaded voltage source   B. Loaded current source

FIGURE 8–8 *Equivalent sources with loads.*

For the current source,

$$I_L = \left(\frac{R_S}{R_S + R_L}\right)\frac{V_S}{R_S} = \frac{V_S}{R_S + R_L}$$

As you see, both expressions for I_L are the same. These equations prove that the sources are equivalent as far as the load or terminals AB are concerned.

Example 8–4

Convert the voltage source in Figure 8–9 to an equivalent current source.

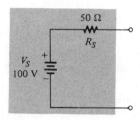

FIGURE 8–9

Solution:

$$I_S = \frac{V_S}{R_S} = \frac{100 \text{ V}}{50 \text{ }\Omega} = 2 \text{ A}$$

$$R_S = 50 \text{ }\Omega$$

The equivalent current source is shown in Figure 8–10.

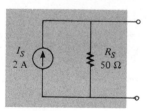

FIGURE 8–10

Converting a Current Source into a Voltage Source

The source current, I_S, multiplied by the source resistance, R_S, gives the value of the equivalent source voltage:

$$V_S = I_S R_S \qquad\qquad (8–2)$$

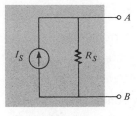

A. Current source

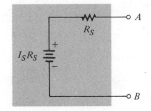

B. Voltage source

FIGURE 8–11   *Conversion of current source to equivalent voltage source.*

Again, R_S remains the same. The polarity of the voltage source is minus to plus in the direction of the current. The equivalent voltage source is the voltage in series with R_S, as illustrated in Figure 8–11.

Example 8–5

Convert the current source in Figure 8–12 to an equivalent voltage source.

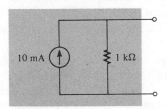

FIGURE 8–12

Solution:

$$V_S = I_S R_S = (10 \text{ mA})(1 \text{ k}\Omega)$$
$$= 10 \text{ V}$$
$$R_S = 1 \text{ k}\Omega$$

The equivalent voltage source is shown in Figure 8–13.

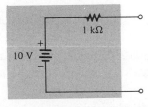

FIGURE 8–13

Review for 8–3

1. Write the formula for converting a voltage source to a current source.
2. Write the formula for converting a current source to a voltage source.
3. Convert the voltage source in Figure 8–14 to an equivalent current source.
4. Convert the current source in Figure 8–15 to an equivalent voltage source.

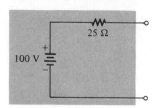

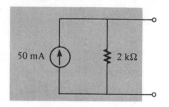

FIGURE 8–14   **FIGURE 8–15**

THE SUPERPOSITION THEOREM

8–4   The *superposition theorem* is useful in the analysis of linear circuits with more than one source. It provides a method of determining the current in any branch of a multiple-source circuit. A statement of the superposition theorem is as follows:

> The current in any given branch of a multiple-source linear circuit can be found by determining the currents in that particular branch produced by each source acting alone, with all other sources replaced by their internal resistances. The total current in the branch is the algebraic sum of the individual source currents in that branch.

Thus, for a circuit with more than one source, proceed as follows to find the total current in any given branch:

1. Leave *one* of the sources in the circuit, and reduce *all* others to zero. Reduce the voltage source to zero by replacing it with a short; any internal series resistance remains (never physically place a short *across* an existing voltage source). Reduce the current source to zero by replacing it with an open circuit; any internal parallel resistance remains.

2. Find the current in the branch of interest due to the one remaining source.

3. Repeat Steps 1 and 2 for each source in turn. When you finish, you will have a number of current values equal to the number of sources in the circuit.

4. Add all of the individual current values algebraically. All currents in one direction will have a plus (+) sign, and all currents in the other direction will have a minus (−) sign.

The following four examples clarify this procedure.

Example 8–6

Find the current in R_2 of Figure 8–16 by using the superposition theorem.

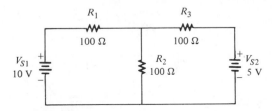

FIGURE 8–16

Solution:

First, by replacing V_{S2} with a short, find the current in R_2 due to voltage source V_{S1}, as shown in Figure 8–17.

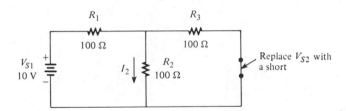

FIGURE 8–17

To find I_2, we can use the current divider formula from Chapter 6. Looking from V_{S1},

$$R_T = R_1 + \frac{R_3}{2} = 100 \ \Omega + 50 \ \Omega$$

$$= 150 \ \Omega$$

$$I_T = \frac{V_{S1}}{R_T} = \frac{10 \ V}{150 \ \Omega} = 0.0667 \ A$$

$$= 66.7 \ mA$$

The current in R_2 due to V_{S1} is

$$I_2 = \left(\frac{R_3}{R_2 + R_3}\right)I_T = \left(\frac{100 \ \Omega}{200 \ \Omega}\right)66.7 \ mA$$

$$= 33.3 \ mA$$

Note that this current flows *downward* through R_2.

Example 8–6 (continued)

Next find the current in R_2 due to voltage source V_{S2} by replacing V_{S1} with a short, as shown in Figure 8–18.

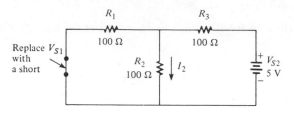

FIGURE 8–18

Looking from V_{S2},

$$R_T = R_3 + \frac{R_1}{2} = 100 \ \Omega + 50 \ \Omega$$

$$= 150 \ \Omega$$

$$I_T = \frac{V_{S2}}{R_T} = \frac{5 \ V}{150 \ \Omega} = 0.0333 \ A$$

$$= 33.3 \ mA$$

The current in R_2 due to V_{S2} is

$$I_2 = \left(\frac{R_1}{R_1 + R_2}\right)I_T = \left(\frac{100 \ \Omega}{200 \ \Omega}\right)33.3 \ mA$$

$$= 16.7 \ mA$$

Note that this current flows *downward* through R_2.

Both component currents are flowing *downward* through R_2, so they have the same algebraic sign. Therefore, we add the values to get the total current through R_2:

$$I_2 \ (\text{total}) = I_2 \ (\text{due to } V_{S1}) + I_2 \ (\text{due to } V_{S2})$$

$$= 33.3 \ mA + 16.7 \ mA = 50 \ mA$$

Example 8–7

Find the current through R_2 in the circuit of Figure 8–19.

Solution:

First find the current in R_2 *due to V_S* by replacing I_S with an *open,* as shown in Figure 8–20. Notice that all of the current produced by V_S flows through R_2.

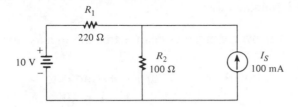

FIGURE 8–19

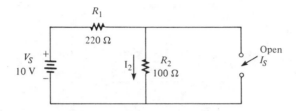

FIGURE 8–20

Looking from V_S,

$$R_T = R_1 + R_2 = 320 \ \Omega$$

The current through R_2 due to V_S is

$$I_2 = \frac{V_S}{R_T} = \frac{10 \ V}{320 \ \Omega} = 0.031 \ A$$

$$= 31 \ mA$$

Note that this current flows *downward* through R_2.

Next find the current through R_2 due to I_S by replacing V_S with a *short*, as shown in Figure 8–21.

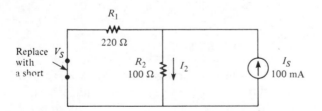

FIGURE 8–21

Using the current divider formula, we get the current through R_2 due to I_S as follows:

$$I_2 = \left(\frac{R_1}{R_1 + R_2}\right)I_S = \left(\frac{220 \ \Omega}{320 \ \Omega}\right)100 \ mA$$

$$= 69 \ mA$$

Note that this current also flows *downward* through R_2.

238

Example 8–7 (continued)

Both currents are in the same direction through R_2; so we add them to get the total:

$$I_2 \text{ (total)} = I_2 \text{ (due to } V_s) + I_2 \text{ (due to } I_s)$$
$$= 31 \text{ mA} + 69 \text{ mA} = 100 \text{ mA}$$

Example 8–8

Find the current through the 100-Ω resistor in Figure 8–22.

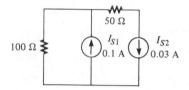

FIGURE 8–22

Solution:

First find the current through the 100-Ω resistor due to current source I_{S1} by replacing source I_{S2} with an *open,* as shown in Figure 8–23. As you can see, the entire 0.1 A from the current source I_{S1} flows *downward* through the 100-Ω resistor.

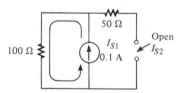

FIGURE 8–23

Next find the current through the 100-Ω resistor due to source I_{S2} by replacing source I_{S1} with an *open,* as indicated in Figure 8–24. Notice that all of the 0.03 A from source I_{S2} flows *upward* through the 100-Ω resistor.

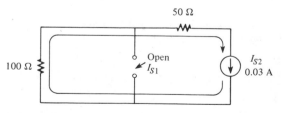

FIGURE 8–24

To get the total current through the 100-Ω resistor, we subtract the smaller current from the larger because they are in opposite directions. The resulting total current flows in the direction of the larger current from source I_{S1}:

$$I_{100\,\Omega} \text{ (total)} = I_{100\,\Omega} \text{ (due to } I_{S1}) - I_{100\,\Omega} \text{ (due to } I_{S2})$$
$$= 0.1 \text{ A} - 0.03 \text{ A} = 0.07 \text{ A}$$

The resulting current flows downward through the resistor.

Example 8–9

Find the total current through R_3 in Figure 8–25.

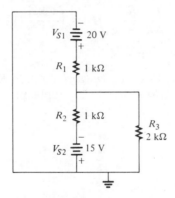

FIGURE 8–25

Solution:

First find the current through R_3 due to source V_{S1} by replacing source V_{S2} with a short, as shown in Figure 8–26.

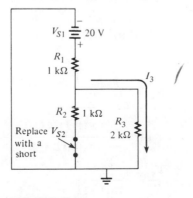

FIGURE 8–26

Example 8–9 (continued)

Looking from V_{S1},

$$R_T = R_1 + \frac{R_2 R_3}{R_2 + R_3}$$

$$= 1 \text{ k}\Omega + \frac{(1 \text{ k}\Omega)(2 \text{ k}\Omega)}{3 \text{ k}\Omega} = 1.67 \text{ k}\Omega$$

$$I_T = \frac{V_{S1}}{R_T} = \frac{20 \text{ V}}{1.67 \text{ k}\Omega}$$

$$= 12 \text{ mA}$$

Now apply the current divider formula to get the current through R_3 due to source V_{S1} as follows:

$$I_3 = \left(\frac{R_2}{R_2 + R_3}\right) I_T = \left(\frac{1 \text{ k}\Omega}{3 \text{ k}\Omega}\right) 12 \text{ mA}$$

$$= 4 \text{ mA}$$

Notice that this current flows *downward* through R_3.

Next find I_3 due to source V_{S2} by replacing source V_{S1} with a short, as shown in Figure 8–27.

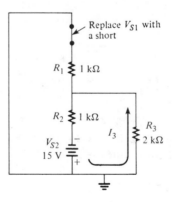

FIGURE 8–27

Looking from V_{S2},

$$R_T = R_2 + \frac{R_1 R_3}{R_1 + R_3}$$

$$= 1 \text{ k}\Omega + \frac{(2 \text{ k}\Omega)(1 \text{ k}\Omega)}{3 \text{ k}\Omega} = 1.67 \text{ k}\Omega$$

$$I_T = \frac{V_{S2}}{R_T} = \frac{15 \text{ V}}{1.67 \text{ k}\Omega}$$

$$= 9 \text{ mA}$$

Now apply the current divider formula to find the current through R_3 due to source V_{S2} as follows:

$$I_3 = \left(\frac{R_1}{R_1 + R_3}\right)I_T = \left(\frac{1 \text{ k}\Omega}{3 \text{ k}\Omega}\right)9 \text{ mA}$$

$$= 3 \text{ mA}$$

Notice that this current flows *upward* through R_3.

Calculation of the total current through R_3 is as follows:

$$I_3 \text{ (total)} = I_3 \text{ (due to } V_{S1}) - I_3 \text{ (due to } V_{S2})$$

$$= 4 \text{ mA} - 3 \text{ mA} = 1 \text{ mA}$$

This current flows downward through R_3.

Review for 8–4

1. State the superposition theorem.

2. Why is the superposition theorem useful for analysis of multiple-source linear circuits?

3. Why is a voltage source shorted and a current source opened when the superposition theorem is applied?

4. Using the superposition theorem, find the current through R_1 in Figure 8–28.

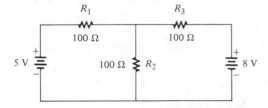

R_1 R_3

100 Ω   100 Ω

5 V   100 Ω R_2 8 V

FIGURE 8–28

5. If two currents are in opposing directions through a branch of a circuit, in what direction does the net current flow?

THEVENIN'S THEOREM

Thevenin's theorem, as applied to linear circuits, provides a method for reducing *any* circuit to an *equivalent* circuit consisting of *an equivalent voltage source in series with an equivalent resistance*. Although we are dealing with dc sources at this point, most circuit theorems, including Thevenin's, apply equally to ac circuits.

The form of Thevenin's equivalent circuit is shown in Figure 8–29. Regardless of how complex the original circuit is, it can always be reduced to this

8–5

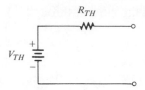

FIGURE 8–29 *Form of Thevenin's equivalent circuit.*

simple series equivalent form. The equivalent voltage source is designated V_{TH}, and the equivalent resistance, R_{TH}.

To apply Thevenin's theorem, you must know how to find the two quantities V_{TH} and R_{TH}. Once you have found them for a given circuit, simply connect them in series to get the complete Thevenin circuit.

Meaning of Equivalency

Figure 8–30A shows a block diagram representing a resistive circuit of any complexity. This circuit has two output terminals A and B. There is a load resistor, R_L, connected across these two terminals. The circuit inside the block produces a certain voltage, V_L, across the load, and a certain current, I_L, through the load as illustrated.

By Thevenin's theorem, the circuit in the block, regardless of how complex it is, can be reduced to an *equivalent* circuit of the form shown in this shaded block of Figure 8–30B. The term *equivalent* means that when the *same value of load* is connected to both the original circuit (block) and Thevenin's equivalent circuit, the voltages across the loads are equal. Also, the currents through the loads are equal. Therefore, *as far as the load is concerned,* there is no difference between the original circuit and Thevenin's circuit—they are equivalent. The load resistance "sees" the same values of V_L and I_L regardless of whether it is connected to the original circuit or to Thevenin's circuit, and therefore it does not "know" the difference.

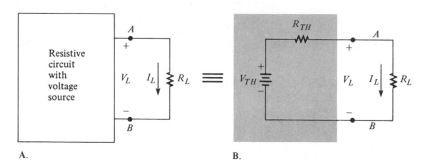

A.   B.

FIGURE 8–30 *Circuit equivalency.*

Thevenin's Equivalent Voltage (V_{TH})

As you have seen, the equivalent voltage, V_{TH}, is one part of the complete Thevenin equivalent circuit. The other part is R_{TH}. V_{TH} *is defined to be the open*

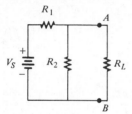 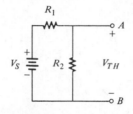

A. Original circuit

B. Remove R_L to open
    the terminals AB
    to get V_{TH}.

FIGURE 8–31

circuit voltage between two points in a circuit. Any component connected between these two points effectively "sees" V_{TH} in series with R_{TH}.

To illustrate, suppose that a resistive circuit of some kind has a resistor connected between two points, as shown in Figure 8–31A. We wish to find the Thevenin circuit that is equivalent to the one shown as "seen" by R_L. V_{TH} is the voltage between points A and B *with R_L removed,* as shown in Figure 8–31B. In other words, we view the rest of the circuit from the viewpoint of the open terminals AB. R_L is considered external to the portion of the circuit to which Thevenin's theorem is applied. The following three examples show how to find V_{TH}.

Example 8–10

Determine V_{TH} for the circuit within the dashed lines in Figure 8–32.

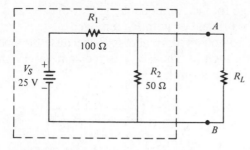

FIGURE 8–32

Solution:

Remove R_L and determine the voltage from A to B, which is V_{TH}. In this case, the voltage from A to B is the same as the voltage across R_2. We determine the voltage across R_2 using the voltage divider formula:

Example 8–10 (continued)

$$V_2 = \left(\frac{R_2}{R_1 + R_2}\right)V_S = \left(\frac{50 \ \Omega}{150 \ \Omega}\right)25 \ V$$

$$= 8.33 \ V$$

$$V_{TH} = V_{AB} = V_2 = 8.33 \ V$$

Example 8–11

For the circuit in Figure 8–33, determine the Thevenin voltage V_{TH} as seen by R_L.

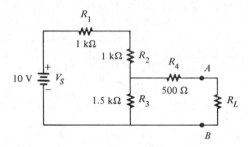

FIGURE 8–33

Solution:

Thevenin's voltage for the circuit between terminals A and B is the voltage that appears across A and B with R_L removed.

There is no voltage drop across R_4 because the open terminals AB prevent current through it. Thus, V_{AB} is the same as V_3 and can be found by the voltage divider formula:

$$V_{AB} = V_3 = \left(\frac{R_3}{R_1 + R_2 + R_3}\right)V_S$$

$$= \left(\frac{1.5 \ k\Omega}{3.5 \ k\Omega}\right)10 \ V = 4.29 \ V$$

V_{TH} is the open terminal voltage from A to B. Therefore,

$$V_{TH} = V_{AB} = 4.29 \ V$$

Example 8–12

For the circuit in Figure 8–34, find V_{TH} as seen by R_L.

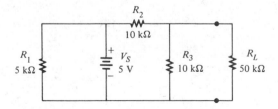

FIGURE 8–34

Solution:

First remove R_L and determine the voltage across the resulting open terminals, which is V_{TH}. We find V_{TH} by applying the voltage divider formula to R_2 and R_3:

$$V_{TH} = V_3 = \left(\frac{R_3}{R_2 + R_3}\right)V_S$$

$$= \left(\frac{10 \text{ k}\Omega}{20 \text{ k}\Omega}\right)5 \text{ V} = 2.5 \text{ V}$$

Notice that R_1 has no effect on the result since 5 V still appear across the R_2 and R_3 combination.

Thevenin's Equivalent Resistance (R_{TH})

The previous examples showed how to find only one part of a Thevenin equivalent circuit: the equivalent voltage, V_{TH}. Now we will illustrate how to find the equivalent resistance, R_{TH}. As defined by Thevenin's theorem, R_{TH} *is the total resistance appearing between two terminals in a given circuit with all sources replaced by their internal resistances.* Thus, if we wish to find R_{TH} between any two terminals in a circuit, first we *short* all voltage sources and *open* all current sources, leaving only their internal resistances, if any. Then we determine the *total* resistance between those two terminals. The following three examples illustrate how to find R_{TH}.

Example 8–13

Find R_{TH} for the circuit within the dashed lines of Figure 8–32 (see Example 8–10).

246

CIRCUIT THEOREMS AND CONVERSIONS

Example 8–13 (continued)

Solution:

First reduce V_S to zero by shorting it, as shown in Figure 8–35.

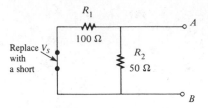

FIGURE 8–35

Looking in at terminals A and B, we see that R_1 and R_2 are in parallel. Thus,

$$R_{TH} = \frac{R_1 R_2}{R_1 + R_2} = \frac{(50\ \Omega)(100\ \Omega)}{150\ \Omega}$$

$$= 33.33\ \Omega$$

Example 8–14

For the circuit in Figure 8–33 (Example 8–11), determine R_{TH} as seen by R_L.

Solution:

First replace the voltage source with a short, as shown in Figure 8–36.

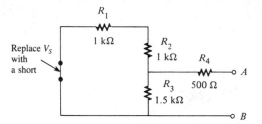

FIGURE 8–36

Looking in at the terminals A and B, we note that R_3 is in parallel with the series combination of R_1 and R_2, and this combination is in series with R_4. The calculation for R_{TH} is as follows:

$$R_{TH} = R_4 + \frac{R_3(R_1 + R_2)}{R_1 + R_2 + R_3}$$

$$= 500\ \Omega + \frac{(1.5\ k\Omega)(2\ k\Omega)}{3.5\ k\Omega} = 1.357\ k\Omega$$

Example 8–15

For the circuit in Figure 8–34 (Example 8–12), determine R_{TH} as seen by R_L.

Solution:

With the voltage source replaced by a short, R_1 is effectively out of the circuit. R_2 and R_3 appear in parallel, as indicated in Figure 8–37. R_{TH} is calculated as follows:

$$R_{TH} = \frac{10\ k\Omega}{2} = 5\ k\Omega$$

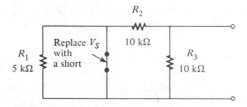

FIGURE 8–37

The previous examples have shown you how to find the two equivalent components of a Thevenin circuit, V_{TH} and R_{TH}. Keep in mind that V_{TH} and R_{TH} can be found for any linear circuit. Once these equivalent values are determined, they must be connected in *series* to form the Thevenin equivalent circuit. The following three examples illustrate this final step.

Example 8–16

Draw the complete Thevenin circuit for the original circuit within the dashed lines of Figure 8–32 (Example 8–10).

Solution:

We found in Examples 8–10 and 8–13 that $V_{TH} = 8.33$ V and $R_{TH} = 33.33\ \Omega$. The Thevenin equivalent circuit is shown in Figure 8–38.

Example 8–16 (continued)

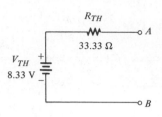

FIGURE 8–38

Example 8–17

For the original circuit in Figure 8–33 (Example 8–11), draw the Thevenin equivalent circuit as seen by R_L.

Solution:

We found in Examples 8–11 and 8–14 that $V_{TH} = 4.29$ V and $R_{TH} = 1.375$ kΩ. Figure 8–39 shows these values combined to form the Thevenin equivalent circuit.

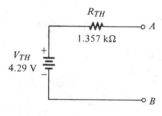

FIGURE 8–39

Example 8–18

For the original circuit in Figure 8–34 (Example 8–12), determine the Thevenin equivalent circuit as seen by R_L.

Solution:

From Examples 8–12 and 8–15, we know that V_{TH} is 2.5 V and R_{TH} is 5 kΩ. Connected in series, they produce the Thevenin equivalent circuit as shown in Figure 8–40.

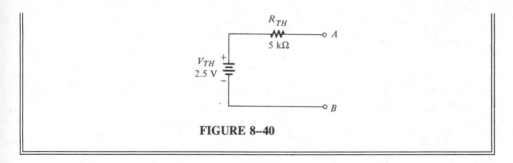

FIGURE 8–40

Summary of Thevenin's Theorem

Remember, the Thevenin equivalent circuit is *always* of the series form regardless of the original circuit that it replaces. The significance of Thevenin's theorem is that the equivalent circuit can replace the original circuit as far as any external load is concerned. Any load resistor connected between the terminals of a Thevenin equivalent circuit will have the same current through it and the same voltage across it as if it were connected to the terminals of the original circuit.

A summary of steps for applying Thevenin's theorem is as follows:

1. Open the two terminals (remove any load) between which you want to find the Thevenin equivalent circuit.

2. Determine the voltage (V_{TH}) across the two open terminals.

3. Determine the resistance (R_{TH}) between the two terminals with all voltage sources shorted and all current sources opened.

4. Connect V_{TH} and R_{TH} in series to produce the complete Thevenin equivalent for the original circuit.

Determining V_{TH} and R_{TH} by Measurement

Thevenin's theorem is largely an analytical tool that is applied theoretically in order to simplify circuit analysis. However, in many cases, Thevenin's equivalent can be found for an actual circuit by the following general measurement methods. These steps are illustrated in Figure 8–41, on the next page.

1. Remove any load from the output terminals of the circuit.

2. Measure the open terminal voltage. The voltmeter used must have an internal resistance much greater (at least 10 times greater) than the R_{TH} of the circuit. (V_{TH} is the open terminal voltage.)

3. Connect a variable resistor (rheostat) across the output terminals. Its maximum value must be greater than R_{TH}.

4. Adjust the rheostat, and measure the terminal voltage. When the terminal voltage equals $\frac{1}{2}V_{TH}$, the resistance of the rheostat is equal to R_{TH}. It should be disconnected from the terminals and measured with an ohmmeter.

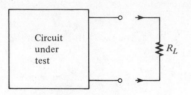

Step 1: Open the terminals.

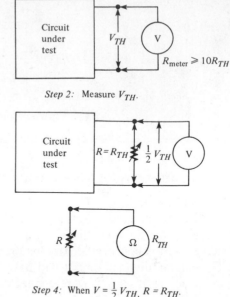

Step 2: Measure V_{TH}.

Step 3: Connect variable R.

Step 4: When $V = \frac{1}{2} V_{TH}$, $R = R_{TH}$.

FIGURE 8–41 *Determination of Thevenin's equivalent by measurement.*

This procedure for determining R_{TH} differs from the theoretical procedure because it is impractical to short voltage sources or open current sources in an actual circuit.

Also, when measuring R_{TH}, be certain that the circuit is capable of providing the required current to the variable resistor load and that the variable resistor can handle the required power. These considerations may make the procedure impractical in some cases.

Review for 8–5

1. State Thevenin's theorem.

2. What are the two components of a Thevenin equivalent circuit?

3. Draw the general form of a Thevenin equivalent circuit.

4. How is V_{TH} determined?

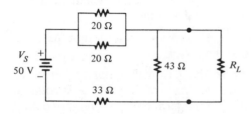

FIGURE 8–42

5. How is R_{TH} determined theoretically?

6. For the original circuit in Figure 8–42, find the Thevenin equivalent circuit as seen by R_L.

NORTON'S THEOREM

8–6

Like Thevenin's theorem, Norton's theorem provides a method of reducing a more complex circuit to a simpler form. The basic difference is that Norton's theorem gives *an equivalent current source in parallel with an equivalent resistance*. The form of Norton's equivalent circuit is shown in Figure 8–43. Regardless of how complex the original circuit is, it can always be reduced to this equivalent form. The equivalent current source is designated I_N, and the equivalent resistance, R_N.

To apply Norton's theorem, you must know how to find the two quantities I_N and R_N. Once you know them for a given circuit, simply connect them in parallel to get the complete Norton circuit.

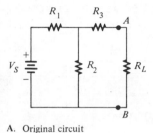

FIGURE 8–43 *Form of Norton's equivalent circuit.*

Norton's Equivalent Current (I_N)

As stated, I_N is one part of the complete Norton equivalent circuit; R_N is the other part. I_N is defined to be the *short circuit current* between two points in a circuit. Any component connected between these two points effectively "sees" a current source of value I_N in parallel with R_N.

To illustrate, suppose that a resistive circuit of some kind has a resistor connected between two points in the circuit, as shown in Figure 8–44A. We wish to find the Norton circuit that is equivalent to the one shown as "seen" by R_L. To

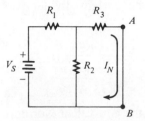

A. Original circuit   B. Short the terminals to get I_N.

FIGURE 8–44 *Determining the Norton equivalent current, I_N.*

find I_N, calculate the current between points A and B with these two points *shorted*, as shown in Figure 8–44B. Example 8–19 demonstrates how to find I_N.

Example 8–19

Determine I_N for the circuit within the dashed lines in Figure 8–45A.

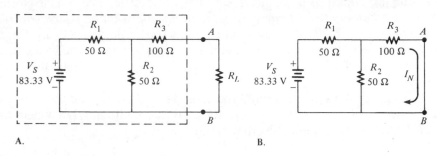

A.   B.

FIGURE 8–45

Solution:

Short terminals A and B as shown in Figure 8–45B. I_N is the current flowing through the short and is calculated as follows: First, the total resistance seen by the voltage source is

$$R_T = R_1 + \frac{R_2 R_3}{R_2 + R_3}$$

$$= 50\ \Omega + \frac{(50\ \Omega)(100\ \Omega)}{150\ \Omega} = 83.33\ \Omega$$

The total current from the source is

$$I_T = \frac{V_S}{R_T} = \frac{83.33\ \text{V}}{83.33\ \Omega}$$

$$= 1\ \text{A}$$

Now apply the current divider formula to find I_N (the current through the short):

$$I_N = \left(\frac{R_2}{R_2 + R_3}\right)I_T = \left(\frac{50\ \Omega}{150\ \Omega}\right)1\ \text{A}$$

$$= 0.33\ \text{A}$$

This is the value for the equivalent Norton current source.

Norton's Equivalent Resistance (R_N)

We define R_N in the same way as R_{TH}: It is the total resistance appearing between two terminals in a given circuit with all sources replaced by their internal resistances. Example 8–20 demonstrates how to find R_N.

Example 8–20

Find R_N for the circuit within the dashed lines of Figure 8–45 (see Example 8–19).

Solution:

First reduce V_S to zero by shorting it, as shown in Figure 8–46.

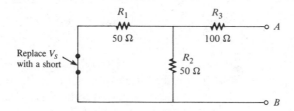

FIGURE 8–46

Looking in at terminals A and B, we see that the parallel combination of R_1 and R_2 is in series with R_3. Thus,

$$R_N = R_3 + \frac{R_1}{2} = 100 \ \Omega + \frac{50 \ \Omega}{2}$$

$$= 125 \ \Omega$$

The last two examples have shown how to find the two equivalent components of a Norton equivalent circuit, I_N and R_N. Keep in mind that these values can be found for any linear circuit. Once these are known, they must be connected in *parallel* to form the Norton equivalent circuit, as illustrated in Example 8–21.

Example 8–21

Draw the complete Norton circuit for the original circuit in Figure 8–45 (Example 8–19).

Example 8–21 (continued)

Solution:

We found in Examples 8–19 and 8–20 that $I_N = 0.33$ A and $R_N = 125 \ \Omega$. The Norton equivalent circuit is shown in Figure 8–47.

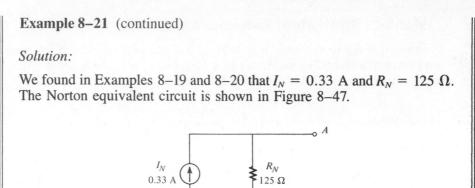

FIGURE 8–47

Summary of Norton's Theorem

Any load resistor connected between the terminals of a Norton equivalent circuit will have the same current through it and the same voltage across it as if it were connected to the terminals of the original circuit. A summary of steps for theoretically applying Norton's theorem is as follows:

1. Short the two terminals between which you want to find the Norton equivalent circuit.

2. Determine the current (I_N) through the shorted terminals.

3. Determine the resistance (R_N) between the two terminals (opened) with all voltage sources shorted and all current sources opened $(R_N = R_{TH})$.

4. Connect I_N and R_N in parallel to produce the complete Norton equivalent for the original circuit.

Norton's equivalent circuit can also be derived from Thevenin's equivalent circuit by use of the source conversion method discussed in Section 8–3.

Review for 8–6

1. State Norton's theorem.
2. What are the two components of a Norton equivalent circuit?
3. Draw the general form of a Norton equivalent circuit.
4. How is I_N determined?
5. How is R_N determined?
6. Find the Norton circuit as seen by R_L in Figure 8–48.

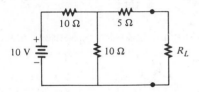

FIGURE 8–48

Millman's theorem allows us to reduce *any number of parallel voltage sources to a single equivalent voltage source*. It simplifies finding the voltage across or current through a load. Millman's theorem gives the same results as Thevenin's theorem for the special case of parallel voltage sources. A conversion by Millman's theorem is illustrated in Figure 8–49.

8–7

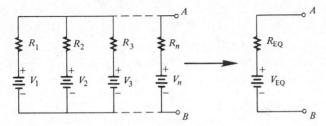

FIGURE 8–49 *Reduction of parallel voltage sources to a single equivalent voltage source.*

Millman's Equivalent Voltage (V_{EQ}) and Equivalent Resistance (R_{EQ})

Millman's theorem gives us a formula for calculating the equivalent voltage, V_{EQ}. To find V_{EQ}, convert each of the parallel voltage sources into current sources, as shown in Figure 8–50.

In Figure 8–50B, the total current from the parallel current sources is

$$I_T = I_1 + I_2 + I_3 + \cdots + I_n$$

The total conductance between terminals A and B is

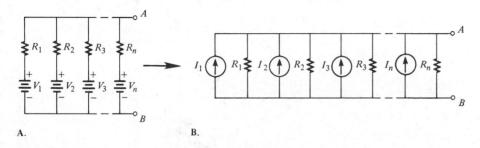

A.   B.

FIGURE 8–50 *Parallel voltage sources converted to current sources.*

$$G_T = G_1 + G_2 + G_3 + \cdots + G_n$$

where $G_T = 1/R_T$, $G_1 = 1/R_1$, and so on. Remember, the current sources are effectively open. Therefore, by Millman's theorem, the *equivalent resistance* is the total resistance R_T.

$$R_{EQ} = \frac{1}{G_T} = \frac{1}{(1/R_1) + (1/R_2) + (1/R_3) + \cdots + (1/R_n)} \qquad (8\text{--}3)$$

By Millman's theorem, the *equivalent voltage* is $I_T R_{EQ}$, where I_T is expressed as follows:

$$I_T = I_1 + I_2 + I_3 + \cdots + I_n$$

$$= \frac{V_1}{R_1} + \frac{V_2}{R_2} + \frac{V_3}{R_3} + \cdots + \frac{V_n}{R_n}$$

The following is the formula for the equivalent voltage:

$$V_{EQ} = \frac{(V_1/R_1) + (V_2/R_2) + (V_3/R_3) + \cdots + (V_n/R_n)}{(1/R_1) + (1/R_2) + (1/R_3) + \cdots + (1/R_n)} \qquad (8\text{--}4)$$

Equations (8–3) and (8–4) are the two Millman formulas. The equivalent voltage source has a polarity such that the total current through a load will flow in the same direction as in the original circuit.

Example 8–22

Use Millman's theorem to find the voltage across R_L and the current through R_L in Figure 8–51.

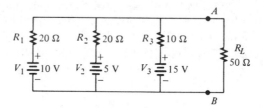

FIGURE 8–51

Solution:

Apply Millman's theorem as follows:

$$R_{EQ} = \frac{1}{(1/R_1) + (1/R_2) + (1/R_3)}$$

$$= \frac{1}{(1/20\ \Omega) + (1/20\ \Omega) + (1/10\ \Omega)}$$

$$= \frac{1}{0.2} = 5 \ \Omega$$

$$V_{EQ} = \frac{(V_1/R_1) + (V_2/R_2) + (V_3/R_3)}{(1/R_1) + (1/R_2) + (1/R_3)}$$

$$= \frac{(10 \text{ V}/20 \ \Omega) + (5 \text{ V}/20 \ \Omega) + (15 \text{ V}/10 \ \Omega)}{(1/20 \ \Omega) + (1/20 \ \Omega) + (1/10 \ \Omega)}$$

$$= \frac{2.25 \text{ A}}{0.2 \text{ S}} = 11.25 \text{ V}$$

The single equivalent voltage source is shown in Figure 8–52.

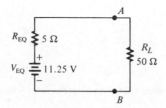

FIGURE 8–52

Now we calculate I_L and V_L for the load resistor.

$$I_L = \frac{V_{EQ}}{R_{EQ} + R_L} = \frac{11.25 \text{ V}}{55 \ \Omega}$$

$$= 0.205 \text{ A}$$

$$V_L = I_L R_L = (0.205 \text{ A}) (50 \ \Omega)$$

$$= 10.25 \text{ V}$$

Review for 8–7

1. To what type of circuit does Millman's theorem apply?
2. Write the Millman theorem formula for R_{EQ}.
3. Write the Millman theorem formula for V_{EQ}.
4. Find the load current and the load voltage in Figure 8–53.

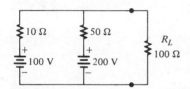

FIGURE 8–53

MAXIMUM POWER TRANSFER THEOREM

8–8

The maximum power transfer theorem states that when a circuit is connected to a load, *maximum power is delivered to the load when the load resistance is equal to the output resistance of the circuit*. The output resistance of a circuit is the equivalent resistance as viewed from the output terminals using Thevenin's theorem. An equivalent circuit with its output resistance and load is shown in Figure 8–54. When $R_L = R_S$, the maximum power possible is transferred from the voltage source to R_L.

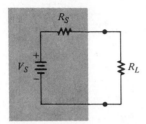

FIGURE 8–54   *Maximum power is transferred to the load when $R_L = R_S$.*

Practical applications of this theorem include audio systems such as stereo, radio, public address, and so on. In these systems the resistance (impedance) of the speaker is the load. The circuit that drives the speaker is a power amplifier. The systems are typically optimized for maximum power to the speakers. Thus, the resistance (impedance) of the speaker must equal the source resistance (impedance) of the amplifier.

Example 8–23 shows that maximum power occurs at the *matched* condition (that is, when $R_L = R_S$).

Example 8–23

The circuit to the left of terminals A and B in Figure 8–55 provides power to the load, R_L. It is the Thevenin equivalent of a more complex circuit. Calculate the power delivered to R_L for each following value of R_L: 1 Ω, 5 Ω, 10 Ω, 15 Ω, and 20 Ω.

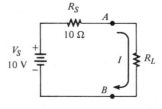

FIGURE 8–55

Solution:

For $R_L = 1 \ \Omega$:

$$I = \frac{V_S}{R_S + R_L} = \frac{10 \text{ V}}{11 \ \Omega}$$

$$= 0.91 \text{ A}$$

$$P_L = I^2 R_L = (0.91 \text{ A})^2 (1 \ \Omega)$$

$$= 0.83 \text{ W}$$

For $R_L = 5 \ \Omega$:

$$I = \frac{10 \text{ V}}{15 \ \Omega}$$

$$= 0.67 \text{ A}$$

$$P_L = I^2 R_L = (0.67 \text{ A})^2 (5 \ \Omega)$$

$$= 2.24 \text{ W}$$

For $R_L = 10 \ \Omega$:

$$I = \frac{10 \text{ V}}{20 \ \Omega}$$

$$= 0.5 \text{ A}$$

$$P_L = I^2 R_L = (0.5 \text{ A})^2 (10 \ \Omega)$$

$$= 2.50 \text{ W}$$

For $R_L = 15 \ \Omega$:

$$I = \frac{10 \text{ V}}{25 \ \Omega}$$

$$= 0.4 \text{ A}$$

$$P_L = I^2 R_L = (0.4 \text{ A})^2 (15 \ \Omega)$$

$$= 2.40 \text{ W}$$

For $R_L = 20 \ \Omega$:

$$I = \frac{10 \text{ V}}{30 \ \Omega} = 0.33 \text{ A}$$

$$P_L = I^2 R_L = (0.33 \text{ A})^2 (20 \ \Omega)$$

$$= 2.18 \text{ W}$$

Notice in Example 8–23 that the maximum load power occurs when $R_L = R_S = 10 \ \Omega$. If R_L is decreased below this value or increased above it, the power falls off, as shown in the graph in Figure 8–56.

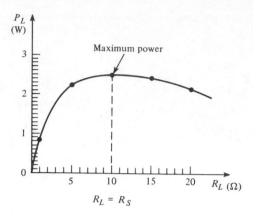

FIGURE 8–56 *Load power versus load resistance.*

Review for 8–8

1. State the maximum power transfer theorem.

2. When is maximum power delivered from a source to a load?

3. A given circuit has a source resistance of 50 Ω. What will be the value of the load to which the maximum power is delivered?

DELTA-WYE (Δ-Y) AND WYE-DELTA (Y-Δ) NETWORK CONVERSIONS

8–9 A resistive delta (Δ) network has the form shown in Figure 8–57A. A wye (Y) network is shown in Figure 8–57B. Notice that letter subscripts are used to designate resistors in the delta network and that numerical subscripts are used to designate resistors in the wye.

Conversion between these two forms of circuits is sometimes helpful. In this section, the conversion formulas and rules for remembering them are given.

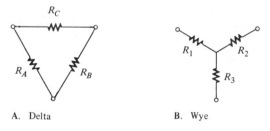

A. Delta   B. Wye

FIGURE 8–57 *Delta and wye networks.*

Δ-to-Y Conversion

It is convenient to think of the wye positioned within the delta, as shown in Figure 8–58. To convert from delta to wye, we need R_1, R_2, and R_3 in terms of R_A, R_B, and R_C. The conversion rule is as follows: *Each resistor in the wye is*

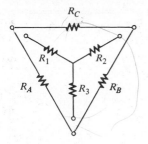

FIGURE 8–58 *"Y within Δ" aid for conversion formulas.*

equal to the product of the resistors in two adjacent delta branches, divided by the sum of all three delta resistors.

In Figure 8–58, R_A and R_C are "adjacent" to R_1:

$$R_1 = \frac{R_A R_C}{R_A + R_B + R_C} \tag{8–5}$$

Also, R_B and R_C are "adjacent" to R_2:

$$R_2 = \frac{R_B R_C}{R_A + R_B + R_C} \tag{8–6}$$

and R_A and R_B are "adjacent" to R_3:

$$R_3 = \frac{R_A R_B}{R_A + R_B + R_C} \tag{8–7}$$

Y-to-Δ Conversion

To convert from wye to delta, we need R_A, R_B, and R_C in terms of R_1, R_2, and R_3. The conversion rule is as follows: *Each resistor in the delta is equal to the sum of all possible products of wye resistors taken two at a time, divided by the opposite wye resistor.*

In Figure 8–58, R_2 is "opposite" to R_A:

$$R_A = \frac{R_1 R_2 + R_1 R_3 + R_2 R_3}{R_2} \tag{8–8}$$

Also, R_1 is "opposite" to R_B:

$$R_B = \frac{R_1 R_2 + R_1 R_3 + R_2 R_3}{R_1} \tag{8–9}$$

and R_3 is "opposite" to R_C:

$$R_C = \frac{R_1 R_2 + R_1 R_3 + R_2 R_3}{R_3} \tag{8–10}$$

The following two examples illustrate conversion between these two forms of circuits.

Example 8–24

Convert the delta network in Figure 8–59 to a wye network.

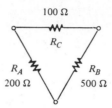

100 Ω
R_C

R_A
200 Ω

R_B
500 Ω

FIGURE 8–59

Solution:

Use Equations (8–5), (8–6), and (8–7):

$$R_1 = \frac{R_A R_C}{R_A + R_B + R_C} = \frac{(200\ \Omega)(100\ \Omega)}{200\ \Omega + 500\ \Omega + 100\ \Omega}$$
$$= 25\ \Omega$$

$$R_2 = \frac{R_B R_C}{R_A + R_B + R_C} = \frac{(500\ \Omega)(100\ \Omega)}{800\ \Omega}$$
$$= 62.5\ \Omega$$

$$R_3 = \frac{R_A R_B}{R_A + R_B + R_C} = \frac{(200\ \Omega)(500\ \Omega)}{800\ \Omega}$$
$$= 125\ \Omega$$

The resulting wye network is shown in Figure 8–60.

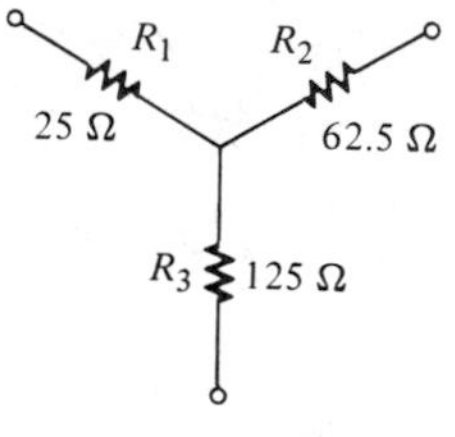

R_1 R_2

25 Ω   62.5 Ω

R_3 125 Ω

FIGURE 8–60

Example 8–25

Convert the wye network in Figure 8–61 to a delta network.

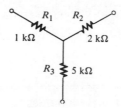

FIGURE 8–61

Solution:

Use Equations (8–8), (8–9), and (8–10):

$$R_A = \frac{R_1 R_2 + R_1 R_3 + R_2 R_3}{R_2}$$

$$= \frac{(1 \text{ k}\Omega)(2 \text{ k}\Omega) + (1 \text{ k}\Omega)(5 \text{ k}\Omega) + (2 \text{ k}\Omega)(5 \text{ k}\Omega)}{2 \text{ k}\Omega} = 8.5 \text{ k}\Omega$$

$$R_B = \frac{R_1 R_2 + R_1 R_3 + R_2 R_3}{R_1}$$

$$= \frac{(1 \text{ k}\Omega)(2 \text{ k}\Omega) + (1 \text{ k}\Omega)(5 \text{ k}\Omega) + (2 \text{ k}\Omega)(5 \text{ k}\Omega)}{1 \text{ k}\Omega} = 17 \text{ k}\Omega$$

$$R_C = \frac{R_1 R_2 + R_1 R_3 + R_2 R_3}{R_3}$$

$$= \frac{(1 \text{ k}\Omega)(2 \text{ k}\Omega) + (1 \text{ k}\Omega)(5 \text{ k}\Omega) + (2 \text{ k}\Omega)(5 \text{ k}\Omega)}{5 \text{ k}\Omega} = 3.4 \text{ k}\Omega$$

The resulting delta network is shown in Figure 8–62.

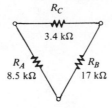

FIGURE 8–62

Review for 8–9

1. Sketch a delta network.
2. Sketch a wye network.

3. Write the formulas for delta-to-wye conversion.
4. Write the formulas for wye-to-delta conversion.

COMPUTER ANALYSIS

8–10   Delta-wye conversion is used in this section as an example for computer analysis. The program listed below converts a specified delta network into a wye network. The resistor labeling conforms to that in Figure 8–57.

```
10 CLS
20 PRINT "THIS PROGRAM CONVERTS A SPECIFIED DELTA NETWORK TO THE"
30 PRINT "CORRESPONDING WYE NETWORK.  REFER TO FIGURE 8-57 FOR"
40 PRINT "THE APPROPRIATE RESISTOR LABELS."
50 PRINT:PRINT:PRINT
60 INPUT "TO CONTINUE PRESS 'ENTER'";X
70 CLS
80 PRINT "PLEASE PROVIDE THE DELTA RESISTOR VALUES WHEN PROMPTED."
90 PRINT:PRINT
100 INPUT "RA IN OHMS";RA
110 INPUT "RB IN OHMS";RB
120 INPUT "RC IN OHMS";RC
130 CLS
140 RT=RA+RB+RC
150 R1=RA*RC/RT
160 R2=RB*RC/RT
170 R3=RA*RB/RT
180 PRINT "THE VALUES FOR THE WYE NETWORK ARE AS FOLLOWS:"
190 PRINT:PRINT "R1=";R1;"OHMS"
200 PRINT "R2=";R2;"OHMS"
210 PRINT "R3=";R3;"OHMS"
```

Review for 8–10

1. Identify the calculation statements.
2. What is the purpose of line 90?

Formulas

$$I_S = \frac{V_S}{R_S} \tag{8–1}$$

$$V_S = I_S R_S \tag{8–2}$$

$$R_{EQ} = \frac{1}{G_T} = \frac{1}{(1/R_1) + (1/R_2) + (1/R_3) + \cdots + (1/R_n)} \tag{8–3}$$

$$V_{EQ} = \frac{(V_1/R_1) + (V_2/R_2) + (V_3/R_3) + \cdots + (V_n/R_n)}{(1/R_1) + (1/R_2) + (1/R_3) + \cdots + (1/R_n)} \tag{8–4}$$

Δ-to-Y conversions:

$$R_1 = \frac{R_A R_C}{R_A + R_B + R_C} \qquad \text{(8–5)}$$

$$R_2 = \frac{R_B R_C}{R_A + R_B + R_C} \qquad \text{(8–6)}$$

$$R_3 = \frac{R_A R_B}{R_A + R_B + R_C} \qquad \text{(8–7)}$$

Y-to-Δ conversions:

$$R_A = \frac{R_1 R_2 + R_1 R_3 + R_2 R_3}{R_2} \qquad \text{(8–8)}$$

$$R_B = \frac{R_1 R_2 + R_1 R_3 + R_2 R_3}{R_1} \qquad \text{(8–9)}$$

$$R_C = \frac{R_1 R_2 + R_1 R_3 + R_2 R_3}{R_3} \qquad \text{(8–10)}$$

Summary

1. An ideal voltage source has zero internal resistance. It provides a constant voltage across its terminals regardless of the load resistance.

2. A practical voltage source has a nonzero internal resistance. Its terminal voltage is essentially constant when $R_L \geq 10R_S$ (rule of thumb).

3. An ideal current source has infinite internal resistance. It provides a constant current regardless of the load resistance.

4. A practical current source has a finite internal resistance. Its current is essentially constant when $10R_L \leq R_S$.

5. The superposition theorem is useful for multiple-source circuits.

6. Thevenin's theorem provides for the reduction of any linear resistive circuit to an equivalent form consisting of an equivalent voltage source in series with an equivalent resistance.

7. The term *equivalency,* as used in Thevenin's and Norton's theorems, means that when a given load resistance is connected to the equivalent circuit, it will have the same voltage across it and the same current through it as when it was connected to the original circuit.

8. Norton's theorem provides for the reduction of any linear resistive circuit to an equivalent form consisting of an equivalent current source in parallel with an equivalent resistance.

9. Millman's theorem provides for the reduction of parallel voltage sources to a single equivalent voltage source consisting of an equivalent voltage and an equivalent series resistance.

10. Maximum power is transferred to a load from a source when the load resistance equals the source resistance.

Self-Test

1. A voltage source has the values $V_S = 25$ V and $R_S = 5$ Ω. What are the values for the equivalent current source?

2. Convert the voltage source in Figure 8–63 to an equivalent current source.

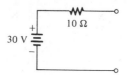

FIGURE 8–63

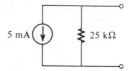

FIGURE 8–64

3. Convert the current source in Figure 8–64 to an equivalent voltage source.

4. In Figure 8–65, use the superposition theorem to find the total current in R_3.

5. In Figure 8–65, what is the total current flowing through R_2?

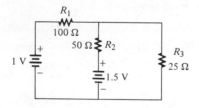

FIGURE 8–65

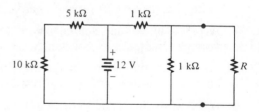

FIGURE 8–66

6. For Figure 8–66, determine the Thevenin equivalent circuit as seen by R.

7. Using Thevenin's theorem, find the current in R_L for the circuit in Figure 8–67.

8. Reduce the circuit in Figure 8–67 to its Norton equivalent as seen by R_L.

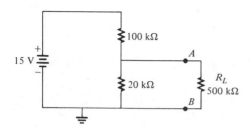

FIGURE 8–67

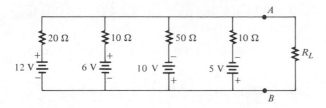

FIGURE 8–68

9. Use Millman's theorem to simplify the circuit in Figure 8–68 to a single voltage source.

10. What value of R_L in Figure 8–68 is required for maximum power transfer?

11. **(a)** Convert the delta network in Figure 8–69A to a wye.
    (b) Convert the wye network in Figure 8–69B to a delta.

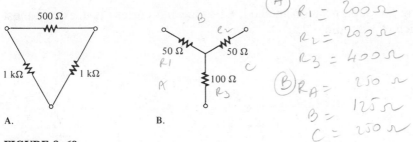

FIGURE 8–69

12. Use the program in Section 8–10 to convert a delta network with R_A = 1.5 kΩ, R_B = 2.2 kΩ, and R_C = 1 kΩ to a wye network.

Problems

Sections 8–1 through 8–3

8–1. A voltage source has the values V_S = 300 V and R_S = 50 Ω. Convert it to an equivalent current source.

8–2. Convert the practical voltage sources in Figure 8–70 to equivalent current sources.

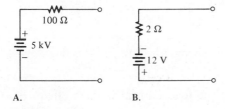

FIGURE 8–70

8–3. A current source has an I_S of 600 mA and an R_S of 1.2 kΩ. Convert it to an equivalent voltage source.

8–4. Convert the practical current sources in Figure 8–71 to equivalent voltage sources.

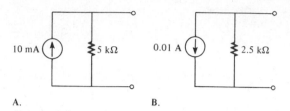

A.   B.

FIGURE 8–71

Section 8–4

8–5. Using the superposition method, calculate the current in the right-most branch of Figure 8–72.

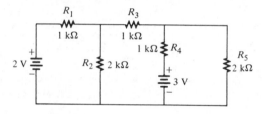

FIGURE 8–72

8–6. Use the superposition theorem to find the current in and the voltage across the R_2 branch of Figure 8–72.

8–7. Using the superposition theorem, solve for the current through R_3 in Figure 8–73.

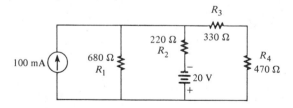

FIGURE 8–73

8–8. Using the superposition theorem, find the load current in each circuit of Figure 8–74.

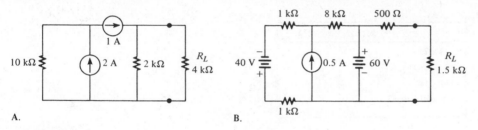

A.

B.

FIGURE 8-74

Section 8-5

8-9. For each circuit in Figure 8-75, determine the Thevenin equivalent as seen by R_L.

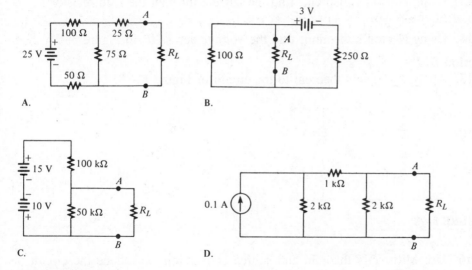

A.

B.

C.

D.

FIGURE 8-75

8-10. Using Thevenin's theorem, determine the current through the load R_L in Figure 8-76.

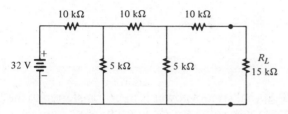

FIGURE 8-76

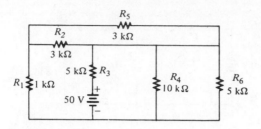

FIGURE 8–77

8–11. Using Thevenin's theorem, find the voltage across R_4 in Figure 8–77.

Section 8–6

8–12. For each circuit in Figure 8–75, determine the Norton equivalent as seen by R_L.

8–13. Using Norton's theorem, find the current through the load resistor R_L in Figure 8–76.

8–14. Using Norton's theorem, find the voltage across R_4 in Figure 8–77.

Section 8–7

8–15. Apply Millman's theorem to the circuit of Figure 8–78.

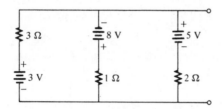

FIGURE 8–78

8–16. Use Millman's theorem and source conversions to reduce the circuit in Figure 8–79 to a single voltage source.

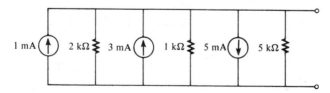

FIGURE 8–79

Section 8–8

8–17. For each circuit in Figure 8–80, maximum power is to be transferred to the load R_L. Determine the appropriate value for R_L in each case.

8–18. Determine R_L for maximum power in Figure 8–81.

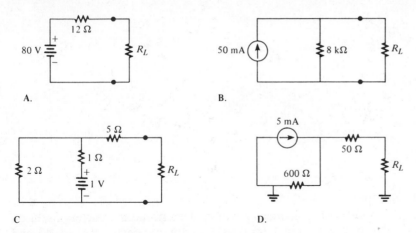

FIGURE 8–80

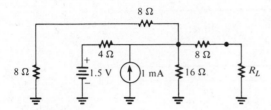

FIGURE 8–81

Section 8–9

8–19. In Figure 8–82, convert each delta network to a wye network.

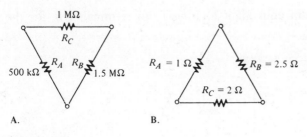

FIGURE 8–82

8–20. In Figure 8–83 (page 272), convert each wye network to a delta network.

Section 8–10

8–21. Change the program in Section 8–10 so that the resistor values can be entered in kilohms and displayed in kilohms.

8–22. Write a program to convert a specified wye network to a corresponding delta network.

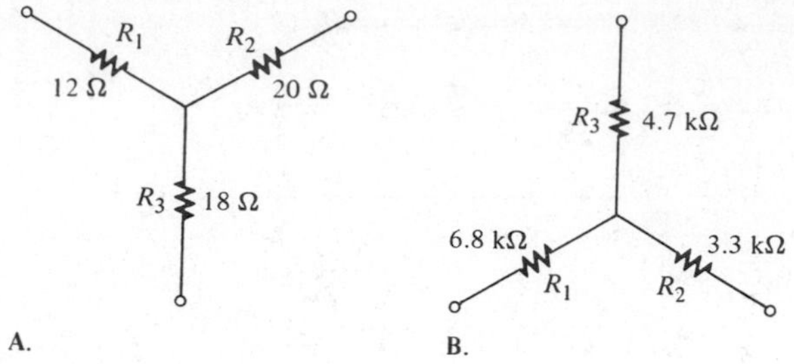

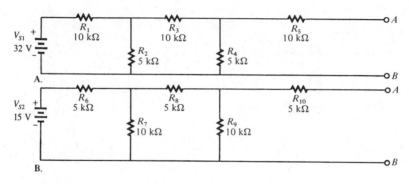

A.   **B.**

FIGURE 8–83

Advanced Problems

8–23. Figure 8–84 shows two ladder networks. Determine the current drain on each of the batteries when terminals A are connected (A to A) and terminals B are connected (B to B).

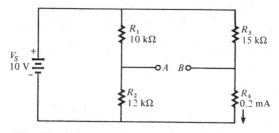

FIGURE 8–84

8–24. Determine the Thevenin equivalent looking from terminals AB for the circuit in Figure 8–85.

FIGURE 8–85

8–25. What are the values of R_4 and R_{TH} when maximum power is transferred from the Thevenized source to the ladder network in Figure 8–86?

8–26. Reduce the circuit between terminals A and B in Figure 8–87 to its Norton equivalent.

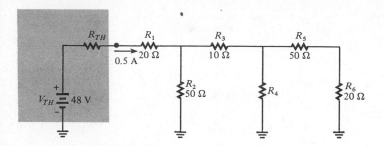

FIGURE 8-86

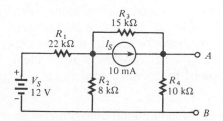

FIGURE 8-87

Answers to Section Reviews

Section 8-1:

1. See Figure 8-88. **2.** See Figure 8-89. **3.** Zero ohms. **4.** Output voltage varies directly with load resistance.

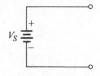

FIGURE 8-88

FIGURE 8-89

Section 8-2:

1. See Figure 8-90. **2.** See Figure 8-91. **3.** Infinite. **4.** Load current varies inversely with load resistance.

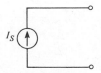

FIGURE 8-90

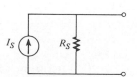

FIGURE 8-91

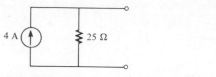

FIGURE 8–92

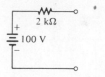

FIGURE 8–93

Section 8–3:
1. $I_S = V_S/R_S$. **2.** $V_S = I_S R_S$. **3.** See Figure 8–92. **4.** See Figure 8–93.

Section 8–4:
1. The total current in any branch of a multiple-source linear circuit is equal to the algebraic sum of the currents due to the individual sources acting alone, with the other sources replaced by their internal resistances. **2.** Because it allows each source to be treated independently. **3.** A short simulates the internal resistance of an ideal voltage source; an open simulates the internal resistance of an ideal current source. **4.** 6.67 mA. **5.** In the direction of the larger current.

Section 8–5:
1. Any linear network can be replaced by an equivalent circuit consisting of an equivalent voltage source and an equivalent series resistance. **2.** V_{TH} and R_{TH}. **3.** See Figure 8–94. **4.** V_{TH} is the open circuit voltage between two terminals in a circuit. **5.** R_{TH} is the resistance as viewed from two terminals in a circuit, with all sources replaced by their internal resistances. **6.** See Figure 8–95.

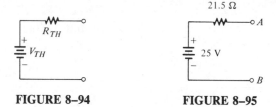

FIGURE 8–94   **FIGURE 8–95**

Section 8–6:
1. Any linear network can be replaced by an equivalent circuit consisting of an equivalent current source and an equivalent parallel resistance. **2.** I_N and R_N. **3.** See Figure 8–96. **4.** I_N is the short circuit current flowing between two terminals in a circuit. **5.** R_N is the resistance as viewed from the two open terminals in a circuit. **6.** See Figure 8–97.

FIGURE 8–96   **FIGURE 8–97**

Section 8–7:
1. Parallel voltage sources. **2.** The equation is as follows:

$$R_{EQ} = \frac{1}{(1/R_1) + (1/R_2) + (1/R_3) + \cdots + (1/R_n)}$$

3. The equation is as follows:

$$V_{EQ} = \frac{(V_1/R_1) + (V_2/R_2) + (V_3/R_3) + \cdots + (V_n/R_n)}{(1/R_1) + (1/R_2) + (1/R_3) + \cdots + (1/R_n)}$$

4. $I_L = 1.08$ A; $V_L = 108$ V.

Section 8–8:
1. Maximum power is transferred from a source to a load when the load resistance is equal to the source resistance. **2.** When $R_L = R_S$. **3.** 50 Ω.

Section 8–9:
1. See Figure 8–98. **2.** See Figure 8–99. **3.** The equations are as follows:

FIGURE 8–98

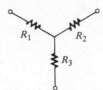

FIGURE 8–99

$$R_1 = \frac{R_A R_C}{R_A + R_B + R_C}$$

$$R_2 = \frac{R_B R_C}{R_A + R_B + R_C}$$

$$R_3 = \frac{R_A R_B}{R_A + R_B + R_C}$$

4. The equations are as follows:

$$R_A = \frac{R_1 R_2 + R_1 R_3 + R_2 R_3}{R_2}$$

$$R_B = \frac{R_1 R_2 + R_1 R_3 + R_2 R_3}{R_1}$$

$$R_C = \frac{R_1 R_2 + R_1 R_3 + R_2 R_3}{R_3}$$

Section 8–10:
1. Lines 140–170. **2.** Double space.

In this chapter, three circuit analysis methods are discussed. These methods are based on Ohm's law and Kirchhoff's voltage and current laws. The methods presented are particularly useful in the analysis of circuits with two or more voltage or current sources and can be used alone or in conjunction with the techniques covered in the previous chapters. Each of these methods can be applied to a given circuit analysis problem. With experience, you will learn which method is best for a particular problem, or you may develop a preference for one of them.

Specifically, you will learn:

1. How to identify loops and nodes in a circuit.
2. How to develop a set of branch current equations and use them to solve for an unknown quantity.
3. How to use determinants in the solution of simultaneous equations.
4. How to develop a set of mesh (loop) equations for a given circuit and use them to solve for an unknown current.
5. How to develop a set of node equations for a given circuit and use them to solve for an unknown voltage.
6. How to use a computer program to find an unknown circuit voltage using the node voltage method.

9
CIRCUIT ANALYSIS METHODS

9–1

In the branch current method, we use Kirchhoff's voltage law and Kirchhoff's current law to solve for the current in each branch of a circuit. Once we know the currents, we can find the voltages.

Loops and Nodes

Figure 9–1 shows a circuit with two voltage sources. It will be used as the basic model throughout the chapter to illustrate each of the circuit analysis methods. In this circuit, there are two *closed loops,* as indicated by arrows 1 and 2. A loop is a complete current path within a circuit. Also, there are four *nodes* in this circuit, as indicated by the letters A, B, C, and D. A node is a junction where two or more current paths come together.

The following are the general steps used in applying the *branch current method*. These steps are demonstrated with aid of Figure 9–2.

1. Assign a current in each circuit branch in an *arbitrary* direction.
2. Show the polarities of the resistor voltages according to the assigned branch current directions.
3. Apply Kirchhoff's voltage law around each closed loop (sum of voltages is equal to zero).
4. Apply Kirchhoff's current law at the *minimum* number of nodes so that *all* branch currents are included (sum of currents at a node equals zero).

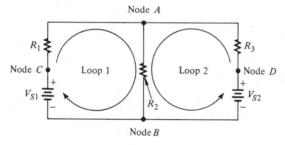

FIGURE 9–1   *Basic multiple-source circuit showing loops and nodes.*

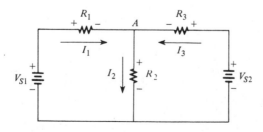

FIGURE 9–2   *Circuit for demonstrating branch current analysis.*

5. Solve the equations resulting from Steps 3 and 4 for the branch current values.

First, the branch currents are assigned in the direction shown. Do not worry about the *actual* current directions at this point.

Second, the polarities of the voltage drops across R_1, R_2, and R_3 are indicated in the figure according to the current directions.

Third, Kirchhoff's voltage law applied to the two loops gives the following equations:

$$R_1 I_1 + R_2 I_2 - V_{S1} = 0 \quad \text{for loop 1}$$

$$R_2 I_2 + R_3 I_3 - V_{S2} = 0 \quad \text{for loop 2}$$

Fourth, Kirchhoff's current law is applied to node A, including all branch currents as follows:

$$I_1 - I_2 + I_3 = 0$$

The negative sign indicates that I_2 flows out of the junction.

Fifth and last, the equations must be solved for I_1, I_2, and I_3. The three equations in the above steps are called *simultaneous equations* and can be solved in two ways: by *substitution* or by *determinants*. Example 9–1 shows how to solve equations by substitution. In the next section, we study the use of determinants and apply them to the later methods.

Example 9–1

Use the branch current method to find each branch current in Figure 9–3.

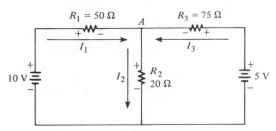

FIGURE 9–3

Solution:

Step 1: Assign branch currents as shown in Figure 9–3. Keep in mind that you can assume any current direction at this point, and the final solution will have a negative sign if the *actual* current is opposite to the assigned current.

Step 2: Mark the polarities of the resistor voltage drops as shown in the figure.

Example 9–1 (continued)

Step 3: Kirchhoff's voltage law around the left loop gives

$$50I_1 + 20I_2 = 10$$

Around the right loop we get

$$20I_2 + 75I_3 = 5$$

Step 4: At node A, the current equation is

$$I_1 - I_2 + I_3 = 0$$

Step 5: The equations are solved by substitution as follows. First find I_1 in terms of I_2 and I_3:

$$I_1 = I_2 - I_3$$

Now substitute $I_2 - I_3$ for I_1 in the first loop equation:

$$50(I_2 - I_3) + 20I_2 = 10$$
$$50I_2 - 50I_3 + 20I_2 = 10$$
$$70I_2 - 50I_3 = 10$$

Next, take the second loop equation and solve for I_2 in terms of I_3:

$$20I_2 = 5 - 75I_3$$
$$I_2 = \frac{5 - 75I_3}{20}$$

Substituting this expression for I_2 into $70I_2 - 50I_3 = 10$, we get the following:

$$70\left(\frac{5 - 75I_3}{20}\right) - 50I_3 = 10$$

$$\frac{350 - 5250I_3}{20} - 50I_3 = 10$$

$$17.5 - 262.5I_3 - 50I_3 = 10$$

$$- 312.5I_3 = -7.5$$

$$I_3 = \frac{7.5}{312.5}$$

$$I_3 = 0.024 \text{ A}$$

Now, substitute this value of I_3 into the second loop equation:

$$20I_2 + 75(0.024) = 5$$

Solve for I_2:

$$I_2 = \frac{5 - 75(0.024)}{20}$$

$$= \frac{3.2}{20} = 0.16 \text{ A}$$

Substituting I_2 and I_3 values into the current equation at node A, we obtain

$$I_1 - 0.16 + 0.024 = 0$$
$$I_1 = 0.16 - 0.024$$
$$= 0.136 \text{ A}$$

Review for 9–1

1. What basic circuit laws are used in the branch current method?
2. When assigning branch currents, you should be careful of the directions (T or F).
3. What is a loop?
4. What is a node?

DETERMINANTS

9–2

When there are several unknown quantities to be found, such as the three currents in the last example, you must have a number of equations equal to the number of unknowns. In this section you will learn how to solve for two unknowns using a systematic method known as *determinants*. This method is an alternate to the substitution method, which we used in the previous section.

Solving Two Simultaneous Equations

To illustrate the method of *second-order determinants*, we will assume two loop equations as follows:

$$10I_1 + 5I_2 = 15$$
$$2I_1 + 4I_2 = 8$$

We want to find the value of I_1 and I_2. To do so, we form a *determinant* with the coefficients of the unknown currents. A *coefficient* is the number associated with an unknown. For example, 10 is the coefficient for I_1 in the first equation.

The first column in the determinant consists of the coefficients of I_1, and the second column consists of the coefficients of I_2. The resulting determinant appears as follows:

$$\begin{vmatrix} 10 & 5 \\ 2 & 4 \end{vmatrix}$$

This is called the *characteristic determinant* for the set of equations.

Next, we form another determinant and use it in conjunction with the characteristic determinant to solve for I_1. We form the determinant for our example by replacing the coefficients of I_1 in the characteristic determinant with the *constants* on the right side of the equations. Doing this, we get the following determinant:

$$\begin{vmatrix} 15 & 5 \\ 8 & 4 \end{vmatrix}$$

We can now solve for I_1 by *evaluating* each determinant and then dividing by the characteristic determinant. To evaluate the determinants, we cross-multiply with appropriate signs and sum the resulting products:

$$\begin{vmatrix} 10 & 5 \\ 2 & 4 \end{vmatrix} = (10)(4) - (2)(5)$$

$$= 40 - 10 = 30$$

The value of this characteristic determinant is 30. Next we cross-multiply the second determinant:

$$\begin{vmatrix} 15 & 5 \\ 8 & 4 \end{vmatrix} = (15)(4) - (8)(5)$$

$$= 60 - 40 = 20$$

The value of this determinant is 20. Now we can solve for I_1 as follows:

$$I_1 = \frac{\begin{vmatrix} 15 & 5 \\ 8 & 4 \end{vmatrix}}{\begin{vmatrix} 10 & 5 \\ 2 & 4 \end{vmatrix}}$$

$$= \frac{20}{30} = 0.667 \text{ A}$$

To find I_2, we form another determinant by substituting the *constants* on the right side of the equations for the coefficients of I_2:

$$\begin{vmatrix} 10 & 15 \\ 2 & 8 \end{vmatrix}$$

We solve for I_2 by dividing this determinant by the characteristic determinant already evaluated:

$$I_2 = \frac{\begin{vmatrix} 10 & 15 \\ 2 & 8 \end{vmatrix}}{30} = \frac{(10)(8) - (2)(15)}{30}$$

$$= \frac{80 - 30}{30} = \frac{50}{30} = 1.67 \text{ A}$$

Three equations with three unknowns can be solved with *third-order determinants*. Since we will limit most of our coverage of analysis methods in this book to those producing only two unknowns, third-order determinants are not covered.

Example 9–2

Solve the following set of equations for the unknown currents:

$$2I_1 - 5I_2 = 10$$
$$6I_1 + 10I_2 = 20$$

Solution:

The characteristic determinant is

$$\begin{vmatrix} 2 & -5 \\ 6 & 10 \end{vmatrix} = (2)(10) - (-5)(6)$$

$$= 20 - (-30) = 20 + 30 = 50$$

Solving for I_1 yields

$$I_1 = \frac{\begin{vmatrix} 10 & -5 \\ 20 & 10 \end{vmatrix}}{50} = \frac{(10)(10) - (-5)(20)}{50}$$

$$= \frac{100 - (-100)}{50} = \frac{200}{50} = 4 \text{ A}$$

Solving for I_2 yields

$$I_2 = \frac{\begin{vmatrix} 2 & 10 \\ 6 & 20 \end{vmatrix}}{50} = \frac{(2)(20) - (6)(10)}{50}$$

$$= \frac{40 - 60}{50} = -0.4 \text{ A}$$

In a circuit problem, the negative sign would indicate that the direction of actual current is opposite to the assigned direction.

Review for 9–2

1. What is a determinant?

2. For what are second-order determinants used?

MESH CURRENT METHOD

9–3   In the mesh current method, we will work with *loop currents* rather than branch currents. As you perhaps realize, a branch current is the *actual* current through a branch. An ammeter in that branch will measure the value of the branch current. Loop currents are abstract or fictitious quantities that are used to make circuit analysis somewhat easier than it is with the branch current method. Keep this in mind as we proceed through this section. The term *mesh* comes from the fact that a multiple-loop circuit, when drawn out, resembles a wire mesh.

A *systematic* method of mesh analysis is listed in the following steps and is illustrated in Figure 9–4, which is the same circuit configuration used in the branch current section. It demonstrates the basic principles well.

1. Assign a current in the *clockwise* (CW) direction around each closed loop. This may not be the actual current direction, but it does not matter. The number of current assignments must be sufficient to include current through all components in the circuit. No *redundant* current assignments should be made.

2. Indicate the voltage drop polarities in each loop based on the *assigned* current directions.

3. Apply Kirchhoff's voltage law around each closed loop. When more than one loop current passes through a component, include its voltage drop.

4. Using substitution or determinants, solve the resulting equations for the loop currents.

First, the loop currents I_1 and I_2 are assigned in the CW direction as shown in the figure. A loop current could be assigned around the outer perimeter of the circuit, but this information would be redundant since I_1 and I_2 already pass through all of the components.

Second, the polarities of the voltage drops across R_1, R_2, and R_3 are shown based on the loop current directions. Notice that I_1 and I_2 flow in opposite directions through R_2 because R_2 is common to both loops. Therefore, two *fictitious* voltage polarities are indicated. In reality, R_2 currents cannot be separated into two parts, but remember that the loop currents are basically abstract quantities used for analysis purposes. The polarities of the voltage sources are fixed and are not affected by the current assignments.

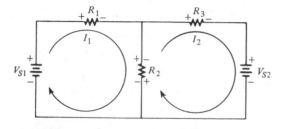

FIGURE 9–4 *Circuit for mesh analysis.*

Third, Kirchhoff's voltage law applied to the two loops results in the following two equations:

$$R_1 I_1 + R_2(I_1 - I_2) = V_{S1} \quad \text{for loop 1}$$
$$R_3 I_2 + R_2(I_2 - I_1) = V_{S2} \quad \text{for loop 2}$$

Fourth, the like terms in the equations are combined and rearranged for convenient solution. The equations are rearranged into the following form. Once the loop currents are evaluated, all of the branch currents can be determined.

$$(R_1 + R_2)I_1 - R_2 I_2 = V_{S1} \quad \text{for loop 1}$$
$$-R_2 I_1 + (R_2 + R_3)I_2 = V_{S2} \quad \text{for loop 2}$$

Notice that only *two* equations are required for the same circuit that required *three* equations in the branch current method. The last two equations follow a certain form which can be used as a *format* to make mesh analysis easier. Referring to these last two equations, notice that for loop 1, the total resistance in the loop, $R_1 + R_2$, is multiplied by I_1 (its loop current). Also in the loop 1 equation, the common resistance R_2 is multiplied by the other loop current I_2 and subtracted from the first term. The same general form is seen in the loop 2 equation. From these observations, a set of rules can be established for applying the same format repeatedly to each loop:

1. Sum the resistances around the loop, and multiply by the loop current.

2. Subtract the common resistance(s) times the adjacent loop current(s).

3. Set the terms in Steps 1 and 2 equal to the total source voltage in the loop. The sign of the source voltage is positive if the assigned loop current flows *out of* its positive terminal. The sign is negative if the loop current flows *into* its positive terminal.

Example 9–3 illustrates the application of these rules to the mesh current analysis of a circuit.

Example 9–3

Using the mesh current method, find the branch currents in Figure 9–5, which is the same circuit as in Example 9–1.

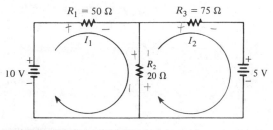

FIGURE 9–5

Example 9–3 (continued)

Solution:

The loop currents are assigned as shown. The format rules are followed for setting up the two equations.

$$(50 + 20)I_1 - 20I_2 = 10 \quad \text{for loop 1}$$
$$70I_1 - 20I_2 = 10$$

$$-20I_1 + (20 + 75)I_2 = -5 \quad \text{for loop 2}$$
$$-20I_1 + \quad\quad 95I_2 = -5$$

Using determinants to find I_1, we obtain

$$I_1 = \frac{\begin{vmatrix} 10 & -20 \\ -5 & 95 \end{vmatrix}}{\begin{vmatrix} 70 & -20 \\ -20 & 95 \end{vmatrix}} = \frac{(10)(95) - (-5)(-20)}{(70)(95) - (-20)(-20)}$$

$$= \frac{950 - 100}{6650 - 400} = 0.136 \text{ A}$$

Solving for I_2 yields

$$I_2 = \frac{\begin{vmatrix} 70 & 10 \\ -20 & -5 \end{vmatrix}}{6250} = \frac{(70)(-5) - (-20)(10)}{6250}$$

$$= \frac{-350 - (-200)}{6250} = -0.024 \text{ A}$$

The negative sign on I_2 means that its direction must be reversed.
Now we find the actual *branch* currents. Since I_1 is the *only* current through R_1, it is also the branch current I_{R1}:

$$I_{R1} = I_1 = 0.136 \text{ A}$$

Since I_2 is the *only* current through R_3, it is also the branch current I_{R3}:

$$I_{R3} = I_2 = 0.024 \text{ A} \quad \text{(opposite direction to } I_2\text{)}$$

Both loop currents I_1 and I_2 flow through R_2 in the *same* direction. Remember, the negative I_2 value told us to reverse its assigned direction.

$$I_{R2} = I_1 - I_2 = 0.136 \text{ A} - (-0.024 \text{ A})$$
$$= 0.16 \text{ A}$$

Keep in mind that once we know the branch currents, we can find the voltages by using Ohm's law.

Circuits with More than Two Loops

The mesh method also can be systematically applied to circuits with any number of loops. Of course, the more loops there are, the more difficult is the solution. However, the basic rules still apply. For example, for a three-loop circuit, three simultaneous equations are required. It is beyond the scope of this book to solve more than two simultaneous equations, but we use Example 9–4 to *set up* the equations for a solution to a three-loop circuit.

Example 9–4

Set up the loop equations for Figure 9–6.

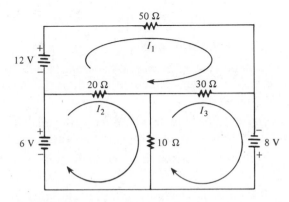

FIGURE 9–6

Solution:

Assign three CW loop currents as shown in the figure. Then use the format rules to write the loop equations. A concise statement of these rules is as follows:

> (Sum of resistors in loop) times (loop current) minus (each common resistor) times (associated adjacent loop current) equals (source voltage in the loop).

$$100I_1 - 20I_2 - 30I_3 = 12 \quad \text{for loop 1}$$
$$-20I_1 + 30I_2 - 10I_3 = 6 \quad \text{for loop 2}$$
$$-30I_1 - 10I_2 + 40I_3 = 8 \quad \text{for loop 3}$$

These three equations can be solved for the currents by substitution or more easily with *third-order determinants*.

Appendix C presents a computer program for solving three equations. The program can also be used to solve for loop currents given the resistor and voltage source values.

Review for 9–3

1. Do the loop currents necessarily represent the actual currents in the branches?
2. When you solve for a loop current and get a negative value, what does it mean?
3. What circuit law is used in the mesh current method?

NODE VOLTAGE METHOD

9–4   Another alternate method of analysis of multiple-source circuits is called the *node voltage method*. It is based on finding the voltages at each node in the circuit using *Kirchhoff's current law*. Remember that a *node* is the junction of two or more current paths.

The general steps for this method are as follows:

1. Determine the number of nodes.
2. Select one node as a *reference*. All voltages will be relative to the reference node. Assign voltage designations to each node where the voltage is unknown.
3. Assign currents at each node where the voltage is unknown, except at the reference node. The directions are arbitrary.
4. Apply Kirchhoff's current law to each node where currents are assigned.
5. Express the current equations in terms of voltages, and solve the equations for the unknown node voltages.

We will use Figure 9–7 to illustrate the general approach to node voltage analysis.

First, establish the nodes. In this case there are *four,* as indicated in the figure.

Second, let us use node B as reference. Think of it as circuit ground. Node voltages C and D are already known to be the source voltages. The voltage at *node A* is the only unknown in this case. It is designated as V_A.

Third, arbitrarily assign the currents at node A as indicated in the figure.

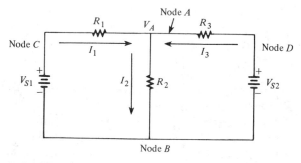

FIGURE 9–7   *Circuit for node voltage analysis.*

Fourth, the Kirchhoff current equation at node A is

$$I_1 - I_2 + I_3 = 0$$

Fifth, express the currents in terms of circuit voltages using Ohm's law as follows:

$$I_1 = \frac{V_1}{R_1} = \frac{V_{S1} - V_A}{R_1}$$

$$I_2 = \frac{V_2}{R_2} = \frac{V_A}{R_2}$$

$$I_3 = \frac{V_3}{R_3} = \frac{V_{S2} - V_A}{R_3}$$

Substituting these into the current equation, we get

$$\frac{V_{S1} - V_A}{R_1} - \frac{V_A}{R_2} + \frac{V_{S2} - V_A}{R_3} = 0$$

The only unknown is V_A; so we can solve the single equation by combining and rearranging terms. Once the voltage is known, all branch currents can be calculated. Example 9–5 illustrates this method further.

Example 9–5

Find the node voltages in Figure 9–8.

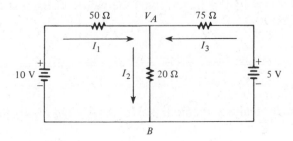

FIGURE 9–8

Solution:

The reference node is chosen at B. The unknown node voltage is V_A, as indicated in the figure. This is the only unknown voltage. Currents are assigned at node A as shown. The current equation is

$$I_1 - I_2 + I_3 = 0$$

Substitution for currents using Ohm's law gives the equation in terms of voltages:

Example 9–5 (continued)

$$\frac{10 - V_A}{50} - \frac{V_A}{20} + \frac{5 - V_A}{75} = 0$$

Solving for V_A yields

$$\frac{10}{50} - \frac{V_A}{50} - \frac{V_A}{20} + \frac{5}{75} - \frac{V_A}{75} = 0$$

$$-\frac{V_A}{50} - \frac{V_A}{20} - \frac{V_A}{75} = -\frac{10}{50} - \frac{5}{75}$$

$$\frac{6V_A + 15V_A + 4V_A}{300} = \frac{30 + 10}{150}$$

$$\frac{25V_A}{300} = \frac{40}{150}$$

$$V_A = \frac{(40)(300)}{(25)(150)}$$

$$= 3.2 \text{ V}$$

Using the same basic procedure, we can solve circuits with more than one unknown node voltage. Example 9–6 illustrates this calculation for two unknown node voltages. Also, the computer program in Appendix C can be used to find two or three unknown node voltages given the values of the resistors and voltage sources in the circuit.

Example 9–6

Using the node analysis method, solve for V_1 and V_2 in the circuit of Figure 9–9.

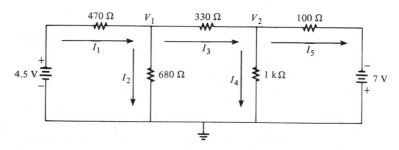

FIGURE 9–9

Solution:

First, the branch currents are assigned as shown in the diagram. Next, Kirchhoff's current law is applied at each node. At node 1,

$$I_1 - I_2 - I_3 = 0$$

Using Ohm's law substitution for the currents, we get

$$\left(\frac{4.5 - V_1}{470}\right) - \left(\frac{V_1}{680}\right) - \left(\frac{V_1 - V_2}{330}\right) = 0$$

$$\frac{4.5}{470} - \frac{V_1}{470} - \frac{V_1}{680} - \frac{V_1}{330} + \frac{V_2}{330} = 0$$

$$\left(\frac{1}{470} + \frac{1}{680} + \frac{1}{330}\right)V_1 - \left(\frac{1}{330}\right)V_2 = \frac{4.5}{470}$$

Now, using the $1/x$ key of the calculator, evaluate the coefficients and constant. The resulting equation for node 1 is

$$0.00663V_1 - 0.00303V_2 = 0.00957$$

At node 2,

$$I_3 - I_4 - I_5 = 0$$

Again using Ohm's law substitution, we get

$$\left(\frac{V_1 - V_2}{330}\right) - \left(\frac{V_2}{1000}\right) - \left(\frac{V_2 - (-7)}{100}\right) = 0$$

$$\frac{V_1}{330} - \frac{V_2}{330} - \frac{V_2}{1000} - \frac{V_2}{100} - \frac{7}{100} = 0$$

$$\left(\frac{1}{330}\right)V_1 - \left(\frac{1}{330} + \frac{1}{1000} + \frac{1}{100}\right)V_2 = \frac{7}{100}$$

Evaluating the coefficients and constant, we obtain the equation for node 2:

$$0.00303V_1 - 0.01403V_2 = 0.07$$

Now, these two node equations must be solved for V_1 and V_2. Using determinants, we get the following solutions:

$$V_1 = \frac{\begin{vmatrix} 0.00957 & -0.00303 \\ 0.07 & -0.01403 \end{vmatrix}}{\begin{vmatrix} 0.00663 & -0.00303 \\ 0.00303 & -0.01403 \end{vmatrix}}$$

$$= \frac{(0.00957)(-0.01403) - (0.07)(-0.00303)}{(0.00663)(-0.01403) - (0.00303)(-0.00303)}$$

$$= -0.928 \text{ V}$$

Example 9–6 (continued)

$$V_2 = \frac{\begin{vmatrix} 0.00663 & 0.00957 \\ 0.00303 & 0.07 \end{vmatrix}}{\begin{vmatrix} 0.00663 & -0.00303 \\ 0.00303 & -0.01403 \end{vmatrix}}$$

$$= \frac{(0.00663)(0.07) - (0.00303)(0.00957)}{(0.00663)(-0.01403) - (0.00303)(-0.00303)}$$

$$= -5.19 \text{ V}$$

Review for 9–4

1. What circuit law is the basis for the node voltage method?

2. What is the reference node?

COMPUTER ANALYSIS

9–5
The program listed below provides for the analysis of a circuit like that in Figure 9–7 for one unknown node voltage. The program in Appendix C provides for a more complete analysis using the methods presented in this chapter.

```
10 CLS
20 PRINT "THIS PROGRAM COMPUTES THE UNKNOWN NODE VOLTAGE"
30 PRINT "(NODE A) FOR A CIRCUIT OF THE GENERAL FORM OF FIGURE"
40 PRINT "9-7."
50 FOR T=1 TO 3500:NEXT:CLS
60 INPUT "VS1 IN VOLTS";V1
70 INPUT "VS2 IN VOLTS";V2
80 INPUT "R1 IN OHMS";R1
90 INPUT "R2 IN OHMS";R2
100 INPUT "R3 IN OHMS";R3
110 CLS
120 VA=((V1/R1)+(V2/R3))/((1/R1)+(1/R2)+(1/R3))
130 PRINT "THE VOLTAGE AT NODE A IS";VA;"VOLTS"
```

Review for 9–5

1. Run the program listed above using the values in Figure 9–8. Compare your result with that in Example 9–5.

Summary

1. A loop is a closed current path in a circuit.

2. A node is the junction of two or more current paths.

3. The branch current method is based on Kirchhoff's voltage law and Kirchhoff's current law.

4. A branch current is an *actual* current in a branch.

5. Determinants are used to solve simultaneous equations.

6. The mesh current method is based on Kirchhoff's voltage law.

7. A loop current is not necessarily the actual current in a branch.

8. The node voltage method is based on Kirchhoff's current law.

Self-Test

1. How many loops and nodes are there in the circuit of Figure 9–10?

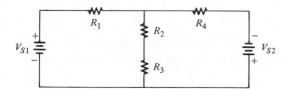

FIGURE 9–10

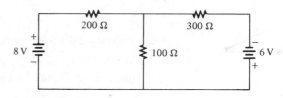

FIGURE 9–11

2. Use the branch current method to find the currents in Figure 9–11.

3. Solve the following two equations by substitution:

$$2I_1 + 5I_2 = 15$$
$$6I_1 - 8I_2 = 10$$

4. Using determinants, solve the following two equations for I_1:

$$15I_1 - 12I_2 = 25$$
$$9I_1 + 3I_2 = 18$$

5. Use the mesh current method to find the current through R_3 in Figure 9–12.

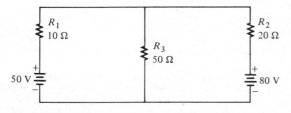

FIGURE 9–12

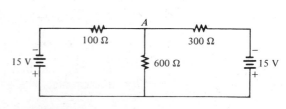

FIGURE 9–13

6. Use the node voltage method to find the voltage at node A in Figure 9–13.

7. Run the program in Section 9–5 using the following circuit values: V_{S1} = 8 V, V_{S2} = 12 V, R_1 = 1 kΩ, R_2 = 680 Ω, and R_3 = 1.5 kΩ.

Problems

Section 9–1

9–1. Identify all *possible* loops in Figure 9–14.

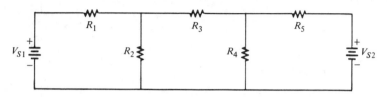

FIGURE 9–14

9–2. Identify all nodes in Figure 9–14. Which ones have a *known* voltage?

9–3. Write the Kirchhoff current equation for the current assignment shown at node A in Figure 9–15.

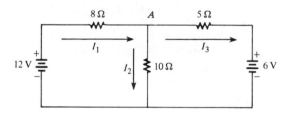

FIGURE 9–15

9–4. Solve for each of the branch currents in Figure 9–15.

9–5. Find the voltage drop across each resistor in Figure 9–15, and indicate its actual polarity.

9–6. Using the substitution method, solve the following set of equations for I_1 and I_2:

$$100I_1 + 50I_2 = 30$$
$$75I_1 + 90I_2 = 15$$

9–7. Using the substitution method, solve the following set of three equations for all currents:

$$5I_1 - 2I_2 + 8I_3 = 1$$
$$2I_1 + 4I_2 - 12I_3 = 5$$
$$10I_1 + 6I_2 + 9I_3 = 0$$

Section 9–2

9–8. Evaluate each determinant.

(a) $\begin{vmatrix} 4 & 6 \\ 2 & 3 \end{vmatrix}$ (b) $\begin{vmatrix} 9 & -1 \\ 0 & 5 \end{vmatrix}$ (c) $\begin{vmatrix} 12 & 15 \\ -2 & -1 \end{vmatrix}$ (d) $\begin{vmatrix} 100 & 50 \\ 30 & -20 \end{vmatrix}$

9–9. Using determinants, solve the following set of equations for both currents:

$$-I_1 + 2I_2 = 4$$
$$7I_1 + 3I_2 = 6$$

Section 9–3

9–10. Using the mesh current method, find the loop currents in Figure 9–16.

9–11. Find the branch currents in Figure 9–16.

9–12. Determine the voltages and their proper polarities for each resistor in Figure 9–16.

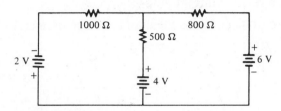

FIGURE 9–16

9–13. Write the loop equations for the circuit in Figure 9–17.

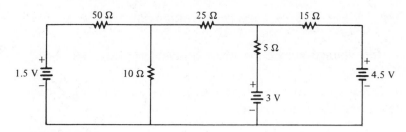

FIGURE 9–17

Section 9–4

9–14. In Figure 9–18, use the node voltage method to find the voltage at point A with respect to point B.

9–15. What are the branch current values in Figure 9–18? Show the actual direction of current in each branch.

9–16. Write the node voltage equations for Figure 9–17.

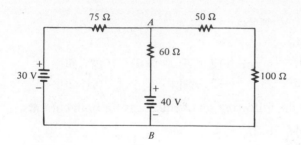

FIGURE 9–18

Section 9–5

9–17. Using the computer program in Appendix C, solve for the loop currents in Problem 9–15.

9–18. Using the computer program in Appendix C, solve for the node voltages in Problem 9–16.

Advanced Problems

9–19. Find the current through each resistor in Figure 9–19.

9–20. In Figure 9–19, determine the voltage across the current source (points A and B).

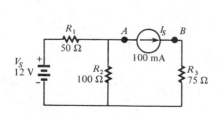

FIGURE 9–19

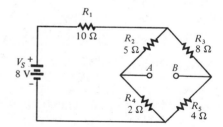

FIGURE 9–20

9–21. Determine the voltage across the open bridge terminals, AB, in Figure 9–20.

9–22. When a 10-Ω resistor is connected from point A to point B in Figure 9–20, how much current flows through it?

Answers to Section Reviews

Section 9–1:

1. Kirchhoff's voltage law and Kirchhoff's current law. **2.** False, but write the equations so that they are consistent with your assigned directions. **3.** A closed path within a circuit. **4.** A junction of two or more current paths.

Section 9–2:
1. The value of a matrix of the coefficients of the unknowns in a set of equations.
2. To solve two equations for two unknowns.

Section 9–3:
1. No. **2.** The direction should be reversed. **3.** Kirchhoff's voltage law.

Section 9–4:
1. Kirchhoff's current law. **2.** The junction to which all circuit voltages are referenced.

Section 9–5:
1. 3.2 V.

Test equipment is indispensable to the electronic technician and technologist. New circuit and systems designs are checked out in the laboratory, and existing equipment is serviced in the shop and in the field, with the aid of a variety of electronic test instruments.

To learn the practical use of test instruments, you must, of course, have actual "hands-on" experience in the laboratory. The intent in this chapter is to present the basics of several types of commonly used dc test instruments and to discuss the fundamentals of troubleshooting circuits.

In this chapter, you will learn:

1. The principles of the moving-coil, the iron-vane, and the electrodynamometer types of meter movements.
2. The definition of current sensitivity.
3. The basics of dc ammeters.
4. The basics of dc voltmeters.
5. How a voltmeter affects a measurement by loading a circuit.
6. How to determine the internal resistance of a voltmeter.
7. The basics of an ohmmeter.
8. The meaning of *back-off scale*.
9. What a multimeter is.
10. What a 3½-digit display is.
11. How to use the various meters in troubleshooting circuits.
12. Some of the common component failures.
13. Precautions to take when making measurements in a circuit.

dc MEASUREMENTS AND TROUBLESHOOTING

10–1

Moving-Coil Movement

In this type of meter movement, known as the *d'Arsonval movement,* the pointer is deflected in proportion to the amount of current through a coil. Figure 10–1A shows a basic d'Arsonval meter movement. It consists of a coil of wire wound on a bearing-mounted assembly that is placed between the poles of a permanent magnet. A pointer is attached to the moving assembly. With no current through the coil, a spring mechanism keeps the pointer at its left-most (zero) position. When current flows through the coil, electromagnetic forces act on the coil, causing a rotation to the right. Electromagnetism is discussed in more detail in a later chapter. The amount of rotation depends on the amount of current. Figure 10–1B shows a construction view of the parts of a typical movement.

Figure 10–2 illustrates how the interaction of magnetic fields produces rotation of the coil assembly. The current flows inward at the "cross" and outward at the "dot" in the single winding shown. The inward current produces a clockwise electromagnetic field that reinforces the permanent magnetic field at the top. The result is a downward force on the right conductor as shown. An upward force is developed on the left side of the coil where the current is outward. These forces produce a clockwise rotation of the coil assembly.

Iron-Vane Movement

This type of movement consists basically of two iron bars placed within a coil. The electromagnetic field produced by the current in the coil induces a north pole and a south pole in the iron bars. The *like* poles *repel* each other, causing the moving element to move away from the stationary element. The attached pointer is deflected by the movement of the element in proportion to current through the coil. A "vane" is attached to the movement and is housed in an air chamber for damping purposes. Figure 10–3A illustrates the basic mechanism, and Part B shows the construction of the basic movement.

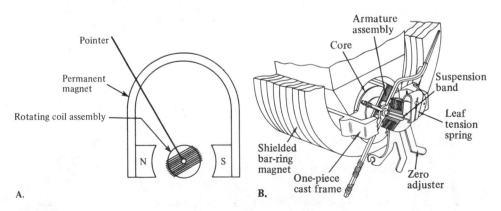

FIGURE 10–1 *Basic d'Arsonval movement.* (**B,** *courtesy of the Triplett Corporation*)

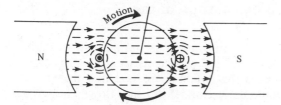

FIGURE 10–2 *Magnetic forces on the coil.*

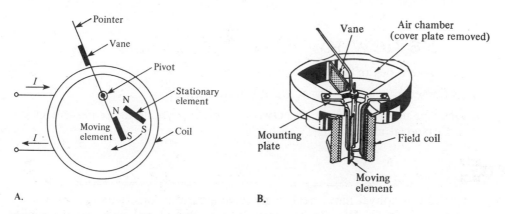

A.

B.

FIGURE 10–3 *Iron-vane movement.* (**B,** *courtesy of the Triplett Corporation*)

Electrodynamometer Movement

Figure 10–4A shows the basic electrodynamometer movement. It differs from the d'Arsonval movement in that it uses an electromagnetic field rather than a permanent magnetic field. The electromagnetic field is produced by current in the stationary coil. The pointer is attached to the moving coil. This type of movement is commonly used in wattmeters. Figure 10–4B shows the construction of a typical movement.

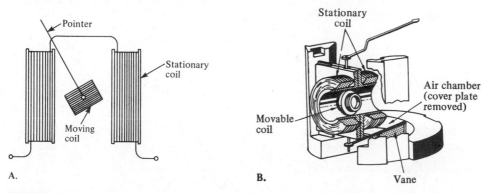

A.

B.

FIGURE 10–4 *Electrodynamometer movement.* (**B,** *courtesy of the Triplett Corporation*)

Current Sensitivity and Resistance of the Meter Movement

The *current sensitivity* of a meter movement is the amount of current required to deflect the pointer full scale (all the way to its right-most position). For example, a 1-mA sensitivity means that when there is 1 mA through the meter coil, the needle is at its maximum deflection. If 0.5 mA flows through the coil, the needle is at the halfway point of its full deflection.

The movement *resistance* is simply the dc resistance of the coil of wire used in the movement.

Review for 10–1

1. List three types of meter movements.

2. Define *current sensitivity* of a meter movement.

3. Describe the basic difference between an electrodynamometer movement and a d'Arsonval movement.

THE AMMETER

10–2   A typical d'Arsonval movement might have a current sensitivity of 1 mA and a resistance of 50 Ω. In order to measure more than 1 mA, additional circuitry must be used with the basic meter movement. Figure 10–5A shows a simple ammeter with a *shunt* (parallel) resistor across the movement. The purpose of the shunt resistor is to bypass current in excess of 1 mA around the meter movement. For example, let us assume that this meter must measure currents of up to 10 mA. Thus, for *full-scale* deflection, the movement must carry 1 mA, and the shunt resistor must carry 9 mA, as indicated in Figure 10–5B.

Determining the Shunt Value

In our example, a proper value of shunt resistance must be used. The following calculations illustrate how this resistance is determined. Since the shunt resistor R_{SH} and the 50-Ω meter movement are in parallel, the voltage drops across them are the same; that is,

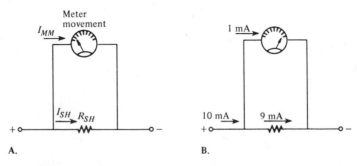

FIGURE 10–5 *Basic ammeter.*

$$V_{SH} = V_{MM}$$

By Ohm's law, $V_{SH} = I_{SH} R_{SH}$ and $V_{MM} = I_{MM} R_{MM}$, and therefore

$$I_{SH} R_{SH} = I_{MM} R_{MM}$$

Solving for R_{SH}, we get

$$R_{SH} = \frac{I_{MM} R_{MM}}{I_{SH}} = \frac{(1 \text{ mA}) (50 \text{ } \Omega)}{9 \text{ mA}} = 5.56 \text{ } \Omega$$

Multiple-Range Ammeter

The example meter just discussed has only one range. It can measure currents from 0 to 10 mA and no higher. However, most practical ammeters have several ranges. Each range must have a different shunt resistance, which is selected with a switch. For example, Figure 10–6 shows a two-range ammeter. A 100-mA range is incorporated with the 10-mA range previously described. When the switch is in the 10-mA position, the meter indicates 10-mA at full-scale deflection of the pointer. When the switch is in the 100-mA position, 100 mA is indicated at full scale.

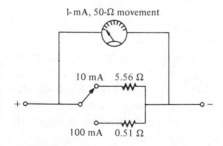

1-mA, 50-Ω movement

10 mA   5.56 Ω

100 mA   0.51 Ω

FIGURE 10–6 *Ammeter with two ranges.*

The value of the 100-mA shunt resistor is determined in the same manner used for the 10-mA shunt. At full scale, the voltage across the movement is $(1 \text{ mA}) (50 \text{ } \Omega) = 50 \text{ mV}$. Therefore, since the shunt must carry 99 mA at full scale, $R_{SH} = 50 \text{ mV}/99 \text{ mA} = 0.51 \text{ } \Omega$. We can obtain other current ranges by switching in appropriate values of shunt resistances.

Example 10–1

Show a three-range basic ammeter that has a 1-A range in addition to the 10-mA and 100-mA ranges of the example meter just discussed.

Solution:

First, we must find the shunt resistance for the 1-A range. Again $V_{SH} = (1 \text{ mA}) (50 \text{ } \Omega) = 50 \text{ mV}$. The shunt resistor must carry all of the 1 A except the 1 mA to operate the movement at full scale. Thus,

Example 10–1 (continued)

$$I_{SH} = 1 \text{ A} - 1 \text{ mA} = 0.999 \text{ A}$$

$$R_{SH} = \frac{50 \text{ mV}}{0.999 \text{ A}} = 0.05 \text{ }\Omega$$

The three-stage meter is shown in Figure 10–7.

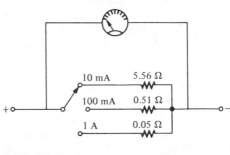

FIGURE 10–7

Effect of the Ammeter on the Circuit

As you know, an ammeter is connected in *series* to measure the current in a circuit. Ideally, the meter should not alter the current that it is intended to measure. In practice, however, the meter unavoidably has some effect on the circuit, because its *internal resistance* is connected in series with the circuit resistance. However, in most cases, the meter's internal resistance is so small compared to the circuit resistance that it can be neglected.

The internal resistance of the ammeter is the shunt resistance in parallel with the coil resistance of the movement. In our example meter, it is approximately 0.05 Ω on the 1-A range. Figure 10–8B shows the meter on the 1-A range and connected to measure the current in the circuit of Figure 10–8A. The 0.05-Ω internal resistance (R_{INT}) of the meter is negligible compared to the 100-Ω circuit resistance. Therefore, the meter does not significantly alter the actual circuit

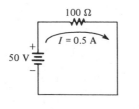

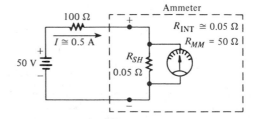

A. Circuit to be measured for current.

B. Ammeter connected in circuit. R_{INT} has negligible effect on I.

FIGURE 10–8 *Measurement of current with an ammeter.*

current. This characteristic is necessary, of course, because we do not want the measuring instrument to change the quantity that is being measured and thus the accuracy of the measurement.

Ammeter Scales

A typical ammeter or milliammeter has more than one scale, each corresponding to different range switch positions. Figure 10–9 shows a two-scale meter as an example. This particular meter has four ranges, as indicated on the range switch in the diagram.

The scales are read in conjunction with the range switch as follows: If the range switch is set at 10 mA or 100 mA, the top scale is used. If the range switch is set at 30 mA or 300 mA, the bottom scale is used. The range switch setting always corresponds to the *full-scale* deflection current. For example, if the range switch is set on 100 mA, the "10" mark on the top scale represents 100 mA.

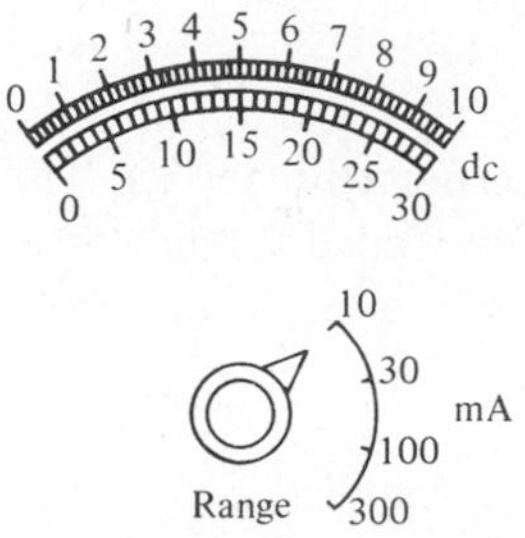

FIGURE 10–9 *Ammeter scale and range switch.*

Review for 10–2

1. What is the purpose of the shunt resistor in an ammeter?

2. A multiple-range ammeter has one shunt for each range (T or F).

3. How is an ammeter connected in a circuit?

THE VOLTMETER

10–3

The voltmeter utilizes the same type of movement as the ammeter. Different external circuitry is added so that the movement will function to measure voltage in a circuit.

As you have seen, the voltage drop across the meter coil is dependent on the current and the coil resistance. For example, a 50-μA, 1000-Ω movement has a full-scale voltage drop of (50 μA)(1000 Ω) = 50 mV. To use the meter to indicate voltages greater than 50 mV, we must add a *series resistance* to drop any additional voltage beyond that which the movement requires for full-scale deflection. This resistance is called the *multiplier resistance* and is designated R_M.

A basic voltmeter is shown in Figure 10–10 with a single multiplier resistor for one range. To make this meter measure 1 V full scale, we must

50-μA, 1000-Ω movement

FIGURE 10–10 *Basic voltmeter with one range.*

determine the value of the multiplier resistance as follows: The movement drops 50 mV at a full-scale current of 50 μA. Therefore, the multiplier resistor R_M must drop the remaining voltage of 1 V − 50 mV = 950 mV. Since R_M is in series with the movement, it also carries 50 μA at full scale. Thus,

$$R_M = \frac{950 \text{ mV}}{50 \text{ } \mu\text{A}} = 19 \text{ k}\Omega$$

Therefore, for 1-V full-scale deflection, the *total resistance* of the voltmeter is 20 kΩ (the multiplier resistance plus the coil resistance).

Voltmeter Sensitivity

Voltmeter sensitivity is defined in terms of resistance per volt (Ω/V). The example meter just discussed has a sensitivity of 20 kΩ/V, because it has a total resistance of 20 kΩ and a full-scale deflection of 1 V. This is a common sensitivity figure for many commercial meters.

Multiple-Range Voltmeter

The meter in Figure 10–10 has only one voltage range (1-V); that is, it can measure voltages from 0 V to 1 V. In order to measure higher voltages with the same movement, additional multiplier resistors must be used. One multiplier resistor is required for each additional range.

For the 50-μA movement, the *total resistance* required is 20 kΩ for each volt of the full-scale reading. In other words, the sensitivity for the 50-μA movement is always 20 kΩ/V *regardless of the range selected*. Thus, the full-scale meter current is 50 μA in *any* range. For any range, we find the *total* meter resistance by multiplying the *sensitivity* by the *full-scale voltage* for that range. For example, for a 10-V range, R_T = (20 kΩ/V)(10 V) = 200 kΩ.

The total resistance for the 1-V range is 20 kΩ; so R_M for the 10-V range must be 200 kΩ − 20 kΩ = 180 kΩ. This *two-range* voltmeter is shown in

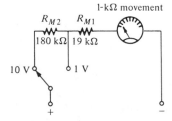

1-kΩ movement

R_{M2} R_{M1}

180 kΩ 19 kΩ

10 V   1 V

+   −

FIGURE 10–11 *Two-range voltmeter.*

Figure 10–11. Additional ranges require the appropriate value of multiplier resistance added in series.

Example 10–2

Show the circuit for a basic voltmeter having 1-V, 10-V, and 100-V ranges.

Solution:

We have already determined R_M for the 1-V and the 10-V ranges. We need only to calculate the additional R_M required for the 100-V range. This calculation is as follows:

$$R_T = (20 \text{ k}\Omega/\text{V})(100 \text{ V}) = 2 \text{ M}\Omega$$

Now, we subtract the meter resistance of the existing two-range meter from 2 MΩ to get the R_M required for the 100-V range:

$$R_{M3} = R_T - R_{M1} - R_{MM}$$
$$= 2 \text{ M}\Omega - 180 \text{ k}\Omega - 19 \text{ k}\Omega - 1 \text{ k}\Omega$$
$$= 2 \text{ M}\Omega - 200 \text{ k}\Omega$$
$$= 1.8 \text{ M}\Omega$$

The schematic for this three-range voltmeter is shown in Figure 10–12.

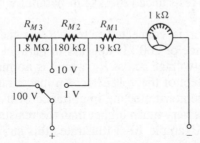

FIGURE 10–12 *Three-range voltmeter.*

Loading Effect of a Voltmeter

As you know, a voltmeter is always connected in parallel with the circuit component across which the voltage is to be measured. Thus, it is much easier to measure voltage than current, because you must break a circuit to insert an ammeter in series. You simply connect a voltmeter across the circuit without disrupting the circuit or breaking a connection.

Since some current is required through the voltmeter to operate the movement, the voltmeter has some effect on the circuit to which it is connected. This effect is called *loading*. However, as long as the meter resistance is much greater

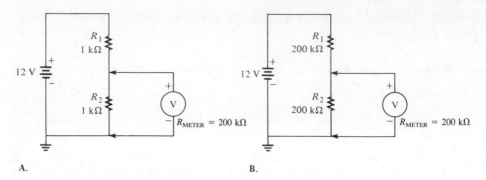

FIGURE 10–13   *Example of loading effect of a voltmeter.*

than the resistance of the circuit across which it is connected, the loading effect is negligible. This characteristic is necessary because we do not want the measuring instrument to change the voltage that it is measuring.

Figure 10–13A shows a simple resistive circuit. Part B of the figure shows the same circuit but with much higher resistor values. Assume that we wish to measure the voltage across R_2 in both circuits with a 20-kΩ/V voltmeter using the 10-V range, because there should be 6 V across the resistor. When the voltmeter is connected across R_2 in the circuit of Part A, the meter's internal resistance of 200 kΩ appears in parallel with the 1 kΩ in the circuit. The voltage across R_2 is still approximately 6 V, because the meter does not significantly affect the circuit resistance (200 kΩ in parallel with 1 kΩ is still approximately 1 kΩ).

When the meter is connected across R_2 in the circuit of Part B, the meter's internal resistance of 200 kΩ appears in parallel with the 200 kΩ in the circuit. In this case, the voltage across R_2, which is normally 6 V, is *reduced* because of the loading effect of the voltmeter. An inaccurate measurement results.

This example demonstrates the importance of using a voltmeter with an internal resistance that is very much higher than the resistance of the circuit across which it is connected. Example 10–3 illustrates this subject further.

Example 10–3

Determine the exact voltage that would be measured with a 20-kΩ/V voltmeter across R_2 in the circuit of Figure 10–13.

Solution:

In the circuit of Part A, the meter is in parallel with the 1-kΩ R_2. The combined resistance of the meter and R_2 is

$$\frac{(200 \text{ k}\Omega)\,(1 \text{ k}\Omega)}{200 \text{ k}\Omega + 1 \text{ k}\Omega} = 0.995 \text{ k}\Omega$$

Using the voltage divider formula, we determine V_2 as follows:

$$V_2 = \frac{0.995 \text{ k}\Omega}{1.995 \text{ k}\Omega}(12 \text{ V}) = 5.985 \text{ V}$$

Without the meter's loading effect, V_2 is 6 V. The meter loading produces an error of 15 mV, which in most cases is not enough to worry about. Figure 10–14 illustrates this situation.

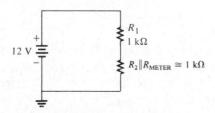

FIGURE 10–14

In the circuit of Part B, the meter is in parallel with the 200-kΩ R_2. The combined resistance of the meter and R_2 is 200 k$\Omega/2$ = 100 kΩ. Using the voltage divider formula, we find V_2 as follows:

$$V_2 = \frac{100 \text{ k}\Omega}{300 \text{ k}\Omega}(12 \text{ V}) = 4 \text{ V}$$

In this situation, the error is 2 V, which is unacceptable. The voltmeter loading is significant, and this voltmeter could not be used in this case. You would have to use a voltmeter with a higher internal resistance (the internal resistance of a meter is usually referred to as *input impedance*). Figure 10–15 illustrates this second case.

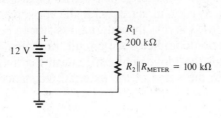

FIGURE 10–15

Voltmeter Scales

Like the ammeter, a typical voltmeter has more than one scale. For example, the voltmeter in Figure 10–16 has four ranges, as indicated on the range switch, and it has two scales. If the range switch is set at 10 V or 100 V, the top

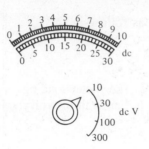

FIGURE 10–16 *Voltmeter scale and range switch.*

scale is used. If the range switch is set at 30 V or 300 V, the bottom scale is used. The range switch setting always corresponds to the *full-scale* voltage. For example, if the range switch is set on 300 V, the "30" mark on the bottom scale represents 300 V.

Review for 10–3

1. What is the purpose of the multiplier resistors in a voltmeter?

2. How many multiplier resistors must a five-scale voltmeter have?

3. The internal resistance (input impedance) of a voltmeter must be large compared to that of the circuit across which it is connected for an accurate measurement (T or F).

THE OHMMETER

10–4
The meter movement used for the ammeter and the voltmeter can also be adapted for use in an ohmmeter. The ohmmeter is used to measure resistance values.

A basic one-range ohmeter is shown in Figure 10–17A. It contains a battery and a variable resistor in series with the movement. To measure resistance, we connect the leads across the external resistor to be measured, as shown in Part B. This connection completes the circuit, allowing the internal battery to produce current through the movement coil, causing a deflection of the pointer (needle) proportional to the value of the external resistance being measured.

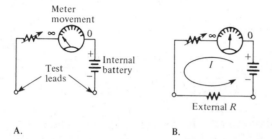

FIGURE 10–17 *Basic ohmmeter circuit.*

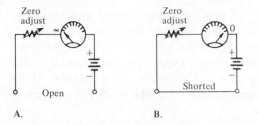

FIGURE 10–18 *Zero adjustment.*

Zero Adjustment

When the ohmmeter leads are open, as in Figure 10–18A, the pointer is at full left scale, indicating *infinite* (∞) resistance (open circuit). When the leads are *shorted,* as in Figure 10–18B, the pointer is at full right scale, indicating *zero* resistance.

The purpose of the variable resistor is to adjust the current so that the pointer is at exactly zero when the leads are shorted. It is used to compensate for changes in the internal battery voltage due to aging.

Ohmmeter Scales

Figure 10–19 shows one type of ohmmeter scale. Between zero and infinity (∞), the scale is marked to indicate various resistor values. Because the values decrease from left to right, this scale is called a *back-off* scale.

Let us assume that a certain ohmmeter uses a 50-μA, 1000-Ω movement and has an internal 1.5-V battery. A current of 50 μA produces a full-scale deflection when the test leads are shorted. To have 50 μA, the *total* ohmmeter resistance is 1.5 V/50 μA = 30 kΩ. Therefore, since the coil resistance is 1 kΩ, the variable zero adjustment resistor must be set at 30 kΩ − 1 kΩ = 29 kΩ.

Suppose that a 120-kΩ resistor is connected to the ohmmeter leads. Combined with the 30-kΩ internal meter resistance, the total R is 150 kΩ. The current is 1.5 V/150 kΩ = 10 μA, which is 20% of the full-scale current and which appears on the scale as shown in Figure 10–19. Now, for example, a 45-kΩ resistor connected to the ohmmeter leads results in a current of

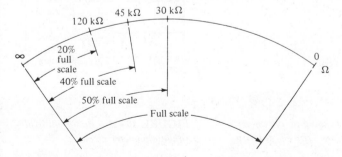

FIGURE 10–19 *Simplified ohmmeter scale illustrating nonlinearity with a few example values.*

1.5 V/75 kΩ = 20 μA, which is 40% of the full-scale current and which is marked on the scale as shown. Additional calculations of this type show that the scale is *nonlinear*. It is more compressed toward the left side than the right side.

The *center scale* point corresponds to an external resistance equal to the internal meter resistance (30 kΩ in this case). The reason is as follows: With 30 kΩ connected to the leads, the current is 1.5 V/60 kΩ = 25 μA, which is half of the full-scale current of 50 μA.

Multiple-Range Ohmmeter

An ohmmeter usually has several ranges. These typically are labeled R×1, R×10, R×100, R×1k, R×10k, R×100k, and R×1M, although some ohmmeters may not have all of the ranges mentioned. These range settings are interpreted differently from those of the ammeter or voltmeter. *The reading on the ohmmeter scale is multiplied by the factor indicated by the range setting.* For example, if the pointer is at 20 on the scale *and* the range switch is set at R×100, the actual resistance measurement is 20×100, or 2 kΩ. This example is illustrated in Figure 10–20 for a typical scale.

To measure small resistance values, you must use a higher ohmmeter current than is needed for measuring large resistance values. Shunt resistors are used to provide multiple ranges on the ohmmeter to measure a range of resistance values from very small to very large. For each range, a different value of shunt resistance is switched in. The shunt resistance increases for higher ohm ranges and is always equal to the center scale reading on any range. In some meters, a higher battery voltage is used for the highest ohm range. A typical circuit is shown in Figure 10–21.

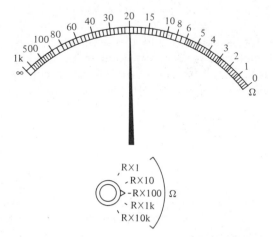

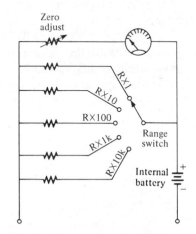

FIGURE 10–20 *Example of reading an ohmmeter scale in conjunction with the range setting (R = 2 kΩ).*

FIGURE 10–21 *Basic multiple-range ohmmeter circuit.*

Review for 10–4

1. Why is an internal battery required in an ohmmeter?
2. What does the term *back-off scale* mean?
3. If the pointer indicates "8" on the ohmmeter scale and the range switch is set at R×1k, what resistance is being measured?

MULTIMETERS

10–5

Generally, the ammeter, voltmeter, and ohmmeter functions are combined into a single instrument for economy and convenience. This instrument is called a *multimeter;* some multifunction meters are called volt-ohm-milliammeters, abbreviated VOM.

Two typical multimeters are shown in Figure 10–22. Notice that both meters have a range switch and a function switch for selecting the desired range of current, voltage, or resistance.

Digital Multimeters

Digital multimeters (DMM) and digital voltmeters (DVM) offer the advantages of easier reading and greater accuracy over conventional analog (needle-type) meters and are, therefore, widely used. Many digital multimeters have a four- or five-digit display with accuracies of ±0.01% or better. Often you will see a reference to a $3\frac{1}{2}$-digit meter. This term means that the fourth (left-most) digit is displayed only as a "1" to handle an overflow. Typical digital multimeters and a digital VOM are shown on the next page in Figure 10–23.

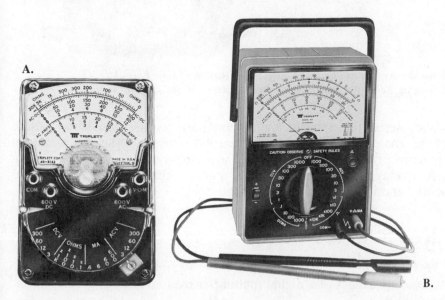

A.

B.

FIGURE 10–22 *Typical portable multimeters. (Courtesy of the Triplett Corporation)*

314

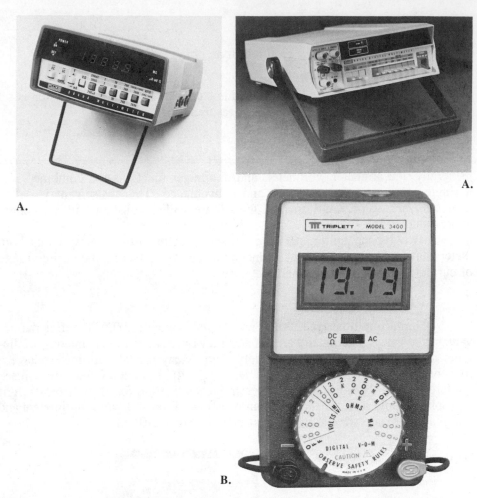

FIGURE 10–23 **A.** *Typical digital multimeters. (Courtesy of John Fluke Mfg. Co.,
Inc.)* **B.** *Typical digital volt-ohm-milliammeter. (Courtesy of the Triplett Corporation)*

The internal principles of digital meters are considerably different from
those of the analog meter previously discussed. However, the use of digital
meters in terms of how they are connected in a circuit is the same as for analog
meters. A coverage of DVMs and DMMs requires some background in digital
fundamentals and is beyond the scope of this book.

Review for 10–5

1. List the measurement functions found on the typical multimeter.

2. Name two advantages of a digital multimeter over an analog (needle) meter.

3. How many display positions does a $3\frac{1}{2}$-digit meter have?

10–6

Troubleshooting has been discussed in several of the previous chapters. We now examine this topic further in the light of what you have learned about the dc meters covered in this chapter.

Definition of Troubleshooting

Troubleshooting can be defined as the *process of recognizing the symptoms of a malfunction, identifying the possible causes, and locating the failed component or components using a systematic procedure.* In order to be an effective troubleshooter, you must understand the basic operation of the circuit or system on which you are working, and you must know how to use the test equipment required to do the job.

Component Failure

Effective troubleshooting requires a familiarity with possible failure modes of the components in an electrical or electronic system. For example, resistors most always become open or their resistance changes to a much higher value when they fail, incandescent lamps open, and dc power supplies or batteries produce either insufficient voltage or no voltage at all.

A resistor can burn out and open if it is dissipating more power than it is rated for. Also, a resistor that is operated well below its power rating is less likely to fail than one operated at or near its rated value.

Checking Power with a Voltmeter

The power in any resistor in a circuit under test can be established by measuring the voltage across the resistor as shown in Figure 10–24A. Using the formula $P = V^2/R$, the power can be calculated from the measured voltage and the known value of the resistor. If the resistance value is not known, the power can be determined from voltage and current measurements, as shown in Figure 10–24B, using the formula $P = VI$. Ideally, a resistor should be operated at no more than one-half of its power rating to assure long life.

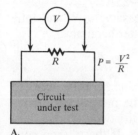

A.

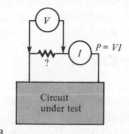

B.

FIGURE 10–24 *Checking the power in a resistor.*

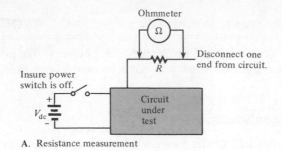

A. Resistance measurement

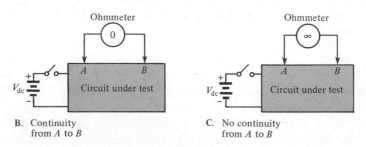

B. Continuity   C. No continuity
   from *A* to *B*   from *A* to *B*

FIGURE 10–25   *Resistance and continuity checks (power to circuit must be off).*

Checking a Resistance Value

If you suspect that the value of a resistor is not the same as its color code or labeling indicates, it can be checked with an ohmmeter or the ohmmeter function of a multimeter.

In order to check accurately the value of a resistor, it *must be disconnected from the circuit*. This prevents possible damage to the ohmmeter due to any voltage source in the circuit under test; it also prevents an inaccurate measurement caused by any other resistive elements that may appear in parallel with the resistor in question.

Two general rules should be followed when making a resistance measurement in a circuit: (1) disconnect the circuit from the power supply; and (2) disconnect the resistor from the circuit by removing at least one of its leads from its circuit connection. This is illustrated in Figure 10–25A.

If a very precise resistance measurement is necessary, a Wheatstone bridge can be used instead of an ohmmeter to provide greater accuracy.

Continuity Check

In addition to resistance measurements, the ohmmeter can be used to check *continuity* from one point to another in a circuit. The purpose of a continuity check is to see if a direct connection exists between the two given points. If there is a direct connection, the ohmmeter indicates zero. If there is no connection or there is a resistive connection, the ohmmeter indicates either infinity or a finite resistance value. This is illustrated in Figure 10–25B and C.

Using the Voltmeter and Ammeter

You already know the basic operation of the voltmeter and ammeter and how they are used to make circuit measurements. The following comments should be noted:

1. When connecting meters in a circuit under test, *always* turn the power off first.

2. Always connect the dc ammeter in *series* with the component through which current is to be measured. This requires breaking the circuit and inserting the meter. The positive terminal must go toward the most positive side of the circuit.

3. Always connect the dc voltmeter in *parallel* with the component across which the voltage is to be measured. The positive terminal must go toward the most positive side of the circuit.

4. Set the range switches to ranges higher than the anticipated current or voltage in order to prevent "pegging" the meter when power is turned back on.

5. Turn the power on and adjust the range switches to get the most accurate measurement.

Example 10–4

This example illustrates a simple troubleshooting procedure on a resistive voltage-divider pc board. A schematic diagram and a pc board layout are shown in Figure 10–26. Test points 1, 2, 3, 4, etc., are labeled on both schematic and board for ease of location.

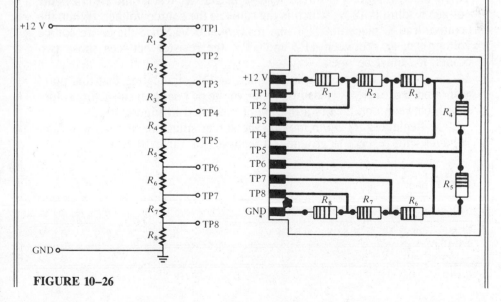

FIGURE 10–26

Example 10–4 (continued)

The circuit has malfunctioned. Determine which component or components have failed based on the observed measurements in Figure 10–27 and your analysis of the circuit.

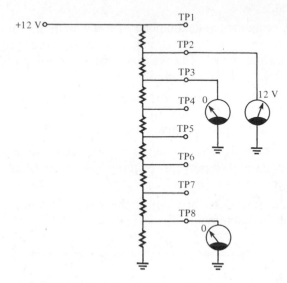

FIGURE 10–27

Solution:

When the voltmeter's positive lead is placed at Test Point 2 (TP2), the voltage reading is 12 V, which is the same as the source voltage. When the positive lead is placed at TP3, the reading is 0 V. Since the entire source voltage appears between TP2 and TP3, the resistor between those two points, R_2, must be open.

Next, a 0-V reading is obtained at TP8, indicating that this point is shorted to ground. A close inspection of the pc board reveals a tiny solder splash between conductors at the point indicated in Figure 10–26.

Replacing R_2 and removing the solder splash results in the proper voltages with respect to ground, as recorded in Table 10–1.

TABLE 10–1

TP1	TP2	TP3	TP4	TP5	TP6	TP7	TP8
12 V	10.93 V	8.57 V	5.04 V	6.96 V	3.45 V	1.1 V	0.5 V

Review for 10–6

1. To check a resistance value with an ohmmeter, the resistor must first be _____.

2. Before connecting any meters to a circuit, one should _____
_____.

3. An open resistor is indicated by an ohmmeter reading of _____.

Summary

1. Three types of meter movements are the d'Arsonval, the iron-vane, and the electrodynamomctcr.

2. A basic ammeter uses a shunt resistor in parallel with the movement for each range setting.

3. An ammeter has a very low internal resistance.

4. An ammeter is always connected in series with the component in which the current is to be measured.

5. A voltmeter uses a multiplier resistor in series with the movement for each range setting.

6. A voltmeter has a very high internal resistance.

7. A voltmeter is always connected in parallel with the element across which the voltage is to be measured.

8. An ohmmeter uses an internal battery to produce current through the meter movement when an external resistance is connected across its terminals.

9. Multimeters combine the voltmeter, ammeter, and ohmmeter functions in one package.

10. Troubleshooting is the process of recognizing the symptoms of a malfunction, identifying possible causes, and locating the faulty component or components by a systematic procedure.

Self-Test

1. If there are 25 μA through a 50-μA meter movement, how much will the pointer be deflected?

2. An ammeter is to measure a full-scale current of 1 mA. What shunt resistance is required for a 1-mA movement? For a 50-μA, 1-kΩ movement?

3. A voltmeter has a sensitivity of 20,000 Ω/V. What is its internal resistance on the 100-V range?

4. What multiplier resistance is used in the meter in Problem 3 on the 100-V range? The resistance of the movement is 1 kΩ.

5. The voltmeter in Figure 10–28 has a sensitivity of 20,000 Ω/V. The range switch is set to the 1-V position in order to measure the voltage across R_2. What voltage is indicated by the meter? What is the actual voltage across R_2 with the voltmeter disconnected? Explain the difference.

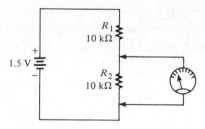

FIGURE 10–28

6. How much resistance is the ohmmeter in Figure 10–29 measuring?

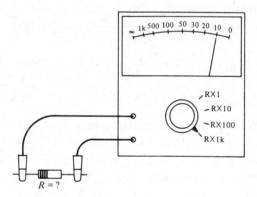

FIGURE 10–29

7. From the indicated measurements in Figure 10–30, locate the faulty component or components, if any.

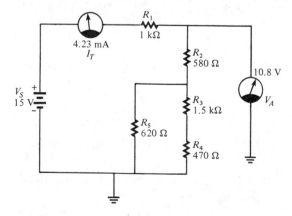

FIGURE 10–30

8. Repeat Problem 7 for Figure 10–31.

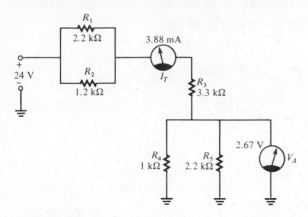

FIGURE 10–31

Problems

Section 10–1

10–1. A certain milliammeter has a 10-mA sensitivity. At what scale position is the pointer with 10 mA through the movement? With 5 mA through the movement?

10–2. How much current is required to produce a full-scale deflection in a meter with a sensitivity of 100 μA?

Section 10–2

10–3. A 50-μA, 1000-Ω movement is used in an ammeter. Determine the full-scale shunt current (I_{SH}) on each of the following ranges:
   (a) 100 μA   **(b)** 1 mA   **(c)** 10 mA   **(d)** 100 mA   **(e)** 1 A

10–4. Repeat Problem 10–3 for a 1-mA, 50-Ω movement.

10–5. Calculate the shunt resistor value (R_{SH}) for each range in Problem 10–3.

10–6. Design a four-range ammeter using a 100-μA, 50-Ω movement. The ranges are to be 100 μA, 1 mA, 10 mA, and 100 mA.

Section 10–3

10–7. Calculate the multiplier resistor values for a 20,000-Ω/V voltmeter with a 1000-Ω movement on the following ranges: 0.1 V, 1 V, 5 V, 10 V, and 50 V.

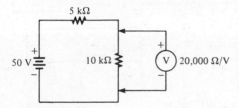

FIGURE 10–32

10–8. What range setting would you use to measure the voltage in Figure 10–32? How much is the error due to the loading effect of the voltmeter?

10–9. What are the voltage readings in Figure 10–33?

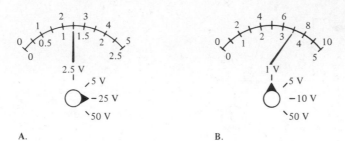

A.   B.

FIGURE 10–33

Section 10–4

10–10. An ohmmeter uses a 50-μA, 1000-Ω movement. It has a 3-V internal battery. What is the value of resistance that it is measuring when the pointer is at center scale, assuming no internal shunt?

10–11. Determine the resistance indicated by each of the following ohmmeter readings and range settings:
(a) Pointer at 2, range setting at R×100.
(b) Pointer at 15, range setting at R×10M.
(c) Pointer at 45, range setting at R×100.

Section 10–5

10–12. A multimeter has the following ranges: 1 mA, 10 mA, 100 mA; 100 mV, 1 V, 10 V; R×1, R×10, R×100. Indicate schematically how you would connect the multimeter in Figure 10–34 to measure the following:
(a) Current through R_1.
(b) Voltage across R_1.
(c) Resistance of R_1.
In each case indicate the *function* on which you would set the meter and the *range* that you would use.

10–13. What voltage reading would you expect on the $3\frac{1}{2}$-digit multimeter in Figure 10–35?

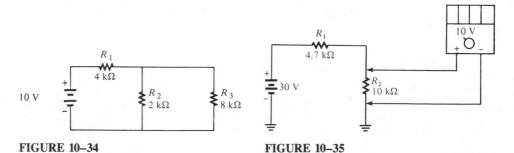

FIGURE 10–34   **FIGURE 10–35**

Section 10–6

10–14. Are the voltage readings in Figure 10–36 correct? If not, what has malfunctioned in the circuit?

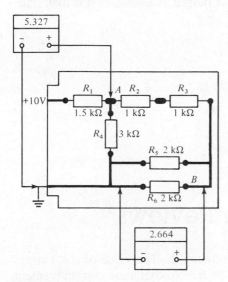

FIGURE 10–36

10–15. A 20,000-Ω/V voltmeter is set on the 10-V range scale. Determine the percent errors for the reading with respect to ground taken at each test point in Figure 10–37 due to the loading effect of the voltmeter.

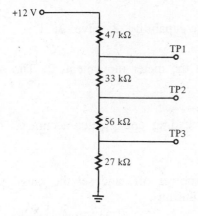

FIGURE 10–37

Advanced Problems

10–16. Using only a voltmeter, show how you would determine by measurement the current, power, and voltage for each resistor on the printed circuit board in Figure 10–36.

10–17. The voltmeter in Figure 10–38 has a sensitivity of 10,000 Ω/V. It is set on the minimum range necessary to measure the voltage at the point to which it is connected. Determine the actual voltage measured by the meter and the current through the meter. The ranges available on this meter are 1 V, 10 V, 25 V, 50 V, and 100 V.

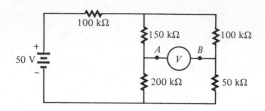

FIGURE 10–38

Answers to Section Reviews

Section 10–1:

1. d'Arsonval, iron-vane, and electrodynamometer. **2.** The value of coil current that produces a full-scale deflection. **3.** The electrodynamometer movement uses an electromagnetic field, whereas the d'Arsonval movement uses a fixed magnetic field.

Section 10–2:

1. The shunt resistor bypasses current (exceeding full-scale current) around the coil. **2.** T. **3.** An ammeter is connected in series.

Section 10–3:

1. The multiplier resistors provide multiple-range capability. **2.** Five. **3.** T.

Section 10–4:

1. The internal battery provides current through the meter movement. **2.** The back-off scale reads from right to left. **3.** 8 kΩ.

Section 10–5:

1. Voltmeter, ammeter, ohmmeter. **2.** Ease of reading and greater accuracy. **3.** Four.

Section 10–6:

1. Disconnected from the circuit. **2.** Turn the power off, and set the range switches to higher than the expected reading. **3.** Infinity.

This chapter provides an introduction to ac circuit analysis in which time-varying electrical signals, particularly the *sine wave,* are presented. An electrical signal, for our purposes, is a voltage or a current that changes in some consistent manner with time. In other words, the voltage or current fluctuates according to a certain pattern called a *wave form.*

Particular emphasis is given to the sine wave because of its basic importance in ac circuit analysis. Other types of wave forms are also introduced, including pulse, triangular, and sawtooth.

Important tools in ac analysis are the phasor concept and complex numbers. These are introduced in this chapter and are used in the following chapters for circuit analysis purposes.

Specifically, in this chapter, you will learn:

1. How to identify a sinusoidal wave form and measure its characteristics.
2. How to relate the frequency and period of a periodic wave form.
3. How to relate the peak, peak-to-peak, rms, and average values of a sine wave.
4. How to use Ohm's law in ac circuits.
5. How to identify points on a sine wave in angular units.
6. How to relate degrees and radians when working in angular units.
7. How to measure the relative phase angle of a sine wave.
8. How to mathematically analyze a sinusoidal wave form.
9. How to use phasors to represent sine waves.
10. How to use the complex number system to represent phasor quantities and to perform arithmetic operations on these quantities.
11. How to identify the characteristics of pulse, triangular, and sawtooth wave forms.

11

ac
CHARACTERISTICS
AND ANALYSIS

11–1 The sine wave is a very common type of alternating current (ac) and alternating voltage. It is also referred to as a sinusoidal wave or, simply, sinusoid. The electrical service provided by the power companies is in the form of sinusoidal voltage and current. In addition, other types of wave forms are composites of many individual sine waves called *harmonics,* as you will see later.

Figure 11–1 shows the general shape of a sine wave, which can be either current or voltage. Notice how the voltage (or current) varies with time. Starting

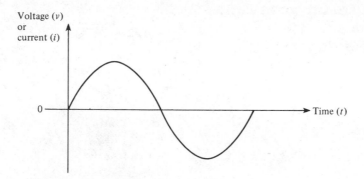

FIGURE 11–1 *Sine wave.*

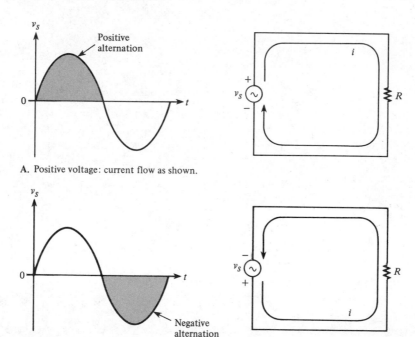

A. Positive voltage: current flow as shown.

B. Negative voltage: current reverses direction.

FIGURE 11–2 *Alternating current and voltage (lower-case v and i represent instantaneous quantities).*

at zero, it *increases* to a positive maximum (peak), returns to zero, and then increases to a *negative* maximum (peak) before returning again to zero.

Polarity

As you have seen, a sine wave changes *polarity* at its zero value; that is, it *alternates* between positive and negative values. When a sine wave voltage is applied to a resistive circuit, as in Figure 11–2, an alternating sine-wave current results. When the voltage changes polarity, the current correspondingly changes direction as indicated.

During the *positive alternation* of the applied voltage V_s, the current is in the direction shown in Figure 11–2A. During a *negative alternation* of the applied voltage, the current is in the opposite direction, as shown in Part B. The combined positive and negative alternations make up one *cycle* of a sine wave.

Review for 11–1

1. Describe one cycle of a sine wave.

2. At what point does a sine wave change polarity?

3. How many maximum points does a sine wave have during one cycle?

4. The maximum points can also be called _____ .

Period

11–2

As you have seen, a sine wave varies with time in a definable manner. Time is designated by t. The time required for a sine wave to complete one full cycle is called the *period,* as indicated in Figure 11–3A. Typically, a sine wave continues to repeat itself in identical cycles, as shown in Part B. Since all cycles of a repetitive sine wave are the same, the period is always a fixed value for a given sine wave.

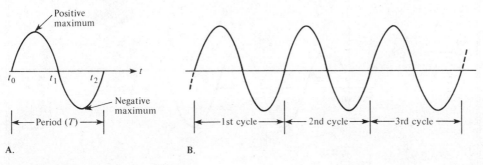

A.

B.

FIGURE 11–3 *Period of a sine wave.*

The period of a sine wave does not necessarily have to be measured between the zero crossings at the beginning and end of a cycle. It can be measured from any point in a given cycle to the *corresponding* point in the next cycle.

Example 11–1

What is the period of the sine wave in Figure 11–4?

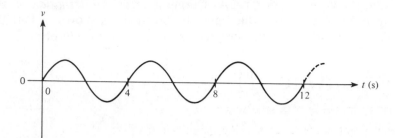

FIGURE 11–4

Solution:

As you can see, it takes four seconds (4 s) to complete each cycle. Therefore, the period is 4 s.

$$T = 4 \text{ s}$$

Example 11–2

Show three possible ways to measure the period of the sine wave in Figure 11–5. How many cycles are shown?

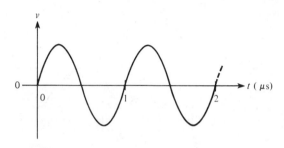

FIGURE 11–5

Solution:

Method 1: The period can be measured from one zero crossing to the corresponding zero crossing in the next cycle.

Method 2: The period can be measured from the positive peak in one cycle to the positive peak in the next cycle.

Method 3: The period can be measured from the negative peak in one cycle to the negative peak in the next cycle.

These measurements are indicated in Figure 11–6, where two cycles of the sine wave are shown. Keep in mind that you obtain the same value for the period no matter which points on the wave form you use.

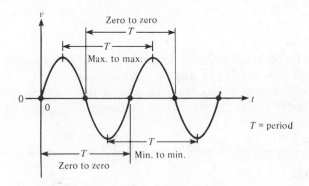

FIGURE 11–6 *Measurement of period.*

Frequency

Frequency is the number of cycles that a sine wave completes in one second. The more cycles completed in one second, the higher the frequency.

Figure 11–7 shows two sine waves. The sine wave in Part A completes two full cycles in one second. The one in Part B completes four cycles in one second. Therefore, the sine wave in Part B has twice the frequency of the one in Part A.

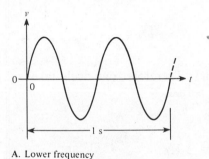

A. Lower frequency

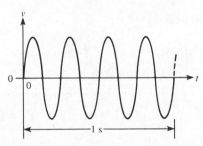

B. Higher frequency

FIGURE 11–7 *Illustration of frequency.*

Frequency is measured in units of *hertz,* abbreviated Hz. One hertz is equivalent to one cycle per second; 60 Hz is 60 cycles per second; and so on. The symbol for frequency is *f.*

Relationship of Frequency and Period

The relationship between frequency and period is very important. The formulas for this relationship are as follows:

$$f = \frac{1}{T} \tag{11-1}$$

$$T = \frac{1}{f} \tag{11-2}$$

There is a *reciprocal* relationship between *f* and *T*. Knowing one, you can calculate the other with the $1/x$ key on your calculator.

This relationship makes sense because a sine wave with a longer period goes through fewer cycles in one second than one with a shorter period.

Example 11–3

Which sine wave in Figure 11–8 has the higher frequency? Determine the period and the frequency of both wave forms.

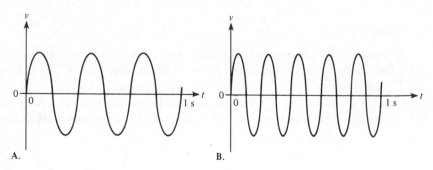

FIGURE 11–8

Solution:

In Part A of the figure, three cycles take 1 s. Therefore, one cycle takes 0.333 s (one-third second), and this is the period.

$$T = 0.333 \text{ s}$$

$$f = \frac{1}{T} = \frac{1}{0.333 \text{ s}} = 3 \text{ Hz}$$

In Part B of the figure, five cycles take 1 s. Therefore, one cycle takes 0.2 s (one-fifth second), and this is the period.

$$T = 0.2 \text{ s}$$

$$f = \frac{1}{T} = \frac{1}{0.2 \text{ s}} = 5 \text{ Hz}$$

Example 11–4

The period of a certain sine wave is 10 milliseconds (10 ms). What is the frequency?

Solution:

Using Equation (11–1), we obtain the following result:

$$f = \frac{1}{T} = \frac{1}{10 \text{ ms}} = \frac{1}{10 \times 10^{-3} \text{ s}}$$

$$= 0.1 \times 10^3 \text{ Hz} = 100 \text{ Hz}$$

Example 11–5

The frequency of a sine wave is 60 Hz. What is the period?

Solution:

Using Equation (11–2), the period is determined as follows:

$$T = \frac{1}{f} = \frac{1}{60 \text{ Hz}} = 0.01667 \text{ s} = 16.67 \text{ ms}$$

Review for 11–2

1. How is the period of a sine wave measured?
2. Define *frequency*, and state its unit.
3. Determine f when $T = 5$ μs.
4. Determine T when $f = 120$ Hz.

VOLTAGE AND CURRENT VALUES

11–3   There are several ways to express the value of a sine wave in terms of its voltage or its current magnitude. These are *instantaneous, peak, peak-to-peak, rms, and average* values.

Instantaneous Value

Figure 11–9 illustrates that at any point in time on a sine wave, the voltage (or current) has an *instantaneous* value. This instantaneous value is different at different points along the curve. Instantaneous values are positive during the positive alternation and negative during the negative alternation. Instantaneous values of voltage and current are symbolized by lower-case v and i, respectively. The curve is shown for voltage only, but it applies equally for current when the v's are replaced with i's.

Peak Value

The peak value of a sine wave is the value of voltage (or current) at the positive or the negative maximum (peaks) with respect to zero. Since the peaks are equal in magnitude, a sine wave is characterized by a single peak value. This is illustrated in Figure 11–10. For a given sine wave, the peak value is constant and is represented by V_p or I_p.

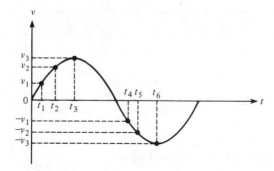

FIGURE 11–9   *Instantaneous values.*

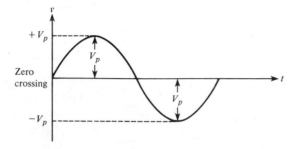

FIGURE 11–10   *Peak values.*

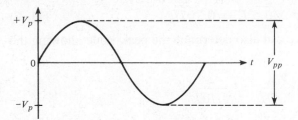

FIGURE 11–11 *Peak-to-peak value.*

Peak-to-Peak Value

The peak-to-peak value of a sine wave, as shown in Figure 11–11, is the voltage or current from the positive peak to the negative peak. Of course, it is always twice the peak value as expressed in the following equations. Peak-to-peak values are represented by the symbols V_{pp} or I_{pp}.

$$V_{pp} = 2V_p \qquad (11\text{--}3)$$

$$I_{pp} = 2I_p \qquad (11\text{--}4)$$

rms Value

The term *rms* stands for *root mean square*. It refers to the mathematical process by which this value is derived (see Appendix D for the derivation). The rms value is also referred to as the *effective value*. Most ac voltmeters display rms voltage. The 120 volts at your wall outlet is an rms value.

The rms value of a sine wave is actually a measure of the *heating effect* of the sine wave. For example, when a resistor is connected across an ac (sine wave) voltage source, as shown in Figure 11–12A, a certain amount of heat is generated by the power in the resistor. Figure 11–12B shows the *same* resistor connected across a dc voltage source. The value of the dc voltage can be adjusted so that the resistor gives off the same amount of heat as it does when connected to the ac source. *The rms value of a sine wave is equal to the dc voltage that produces the same amount of heat as the sinusoidal voltage.*

The peak value of a sine wave can be converted to the corresponding rms value using the following relationships for either voltage or current:

$$V_{rms} = \sqrt{0.5}\ V_p \cong 0.707V_p \qquad (11\text{--}5)$$

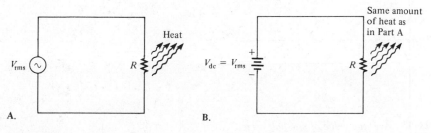

FIGURE 11–12 *rms value.*

$$I_{rms} = \sqrt{0.5}\, I_p \cong 0.707 I_p \qquad (11-6)$$

Using these formulas, we can also determine the peak value knowing the rms value as follows:

$$V_p = \frac{V_{rms}}{0.707} = \left(\frac{1}{0.707}\right) V_{rms}$$

$$V_p = \sqrt{2}\, V_{rms} \cong 1.414 V_{rms} \qquad (11-7)$$

Similarly,

$$I_p = \sqrt{2}\, I_{rms} \cong 1.414 I_{rms} \qquad (11-8)$$

To get the peak-to-peak value, simply double the peak value:

$$V_{pp} = 2.828 V_{rms} \qquad (11-9)$$

and

$$I_{pp} = 2.828 I_{rms} \qquad (11-10)$$

Average Value

The average value of a sine wave taken over *one complete cycle* is always zero, because the positive values (above the zero crossing) offset the negative values (below the zero crossing).

To be useful for comparison purposes, the average value of a sine wave is defined over the positive *half-cycle* rather than over a full cycle. The average value is the total area under the half-cycle curve divided by the distance of the curve along the horizontal axis. Since the derivation is quite complex, it is reserved for Appendix E. The result is expressed in terms of the peak value as follows for both voltage and current:

$$V_{avg} = \left(\frac{2}{\pi}\right) V_p \cong 0.637 V_p \qquad (11-11)$$

$$I_{avg} = \left(\frac{2}{\pi}\right) I_p \cong 0.637 I_p \qquad (11-12)$$

The average value of a sine wave voltage is illustrated in Figure 11–13.

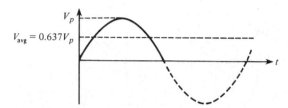

FIGURE 11–13 *Average value of one-half cycle.*

Example 11–6

Determine V_p, V_{pp}, V_{rms}, and V_{avg} for the sine wave in Figure 11–14.

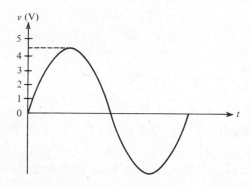

FIGURE 11–14

Solution:

V_p = 4.5 V as taken directly from the graph. From this, the other values are calculated.

$$V_{pp} = 2V_p = 2(4.5 \text{ V}) = 9 \text{ V}$$

$$V_{rms} = 0.707V_p = 0.707(4.5 \text{ V}) = 3.182 \text{ V}$$

$$V_{avg} = 0.637V_p = 0.637(4.5 \text{ V}) = 2.867 \text{ V}$$

Ohm's Law in ac Analysis

When a sinusoidal voltage is applied to a resistive circuit, as shown in Figure 11–15, a sinusoidal current is generated. *Ohm's law can be used in ac circuits such as this just as in dc circuits.* The sine wave values must be consistent; for example, when voltage is expressed in rms, the current must also be rms. Equations (11–13) through (11–17) express Ohm's law in terms of the sine wave values of instantaneous, peak, peak-to-peak, rms, and average:

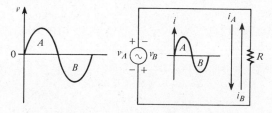

FIGURE 11–15 *Sinusoidal voltage applied to a resistive circuit.*

$$v = iR \qquad\qquad (11\text{--}13)$$

$$V_p = I_p R \qquad\qquad (11\text{--}14)$$

$$V_{pp} = I_{pp} R \qquad\qquad (11\text{--}15)$$

$$V_{\text{rms}} = I_{\text{rms}} R \qquad\qquad (11\text{--}16)$$

$$V_{\text{avg}} = I_{\text{avg}} R \qquad\qquad (11\text{--}17)$$

Review for 11–3

1. Determine V_{pp} when (a) $V_p = 1$ V, (b) $V_{\text{rms}} = 1.414$ V, (c) $V_{\text{avg}} = 3$ V.

2. Determine V_{rms} when (a) $V_p = 2.5$ V, (b) $V_{pp} = 10$ V, (c) $V_{\text{avg}} = 1.5$ V.

3. A sinusoidal voltage with an rms value of 5 V is applied to a circuit with a resistance of 10 Ω. What is the rms value of the current? The peak value of the current?

ANGULAR RELATIONSHIPS OF A SINE WAVE

11–4   As you have seen, sine waves can be measured along the horizontal axis on a time basis; however, since the time for completion of one full cycle or any portion of cycle is frequency-dependent, it is often useful to specify points on the sine wave in terms of an *angular measurement* expressed in degrees or radians. Angular measurement is independent of frequency.

A sine wave voltage can be produced by rotating electromechanical machines. As the rotor of the ac generator goes through a full 360 degrees of rotation, the resulting voltage output is one full cycle of a sine wave. Thus the angular measurement of a sine wave can be related to the angular rotation of a generator (ac generators are discussed in Chapter 20).

Angular Measurement

A *radian* (rad) is defined as the angular distance along the circumference of a circle equal to the radius of the circle. One radian is equivalent to 57.3 degrees, as illustrated in Figure 11–16.

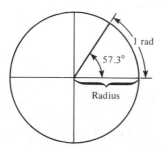

FIGURE 11–16   *Angular measurement showing relationship of radian to degrees.*

TABLE 11–1

Degrees	Radians (rad)
0	0
45	$\pi/4$
90	$\pi/2$
135	$3\pi/4$
180	π
225	$5\pi/4$
270	$3\pi/2$
315	$7\pi/4$
360	2π

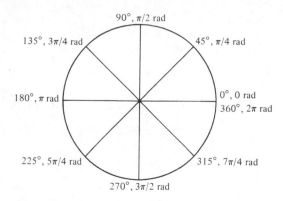

FIGURE 11–17 *Angular measurements.*

In a 360-degree revolution, there are 2π radians; *π is the ratio of the circumference of any circle to its diameter and has a constant value of 3.1416.* Most calculators have a π key so that the actual numerical value does not have to be entered.

Table 11–1 lists several values of degrees and the corresponding radian values. These angular measurements are illustrated in Figure 11–17.

Radian/Degree Conversion

Radians can be converted to degrees using Equation (11–18).

$$\text{rad} = \left(\frac{\pi \text{ rad}}{180°}\right) \text{degrees} \qquad \textbf{(11–18)}$$

Similarly, degrees can be converted to radians with Equation (11–19).

$$\text{degrees} = \left(\frac{180°}{\pi \text{ rad}}\right) \text{rad} \qquad \textbf{(11–19)}$$

Example 11–7

(a) Convert 60 degrees to radians.
(b) Convert $\pi/6$ radians to degrees.

Solution:

(a) $\text{Rad} = \left(\frac{\pi \text{ rad}}{180°}\right)60° = \frac{\pi}{3} \text{ rad} = 1.047 \text{ rad}$

(b) $\text{Degrees} = \left(\frac{180°}{\pi \text{ rad}}\right)\left(\frac{\pi \text{ rad}}{6}\right) = 30°$

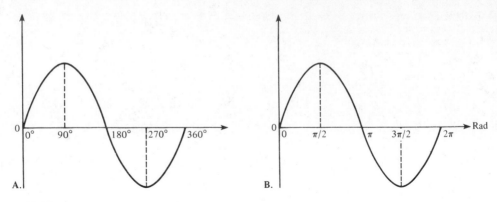

FIGURE 11–18   *Sine wave angles.*

Sine Wave Angles

The angular measurement of a sine wave is based on 360 degrees or 2π radians for a complete cycle. A half-cycle is 180 degrees or π radians; a quarter-cycle is 90 degrees or $\pi/2$ radians; and so on. Figure 11–18A shows angles in degrees for a full cycle of a sine wave; Part B shows the same points in radians.

Phase

The phase of a sine wave is an angular measurement that specifies the position of that sine wave *relative* to a reference. Figure 11–19 shows one cycle of a sine wave to be used as the *reference*. Note that the first positive-going crossing of the horizontal axis (zero crossing) is at 0 degrees (0 rad), and the positive peak is at 90 degrees ($\pi/2$ rad). The negative-going zero crossing is at 180 degrees (π rad), and the negative peak is at 270 degrees ($3\pi/2$ rad). The cycle is completed at 360 degrees (2π rad). When the sine wave is shifted left or right with respect to this reference, there is a *phase shift*.

Figure 11–20 illustrates phase shifts of a sine wave. In Part A of the figure, sine wave *B* is shifted to the right by 90 degrees ($\pi/2$ rad). Thus, there is a *phase angle* of 90 degrees between sine wave *A* and sine wave *B*. In terms of time, the positive peak of sine wave *B* occurs *later* than the positive peak of

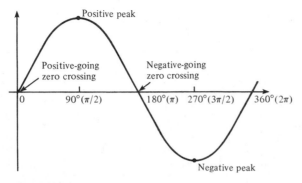

FIGURE 11–19   *Phase reference.*

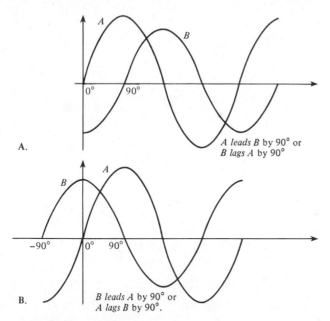

FIGURE 11–20 *Phase shifts.*

sine wave A, because time increases to the right along the horizontal axis. In this case, sine wave B is said to *lag* sine wave A by 90 degrees or $\pi/2$ radians. Stated another way, sine wave A *leads* sine wave B by 90 degrees.

    In Figure 11–20B, sine wave B is shown shifted left by 90 degrees; so, again, there is a phase angle of 90 degrees between sine wave A and sine wave B. In this case, the positive peak of sine wave B occurs earlier in time than that of sine wave A; therefore, sine wave B is said to *lead* by 90 degrees. A sine wave shifted 90 degrees to the left with respect to the reference is called a *cosine* wave.

Example 11–8

What are the phase angles between the two sine waves in Figure 11–21A and B?

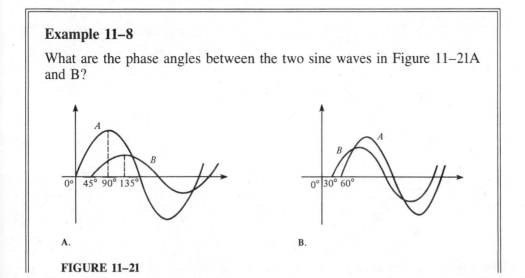

FIGURE 11–21

Example 11–8 (continued)

Solution:

In Part A of the figure, the phase angle is 45 degrees. Sine wave *A* leads sine wave *B* by 45 degrees.

       In Part B, the phase angle is 30 degrees. Sine wave *A* lags sine wave *B* by 30 degrees.

Review for 11–4

1. When the positive-going zero crossing of a sine wave occurs at 0 degrees, at what angle does each of the following points occur?
   (a) Positive peak   **(b)** Negative-going zero crossing
   (c) Negative peak   **(d)** End of first complete cycle

2. A half-cycle is completed in _____ degrees or _____ radians.

3. A full cycle is completed in _____ degrees or _____ radians.

4. Determine the phase angle between the two sine waves in Figure 11–22.

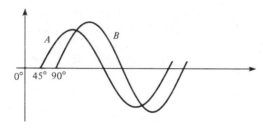

FIGURE 11–22

THE SINE WAVE EQUATION

11–5
    As you have seen, a sine wave can be graphically represented by voltage (current) values on the vertical axis and by angular measurement (degrees or radians) along the horizontal axis. A generalized graph of one cycle of a sine wave is shown in Figure 11–23. The *amplitude, A,* is the maximum value of the voltage or current on the vertical axis, and angular values run along the horizontal axis.

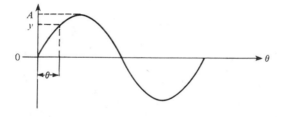

FIGURE 11–23 *One cycle of a sine wave showing amplitude and phase.*

A sine wave curve follows a specific mathematical formula. The general expression for the sine wave curve in Figure 11–23 is

$$y = A \sin \theta \qquad\qquad (11–20)$$

This expression states that any point on the sine wave, represented by an instantaneous value y, is equal to the maximum value times the sine (sin) of the angle θ at that point. For example, a certain voltage sine wave has a peak value of 10 V. The instantaneous voltage at a point 60 degrees along the horizontal axis can be calculated as follows:

$$v = V_p \sin \theta = 10 \sin 60° = 10(0.866) = 8.66 \text{ V}$$

Figure 11–24 shows this particular instantaneous value on the curve. You can find the sine of any angle on your calculator by first entering the value of the angle and then pressing the *sin* key.

Expressions for Shifted Sine Waves

When a sine wave is shifted to the right of the reference by a certain angle, ϕ, as illustrated in Figure 11–25A, the general expression is

$$y = A \sin(\theta - \phi) \qquad\qquad (11–21)$$

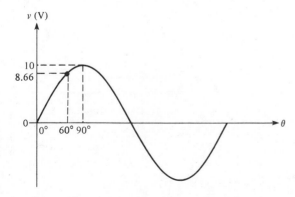

FIGURE 11–24 *Illustration of the instantaneous value at $\theta = 60°$.*

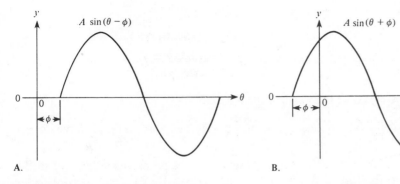

FIGURE 11–25 *Shifted sine waves.*

When a sine wave is shifted to the left of the reference by a certain angle, ϕ, as shown in Figure 11–25B, the general expression is

$$y = A \sin(\theta + \phi) \tag{11–22}$$

Example 11–9

Determine the instantaneous value at the 90-degree reference point on the horizontal axis for each sine wave voltage in Figure 11–26.

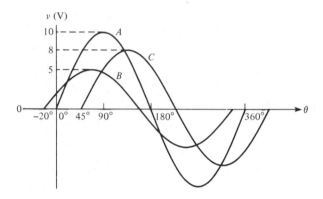

FIGURE 11–26

Solution:

Sine wave A is the reference. Sine wave B is shifted left 20 degrees with respect to A; so it leads. Sine wave C is shifted right 45 degrees with respect to A; so it lags.

$$v_A = V_p \sin \theta$$

$$v_A = 10 \sin 90° = 10(1) = 10 \text{ V}$$

$$v_B = V_p \sin(\theta + \phi_B)$$

$$v_B = 5 \sin(90° + 20°) = 5 \sin 110° = 5(0.9397) = 4.7 \text{ V}$$

$$v_C = V_p \sin(\theta - \phi_C)$$

$$v_C = 8 \sin(90° - 45°) = 8 \sin 45° = 8(0.7071) = 5.66 \text{ V}$$

Review for 11–5

1. Calculate the instantaneous value at 120 degrees for the sine wave in Figure 11–24.

2. Determine the instantaneous value at the 45-degree point of a sine wave shifted 10 degrees to the left of the zero reference ($V_p = 10$ V).

3. Find the instantaneous value at the 90-degree point of a sine wave shifted 25 degrees to the right from the zero reference ($V_p = 5$ V).

Phasors can be used to represent time-varying quantities, such as sine waves, in terms of their magnitude and angular position (phase angle).

Examples of phasors are shown in Figure 11–27. The *length* of the phasor "arrow" represents the magnitude. The angle, θ (relative to 0 degrees), represents the angular position, as shown in Part A. The specific phasor example in Part B has a magnitude of 2 and a phase angle of 60 degrees. The phasor in Part C has a magnitude of 3 and a phase angle of 180 degrees. The phasor in Part D has a magnitude of 1 and a phase angle of −45 degrees.

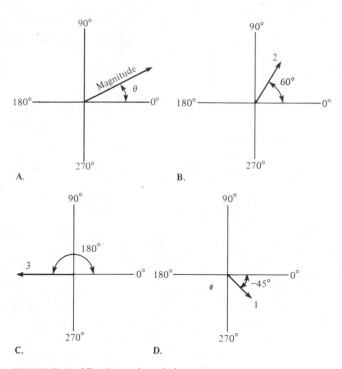

FIGURE 11–27 *Examples of phasors.*

Phasor Representation of a Sine Wave

A full cycle of a sine wave can be represented by rotation of a phasor through 360 degrees. *The instantaneous value of the sine wave at any point is equal to the vertical distance from the tip of the phasor to the horizontal axis.* Figure 11–28 shows how the phasor "traces out" the sine wave as it goes from 0 to 360 degrees. You can relate this concept to the rotation in an ac generator.

Notice in Figure 11–28 that the length of the phasor is equal to the *peak* value of the sine wave (observe the 90-degree and the 270-degree points). The

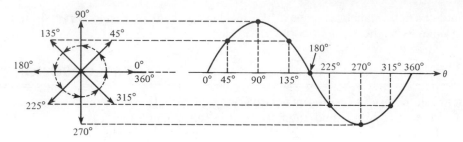

FIGURE 11–28   *Sine wave represented by rotational phasor motion.*

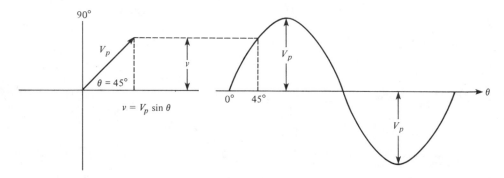

FIGURE 11–29   *Right triangle derivation of sine wave formula.*

angle of the phasor measured from 0 degrees is the corresponding angular point on the sine wave.

Phasors and the Sine Wave Formula

Let's examine a phasor representation at one specific angle. Figure 11–29 shows a voltage phasor at an angular position of 45 degrees and the corresponding point on the sine wave. The instantaneous value of the sine wave at this point is related to both the position and the length of the phasor. As previously mentioned, the vertical distance from the phasor tip down to the horizontal axis represents the instantaneous value of the sine wave at that point.

Notice that when a vertical line is drawn from the phasor tip down to the horizontal axis, a *right triangle* is formed as shown in the figure. The length of the phasor is the *hypotenuse* of the triangle, and the vertical projection is the *opposite side*. From trigonometry, *the opposite side of a right triangle is equal to the hypotenuse times the sine of the angle θ*. In this case, the length of the phasor is the *peak* value of the sine wave voltage, V_p. Thus, the opposite side of the triangle, which is the instantaneous value, can be expressed as $v = V_p \sin \theta$. Recall that this formula is the one stated earlier for calculating instantaneous sine wave values. Of course, this also applies to a current sine wave.

Positive and Negative Phasor Angles

The position of a phasor at any instant can be expressed as a *positive* angle, as you have seen, or as an *equivalent negative angle*. Positive angles are

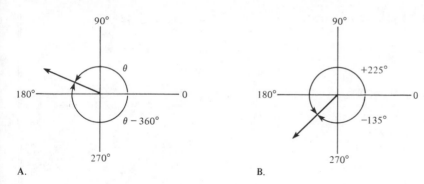

FIGURE 11–30 *Positive and negative phasor angles.*

measured counterclockwise from 0 degrees. Negative angles are measured clock-wise from 0 degrees. For a given positive angle θ, the corresponding negative angle is $\theta - 360°$, as illustrated in Figure 11–30A. In Part B, a specific example is shown. The angle of the phasor in this case can be expressed as +225 degrees or −135 degrees.

Example 11–10

For each phasor in Figure 11–31, determine the instantaneous sine wave value. Also express each positive angle shown as an equivalent negative angle. The length of each phasor represents the peak value of the sine wave.

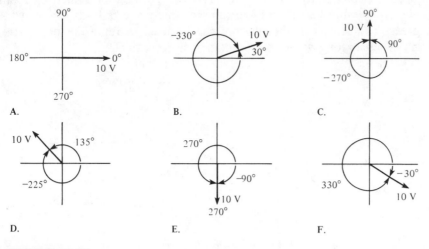

FIGURE 11–31

Solution:

Part A: $\theta = 0°$

$v = 10 \sin 0° = 10(0) = 0$ V

Example 11–10 (continued)

Part B: $\theta = 30° = -330°$

$v = 10 \sin 30° = 10(0.5) = 5$ V

Part C: $\theta = 90° = -270°$

$v = 10 \sin 90° = 10(1) = 10$ V

Part D: $\theta = 135° = -225°$

$v = 10 \sin 135° = 10(0.707) = 7.07$ V

Part E: $\theta = 270° = -90°$

$v = 10 \sin 270° = 10(-1) = -10$ V

Part F: $\theta = 330° = -30°$

$v = 10 \sin 330° = 10(-0.5) = -5$ V

The equivalent negative angles are shown in Figure 11–31.

Phasor Diagrams

A phasor diagram shows the *relative relationship* of two or more sine waves of the same frequency. A phasor in a *fixed* position represents a *complete* sine wave, because once the phase angle between two or more sine waves of the same frequency is established, it remains constant throughout the cycles. For example, the two sine waves in Figure 11–32A can be represented by a phasor diagram, as shown in Part B. As you can see, sine wave *B* leads sine wave *A* by 30 degrees. The length of the phasors can be used to represent peak, rms, or average values as long as the representation is consistent.

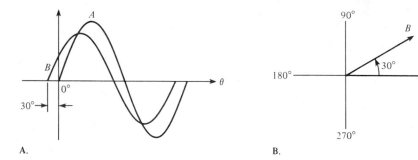

A.   B.

FIGURE 11–32 *Example of a phasor diagram.*

Example 11–11

Use a phasor diagram to represent the sine waves in Figure 11–33.

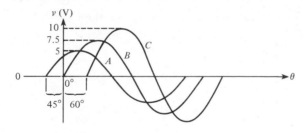

FIGURE 11-33

Solution:

The phasor diagram representing the sine waves is shown in Figure 11-34. In this case, the length of each phasor represents the peak value of the sine wave.

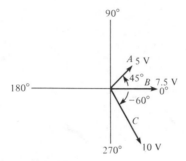

FIGURE 11-34

Angular Velocity of a Phasor

As you have seen, one cycle of a sine wave is traced out when a phasor is rotated through 360 degrees. The faster it is rotated, the faster the sine wave cycle is traced out. Thus, the *period* and *frequency* are related to the *velocity of rotation* of the phasor. The velocity of rotation is called the *angular velocity* and is designated ω (the Greek letter omega).

When a phasor rotates through 360 degrees or 2π radians, one complete cycle is traced out. Therefore, the time required for the phasor to go through 2π radians is the *period* of the sine wave. Because the phasor rotates through 2π radians in a time equal to the period T, the angular velocity can be expressed as

$$\omega = \frac{2\pi}{T}$$

Since $f = 1/T$,

$$\omega = 2\pi f \qquad\qquad \textbf{(11-23)}$$

When a phasor is rotated at a velocity ω, then ωt is the *angular distance through which the phasor has passed at any instant*. Therefore, the following relationship can be stated:

$$\theta = \omega t \qquad\qquad (11\text{--}24)$$

Review for 11–6

1. What is a *phasor*?

2. What is the angular velocity of a phasor representing a sine wave with a frequency of 1500 Hz?

3. A certain phasor has an angular velocity of 628 rad/s. To what frequency does this correspond?

4. Sketch a phasor diagram to represent the two sine waves in Figure 11–35. Use peak values.

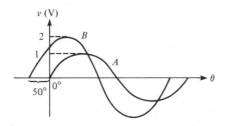

FIGURE 11–35

THE COMPLEX NUMBER SYSTEM

11–7   Complex numbers allow mathematical operations with phasor quantities and are very useful in the analysis of ac circuits. With the complex number system, we can add, subtract, multiply, and divide quantities that have both magnitude and angle, such as sine waves and other ac circuit quantities.

Positive and Negative Numbers

Positive numbers can be represented by points to the right of the origin on the horizontal axis of a graph, and negative numbers can be represented by points to the left of the origin, as illustrated in Figure 11–36A. Also, positive numbers can be represented by points on the vertical axis above the origin, and negative numbers can be represented by points below the origin, as shown in Figure 11–36B.

The Complex Plane

To distinguish between values on the horizontal axis and values on the vertical axis, a *complex plane* is used. In the complex plane, the horizontal axis

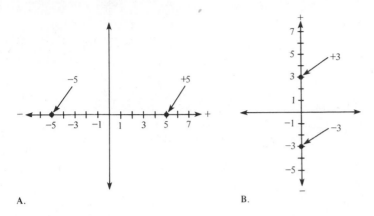

A.   B.

FIGURE 11–36   *Graphical representation of positive and negative numbers.*

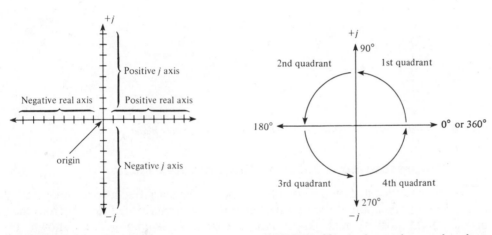

FIGURE 11–37   *The complex plane.*   **FIGURE 11–38**   *Angles on the complex plane.*

is called the *real axis,* and the vertical axis is called the *imaginary axis,* as shown in Figure 11–37.

In electrical work, a $\pm j$ prefix is used to designate numbers that lie on the imaginary axis in order to distinguish them from numbers lying on the real axis. This prefix is known as the *j operator.* In mathematics, an *i* is used instead of a *j,* but in electric circuits, the *i* can be confused with instantaneous current, so *j* is used.

Angular Position on the Complex Plane

Angular positions can be represented on the complex plane, as shown in Figure 11–38. The positive real axis represents 0 degrees. Proceeding counterclockwise, the $+j$ axis represents 90 degrees, the negative real axis represents 180 degrees, the $-j$ axis is the 270-degree point, and, after a full rotation of 360 degrees, we are back to the positive real axis. Notice that the plane is sectioned into four *quadrants.*

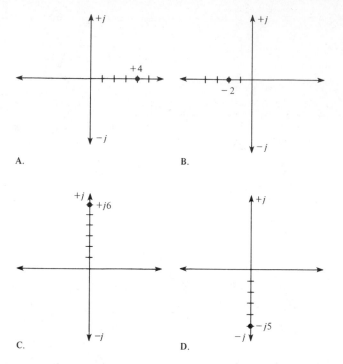

FIGURE 11–39 *Real and imaginary (j) numbers on the complex plane.*

Representing a Point on the Complex Plane

A point located on the complex plane can be classified as real, imaginary ($\pm j$), or a combination of the two. For example, a point located 4 units from the origin on the positive real axis is the *positive real number* $+4$, as shown in Figure 11–39A. A point 2 units from the origin on the negative real axis is the *negative real number,* -2, as shown in Part B. A point on the $+j$ axis 6 units from the origin, as in Part C, is designated $+j6$. Finally, a point 5 units along the $-j$ axis, is designated $-j5$, as in Part D.

When a point lies not on any axis but somewhere in one of the four quadrants, it is a *complex number* and can be defined by its *coordinates*. For example, in Figure 11–40, the point located in the first quadrant has a real value of $+4$ and a j value of $+j4$. The point located in the second quadrant has coordinates -3 and $+j2$. The point located in the third quadrant has coordinates -3 and $-j5$. The point located in the fourth quadrant has coordinates of $+6$ and $-j4$.

Value of j

If we multiply the positive real value of $+2$ by j, the result is $+j2$. This multiplication has effectively moved the $+2$ through a 90-degree angle to the $+j$ axis. Similarly, multiplying $+2$ by $-j$ rotates it -90 degrees to the $-j$ axis.

Mathematically, the j operator has a value of $\sqrt{-1}$. If $+j2$ is multiplied by j, we get

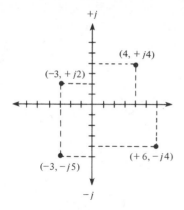

FIGURE 11–40 *Points on the complex plane.*

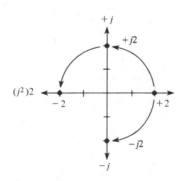

FIGURE 11–41 *Effect of the j operator on location of a number on the complex plane.*

$$j^2 2 = (\sqrt{-1})(\sqrt{-1})(2) = (-1)(2) = -2$$

This calculation effectively places the value on the negative real axis. Therefore, multiplying a positive real number by j^2 converts it to a negative real number, which, in effect, is a rotation of 180 degrees on the complex plane. These operations are illustrated in Figure 11–41.

Review for 11–7

1. Locate the following points on the complex plane:
   (a) $+3$ (b) -4 (c) $+j1$
2. What is the angular difference between the following numbers:
   (a) $+4$ and $+j4$ (b) $+j6$ and -6 (c) $+j2$ and $-j2$

RECTANGULAR AND POLAR FORMS OF COMPLEX NUMBERS

11–8

There are two forms of complex numbers: the *rectangular form* and the *polar form*. Each has certain advantages when used in circuit analysis, depending on the particular application.

As you know, a phasor quantity contains both *magnitude* and *phase*. In this text, italic letters such as V and I are used to represent magnitude only, and boldface letters such as **V** and **I** are used to represent complete phasor quantities. Other circuit quantities that can be expressed in phasor form will be introduced later.

Rectangular Form

A phasor quantity is represented in *rectangular form* by the algebraic sum of the *real value* of the coordinate and the *j value* of the coordinate. An "arrow" drawn from the origin to the coordinate point in the complex plane is used to represent *graphically* the phasor quantity. Examples are $1 + j2$, $5 - j3$,

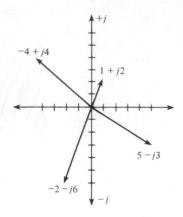

FIGURE 11–42   *Examples of phasors specified by rectangular coordinates.*

$-4 + j4$, and $-2 - j6$, which are shown on the complex plane in Figure 11–42. As you can see, the rectangular coordinates describe the phasor in terms of its values projected onto the real axis and the j axis.

Polar Form

Phasor quantities can also be expressed in *polar form,* which consists of the phasor magnitude and the angular position relative to the positive real axis. Examples are $2\angle45°$, $5\angle120°$, and $8\angle-30°$. The first number is the magnitude, and the symbol $\angle$ precedes the value of the angle. Figure 11–43 shows these phasors on the complex plane. The length of the phasor, of course, represents the magnitude of the quantity. Keep in mind that for every phasor expressed in polar form, there is also an equivalent expression in rectangular form.

Conversion from Rectangular to Polar Form

Most scientific calculators have provisions for conversion between rectangular and polar forms (see Appendix G). However, we discuss the conversion method here so that you will understand the mathematical procedure.

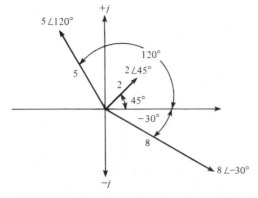

FIGURE 11–43   *Examples of phasors specified by polar values.*

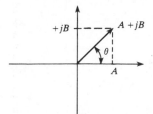

FIGURE 11–44 *Right triangle relationship in the complex plane.*

First, the phasor can be visualized as forming a *right triangle* in the complex plane, as indicated in Figure 11–44. The horizontal side of the triangle is the real value, A, and the vertical side is the j value, B. The hypotenuse of the triangle is the length of the phasor, C, representing the magnitude, and can be expressed as

$$C = \sqrt{A^2 + B^2} \qquad (11\text{--}25)$$

using the Pythagorean theorem.

Next, the angle, θ, indicated in Figure 11–44 is expressed as

$$\theta = \arctan\left(\frac{B}{A}\right) \qquad (11\text{--}26)$$

Note that arctan is INV TAN on a calculator. The general formula for converting from rectangular to polar is as follows:

$$A \pm jB = \sqrt{A^2 + B^2}\angle\pm\arctan\left(\frac{B}{A}\right) = C\angle\pm\theta \qquad (11\text{--}27)$$

The following example illustrates this conversion procedure.

Example 11–12

Convert the following complex numbers from rectangular form to polar form:
(a) $8 + j6$
(b) $10 - j5$
(c) $12 - j18$

Solution:

(a) The magnitude of the phasor represented by $8 + j6$ is

$$C = \sqrt{8^2 + 6^2} = \sqrt{100} = 10$$

The angle is

$$\theta = \arctan\left(\frac{6}{8}\right) = 36.87°$$

Example 11–12 (continued)

The complete polar expression for this phasor is

$$\mathbf{C} = 10\angle 36.87°$$

(b) The magnitude of the phasor represented by $10 + j5$ is

$$\mathbf{C} = \sqrt{10^2 + 5^2} = \sqrt{125} = 11.18$$

The angle is

$$\theta = \arctan\left(\frac{5}{10}\right) = 26.57°$$

The complete polar expression for this phasor is

$$\mathbf{C} = 11.18\angle 26.57°$$

(c) The magnitude of the phasor represented by $12 - j18$ is

$$\mathbf{C} = \sqrt{12^2 + (-18)^2} = \sqrt{468} = 21.63$$

The angle is

$$\theta = \arctan\left(\frac{-18}{12}\right) = -56.31°$$

The complete polar expression for this phasor is

$$\mathbf{C} = 21.63\angle -56.31°$$

Conversion from Polar to Rectangular Form

The polar form gives the magnitude and angle of a phasor quantity, as indicated in Figure 11–45. To get the rectangular form, sides A and B of the triangle must be found using the rules from trigonometry stated below:

$$A = C \cos \theta \qquad\qquad\qquad \textbf{(11–28)}$$

$$B = C \sin \theta \qquad\qquad\qquad \textbf{(11–29)}$$

The general polar-to-rectangular conversion formula is as follows:

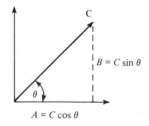

FIGURE 11–45 *Polar components of a phasor.*

$$C\angle\theta = C \cos \theta \pm jC \sin \theta = A \pm jB \qquad (11\text{--}30)$$

The following example demonstrates this conversion.

Example 11–13

Convert the following polar quantities to rectangular form:
(a) $10\angle30°$
(b) $200\angle-45°$
(c) $4\angle135°$

Solution:

(a) The real part of the phasor represented by $10\angle30°$ is

$$A = 10 \cos 30° = 10(0.866) = 8.66$$

The j part of this phasor is

$$jB = j10 \sin 30° = j10(0.5) = j5$$

The complete rectangular expression is

$$A + jB = 8.66 + j5$$

(b) The real part of the phasor represented by $200\angle-45°$ is

$$A = 200 \cos(-45°) = 200(0.707) = 141.4$$

The j part is

$$jB = j200 \sin(-45°) = j200(-0.707) = -j141.4$$

The complete rectangular expression is

$$A + jB = 141.4 - j141.4$$

(c) The real part of the phasor represented by $4\angle135°$ is

$$A = 4 \cos 135° = 4(-0.707) = -2.828$$

The j part is

$$jB = j4 \sin 135° = 4(0.707) = 2.828$$

The complete rectangular expression is

$$A + jB = -2.828 + j2.828$$

Review for 11–8

1. Name the two parts of a complex number in rectangular form.
2. Name the two parts of a complex number in polar form.

3. Convert $2 + j2$ to polar form.

4. Convert $5\angle45°$ to rectangular form.

MATHEMATICAL OPERATIONS WITH COMPLEX NUMBERS

11–9   Addition

Complex numbers must be in rectangular form in order to add them. The rule is: *Add the real parts of each complex number to get the real part of the sum. Then add the j parts of each complex number to get the j part of the sum.*

Example 11–14

Add the following sets of complex numbers.
(a) $8 + j5$ and $2 + j1$
(b) $20 - j10$ and $12 + j6$

Solution:

(a) $(8 + j5) + (2 + j1) = (8 + 2) + j(5 + 1) = 10 + j6$
(b) $(20 - j10) + (12 + j6) = (20 + 12) + j(-10 + 6)$
$$= 32 + j(-4) = 32 - j4$$

Subtraction

As in addition, the numbers must be in rectangular form to be subtracted. The rule is: *Subtract the real parts of the numbers to get the real part of the difference, and subtract the j parts of the numbers to get the j part of the difference.*

Example 11–15

(a) Subtract $1 + j2$ from $3 + j4$.
(b) Subtract $10 - j8$ from $15 + j15$.

Solution:

(a) $(3 + j4) - (1 + j2) = (3 - 1) + j(4 - 2) = 2 + j2$
(b) $(15 + j15) - (10 - j8) = (15 - 10) + j[15 - (-8)] = 5 + j23$

Multiplication

Multiplication of two complex numbers is most easily performed with both numbers in polar form. The rule is: *Multiply the magnitudes, and add the angles algebraically.*

Example 11–16

Perform the following multiplications.
(a) $10\angle 45°$ times $5\angle 20°$.
(b) $2\angle 60°$ times $4\angle -30°$

Solution:

(a) $(10\angle 45°)(5\angle 20°) = (10)(5)\angle(45° + 20°) = 50\angle 65°$
(b) $(2\angle 60°)(4\angle -30°) = (2)(4)\angle[60° + (-30°)] = 8\angle 30°$

Division

Like multiplication, division is done most easily when the numbers are in polar form. The rule is: *Divide the magnitude of the numerator by the magnitude of the denominator to get the magnitude of the quotient, and subtract the denominator angle from the numerator angle to get the angle of the quotient.*

Example 11–17

Perform the following divisions.
(a) Divide $100\angle 50°$ by $25\angle 20°$.
(b) Divide $15\angle 10°$ by $3\angle -30°$.

Solution:

(a) $\dfrac{100\angle 50°}{25\angle 20°} = \left(\dfrac{100}{25}\right)\angle(50° - 20°) = 4\angle 30°$

(b) $\dfrac{15\angle 10°}{3\angle -30°} = \left(\dfrac{15}{3}\right)\angle[10° - (-30°)] = 5\angle 40°$

Application of Complex Numbers to Sine Waves

Since sine waves can be represented by phasors, they can be described in terms of complex numbers in rectangular or polar form. For example, a sine wave with an rms value of 10 V and an angle of 30 degrees with respect to the 0-degree reference can be expressed in polar form as $10\angle 30°$ V. Converting this value to rectangular form, we get 8.66 V $+ j5$ V.

Because sine waves can be represented in complex form, they can be added, subtracted, multiplied, and divided using the rules that we have discussed. Also, as you will learn later, other electrical quantities can be described in complex form to ease many circuit analysis problems.

Review for 11–9

1. Add $1 + j2$ and $3 - j1$.

2. Subtract $12 + j18$ from $15 + j25$.

3. Multiply $8\angle 45°$ times $2\angle 65°$.

4. Divide $30\angle 75°$ by $6\angle 60°$.

NONSINUSOIDAL WAVE FORMS

11–10 Sine waves are very important in electronics, but they are by no means the only type of ac or time-varying wave form. Two other categories are discussed next; these are the *pulse wave form* and the *triangular wave form*.

Pulse Wave Forms

Basically, a pulse can be described as a very rapid transition from one voltage or current level (baseline) to another, and then, after an interval of time, a very rapid transition back to the original baseline level. The transitions in level are called *steps*. An *ideal* pulse consists of two opposite-going steps of equal amplitude. Figure 11–46A shows an ideal positive-going pulse consisting of two equal but opposite instantaneous steps separated by an interval of time called the *pulse width*. Part B of the figure shows an ideal negative-going pulse. The height of the pulse measured from the baseline is its voltage (or current) amplitude.

In many applications, analysis is simplified by treating all pulses as ideal (composed of instantaneous steps and perfectly rectangular in shape). Actual pulses, however, are never ideal. All pulses possess certain characteristics that cause them to be different from the ideal.

In practice, pulses cannot change from one level to another instantaneously. Time is always required for a transition (step), as illustrated in Figure 11–47A. As you can see, there is an interval of time during which the pulse is rising from its lower value to its higher value. This interval is called the *rise time, t_r*. The most widely accepted definition of rise time is *the time required for the pulse to go from 10% of its full amplitude to 90% of its full amplitude*.

The interval of time during which the pulse is falling from its higher value to its lower value is called the *fall time, t_f*. The accepted definition of fall

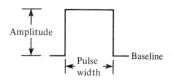

A. Positive-going pulse   B. Negative-going pulse

FIGURE 11–46 *Ideal pulses.*

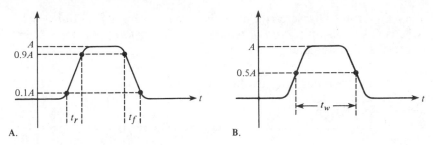

A.   B.

FIGURE 11–47 *Nonideal pulse.*

time is *the time required for the pulse to go from 90% of its full amplitude to 10% of its full amplitude.*

Pulse width also requires a precise definition for the nonideal pulse, because the rising and falling edges are not vertical. The most widely accepted definition of pulse width (t_w) is *the time between the point on the rising edge, where the value is 50% of full amplitude, to the point on the falling edge, where the value is 50% of full amplitude.* Pulse width is shown in Figure 11–47B.

Repetitive Pulses

Any wave form that repeats itself at fixed intervals is *periodic*. Some examples of periodic pulse wave forms are shown in Figure 11–48. Notice that in each case, the pulses repeat at regular intervals. The rate at which the pulses repeat is the pulse repetition rate (PRR) or pulse repetition frequency (PRF), which is the *fundamental frequency* of the wave form. The frequency can be expressed in *hertz* or in *pulses per second*. The time from one pulse to the corresponding point on the next pulse is the period T. The relationship between frequency and period is the same as with the sine wave.

$$\text{PRF} = \frac{1}{T} \qquad\qquad (11\text{--}31)$$

$$T = \frac{1}{\text{PRF}} \qquad\qquad (11\text{--}32)$$

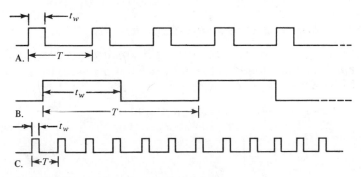

A.

B.

C.

FIGURE 11–48 *Repetitive pulse wave forms.*

362

ac CHARACTERISTICS AND ANALYSIS

A very important characteristic of pulse wave forms is the *duty cycle*. The duty cycle is defined as the *ratio of the pulse width (t_w) to the period T* and is usually expressed as a percentage:

$$\text{percent duty cycle} = \left(\frac{t_w}{T}\right)100\% \qquad (11\text{--}33)$$

Example 11–18

Determine the period, PRF, and duty cycle for the pulse wave form in Figure 11–49.

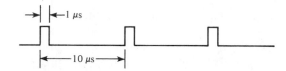

FIGURE 11–49

Solution:

$$T = 10 \ \mu s$$

$$\text{PRF} = \frac{1}{T} = \frac{1}{10 \ \mu s} = 0.1 \text{ MHz}$$

$$\text{percent duty cycle} = \left(\frac{1 \ \mu s}{10 \ \mu s}\right)100\% = 10\%$$

Square Waves

A square wave is a pulse wave form with a duty cycle of 50%. Thus, the pulse width is equal to one-half of the period. A square wave is shown in Figure 11–50.

The Average Value of a Pulse Wave Form

The average value (V_{avg}) of a pulse wave form is equal to its baseline value plus its duty cycle times its amplitude. The lower level of the wave form is taken as the baseline. The formula is as follows:

$$V_{avg} = \text{baseline} + (\text{duty cycle})(\text{amplitude}) \qquad (11\text{--}34)$$

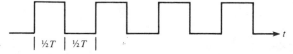

FIGURE 11–50 *Square wave.*

The following example illustrates the calculation of the average value.

Example 11–19

Determine the average value of each of the wave forms in Figure 11–51.

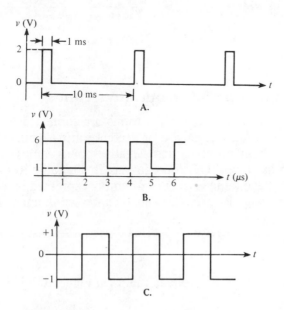

FIGURE 11–51

Solution:

(a) Here the baseline is at 0 V, the amplitude is 2 V, and the duty cycle is 10%. The average value is as follows:

$$V_{avg} = \text{baseline} + (\text{duty cycle})(\text{amplitude})$$
$$= 0 \text{ V} + (0.1)(2 \text{ V}) = 0.2 \text{ V}$$

(b) This wave form has a baseline of +1 V, an amplitude of 5 V, and a duty cycle of 50%. The average value is

$$V_{avg} = \text{baseline} + (\text{duty cycle})(\text{amplitude})$$
$$= 1 \text{ V} + (0.5)(5 \text{ V}) = 1 \text{ V} + 2.5 \text{ V} = 3.5 \text{ V}$$

(c) This is a square wave with a baseline of −1 V and an amplitude of 2 V. The average value is

$$V_{avg} = \text{baseline} + (\text{duty cycle})(\text{amplitude})$$
$$= -1 \text{ V} + (0.5)(2 \text{ V}) = -1 \text{ V} + 1 \text{ V} = 0 \text{ V}$$

This is an *alternating* square wave, and, as with an alternating sine wave, it has an average of zero.

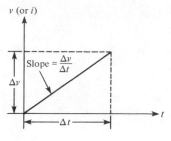

A. Positive ramp

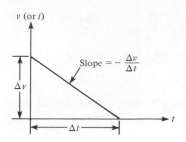

B. Negative ramp

FIGURE 11–52 *Ramps.*

Triangular and Sawtooth Wave Forms

Triangular and sawtooth wave forms are formed by voltage or current *ramps*. A ramp is a *linear* increase or decrease in the voltage or current. Figure 11–52 shows both positive- and negative-going ramps. In Part A of the figure, the ramp has a positive slope; in Part B, the ramp has a negative slope. The *slope* of a voltage ramp is $\pm \Delta v / \Delta t$ and is expressed in units of V / s. The slope of a current ramp is $\pm \Delta i / \Delta t$ and is expressed in units of A / s.

Example 11–20

What are the slopes of the voltage ramps in Figure 11–53?

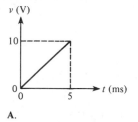

A.

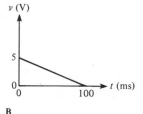

B.

FIGURE 11–53

Solution:

In Part A, the voltage increases from 0 V to +10 V in 5 ms. Thus, $\Delta v = 10$ V and $t = 5$ ms. The slope is

$$\frac{\Delta v}{\Delta t} = \frac{10 \text{ V}}{5 \text{ ms}} = 2 \text{ V/ms} = 2 \text{ kV/s}$$

In Part B, the voltage decreases from +5 V to 0 V in 100 ms. Thus, $\Delta v = -5$ V and $t = 100$ ms. The slope is

$$\frac{\Delta v}{\Delta t} = \frac{-5 \text{ V}}{100 \text{ ms}} = -0.05 \text{ V/ms} = -50 \text{ V/s}$$

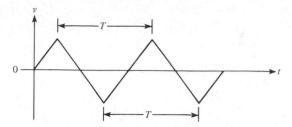

FIGURE 11–54 *Alternating triangular wave form.*

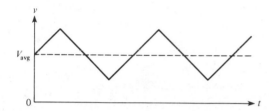

FIGURE 11–55 *Triangular wave form with a nonzero average value.*

Triangular Wave Forms

Figure 11–54 shows that a triangular wave form is composed of positive- and negative-going ramps having equal slopes. The period of this wave form is measured from one peak to the next corresponding peak, as illustrated. This particular triangular wave form is alternating and has an average value of zero.

Figure 11–55 depicts a triangular wave form with a nonzero average value. The frequency for triangular waves is determined in the same way as for sine waves, that is, $f = 1/T$.

Sawtooth Wave Forms

The sawtooth wave is actually a special case of the triangular wave consisting of two ramps, one of much longer duration than the other. Sawtooth wave forms are commonly used in many electronic systems. For example, the electron beam that sweeps across the screen of your TV receiver, creating the picture, is controlled by sawtooth voltages and currents. One sawtooth wave produces the horizontal beam movement, and the other produces the vertical beam movement. The sawtooth is sometimes called a *sweep*.

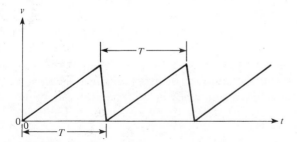

FIGURE 11–56 *Sawtooth wave form.*

Figure 11–56 is an example of a sawtooth wave. Notice that it consists of a positive-going ramp of relatively long duration, followed by a negative-going ramp of relatively short duration.

Review for 11–10

1. Define the following parameters:
(a) rise time   (b) fall time   (c) pulse width

2. In a certain repetitive pulse wave form, the pulses occur once every millisecond. What is the PRF of this wave form?

3. Determine the duty cycle, amplitude, and average value of the wave form in Figure 11–57A.

4. What is the period of the triangular wave in Figure 11–57B?

5. What is the frequency of the sawtooth wave in Figure 11–57C?

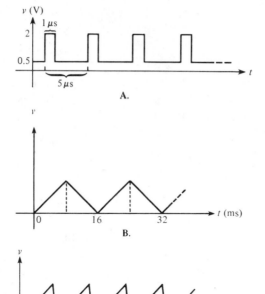

A.

B.

C.

FIGURE 11–57

HARMONICS

11–11 A repetitive nonsinusoidal wave form is composed of a *fundamental frequency* and *harmonic frequencies*. The fundamental frequency is the repetition rate of the wave form, and the harmonics are higher-frequency sine waves that are multiples of the fundamental.

Odd Harmonics

Odd harmonics are frequencies that are *odd multiples* of the fundamental frequency of a wave form. For example, a 1-kHz square wave consists of a fundamental of 1 kHz and odd harmonics of 3 kHz, 5 kHz, 7 kHz, and so on. The 3-kHz frequency in this case is called the *third harmonic;* the 5-kHz frequency is the *fifth harmonic;* and so on.

Even Harmonics

Even harmonics are frequencies that are *even multiples* of the fundamental frequency. For example, if a certain wave has a fundamental of 200 Hz, the second harmonic is 400 Hz, the fourth harmonic is 800 Hz, the sixth harmonic is 1200 Hz, and so on. These are even harmonics.

Composite Wave Form

Any variation from a pure sine wave produces harmonics. A non-sinusoidal wave is a composite of the fundamental and the harmonics. Some types of wave forms have only odd harmonics, some have only even harmonics, and some contain both. The shape of the wave is determined by its harmonic content. Generally, only the fundamental and the first few harmonics are of significant importance in determining the wave shape.

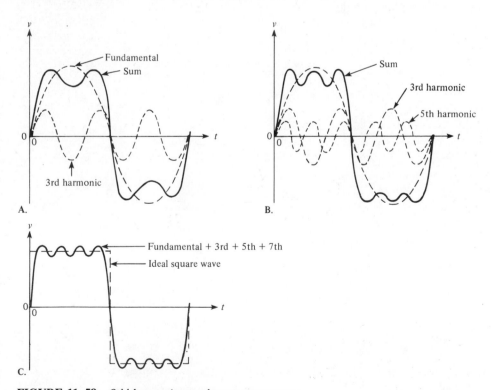

FIGURE 11–58 *Odd harmonics produce a square wave.*

A *square wave* is an example of a wave form that consists of a fundamental and only odd harmonics. When the instantaneous values of the fundamental and each odd harmonic are added algebraically at each point, the resulting
curve will have the shape of a square wave, as illustrated in Figure 11–58. In
Part A of the figure, the fundamental and the third harmonic produce a wave
shape that begins to resemble a square wave. In Part B, the fundamental, third,
and fifth harmonics produce a closer resemblance. When the seventh harmonic is
included, as in Part C, the resulting wave shape becomes even more like a square
wave. As more harmonics are included, a square wave is approached.

Review for 11–11

1. Define *fundamental frequency.*

2. What is the second harmonic of a fundamental frequency of 1 kHz?

3. What is the fundamental frequency of a square wave having a period of 10 μs?

COMPUTER ANALYSIS

11–12

The program in this section computes the values of a sinusoidal voltage given the
peak voltage and the period. In addition, the instantaneous value is calculated for
any specified phase angle.

```
10 CLS
20 PRINT "THIS PROGRAM COMPUTES ALL SINE WAVE VOLTAGE VALUES WHEN THE"
30 PRINT "PEAK VALUE AND PERIOD ARE PROVIDED.  THE INSTANTANEOUS"
40 PRINT "VALUE FOR A SPECIFIED PHASE ANGLE IS ALSO COMPUTED."
50 PRINT:PRINT:PRINT
60 INPUT "TO CONTINUE PRESS 'ENTER'";X
70 CLS
80 INPUT "THE PEAK VALUE IN VOLTS";VM
90 INPUT "THE PERIOD IN SECONDS";T
100 INPUT "INSTANTANEOUS PHASE ANGLE IN DEGREES";THETA
110 CLS
120 VPP=2*VM
130 VRMS=SQR(0.5)*VM
140 VAVG=(2/3.1415927)*VM
150 F=1/T
160 V=VM*SIN(THETA/57.29577786)
170 PRINT "VP =";VM;"V"
180 PRINT "0 =";THETA;"DEGREES"
190 PRINT "T =";T;"SECONDS"
200 PRINT "VPP =";VPP;"V"
210 PRINT "VRMS =";VRMS;"V"
220 PRINT "VAVG =";VAVG;"V"
230 PRINT "F =";F;"HZ"
240 PRINT "INSTANTANEOUS VOLTAGE AT";THETA;"DEGREES =";V;"V"
```

Review for 11–12

1. Run the program listed above for the following values: $V_p = 10$ V, $T =$
0.001 s, and $\theta = 60°$.

Formulas

$$f_{HZ} = \frac{1}{T_S} \qquad (11\text{--}1)$$

$$T = \frac{1}{f} \qquad (11\text{--}2)$$

$$V_{pp} = 2V_p \qquad (11\text{--}3)$$

$$I_{pp} = 2I_p \qquad (11\text{--}4)$$

$$V_{rms} = \sqrt{0.5}\, V_p \cong 0.707V_p \qquad (11\text{--}5)$$

$$I_{rms} = \sqrt{0.5}\, I_p \cong 0.707I_p \qquad (11\text{--}6)$$

$$V_p = \sqrt{2}\, V_{rms} \cong 1.414V_{rms} \qquad (11\text{--}7)$$

$$I_p = \sqrt{2}\, I_{rms} \cong 1.414I_{rms} \qquad (11\text{--}8)$$

$$V_{pp} = 2.828V_{rms} \qquad (11\text{--}9)$$

$$I_{pp} = 2.828I_{rms} \qquad (11\text{--}10)$$

$$V_{avg} = \left(\frac{2}{\pi}\right)V_p \cong 0.637V_p \qquad (11\text{--}11)$$

$$I_{avg} = \left(\frac{2}{\pi}\right)I_p \cong 0.637I_p \qquad (11\text{--}12)$$

$$v = iR \qquad (11\text{--}13)$$

$$V_p = I_pR \qquad (11\text{--}14)$$

$$V_{pp} = I_{pp}R \qquad (11\text{--}15)$$

$$V_{rms} = I_{rms}R \qquad (11\text{--}16)$$

$$V_{avg} = I_{avg}R \qquad (11\text{--}17)$$

$$\text{rad} = \left(\frac{\pi\ \text{rad}}{180°}\right)\text{degrees} \qquad (11\text{--}18)$$

$$\text{degrees} = \left(\frac{180°}{\pi\ \text{rad}}\right)\text{rad} \qquad (11\text{--}19)$$

$$y = A\sin\theta \qquad (11\text{--}20)$$

$$y = A\sin(\theta - \phi) \qquad (11\text{--}21)$$

$$y = A\sin(\theta + \phi) \qquad (11\text{--}22)$$

$$\omega = 2\pi f \qquad (11\text{--}23)$$

$$\theta = \omega t \qquad (11\text{--}24)$$

$$C = \sqrt{A^2 + B^2} \qquad (11\text{--}25)$$

$$\theta = \arctan\left(\frac{B}{A}\right) \tag{11-26}$$

$$A \pm jB = \sqrt{A^2 + B^2} \angle \pm \arctan\left(\frac{B}{A}\right) = C \angle \pm \theta \tag{11-27}$$

$$A = C \cos \theta \tag{11-28}$$

$$B = C \sin \theta \tag{11-29}$$

$$C \angle \theta = C \cos \theta \pm jC \sin \theta = A \pm jB \tag{11-30}$$

$$\text{PRF} = \frac{1}{T} \tag{11-31}$$

$$T = \frac{1}{\text{PRF}} \tag{11-32}$$

$$\text{percent duty cycle} = \left(\frac{t_w}{T}\right)100\% \tag{11-33}$$

$$V_{\text{avg}} = \text{baseline} + (\text{duty cycle})(\text{amplitude}) \tag{11-34}$$

Summary

1. The sine wave is a time-varying periodic wave form. Voltage or current can vary sinusoidally.

2. The sine wave is a common form of ac (alternating current).

3. Alternating current reverses direction in response to changes in the voltage polarity.

4. A periodic wave form is one that repeats at fixed intervals.

5. The time required for a sine wave to repeat is called the *period*.

6. The period is the time for one complete cycle.

7. Frequency is the rate of change of the sine wave in cycles per second. The unit of frequency is the *hertz*. One hertz is one cycle per second.

8. The frequency is the reciprocal of the period, and vice versa.

9. The instantaneous rate of change for a sine wave is maximum at the zero crossings and minimum at the peaks.

10. The instantaneous value of a sine wave is its value at any point in time.

11. The peak value of a sine wave is its maximum positive value or its maximum negative value measured from the zero crossing.

12. The peak-to-peak value of a sine wave is measured from the positive peak to the negative peak and is equal to twice the peak value.

13. The term *rms* stands for *root mean square*. The rms value of a sine wave, also known as the *effective* value, is a measure of the heating effect of a sine wave. It is 0.707 times the peak value.

14. The average value of a sine wave is defined over a half-cycle. It is 0.637 times the peak value. The average value over a full sine wave cycle is zero.

15. A full cycle of a sine wave is 360 degrees or 2π radians. A half-cycle is 180 degrees or π radians. A quarter cycle is 90 or $\pi/2$ radians.

16. A phase angle is the difference in degrees or radians between two sine waves.

17. A phasor represents a time-varying quantity in terms of both magnitude and direction.

18. The angular position of a phasor represents the angle of the sine wave, and the length of a phasor represents the amplitude.

19. A complex number represents a phasor quantity.

20. Complex numbers can be added, subtracted, multiplied, and divided.

21. The rectangular form of a complex number consists of a real part and a j part.

22. The polar form of a complex number consists of a magnitude and an angle.

23. A pulse consists of a transition from a baseline level to an amplitude level followed by a transition back to the baseline level.

24. The rise time of a pulse is the time required for the pulse to change from the 10% point to the 90% point on its rising edge.

25. The fall time of a pulse is the time required for the pulse to change from the 90% point to the 10% point on its falling edge.

26. Pulse width is the time between the 50% points on the rising and falling edges.

27. The duty cycle is the ratio of the pulse width to the period.

28. Harmonic frequencies are odd or even multiples of the repetition rate (fundamental frequency) of a wave form.

Self-Test

1. A certain sine wave has a period of 200 ms. What is its frequency?

2. A certain sine wave has a frequency of 25 kHz. What is its period?

3. A given sine wave takes 5 μs to complete one cycle. Determine its frequency.

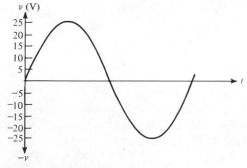

FIGURE 11–59

4. For the sine wave in Figure 11–59, determine the peak, peak-to-peak, rms, and average values.

5. The rms value for a given sine wave is 115 V. What is its peak value?

6. A 10-V rms sine wave is applied to a 100-Ω resistive circuit. Determine the rms value of the current.

7. Make a sketch of two sine waves as follows: Sine wave A is the reference, and sine wave B lags A by 90 degrees. Both have equal amplitudes.

8. A certain sine wave has a peak value of 20 V and begins at 0-degree reference. Calculate its instantaneous value at each of the following points: 10 degrees, 25 degrees, 30 degrees, 90 degrees, 180 degrees, 210 degrees, and 300 degrees.

9. Calculate the instantaneous value of the sine wave in Figure 11–60 at the following points on the horizontal axis:
   (a) 30 degrees   (b) 60 degrees   (c) 100 degrees   (d) 225 degrees

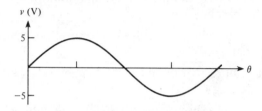

FIGURE 11–60

10. Determine the instantaneous value represented by each of the phasors in Figure 11–61.

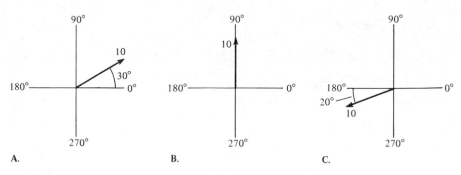

FIGURE 11–61

11. Express each of the following angles as an equivalent negative angle:
    (a) 20 degrees   (b) 60 degrees   (c) 135 degrees
    (d) 200 degrees   (e) 315 degrees   (f) 330 degrees

12. If the angular velocity, ω, is 1000 rad/s, what is the frequency?

13. Locate the following points in the complex plane.
    (a) $1 + j2$ (b) $3 + j4$ (c) $5 - j3$ (d) $-2 + j3$ (e) $-1 - j2$

14. Sketch the phasors specified by the following polar coordinates:
    (a) $2\angle45°$ (b) $4\angle0°$ (c) $5\angle30°$ (d) $1\angle90°$ (e) $3\angle-90°$

15. Convert the following rectangular numbers into polar form:
    (a) $5 + j5$ (b) $12 + j9$ (c) $8 - j10$ (d) $100 - j50$

16. Convert the following polar numbers into rectangular form:
    (a) $1\angle45°$ (b) $12\angle60°$ (c) $100\angle-80°$ (d) $40\angle125°$

17. Determine the approximate rise time, fall time, pulse width, and amplitude of the pulse in Figure 11–62.

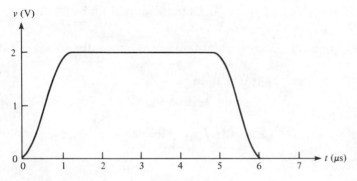

FIGURE 11–62

18. The repetition frequency of a pulse wave form is 2 kHz, and the pulse width is 1 μs. What is the percent duty cycle?

19. Calculate the average value of the pulse wave form in Figure 11–63.

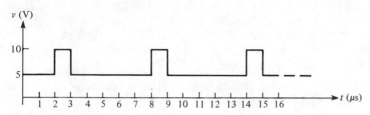

FIGURE 11–63

20. What is the fundamental frequency of the pulse wave form in Figure 11–63?

21. What are the second and third harmonics of 25 kHz?

Problems

Sections 11–1 through 11–2

11–1. Calculate the frequency for each of the following values of period:

    (a) 1 s   **(b)** 0.2 s   **(c)** 50 ms

    (d) 1 ms   **(e)** 500 μs   **(f)** 10 μs

11–2. Calculate the period for each of the following values of frequency:

    (a) 1 Hz   **(b)** 60 Hz   **(c)** 500 Hz

    (d) 1 kHz   **(e)** 200 kHz   **(f)** 5 MHz

11–3. A sine wave goes through 5 cycles in 10 μs. What is its period?

11–4. A sine wave has a frequency of 50 kHz. How many cycles does it complete in 10 ms?

Section 11–3

11–5. A sine wave has a peak value of 12 V. Determine the following values:

    (a) rms   **(b)** peak-to-peak   **(c)** average

11–6. A sinusoidal current has an rms value of 5 mA. Determine the following values:

    (a) peak   **(b)** average   **(c)** peak-to-peak

11–7. A sinusoidal voltage is applied to the resistive circuit in Figure 11–64. Determine the following:

    (a) I_{rms} **(b)** I_{avg} **(c)** I_p **(d)** I_{pp} **(e)** i at the positive peak

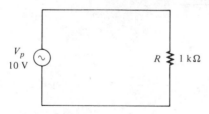

FIGURE 11–64

Section 11–4

11–8. Convert the following angular values from degrees to radians:

    (a) 30 degrees   **(b)** 45 degrees   **(c)** 78 degrees

    (d) 135 degrees   **(e)** 200 degrees   **(f)** 300 degrees

11–9. Convert the following angular values from radians to degrees:

    (a) $\pi/8$ **(b)** $\pi/3$ **(c)** $\pi/2$

    (d) $3\pi/5$ **(e)** $6\pi/5$ **(f)** 1.8π

11–10. Sine wave A has a positive-going zero crossing at 30 degrees. Sine wave B has a positive-going zero crossing at 45 degrees. Determine the phase angle between the two signals. Which signal leads?

11–11. One sine wave has a positive peak at 75 degrees, and another has a positive peak at 100 degrees. How much is each sine wave shifted in phase from the 0-degree reference? What is the phase angle between them?

Section 11–5

11–12. A certain sine wave has a positive-going zero crossing at 0 degrees and an rms value of 20 V. Calculate its instantaneous value at each of the following angles:
    (a) 15 degrees   **(b)** 33 degrees   **(c)** 50 degrees
    (d) 110 degrees   **(e)** 70 degrees   **(f)** 145 degrees
    (g) 250 degrees   **(h)** 325 degrees

11–13. For a particular 0-degree reference sinusoidal current, the peak value is 100 mA. Determine the instantaneous value at each of the following points:
    (a) 35 degrees   **(b)** 95 degrees   **(c)** 190 degrees
    (d) 215 degrees   **(e)** 275 degrees   **(f)** 360 degrees

11–14. For a 0-degree reference sine wave with an average value of 6.37 V, determine its instantaneous value at each of the following points:
    (a) $\pi/8$ radians   **(b)** $\pi/4$ radians   **(c)** $\pi/2$ radians
    (d) $3\pi/4$ radians   **(e)** π radians   **(f)** $3\pi/2$ radians
    (g) 2π radians

11–15. Sine wave A lags sine wave B by 30 degrees. Both have peak values of 15 V. Sine wave A is the reference with a positive-going crossing at 0 degrees. Determine the instantaneous value of sine wave B at 30 degrees, 45 degrees, 90 degrees, 180 degrees, 200 degrees, and 300 degrees.

11–16. Repeat Problem 11–15 for the case when sine wave A *leads* sine wave B by 30 degrees.

Section 11–6

11–17. Draw a phasor diagram to represent the sine waves in Figure 11–65.

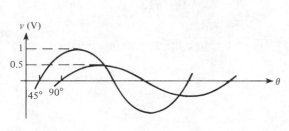

FIGURE 11–65   **FIGURE 11–66**

11–18. Sketch the sine waves represented by the phasor diagram in Figure 11–66. The phasor lengths represent peak values.

11–19. Determine the frequency for each angular velocity.
    (a) 60 rad/s   **(b)** 360 rad/s   **(c)** 2 rad/s   **(d)** 1256 rad/s

Sections 11–7 through 11–8

11–20. Points on the complex plane are described below. Express each point as a complex number in rectangular form.
   (a) 3 units to the right of the origin on the real axis, and up 5 units on the j axis.
   (b) 2 units to the left of the origin on the real axis, and 1.5 units up on the j axis.
   (c) 10 units to the left of the origin on the real axis, and down 14 units on the $-j$ axis.

11–21. What is the value of the hypotenuse of a right triangle whose sides are 10 and 15?

11–22. Convert each of the following rectangular numbers to polar form.
   (a) $40 - j40$ **(b)** $50 - j200$ **(c)** $35 - j20$ **(d)** $98 + j45$

11–23. Convert each of the following polar numbers to rectangular form.
   (a) $1000\angle -50°$ **(b)** $15\angle 160°$ **(c)** $25\angle -135°$ **(d)** $3\angle 180°$

Section 11–9

11–24. Add the following sets of complex numbers:
   (a) $9 + j3$ and $5 + j8$ **(b)** $3.5 - j4$ and $2.2 + j6$
   (c) $-18 + j23$ and $30 - j15$ **(d)** $12\angle 45°$ and $20\angle 32°$
   (e) $3.8\angle 75°$ and $1 + j1.8$ **(f)** $50 - j39$ and $60\angle -30°$

11–25. Perform the following subtractions:
   (a) $(2.5 + j1.2) - (1.4 + j0.5)$ **(b)** $(-45 - j23) - (36 + j12)$
   (c) $(8 - j4) - 3\angle 25°$ **(d)** $48\angle 135° - 33\angle -60°$

11–26. Multiply the following numbers:
   (a) $4.5\angle 48°$ and $3.2\angle 90°$ **(b)** $120\angle -220°$ and $95\angle 200°$
   (c) $-3\angle 150°$ and $4 - j3$ **(d)** $67 + j84$ and $102\angle 40°$
   (e) $15 - j10$ and $-25 - j30$ **(f)** $0.8 + j0.5$ and $1.2 - j1.5$

11–27. Perform the following divisions:
   (a) $\dfrac{8\angle 50°}{2.5\angle 39°}$ **(b)** $\dfrac{63\angle -91°}{9\angle 10°}$ **(c)** $\dfrac{28\angle 30°}{14 - j12}$ **(d)** $\dfrac{40 - j30}{16 + j8}$

Section 11–10

11–28. From the graph in Figure 11–67, determine the approximate values of t_r, t_f, t_w, and amplitude.

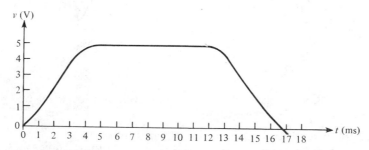

FIGURE 11–67

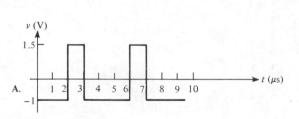
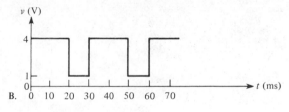

FIGURE 11–68

11–29. Determine the duty cycle for each wave form in Figure 11–68.

11–30. Find the average value of each pulse wave form in Figure 11–68.

11–31. What is the frequency of each wave form in Figure 11–68?

11–32. What is the frequency of each sawtooth wave form in Figure 11–69?

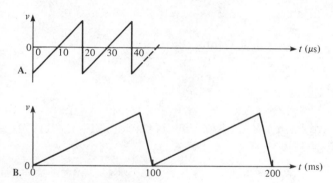

FIGURE 11–69

Section 11–11

11–33. A square wave has a period of 40 μs. List the first six odd harmonics.

11–34. What is the fundamental frequency of the square wave mentioned in Problem 11–33?

Section 11–12

11–35. Modify the program in Section 11–12 to compute sinusoidal current values instead of voltage.

11–36. Write a program to convert degrees to radians.

11–37. Develop a program to convert complex numbers in rectangular form into polar form.

Advanced Problems

11–38. A certain sine wave has a frequency of 2.2 kHz and an rms value of 25 V. Assuming a given cycle begins (zero crossing) at $t = 0$ s, what is the change in voltage from 0.12 ms to 0.2 ms?

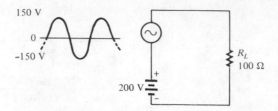

FIGURE 11–70

11–39. Figure 11–70 shows a sinusoidal voltage source in series with a dc source. Effectively, the two voltages are superimposed. Determine the average power dissipation in the load resistor.

11–40. Perform the following operations:

(a) $\dfrac{2.5\angle 65° - 1.8\angle -23°}{1.2\angle 37°}$

(b) $\dfrac{(100\angle 15°)\,(85 - j150)}{25 + j45}$

(c) $\dfrac{(250\angle 90° + 175\angle 75°)\,(50 - j100)}{(125 + j90)\,(35\angle 50°)}$

(d) $\dfrac{(1.5)^2\,(3.8)}{1.1} + j\!\left(\dfrac{8}{4} - j\dfrac{4}{2}\right)$

11–41. A nonsinusoidal wave form called a *stairstep* is shown in Figure 11–71. Determine its average value.

FIGURE 11–71

Answers to Section Reviews

Section 11–1:
1. From the zero crossing through a positive peak, then through zero to a negative peak and back to the zero crossing. **2.** At the zero crossings. **3.** 2. **4.** Peaks.

Section 11–2:
1. From one zero crossing to the next *corresponding* zero crossing, or from one peak to the next *corresponding* peak. **2.** The number of cycles completed in one second; hertz. **3.** 200 kHz. **4.** 8.33 ms.

Section 11–3:
1. (a) 2 V, (b) 4 V, (c) 9.42 V. **2.** (a) 1.77 V, (b) 3.54 V, (c) 1.66 V. **3.** 0.5 A; 0.707 A

Section 11–4:
1. (a) 90 degrees, (b) 180 degrees, (c) 270 degrees, (d) 360 degrees. **2.** 180; π. **3.** 360; 2π. **4.** 45 degrees.

Section 11–5:
1. 8.66 V. **2.** 8.19 V. **3.** 4.53 V.

Section 11–6:
1. A graphic representation of the magnitude and angular position of a time-varying quantity. **2.** 9425 rad/s. **3.** 99.95 Hz. **4.** See Figure 11–72.

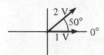

FIGURE 11–72

Section 11–7:
1. (a) 3 units right of the origin on real axis, (b) 4 units left of the origin on real axis, (c) 1 unit above origin on j axis. **2.** (a) 90 degrees, (b) 90 degrees, (c) 180 degrees.

Section 11–8:
1. Real part and j part. **2.** Magnitude and angle. **3.** $2.828\angle 45°$. **4.** $3.54 + j3.54$.

Section 11–9:
1. $4 + j1$. **2.** $3 + j7$. **3.** $16\angle 110°$. **4.** $5\angle 15°$.

Section 11–10:
1. (a) The time interval from 10% to 90% of the rising pulse edge, (b) The time interval from 90% to 10% of the falling pulse edge, (c) The time interval from 50% of the leading pulse edge to 50% of the trailing pulse edge. **2.** 1 kHz. **3.** 20%, 1.5 V, 0.8 V. **4.** 16 ms. **5.** 1 MHz.

Section 11–11:
1. The repetition rate of the wave form. **2.** 2 kHz. **3.** 100 kHz.

Section 11–12:
1. $V_{pp} = 20$ V, $V_{rms} = 7.07107$ V, $V_{avg} = 6.3662$ V, $F = 1000$ Hz, $v = 8.66025$ V.

The capacitor is an electrical device that has the ability to store electrical charge; the measure of that ability is called *capacitance*. In this chapter, the basic capacitor is introduced, and its characteristics are studied. Various types of capacitors are covered in terms of their physical construction and their electrical properties.

The basic behavior of capacitors in both dc and ac circuits is studied, and series and parallel combinations are analyzed. Methods for testing capacitors are also discussed.

In this chapter, you will learn:

1. How capacitance is measured.
2. The conversions for farad, microfarad, and picofarad units of capacitance.
3. How a capacitor stores energy.
4. The definition of Coulomb's law, and how it relates to an electric field.
5. How the various physical characteristics affect the capacitance.
6. The classifications of capacitors.
7. The definition of time constant, and how to determine the charging or discharging time of a capacitor.
8. Why a capacitor blocks dc.
9. How to calculate instantaneous values on a charging or discharging curve.
10. How to obtain a desired capacitance value by combining capacitors in series.
11. How to obtain a desired capacitance value by combining capacitors in parallel.
12. How a capacitor introduces a phase shift between current and voltage.
13. The definition of capacitive reactance, and how to determine its value in any given circuit condition.
14. Why a capacitor does not dissipate any power.
15. The meaning of reactive power in a capacitive circuit.

12

THE CAPACITOR

12–1

In its simplest form, a capacitor is constructed of two parallel conductive plates separated by an insulating material called the *dielectric*. A basic capacitor is shown in Figure 12–1.

Capacitance

Basically, a capacitor is a device that stores electrical charge; a measure of a capacitor's ability to store charge is its *capacitance,* symbolized by C. Charge is stored when an excess of electrons accumulate on one of the plates, creating a net negative charge with respect to the other plate. To charge a capacitor, a voltage source must be connected to the plates, as shown in Figure 12–2.

Electrical Charge

Electrical charge, symbolized by Q, is measured in units of *coulombs*. The coulomb is symbolized by C, as with capacitance. This can be confusing, so keep the double usage of C in mind (italic C is used to denote capacitance). One coulomb of charge is defined as the charge possessed by 6.28×10^{18} electrons.

The Unit of Capacitance

The *farad,* symbolized by F, is the basic unit of capacitance. By definition, the capacitance is *one farad* when *one coulomb* of charge is stored with *one volt* across the plates of a capacitor.

The formula for capacitance in terms of *charge* and *voltage* is

$$C = \frac{Q}{V} \tag{12–1}$$

where C is for capacitance, Q is for charge, and V is for voltage. When charge is expressed in *coulombs* and voltage in *volts,* capacitance is in *farads*. By rearranging Equation (12–1), we obtain two other forms as follows:

$$Q = CV \tag{12–2}$$

$$V = \frac{Q}{C} \tag{12–3}$$

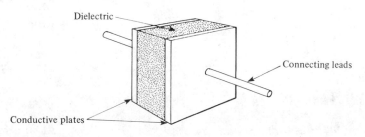

Dielectric

Connecting leads

Conductive plates

FIGURE 12–1 *Basic capacitor.*

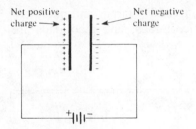

FIGURE 12–2 *Charge on a capacitor.*

Example 12–1

(a) A certain capacitor stores 50 coulombs (50 C) with 10 V across its plates. What is its *capacitance* in units of farads?

(b) A 2-F capacitor has 100 V across its plates. How much charge does it store?

(c) Determine the voltage across a 1-F capacitor that is storing 20 coulombs (20 C) of charge.

Solution:

(a) $C = \dfrac{Q}{V} = \dfrac{50 \text{ C}}{10 \text{ V}} = 5 \text{ F}$

(b) $Q = CV = (2 \text{ F})(100 \text{ V}) = 200 \text{ C}$

(c) $V = \dfrac{Q}{C} = \dfrac{20 \text{ C}}{1 \text{ F}} = 20 \text{ V}$

Smaller Units of Capacitance

A farad represents a relatively large amount of capacitance. In practice, particularly in electronics, smaller units such as the *microfarad* (μF) and the

TABLE 12–1 *Conversions for farads, microfarads, and picofarads.*

To convert from	to	multiply
farads	microfarads	farads by 10^6
farads	picofarads	farads by 10^{12}
microfarads	farads	microfarads by 10^{-6}
microfarads	picofarads	microfarads by 10^6
picofarads	farads	picofarads by 10^{-12}
picofarads	microfarads	picofarads by 10^{-6}

picofarad (pF) are very common. One microfarad is 10^{-6} farad, and one pico-farad is 10^{-12} farad. Conversions for farads, microfarads, and picofarads are given in Table 12–1 on page 383.

Example 12–2

Convert the following values to microfarads:
(a) 0.00001 F   (b) 0.005 F   (c) 1000 pF   (d) 200 pF

Solution:

(a) $0.00001 \text{ F} \times 10^6 = 10 \ \mu\text{F}$
(b) $0.005 \text{ F} \times 10^6 = 5000 \ \mu\text{F}$
(c) $1000 \text{ pF} \times 10^{-6} = 0.001 \ \mu\text{F}$
(d) $200 \text{ pF} \times 10^{-6} = 0.0002 \ \mu\text{F}$

Example 12–3

Convert the following values to picofarads:
(a) 0.1×10^{-8} F   (b) 0.000025 F   (c) 0.01 μF   (d) 0.005 μF

Solution:

(a) $0.1 \times 10^{-8} \text{ F} \times 10^{12} = 1000 \text{ pF}$
(b) $0.000025 \text{ F} \times 10^{12} = 25 \times 10^6 \text{ pF}$
(c) $0.01 \ \mu\text{F} \times 10^6 = 10,000 \text{ pF}$
(d) $0.005 \ \mu\text{F} \times 10^6 = 5000 \text{ pF}$

Voltage Rating

Every capacitor has a limit on the amount of voltage that it can withstand across its plates. The *voltage rating* specifies the maximum dc voltage that can be applied without risk of damage to the device. If this maximum voltage, commonly called the *breakdown voltage* or *working voltage,* is exceeded, permanent damage to the capacitor can result.

Both the capacitance and the voltage rating must be taken into consideration before a capacitor is used in a circuit application. The choice of capacitance value is based on particular circuit requirements (and on factors that are studied later). The voltage rating should always be well above the maximum voltage expected in a particular application.

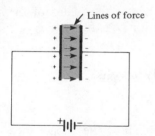

Lines of force

FIGURE 12–3   *The electric field stores energy in a capacitor.*

How a Capacitor Stores Energy

A capacitor stores energy in the form of an *electric field* that is established by the opposite charges on the two plates. The electric field is represented by *lines of force* between the positive and negative charges and concentrated within the dielectric, as shown in Figure 12–3.

Coulomb's law states: *A force exists between two charged bodies that is directly proportional to the product of the two charges and inversely proportional to the square of the distance between the bodies.* This relationship is expressed in Equation (12–4).

$$F = \frac{kQ_1 Q_2}{d^2} \qquad (12\text{–}4)$$

where F is the force in newtons, Q_1 and Q_2 are the charges in coulombs, d is the distance between the charges in meters, and k is a proportionality constant equal to 9×10^9. Figure 12–4A illustrates the line of force between a positive and a negative charge; Part B shows that many opposite charges on the plates of a capacitor create many lines of force, which form an *electric field* that stores energy within the dielectric.

The greater the forces between the charges on the plates of a capacitor, the more energy is stored. The amount of energy stored therefore is directly proportional to the capacitance, because, from Coulomb's law, the more charge stored, the greater the force.

Also, from Equation (12–2), the amount of charge stored is directly related to the voltage as well as the capacitance. Therefore, the amount of energy

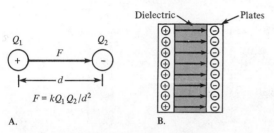

Dielectric   Plates

A.

Q_1 Q_2

F

d

$F = kQ_1 Q_2 / d^2$

B.

FIGURE 12–4   *Lines of force are created by opposite charges.*

stored is also dependent on the square of the voltage across the plates of the capacitor. The formula for the energy stored by a capacitor is as follows:

$$\mathscr{E} = \tfrac{1}{2}CV^2 \tag{12-5}$$

The energy, $\mathscr{E}$, is in joules when C is in farads and V is in volts.

Capacitor Characteristics

The following parameters are important in establishing the capacitance and the voltage rating of a capacitor.

Plate Area: *Capacitance is directly proportional to the physical size of the plates as determined by the plate area, A.* A larger plate area produces a larger capacitance, and vice versa. Figure 12–5A shows that the plate area of a parallel plate capacitor is the area of one of the plates. If the plates are moved in relation to each other, as shown in Part B, the *overlapping area* determines the effective plate area. This variation in effective plate area is the basis for a certain type of variable capacitor. $A \left(m^2 \right)$

Plate Separation: *Capacitance is inversely proportional to the distance between the plates.* The plate separation is designated *d*, as shown in Figure 12–6. A greater separation of the plates produces a smaller capacitance, as illustrated in the figure. The breakdown voltage is directly proportional to the plate separation. The further the plates are separated, the greater the breakdown voltage.

Dielectric Constant: As you know, the insulating material between the plates of a capacitor is called the *dielectric*. Every dielectric material has the ability

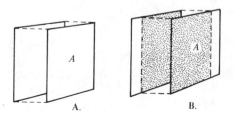

FIGURE 12–5   *Plate area.*

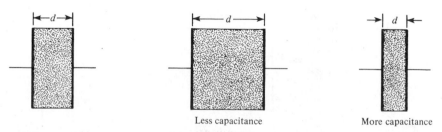

Less capacitance   More capacitance

FIGURE 12–6   *Plate separation.*

to concentrate the lines of force of the electric field existing between the oppositely charged plates of a capacitor and thus increase the capacity for energy storage. The measure of a material's ability to establish an electric field is called the *dielectric constant* or *relative permittivity,* symbolized by ϵ_r (the Greek letter epsilon).

Capacitance is directly proportional to the dielectric constant. The dielectric constant of a vacuum is defined as 1, and that of air is very close to 1. These values are used as a reference, and all other materials have values of ϵ_r specified with respect to that of a vacuum or air. For example, a material with $\epsilon_r = 8$ can result in a capacitance eight times greater than that of air with all other factors being equal.

Table 12–2 lists several common dielectric materials and typical dielectric constants for each. Values can vary because they depend on the specific composition of the material.

The dielectric constant (relative permittivity) is dimensionless, because it is a relative measure and is a ratio of the *absolute permittivity,* ϵ, of a material to the *absolute permittivity,* ϵ_0, of a vacuum, as expressed by the following formula:

$$\epsilon_r = \frac{\epsilon}{\epsilon_0} \quad \text{epsilon sub o} \tag{12–6}$$

The value of ϵ_0 is 8.85×10^{-12} F/m (farads per meter). *for air*

Dielectric Strength: The breakdown voltage of a capacitor is determined by the dielectric strength of the material used. The dielectric strength is expressed in volts/mil (1 mil = 0.001 in). Table 12–3 lists typical values for several materials. Exact values vary depending on the specific composition of the material.

The dielectric strength can best be explained by an example. Assume that a certain capacitor has a plate separation of 1 mil and that the dielectric material is ceramic. This particular capacitor can withstand a maximum voltage of 1000 V, because its dielectric strength is 1000 V/mil. If the maximum voltage is exceeded, the dielectric may break down and conduct current, causing permanent damage to the capacitor. Similarly, if the ceramic capacitor has a plate separation of 2 mils, its breakdown voltage is 2000 V.

TABLE 12–2 *Some common dielectric materials and their dielectric constants.*

Material	Typical ϵ_r Values
Air (vacuum)	1.0
Teflon®	2.0
Paper (paraffined)	2.5
Oil	4.0
Mica	5.0
Glass	7.5
Ceramic	1200

TABLE 12–3 *Some common dielectric materials and their dielectric strengths.*

Material	Dielectric Strength (volts/mil)
Air	80
Oil	375
Ceramic	1000
Paper (paraffined)	1200
Teflon®	1500
Mica	1500
Glass	2000

Formula for Capacitance in Terms of Physical Parameters

You have seen how capacitance is directly related to plate area A and the dielectric constant ϵ_r, and inversely related to plate separation d. An exact formula for calculating the capacitance in terms of these three quantities is as follows:

$$C = \frac{A\epsilon_r(8.85 \times 10^{-12} \text{ F/m})}{d} \qquad (12\text{--}7)$$

A is in square meters (m^2), d is in meters (m), and C is in farads (F). Recall that 8.85×10^{-12} F/m is the absolute permittivity, ϵ_0, of a vacuum and that $\epsilon_r(8.85 \times 10^{-12})$ F/m is the absolute permittivity of a dielectric, as derived from Equation (12--6).

Example 12--4

Determine the capacitance of a parallel plate capacitor having a plate area of 0.01 m^2 and a plate separation of 0.02 m. The dielectric is mica.

Solution:

Use Equation (12--6):

$$C = \frac{A\epsilon_r(8.85 \times 10^{-12} \text{ F/m})}{d}$$

$$= \frac{(0.01 \text{ m}^2)(5.0)(8.85 \times 10^{-12} \text{ F/m})}{0.02 \text{ m}}$$

$$= 22.13 \text{ pF}$$

Temperature Coefficient

The temperature coefficient indicates the amount and direction of a change in capacitance value with temperature. A *positive* temperature coefficient means that the capacitance increases with an increase in temperature or decreases with a decrease in temperature. A *negative* coefficient means that the capacitance decreases with an increase in temperature or increases with a decrease in temperature.

Temperature coefficients are typically specified in *parts per million per degree Celsius* (ppm/°C). For example, a negative temperature coefficient of 150 ppm/°C for a 1-μF capacitor means that for every degree rise in temperature, the capacitance decreases by 150 pF (there are one million picofarads in one microfarad).

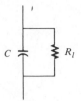

FIGURE 12–7 *Equivalent circuit for a nonideal capacitor.*

Leakage

No insulating material is perfect. The dielectric of any capacitor will conduct some very small amount of current. Thus, the charge on a capacitor will eventually leak off. Some types of capacitors have higher leakages than others. An equivalent circuit for a nonideal capacitor is shown in Figure 12–7. The parallel resistor represents the extremely high resistance of the dielectric material through which leakage current flows.

Review for 12–1

1. Define *capacitance*.
2. (a) How many microfarads in a farad?
   (b) How many picofarads in a farad?
   (c) How many picofarads in a microfarad?
3. Convert 0.0015 μF to picofarads. To farads.
4. How much energy in joules is stored by a 0.01-μF capacitor with 15 V across its plates?
5. (a) When the plate area of a capacitor is increased, does the capacitance increase or decrease?
   (b) When the distance between the plates is increased, does the capacitance increase or decrease?
6. The plates of a ceramic capacitor are separated by 10 mils. What is the typical breakdown voltage?
7. A ceramic capacitor has a plate area of 0.2 m². The thickness of the dielectric is 0.005 m. What is the capacitance?
8. A capacitor with a value of 2 μF at 25°C has a positive temperature coefficient of 50 ppm/°C. What is the capacitance value when the temperature increases to 125°C?

TYPES OF CAPACITORS

12–2

There are many types of capacitors available, and they come in many shapes and sizes, ranging from huge oil-filled capacitors used in power applications to inte-

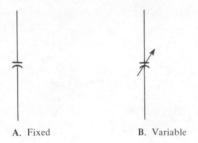

A. Fixed   **B.** Variable

FIGURE 12–8 *Schematic symbols for fixed and variable capacitors.*

grated circuit capacitors used in microelectronic applications. We discuss some of the most common ones in this section.

First, capacitors can be either *fixed* or *variable*. The standard schematic symbols for both fixed and variable capacitors are shown in Figure 12–8.

Fixed Capacitors

A fixed capacitor has a capacitance value that cannot be altered. Types of fixed capacitors, which are generally classified by their dielectric material, are discussed in the following subsections.

Mica Capacitors: Figure 12–9 shows the typical construction of a mica capacitor. Thin sheets of mica form the dielectric. They are stacked alternately with sheets of metal foil, which form the plates. Alternate foil sections are connected together to form effectively a larger single plate. More sections are used to increase the plate area, thus increasing the capacitance. A reduction in the number of foil sections or in the size of each section reduces the capacitance. The mica and foil assembly is encapsulated in Bakelite® or other protective material.

Mica capacitors are most commonly used for low capacitance values ranging from a few picofarads to several hundred picofarads. Several typical mica capacitors are shown in Figure 12–10.

Ceramic Capacitors: Ceramic dielectrics provide very high dielectric constants. As a result, comparatively high capacitance values can be achieved in

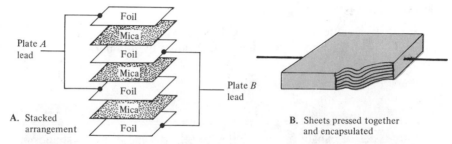

A. Stacked arrangement   **B.** Sheets pressed together and encapsulated

FIGURE 12–9 *Construction of a typical mica capacitor.*

FIGURE 12–10 *Typical mica capacitors.*

a small physical size. Typical values range from a few picofarads to a few microfarads and with very high breakdown voltages (several thousand volts are common). Some typical ceramic capacitors are shown in Figure 12–11.

Paper Capacitors: The typical construction of a paper capacitor is illustrated in Figure 12–12. The paper dielectric is commonly permeated with paraffin. The tubular construction shown in the figure consists of a strip of paper dielectric between two strips of foil that act as plates. A wire lead is connected to each foil sheet to provide for external connections to a circuit. The assembly is rolled into a tubular shape and encapsulated in a molded case. (Other varieties of paper capacitors have a flat rectangular shape.) One lead is connected to the inner foil

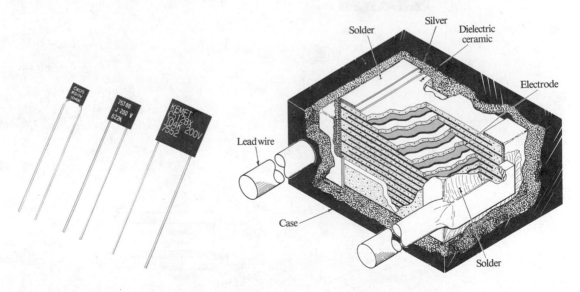

A.   **B.**

FIGURE 12–11 *Ceramic capacitors.* **A.** *Typical capacitors.* **B.** *Construction view (***A** *and* **B,** *courtesy of Union Carbide Corporation, Electronics Division)*

Lead connected
to outer foil sheet

Inner foil
Paper dielectric
Outer foil

Lead connected to
inner foil sheet

FIGURE 12-12 *Construction of a paper tubular capacitor.*

and one to the outer foil, as indicated. The outer foil connection is often marked with a band around the case; other markings are also used. When the capacitor is connected to a circuit, the banded end should be grounded to take advantage of the shielding effect of the outer foil. Figure 12-13 shows typical paper capacitors.

Electrolytic Capacitors: Electrolytic capacitors are polarized so that one plate is positive and the other negative. These capacitors are used for comparatively high capacitance values ranging from a few microfarads to several thousand microfarads. However, they have relatively low breakdown voltages and exhibit fairly high leakage currents compared to the other types of capacitors. The basic construction of an electrolytic capacitor is illustrated in Figure 12-14. As shown in Part A, the capacitor consists of a roll of foil with a layer of aluminum oxide deposited on its inner surface by electrolytic action, with the paper or gauze

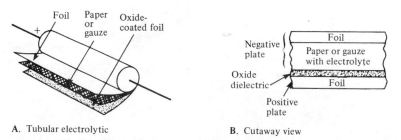

FIGURE 12-13 *Typical paper capacitors.*

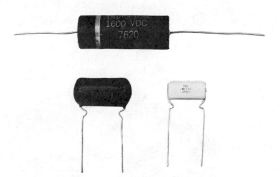

Foil   Paper   Oxide-
        or   coated foil
        gauze

A. Tubular electrolytic

Negative
plate

Oxide
dielectric

Positive
plate

Foil
Paper or gauze
with electrolyte
Foil

B. Cutaway view

FIGURE 12-14 *Construction of an electrolytic capacitor.*

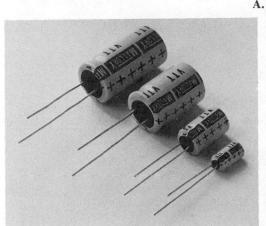

FIGURE 12–15 *Electrolytic capacitors.* **A.** *Typical examples. (Courtesy of Mallory Capacitor Division)* **B.** *Symbol.*

separator saturated with an electrolyte. This foil is the *positive plate,* and the oxide layer acts as the dielectric. The electrolyte in the paper and an inner foil layer act as the *negative plate.*

Since the electrolytic capacitor is polarized, *the positive plate must always be connected to the more positive side of a circuit.* Always be certain that this connection is correct. The positive end is usually indicated by plus signs or some other obvious markings.

Another type of electrolytic capacitor uses *tantalum* rather than aluminum. These capacitors are commonly used in transistor circuit applications. Figure 12–15 shows several typical electrolytic capacitors.

Film Capacitors: There are several basic types of film capacitors. Various types of materials are used as the dielectric, including Teflon®, Mylar®, Parylene®, and metalized film. These capacitors exhibit very high dielectric resistances and are available in capacitance values ranging from about 0.001 μF to 1 μF. Figure 12–16 (page 394) illustrates the construction of typical film capacitors.

Integrated Circuit Capacitors: An integrated circuit is constructed on a single tiny silicon chip. All of the circuit components, including transistors, diodes, resistors, and capacitors, are *integrated* onto this one piece of semiconductor material. In a later course, you will study the total integrated circuit and examine the construction of this type of capacitor.

Variable Capacitors

Variable capacitors are used in circuits when there is a need to adjust the capacitance value either manually or automatically.

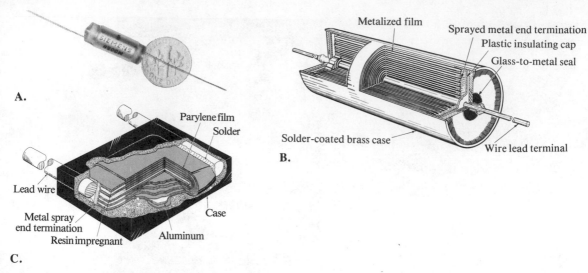

A.

B.

C.

FIGURE 12–16   *Film capacitors.* **A.** *Typical example. (Courtesy of Siemens Corporation)* **B.** *Construction view of metalized film capacitor.* **C.** *Construction view of molded film capacitor. (Courtesy of Union Carbide Corporation, Electronics Division)*

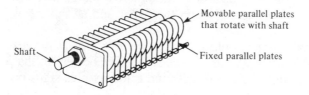

FIGURE 12–17   *Typical variable air capacitor.*

Air Capacitors: Variable air capacitors, such as the one shown in Figure 12–17, are sometimes used in radio receivers as tuning capacitors to provide frequency selection. This type of capacitor is constructed of several plates that mesh together. One set of plates can be moved relative to the other, thus changing the effective plate area and the capacitance. The plates are linked together mechanically so that all of the plates are moved by rotation of a shaft.

Trimmers and Padders: These adjustable capacitors normally have screwdriver-type adjustments and are used for very fine adjustments in circuit applications. They are usually ceramic or mica capacitors in which the plate separation is adjusted. Many have an appearance similar to that shown in Figure 12–18.

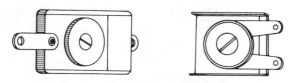

FIGURE 12–18   *Mica trimmer capacitors.*

Varactors: The varactor is a semiconductor device that exhibits a capacitance that can be varied by adjusting the voltage across its terminals. This device is usually covered in detail in an electronic devices course.

Review for 12–2

1. How are capacitors commonly classified?
2. What is the difference between a fixed and a variable capacitor?
3. What type of capacitor is polarized?
4. What precautions must be taken when installing a polarized capacitor in a circuit?

CAPACITORS IN dc CIRCUITS

Charging a Capacitor

12–3

A capacitor charges when it is connected to a dc voltage source, as shown in Figure 12–19. The capacitor in Part A of the figure is uncharged; that is, plate *A* and plate *B* have equal numbers of free electrons. When the switch is closed, as shown in Part B, the source moves electrons away from plate *A* through the circuit to plate *B*. As plate *A* loses electrons and plate *B* gains electrons, plate *A* becomes positive with respect to plate *B*. As this charging process continues, the voltage across the plates builds up rapidly until it is equal to the applied voltage, V_S, but opposite in polarity, as shown in Part C. *When the capacitor is fully charged, there is no current.* A capacitor *blocks* constant dc.

When the charged capacitor is disconnected from the source, as shown in Figure 12–19D, it remains charged for long periods of time, depending on its leakage resistance, and can cause severe electrical shock. The charge on an electrolytic capacitor generally leaks off more rapidly than in other types of capacitors.

V_C does not change instantaneously

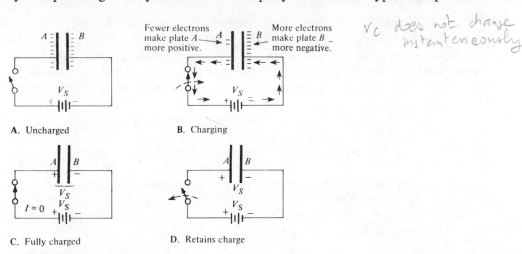

A. Uncharged   B. Charging

C. Fully charged   D. Retains charge

FIGURE 12–19   *Charging a capacitor.*

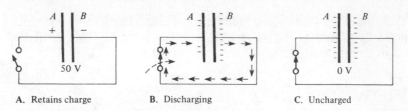

A. Retains charge   B. Discharging   C. Uncharged

FIGURE 12–20 *Discharging a capacitor.*

Discharging a Capacitor

When a wire is connected across a charged capacitor, as shown in Figure 12–20, the capacitor will discharge. In this particular case, a very low resistance path (the wire) is connected across the capacitor with a switch. Before the switch is closed, the capacitor is charged to 50 V, as indicated in Part A. When the switch is closed, as shown in Part B, the excess electrons on plate B move through the circuit to plate A; as a result of the current through the low resistance of the wire, the energy stored by the capacitor is dissipated in the wire. The charge is neutralized when the numbers of free electrons on both plates are again equal. At this time, the voltage across the capacitor is zero, and the capacitor is completely discharged, as shown in Part C.

Current during Charging and Discharging

Notice in Figures 12–19 and 12–20 that the direction of the current during discharge is opposite to that of the charging current. It is important to understand that *that there is no current through the dielectric of the capacitor during charging or discharging, because the dielectric is an insulating material.* Current flows from one plate to the other only through the external circuit.

The *RC* Time Constant

In a practical situation, there cannot be capacitance without some resistance in a circuit. It may simply be the small resistance of a wire, or it may be a designed-in resistance. Because of this, the charging and discharging characteristics of a capacitor must always be considered in light of the associated resistance. The resistance introduces the element of *time* in the charging and discharging of a capacitor.

When a capacitor charges or discharges through a resistance, a certain time is required for the capacitor to charge fully or discharge fully. *The voltage across a capacitor cannot change instantaneously,* because a finite time is required to move charge from one point to another. The rate at which the capacitor charges or discharges is determined by the *time constant* of the circuit. *The time constant of a series RC circuit is a time interval that equals the product of the resistance and the capacitance.* The time constant is symbolized by τ (Greek letter tau), and the formula is as follows:

$$\tau = RC \qquad\qquad\qquad \textbf{(12–8)}$$

Recall that $I = Q/t$. The current depends on the amount of charge moved in a given time. When the resistance is increased, the charging current is reduced, thus increasing the charging time of the capacitor. When the capacitance is increased, the amount of charge increases; thus, for the same current, more time is required to charge the capacitor.

Transient time = 5T

Example 12–5

A series RC circuit has a resistance of 1 MΩ and a capacitance of 5 μF. What is the time constant?

Solution:

$$\tau = RC = (1 \times 10^6 \ \Omega)(5 \times 10^{-6} \ \text{F}) = 5 \ \text{seconds}$$

During one time constant interval, the charge on a capacitor changes approximately 63%. Therefore, an uncharged capacitor charges to 63% of its fully charged voltage in one time constant. When discharging, the capacitor voltage drops to approximately 37% (100% − 63%) of its initial value in one time constant. This change also corresponds to a 63% change.

37% − 63%

The Charging and Discharging Curves

A capacitor charges and discharges following a nonlinear curve, as shown in Figure 12–21. In these graphs, the percentage of full charge is shown at each time-constant interval. This type of curve follows a precise mathematical formula and is called an *exponential curve.* The charging curve is an *increasing exponential,* and the discharging curve is a *decreasing exponential.* As you can see, it takes *five time constants* to approximately reach the final value.

General Formula: The general expressions for either increasing or decreasing exponential curves are given in the following equations for both voltage and current.

100 − 37
63

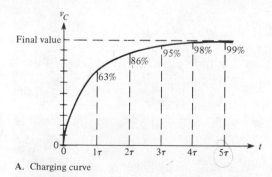

A. Charging curve

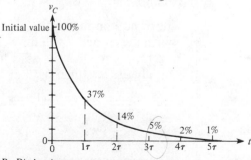

B. Discharging curve

FIGURE 12–21 *Charging and discharging exponential curves for an RC circuit.*

$\ln e^x = x$

$$v = V_F + (V_i - V_F)e^{-t/\tau} \tag{12-9}$$

$$i = I_F + (I_i - I_F)e^{-t/\tau} \tag{12-10}$$

$\dfrac{N}{V_F + (V_i - V_F)} = e^{-\frac{t}{\tau}}$

where V_F and I_F are the *final* values, and V_i and I_i are the *initial* values. v and i are the instantaneous values of the capacitor voltage or current at time t, and e is the base of natural logarithms with a value of 2.718. The e^x key or the INV and $\ln x$ keys on your calculator provide ease in evaluating this exponential term.

$\ln\left(\dfrac{N}{V_F + V_i - V_F}\right) = -\dfrac{t}{\tau}$

$-\tau \ln\left(\dfrac{N}{V_F + V_i - V_F}\right) = t$

The Charging Curve: The formula for the special case in which an increasing exponential voltage curve begins at zero ($V_i = 0$) is given in Equation (12-11). It is developed as follows, starting with the general formula.

$$v = V_F + (V_i - V_F)e^{-t/\tau}$$
$$= V_F + (0 - V_F)e^{-t/RC}$$
$$v = V_F(1 - e^{-t/RC}) \tag{12-11}$$

Using Equation (12-11), we can calculate the value of the charging voltage of a capacitor at any instant of time. The same is true for an increasing current.

$N = .99 V_F$

inductor

$i = I_F^+$

θ_{t_1}
v_{t_2}
dv

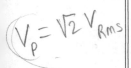

$A \to V_p$

$V_p = \sqrt{2} V_{RMS}$

Example 12-6

In Figure 12-22, determine the capacitor voltage 50 microseconds (μs) after the switch is closed if the capacitor is initially uncharged. Sketch the charging curve.

8 kΩ

50 V

0.01 μF

FIGURE 12-22

Solution:

The time constant is $RC = (8\ \text{k}\Omega)(0.01\ \mu\text{F}) = 80\ \mu\text{s}$. The voltage to which the capacitor will fully charge is 50 V (this is V_F). The initial voltage is zero. Notice that 50 μs is less than one time constant; so the capacitor will charge less than 63% of the full voltage in that time.

$$v_C = V_F(1 - e^{-t/RC}) = 50\ \text{V}(1 - e^{-50\,\mu\text{s}/80\,\mu\text{s}})$$
$$= 50\ \text{V}(1 - e^{-0.625}) = 50\ \text{V}(1 - 0.535)$$
$$= 23.2\ \text{V}$$

We determine the value of $e^{-0.625}$ on the calculator by entering -0.625 and then pressing the e^x key (or INV and then $\ln x$).

The charging curve for the capacitor is shown in Figure 12–23.

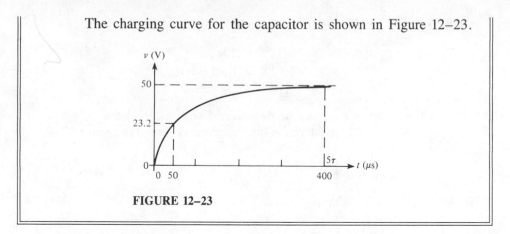

FIGURE 12–23

The Discharging Curve: The formula for the special case in which a decreasing exponential voltage curve ends at zero is derived from the general formula as follows:

$$v = V_F + (V_i - V_F)e^{-t/\tau}$$
$$= 0 + (V_i - 0)e^{-t/RC}$$

$$v = V_i e^{-t/RC} \qquad (12\text{--}12)$$

where V_i is the voltage at the beginning of the discharge. We can use this formula to calculate the discharging voltage at any instant, as Example 12–7 illustrates.

Example 12–7

Determine the capacitor voltage in Figure 12–24 at a point in time 6 milliseconds (ms) after the switch is closed. Sketch the discharging curve.

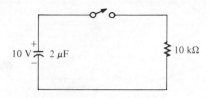

FIGURE 12–24

Solution:

The discharge time constant is $RC = (10 \text{ k}\Omega)(2 \text{ }\mu\text{F}) = 20$ ms. The initial capacitor voltage is 10 V. Notice that 6 ms is less than one time constant, so the capacitor will discharge less than 63%. Therefore, it will have a voltage greater than 37% of the initial voltage at 6 ms.

Example 12–7 (continued)

$$v_C = V_i e^{-t/RC} = 10e^{-6\,\text{ms}/20\,\text{ms}}$$
$$= 10e^{-0.3} = 10(0.741)$$
$$= 7.41 \text{ V}$$

Again, the value of $e^{-0.3}$ can be determined with a calculator.
The discharging curve for the capacitor is shown in Figure 12–25.

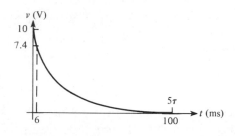

FIGURE 12–25

Universal Exponential Curves: The universal curves in Figure 12–26 provide a graphical solution of the charge and discharge of capacitors. Example 12–8 illustrates this graphical method.

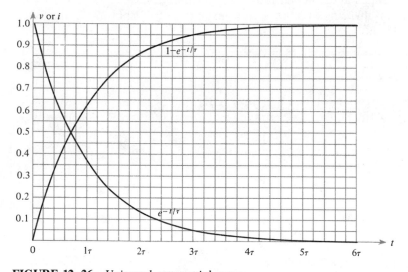

FIGURE 12–26 *Universal exponential curves.*

Example 12–8

How long will it take the capacitor in Figure 12–27 to charge to 75 V? What is the capacitor voltage 2 ms after the switch is closed? Use the universal curves in Figure 12–26 to determine the answers.

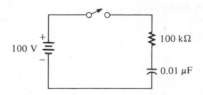

100 V

100 kΩ

0.01 μF

FIGURE 12–27

Solution:

The full charge voltage is 100 V, which is the 100% level on the graph. Since 75 V is 75% of the maximum, you can see that this value occurs at 1.4 time constants. One time constant is 1 ms. Therefore, the capacitor voltage reaches 75 V at 1.4 ms after the switch is closed.

The capacitor is at approximately 87 V in 2 ms. These graphical solutions are shown in Figure 12–28.

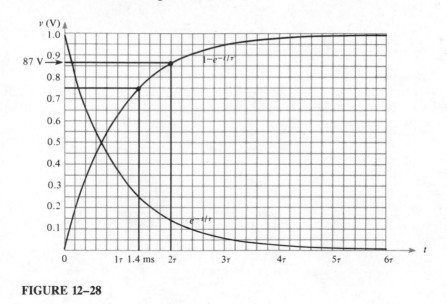

FIGURE 12–28

Time-Constant Percentage Tables: The percentages of full charge or discharge at each time-constant interval can be calculated using the exponential formulas, or they can be extracted from the universal graphs. The results are summarized in Tables 12–4 and 12–5 (page 402).

Solving for Time

Occasionally, it is necessary to determine how long it will take a capacitor to charge or discharge to a specified voltage. Equations (12–10) and (12–11) can be solved for t if v is specified. The natural logarithm (abbreviated ln) of

TABLE 12–4 *Percentage of final charge after each charging time-constant interval.*

Number of Time Constants	% Final Charge
1	63
2	86
3	95
4	98
5	99 (considered 100%)

TABLE 12–5 *Percentage of initial charge after each discharging time-constant interval.*

Number of Time Constants	% Initial Charge
1	37
2	14
3	5
4	2
5	1 (considered 0)

$e^{-t/RC}$ is the exponent $-t/RC$. Therefore, taking the natural logarithm of both sides of the equation allows us to solve for time. This calculation is done as follows with Equation (12–12):

$$v = V_i e^{-t/RC} \qquad V =$$

$$\frac{v}{V_i} = e^{-t/RC}$$

$$\ln\left(\frac{v}{V_i}\right) = \ln e^{-t/RC}$$

$$\ln\left(\frac{v}{V_i}\right) = \frac{-t}{RC}$$

$$t = -RC \ln\left(\frac{v}{V_i}\right) \qquad t = -RC \ln\left(1 - \frac{v}{V_F}\right)$$

The same procedure can be used for the increasing exponential formula in Equation (12–11).

Example 12–9

In Figure 12–29, how long will it take the capacitor to discharge to 25 V when the switch is closed?

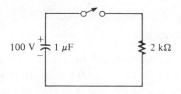

FIGURE 12–29

Solution:

$$t = -RC \ln\left(\frac{v}{V_i}\right) = -(2 \text{ ms}) \ln\left(\frac{25 \text{ V}}{100 \text{ V}}\right)$$

$$= -(2 \text{ ms}) \ln(0.25) = -(2 \text{ ms})(-1.39)$$

$$= 2.77 \text{ ms}$$

We can determine $\ln(0.25)$ with a calculator by first entering 0.25 and then pressing the $\ln x$ key.

Review for 12–3

1. Determine the time constant when $R = 1.2 \text{ k}\Omega$ and $C = 1000 \text{ pF}$.

2. If the circuit mentioned in Problem 1 is charged with a 5-V source, how long will it take the capacitor to reach full charge? At full charge, what is the capacitor voltage?

3. A certain circuit has a time constant of 1 ms. If it is charged with a 10-V battery, what will the capacitor voltage be at each of the following intervals: 2 ms, 3 ms, 4 ms, and 5 ms?

4. A capacitor is charged to 100 V. If it is discharged through a resistor, what is the capacitor voltage at one time constant?

5. In Figure 12–30, determine the voltage across the capacitor at 2.5 time constants after the switch is closed.

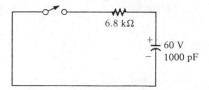

FIGURE 12–30

6. In Figure 12–30, how long will it take the capacitor to discharge to 15 V?

SERIES CAPACITORS

12-4

Total Capacitance

When capacitors are connected in series, the *effective* plate separation increases, and *the total capacitance is less than that of the smallest capacitor.* The reason is as follows: Consider the generalized circuit in Figure 12–31A, which has n capacitors in series with a voltage source and a switch. When the switch is closed, the capacitors charge as current is established through the circuit. Since this is a series circuit, the current must be the same at all points, as illustrated. Since current is the rate of flow of charge, *the amount of charge stored by each capacitor is equal to the total charge,* expressed as follows:

$$Q_T = Q_1 = Q_2 = Q_3 = \cdots = Q_n \tag{12–13}$$

Next, according to *Kirchhoff's voltage law,* the *sum* of the voltages across the charged capacitors must equal the total voltage, V_T, as shown in Figure 12–31B. This is expressed in equation form as

$$V_T = V_1 + V_2 + V_3 + \cdots + V_n$$

From Equation (12–3), $V = Q/C$. Substituting this relationship into each term of the voltage equation, the following result is obtained:

$$\frac{Q_T}{C_T} = \frac{Q_1}{C_1} + \frac{Q_2}{C_2} + \frac{Q_3}{C_3} + \cdots + \frac{Q_n}{C_n} \tag{12–14}$$

Since the charges on all the capacitors are equal, the Q terms can be factored from Equation (12–14) and canceled, resulting in

$$\frac{1}{C_T} = \frac{1}{C_1} + \frac{1}{C_2} + \frac{1}{C_3} + \cdots + \frac{1}{C_n}$$

Taking the reciprocal of both sides

$$C_T = \frac{1}{\dfrac{1}{C_1} + \dfrac{1}{C_2} + \dfrac{1}{C_3} + \cdots + \dfrac{1}{C_n}} \tag{12–15}$$

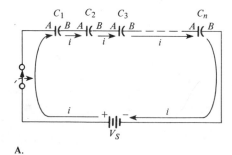

A.

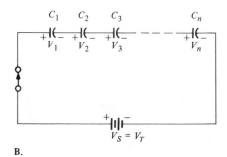

B.

FIGURE 12–31 *Series capacitive circuit.*

Special Case of Two Capacitors in Series

When only two capacitors are in series, a special form of Equation (12–14) can be used.

$$\frac{1}{C_T} = \frac{1}{C_1} + \frac{1}{C_2}$$

$$= \frac{C_1 + C_2}{C_1 C_2}$$

Taking the reciprocal of both sides we get

$$C_T = \frac{C_1 C_2}{C_1 + C_2} \qquad\qquad (12\text{–}16)$$

Capacitors of Equal Value in Series

This special case is another in which a formula can be developed from Equation (12–15). When all n values are the same and equal to C, we get

$$\frac{1}{C_T} = \frac{1}{C} + \frac{1}{C} + \frac{1}{C} + \cdots + \frac{1}{C}$$

Adding all the terms on the right, we get

$$\frac{1}{C_T} = \frac{n}{C}$$

Taking the reciprocal of both sides:

$$C_T = \frac{C}{n} \qquad\qquad (12\text{–}17)$$

The capacitance value of the equal capacitors divided by the number of equal series capacitors gives the total capacitance. Notice that the total series capacitance is calculated in the same manner as *total parallel resistance*.

Example 12–10

Determine the total capacitance in Figure 12–32.

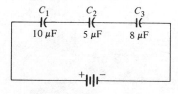

FIGURE 12–32

Example 12–10 (continued)

Solution:

$$\frac{1}{C_T} = \frac{1}{C_1} + \frac{1}{C_2} + \frac{1}{C_3}$$

$$= \frac{1}{10 \ \mu F} + \frac{1}{5 \ \mu F} + \frac{1}{8 \ \mu F}$$

Taking the reciprocal of both sides:

$$C_T = \frac{1}{\dfrac{1}{10 \ \mu F} + \dfrac{1}{5 \ \mu F} + \dfrac{1}{8 \ \mu F}}$$

$$= \frac{1}{0.425} \ \mu F$$

$$= 2.35 \ \mu F$$

Example 12–11

Find C_T in Figure 12–33.

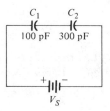

FIGURE 12–33

Solution:

$$C_T = \frac{C_1 C_2}{C_1 + C_2}$$

$$= \frac{(100 \ pF) \ (300 \ pF)}{400 \ pF}$$

$$= 75 \ pF$$

Example 12–12

Determine C_T for the series capacitors in Figure 12–34.

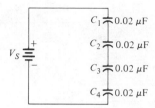

$C_1 \gtrless 0.02 \ \mu\text{F}$

$C_2 \gtrless 0.02 \ \mu\text{F}$

V_S

$C_3 \gtrless 0.02 \ \mu\text{F}$

$C_4 \gtrless 0.02 \ \mu\text{F}$

FIGURE 12–34

Solution:

$$C_1 = C_2 = C_3 = C_4 = C$$

$$C_T = \frac{C}{n} = \frac{0.02 \ \mu\text{F}}{4}$$

$$= 0.005 \ \mu\text{F}$$

Capacitor Voltages

A series connection of charged capacitors acts as a *voltage divider.* The voltage across each capacitor in series is inversely proportional to its capacitance value, as shown by the formula $V = Q/C$.

The voltage across any capacitor in series can be calculated as follows:

$$V_X = \left(\frac{C_T}{C_X}\right) V_T \qquad \textbf{(12–18)}$$

where C_X is C_1, or C_2, or C_3, and so on. The derivation is as follows: Since the charge on any capacitor in series is the same as the total charge ($Q_X = Q_T$), and since $Q_X = V_X C_X$ and $Q_T = V_T C_T$, then

$$V_X C_X = V_T C_T$$

Solving for V_X, we get

$$V_X = \frac{C_T V_T}{C_X}$$

The largest capacitor in series will have the smallest voltage, and the smallest capacitor will have the largest voltage.

Example 12–13

Find the voltage across each capacitor in Figure 12–35.

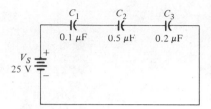

FIGURE 12–35

Solution:

$$\frac{1}{C_T} = \frac{1}{C_1} + \frac{1}{C_2} + \frac{1}{C_3}$$

$$= \frac{1}{0.1\ \mu F} + \frac{1}{0.5\ \mu F} + \frac{1}{0.2\ \mu F}$$

$$C_T = \frac{1}{17}\ \mu F = 0.0588\ \mu F$$

$$V_S = V_T = 25\ V$$

$$V_1 = \left(\frac{C_T}{C_1}\right)V_T = \left(\frac{0.0588\ \mu F}{0.1\ \mu F}\right)25\ V = 14.71\ V$$

$$V_2 = \left(\frac{C_T}{C_2}\right)V_T = \left(\frac{0.0588\ \mu F}{0.5\ \mu F}\right)25\ V = 2.94\ V$$

$$V_3 = \left(\frac{C_T}{C_3}\right)V_T = \left(\frac{0.0588\ \mu F}{0.2\ \mu F}\right)25\ V = 7.35\ V$$

Review for 12–4

1. Is the total capacitance of a series connection less than or greater than the value of the smallest capacitor?

2. The following capacitors are in series: 100 pF, 250 pF, and 500 pF. What is the total capacitance?

3. A 0.01-μF and a 0.015-μF capacitor are in series. Determine the total capacitance.

4. Five 100-pF capacitors are connected in series. What is C_T?

5. Determine the voltage across C_1 in Figure 12–36.

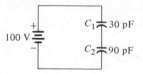

FIGURE 12–36

When capacitors are connected in parallel, the *effective* plate area increases, and *the total capacitance is the sum of the individual capacitances.* To understand this, consider what happens when the switch in Figure 12–37 is closed.

The total charging current from the source divides at the junction of the parallel branches. There is a separate charging current through each branch so that a different charge can be stored by each capacitor. By Kirchhoff's current law, the *sum* of all of the charging currents is equal to the total current. Therefore, the sum of the charges on the capacitors is equal to the total charge. Also, the voltages across all of the parallel branches are equal. These observations are used to develop a formula for total parallel capacitance as follows for the general case of n capacitors in parallel.

$$Q_T = Q_1 + Q_2 + Q_3 + \cdots + Q_n \qquad (12\text{–}19)$$

Since $Q = CV$,

$$C_T V_T = C_1 V_1 + C_2 V_2 + C_3 V_3 + \cdots + C_n V_n$$

Since $V_T = V_1 = V_2 = V_3 = \cdots = V_n$, the voltages can be factored and canceled, giving

$$C_T = C_1 + C_2 + C_3 + \cdots + C_n \qquad (12\text{–}20)$$

Equation (12–20) is the general formula for total parallel capacitance where n is the number of capacitors. Remember that *capacitors add in parallel.*

For the special case when all of the capacitors have the same value, C, multiply the value by the number of capacitors in parallel:

$$C_T = nC \qquad (12\text{–}21)$$

Notice that in all cases, the total parallel capacitance is calculated in the same manner as *total series resistance.*

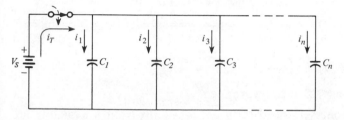

FIGURE 12–37 *Capacitors in parallel.*

Example 12–14

What is the total capacitance in Figure 12–38? What is the voltage across each capacitor?

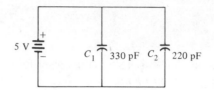

FIGURE 12–38

Solution:

$$C_T = C_1 + C_2 = 330 \text{ pF} + 220 \text{ pF}$$
$$= 550 \text{ pF}$$
$$V_1 = V_2 = 5 \text{ V}$$

Example 12–15

Determine C_T in Figure 12–39.

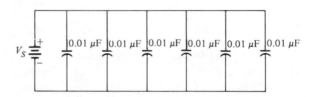

FIGURE 12–39

Solution:

There are six equal-valued capacitors in parallel, so $n = 6$.

$$C_T = nC = (6)(0.01 \ \mu\text{F}) = 0.06 \ \mu\text{F}$$

Review for 12–5

1. How is total parallel capacitance determined?
2. In a certain application, you need 0.05 μF. The only values available are 0.01 μF, which are available in large quantities. How can you get the total capacitance that you need?

3. The following capacitors are in parallel: 10 pF, 5 pF, 33 pF, and 50 pF. What is C_T?

12-6

In order to understand fully the action of capacitors in ac circuits, the concept of the *derivative* must be introduced. *The derivative of a time-varying quantity is the instantaneous rate of change of that quantity.*

Recall that current is the *rate of flow of charge (electrons)*. Therefore, instantaneous current, i, can be expressed as the instantaneous rate of change of charge, q, with respect to time, t:

$$i = \frac{dq}{dt} \tag{12-22}$$

The term dq/dt is the *derivative* of q with respect to time and represents the instantaneous rate of change of q. Also, in terms of instantaneous quantities, $q = Cv$. Therefore, from a basic rule of differential calculus, the derivative of q is $dq/dt = C(dv/dt)$. Since $i = dq/dt$, we get the following relationship:

$$i = C\left(\frac{dv}{dt}\right) \tag{12-23}$$

This equation says that *the instantaneous capacitor current is equal to the capacitance times the instantaneous rate of change of the voltage across the capacitor.* From this, you can see that the faster the voltage across a capacitor changes, the greater the current.

Phase Relationship of Current and Voltage in a Capacitor

Now consider what happens when a sinusoidal voltage is applied across a capacitor, as shown in Figure 12–40. The voltage wave form has a maximum rate of change (dv/dt = max) at the zero crossings and a zero rate of change ($dv/dt = 0$) at the peaks, as indicated in Figure 12–41A on page 412.

Using Equation (12–23), the phase relationship between the current and the voltage for the capacitor can be established. When $dv/dt = 0$, i is also zero, because $i = C(dv/dt) = C(0) = 0$. When dv/dt is a positive-going maximum, i is a positive maximum; when dv/dt is a negative-going maximum, i is a negative maximum.

A sinusoidal voltage always produces a sinusoidal current in a capacitive circuit. Therefore, the current can be plotted with respect to the voltage by

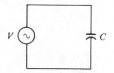

FIGURE 12–40   *Sine wave applied to a capacitor.*

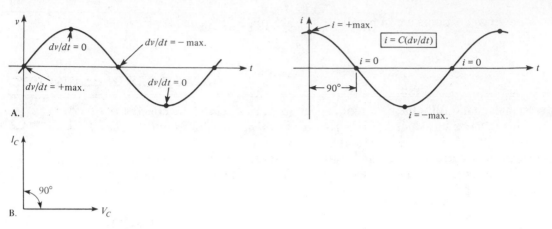

FIGURE 12–41   *Phase relation of v and i in a capacitor. Current leads voltage by 90 degrees.*

knowing the points on the voltage curve at which the current is zero and those at which it is maximum. This relationship is shown in Figure 12–41A. Notice that *the current leads the voltage in phase by 90 degrees*. This is always true in a purely capacitive circuit. A phasor diagram of this relationship is shown in Figure 12–41B.

Capacitive Reactance

Capacitive reactance is the opposition to sinusoidal current, expressed in ohms. The symbol for capacitive reactance is X_C.

To develop a formula for X_C, we use the relationship $i = C(dv/dt)$ and the curves in Figure 12–42. The rate of change of voltage is directly related to *frequency*. The faster the voltage changes, the higher the frequency. For example, you can see that in Figure 12–42, the slope of sine wave A at the zero crossings is greater than that of sine wave B. The *slope* of a curve at a point indicates the rate of change at that point. Sine wave A has a higher frequency than sine wave B, as indicated by a greater maximum rate of change (dv/dt is greater at the zero crossings).

When frequency increases, dv/dt increases, and thus i increases. When frequency decreases, dv/dt decreases, and thus i decreases:

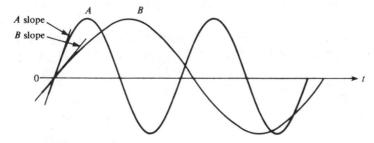

FIGURE 12–42   *A higher-frequency wave has a greater slope at its zero crossings, corresponding to a higher rate of change.*

$$\overset{\uparrow}{i} \;=\; C(d\overset{\uparrow}{v}/dt) \qquad \text{and} \qquad \underset{\downarrow}{i} \;=\; C(d\underset{\downarrow}{v}/dt)$$

An increase in i means that there is less opposition to current (X_C is less), and a decrease in i means a greater opposition (X_C is greater). Therefore, X_C *is inversely proportional to i and thus inversely proportional to frequency.*

$$X_C \text{ is proportional to } \frac{1}{f}$$

Again, from the relationship $i = C(dv/dt)$, you can see that if dv/dt is constant and C is varied, an increase in C produces an increase in i, and a decrease in C produces a decrease in i.

$$\overset{\uparrow}{i} \;=\; \overset{\uparrow}{C}(dv/dt) \qquad \text{and} \qquad \underset{\downarrow}{i} \;=\; \underset{\downarrow}{C}(dv/dt)$$

Again, an increase in i means less opposition (X_C is less), and a decrease in i means greater opposition (X_C is greater). Therefore, X_C *is inversely proportional to i and thus inversely proportional to capacitance.*

The capacitive reactance is inversely proportional to both f and C.

$$X_C \text{ is proportional to } \frac{1}{fC}$$

Thus far, we have determined a proportional relationship between X_C and $1/fC$. What is needed is a formula that tells us what X_C is *equal* to so that it can be calculated. This formula is derived in Appendix F and is stated as follows:

$$X_C = \frac{1}{2\pi fC} \qquad\qquad (12\text{–}24)$$

X_C is in *ohms* when f is in *hertz* and C is in *farads*. Notice that 2π appears in the denominator as a constant of proportionality. This term is derived from the relationship of a sine wave to rotational motion, as you can see in Appendix F.

Example 12–16

A sinusoidal voltage is applied to a capacitor, as shown in Figure 12–43. The frequency of the sine wave is 1 kHz. Determine the capacitive reactance.

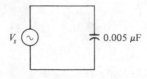

V_s 0.005 μF

FIGURE 12–43

Example 12–16 (continued)

Solution:

$$X_C = \frac{1}{2\pi f C} = \frac{1}{2\pi(1 \times 10^3 \text{ Hz})(0.005 \times 10^{-6} \text{ F})}$$

$$= 31.83 \text{ k}\Omega$$

Ohm's Law as Applied to X_C: Ohm's law applies to reactance as well as to resistance. The three equivalent forms are as follows:

$$V = IX_C \tag{12–25}$$

$$I = \frac{V}{X_C} \tag{12–26}$$

$$X_C = \frac{V}{I} \tag{12–27}$$

where X_C is used in the place of R. Both current and voltage must be expressed in the same units of rms, peak, peak-to-peak, or average values.

Example 12–17

Determine the rms current in Figure 12–44.

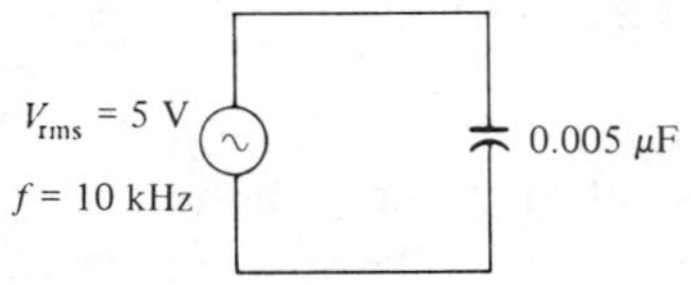

FIGURE 12–44

Solution:

$$X_C = \frac{1}{2\pi f C} = \frac{1}{2\pi(10 \times 10^3 \text{ Hz})(0.005 \times 10^{-6} \text{ F})}$$

$$= 3.18 \text{ k}\Omega$$

$$I_{rms} = \frac{V_{rms}}{X_C} = \frac{5 \text{ V}}{3.18 \text{ k}\Omega}$$

$$= 1.57 \text{ mA}$$

Complex Form of Capacitive Reactance: X_C can be treated as a complex quantity, because it introduces a phase angle of 90 degrees between the current and the voltage. Italic X_C represents just the magnitude; boldface $\mathbf{X}_C$ is a phasor quantity representing both magnitude and angle. $\mathbf{X}_C$ causes the voltage to lag the current by 90 degrees; so in polar form it is assigned a -90-degree angle and is expressed as $X_C\angle-90°$. In rectangular form, it is expressed as $-jX_C$, where the $-j$ indicates a -90-degree phase shift.

We make use of these complex forms for circuit analysis in later chapters.

Power in a Capacitor

As discussed earlier in this chapter, a charged capacitor stores energy in the electric field within the dielectric. An ideal capacitor does not dissipate energy; it only stores it. When an ac voltage is applied to a capacitor, energy is stored by the capacitor during a portion of the voltage cycle; then the stored energy is *returned to the source* during another portion of the cycle. *There is no net energy loss.* Figure 12–45 shows the power curve that results from one cycle of capacitor voltage and current.

Instantaneous Power (p): *The product of v and i gives instantaneous power, p.* At points where v or i is zero, p is also zero. When both v and i are positive, p is also positive. When either v or i is positive and the other negative, p is negative. When both v and i are negative, p is positive. As you can see, the power follows a sinusoidal curve. Positive values of power indicate that energy is stored by the capacitor. Negative values of power indicate that energy is returned from the capacitor to the source. Note that the power fluctuates at a frequency twice that of the voltage or current as energy is alternately stored and returned to the source.

Average Power (P_{avg}): Ideally, all of the energy stored by a capacitor during the positive portion of the power cycle is returned to the source during the negative

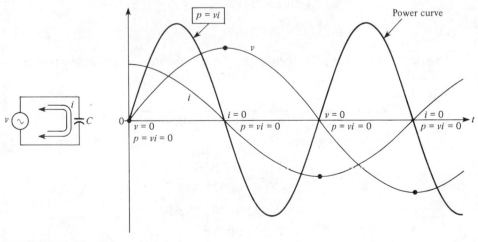

FIGURE 12–45 *Power curve.*

portion. *No net energy is consumed in the capacitor,* so the *average power is zero.* Actually, because of leakage and foil resistance in a practical capacitor, a small percentage of the total power is dissipated.

Reactive Power (P_r): The rate at which a capacitor stores or returns energy is called its *reactive power,* P_r. The reactive power is a nonzero quantity, because at any instant in time, the capacitor is actually taking energy from the source or returning energy to it. Reactive power does not represent an energy loss. The following formulas apply:

$$P_r = V_{rms} I_{rms} \qquad\qquad (12\text{--}28)$$

$$P_r = \frac{V_{rms}^2}{X_C} \qquad\qquad (12\text{--}29)$$

$$P_r = I_{rms}^2 X_C \qquad\qquad (12\text{--}30)$$

Notice that these equations are of the same form as those for average power in a resistor. The voltage and current are expressed in rms. The unit of reactive power is *volt-amperes reactive* (VAR).

Example 12–18

Determine the average power and the reactive power in Figure 12–46.

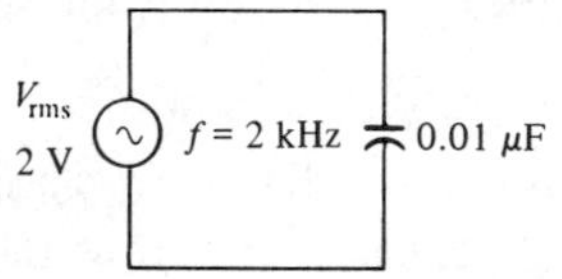

V_{rms}
2 V $f = 2$ kHz 0.01 μF

FIGURE 12–46

Solution:

The average power, P_{avg}, is *always zero for a capacitor.* The reactive power is as follows:

$$X_C = \frac{1}{2\pi f C} = \frac{1}{2\pi(2 \times 10^3 \text{ Hz})(0.01 \times 10^{-6} \text{ F})}$$

$$= 7.958 \text{ k}\Omega$$

$$P_r = \frac{V_{rms}^2}{X_C} = \frac{(2 \text{ V})^2}{7.958 \text{ k}\Omega}$$

$$= 0.503 \times 10^{-3} \text{ VAR}$$

$$= 0.503 \text{ mVAR}$$

Review for 12–6

1. State the phase relationship between current and voltage in a capacitor.
2. Calculate X_C for $f = 5$ kHz and $C = 50$ pF.
3. At what frequency is the reactance of a 0.1-μF capacitor equal to 2 kΩ?
4. Calculate the rms current in Figure 12–47.

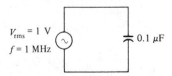

$V_{rms} = 1$ V
$f = 1$ MHz
$\doteqdot 0.1~\mu$F

FIGURE 12–47

5. A 1-μF capacitor is connected to an ac voltage source of 12 V rms. What is the average power?
6. In Problem 5, determine reactive power at a frequency of 500 Hz.

TESTING CAPACITORS

12–7

Capacitors are generally very reliable devices. Their useful life can be extended significantly by operating well within the voltage rating and at moderate temperatures.

Failures can be categorized into two areas: *catastrophic* and *degradation*. The catastrophic failures are usually a short circuit caused by dielectric breakdown or an open circuit caused by connection failure. Degradation usually results in a gradual decrease in leakage resistance, hence an increase in leakage current.

Ohmmeter Check

When there is a suspected problem, the capacitor can be removed from the circuit and checked with an ohmmeter. First, to be sure that the capacitor is discharged, short its leads, as indicated in Figure 12–48A. Connect

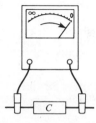
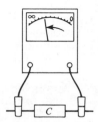

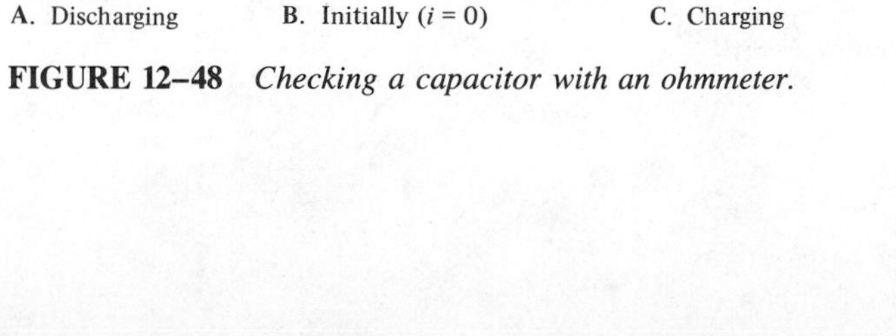

A. Discharging   B. Initially ($i = 0$)   C. Charging

FIGURE 12–48 *Checking a capacitor with an ohmmeter.*

the meter, set on a high ohms range such as R×1M, to the capacitor, as shown in Figure 12–48B, and observe the needle. It should initially indicate near zero ohms. Then it should begin to move toward the high-resistance end of the scale as the capacitor charges from the ohmmeter's battery, as shown in Figure 12–48C. When the capacitor is fully charged, the meter will indicate an extremely high resistance.

As mentioned, the capacitor charges from the internal battery of the ohmmeter, and the meter responds to the charging current. The larger the capacitance value, the more slowly the capacitor will charge, as indicated by the needle movement. For very small pF values, the meter response may be insufficient to indicate the fast charging action.

If the capacitor is internally shorted, the meter will go to zero and stay there. If it is leaky, the final meter reading will be much less than normal. Most capacitors have a resistance of several hundred megohms. The exception is the electrolytic, which may normally have less than one megohm of leakage resistance. If the capacitor is open, no charging action will be observed, and the meter will indicate an infinite resistance. The actual capacitance value can be measured on a capacitance bridge.

Review for 12–7

1. How can a capacitor be discharged after removal from the circuit?

2. Describe how the needle of an ohmmeter responds when a good capacitor is checked.

COMPUTER ANALYSIS

12–8   This program computes the instantaneous charging voltage for a capacitor after switch closure in a series *RC* circuit connected to a dc source. The program requires that you input the dc source voltage, the resistance value in ohms, the capacitance value in farads, and the number of time intervals for which you want the voltage computed during the full charging interval. The program output is a tabulation of the instantaneous voltages and percent of full charge at each point in time during the charging interval. Note that the exponentiation symbol ([) in line 190 is the ↑ symbol on the keyboard.

```
10 CLS
20 PRINT "THIS PROGRAM COMPUTES AND TABULATES THE INSTANTANEOUS"
30 PRINT "CAPACITOR VOLTAGE AND PERCENT OF FULL CHARGE AT"
40 PRINT "NUMBER OF SPECIFIED TIME INTERVALS DURING CHARGING"
50 PRINT "IN A SPECIFIED RC CIRCUIT."
60 PRINT
70 PRINT "ENTER THE DC SOURCE VOLTAGE, RESISTANCE IN OHMS,"
80 PRINT "CAPACITANCE IN FARADS, AND DESIRED NUMBER OF TIME"
90 PRINT "INTERVALS."
100 FOR T=0 TO 4000:NEXT:CLS
110 INPUT "DC SOURCE VOLTAGE";VDC
120 INPUT "CAPACITANCE IN FARADS";C
130 INPUT "RESISTANCE IN OHMS";R
```

```
140 INPUT "NUMBER OF TIME INTERVALS ";TI
150 CLS
160 PRINT "TIME (SEC)","CAPACITOR VOLTAGE","% OF FULL CHARGE"
170 FOR T=0 TO 5*R*C STEP 5*R*C/TI
180 X=T/(R*C)
190 V=VDC*(1-(2.718)[-X)
200 P=(V/VDC)*100
210 PRINT TAB(0) T "S";TAB(20) V "V";TAB(50) P "%"
220 NEXT
```

Review for 12–8

1. Identify by line number each FOR/NEXT loop.

2. State the purpose of each FOR/NEXT loop.

Formulas

$$C = \frac{Q}{V} \tag{12-1}$$

$$Q = CV \tag{12-2}$$

$$V = \frac{Q}{C} \tag{12-3}$$

$$F = \frac{kQ_1 Q_2}{d^2} \tag{12-4}$$

$$\mathcal{E} = \tfrac{1}{2}CV^2 \tag{12-5}$$

$$\epsilon_r = \frac{\epsilon}{\epsilon_0} \tag{12-6}$$

$$C = \frac{A\epsilon_r(8.85 \times 10^{-12} \text{ F/m})}{d} \tag{12-7}$$

$$\tau = RC \tag{12-8}$$

$$v = V_F + (V_i - V_F)e^{-t/\tau} \tag{12-9}$$

$$i = I_F + (I_i - I_F)e^{-t/\tau} \tag{12-10}$$

$$v = V_F(1 - e^{-t/RC}) \tag{12-11}$$

$$v = V_i e^{-t/RC} \tag{12-12}$$

$$Q_T = Q_1 = Q_2 = Q_3 = \cdots = Q_n \tag{12-13}$$

$$\frac{Q_T}{C_T} = \frac{Q_1}{C_1} + \frac{Q_2}{C_2} + \frac{Q_3}{C_3} + \cdots + \frac{Q_n}{C_n} \tag{12-14}$$

$$C_T = \frac{1}{(1/C_1) + (1/C_2) + (1/C_3) + \cdots + (1/C_n)} \tag{12-15}$$

$$C_T = \frac{C_1 C_2}{C_1 + C_2} \tag{12–16}$$

$$C_T = \frac{C}{n} \tag{12–17}$$

$$V_X = \left(\frac{C_T}{C_X}\right) V_T \tag{12–18}$$

$$Q_T = Q_1 + Q_2 + Q_3 + \cdots + Q_n \tag{12–19}$$

$$C_T = C_1 + C_2 + C_3 + \cdots + C_n \tag{12–20}$$

$$C_T = nC \tag{12–21}$$

$$i = \frac{dq}{dt} \tag{12–22}$$

$$i = C\left(\frac{dv}{dt}\right) \tag{12–23}$$

$$X_C = \frac{1}{2\pi f C} \tag{12–24}$$

$$V = I X_C \tag{12–25}$$

$$I = \frac{V}{X_C} \tag{12–26}$$

$$X_C = \frac{V}{I} \tag{12–27}$$

$$P_r = V_{rms} I_{rms} \tag{12–28}$$

$$P_r = \frac{V_{rms}^2}{X_C} \tag{12–29}$$

$$P_r = I_{rms}^2 X_C \tag{12–30}$$

Summary

1. A capacitor is composed of two parallel conducting plates separated by a dielectric insulator.
2. Capacitance is the measure of a capacitor's ability to store electrical charge.
3. Energy is stored by a capacitor in the electric field concentrated in the dielectric.
4. The farad is the unit of capacitance.
5. *One farad* is the amount of capacitance when *one coulomb* of charge is stored with *one volt* across the plates.

6. Capacitance is directly proportional to the plate area and inversely proportional to the plate separation.

7. The dielectric constant is an indication of the ability of a material to establish an electric field.

8. The dielectric strength is one factor that determines the breakdown voltage of a capacitor.

9. The time constant for a series RC circuit is the resistance times the capacitance.

10. In an RC circuit, the voltage and current in a charging or discharging capacitor make a 63% change during each time-constant interval.

11. Five time constants are required for a capacitor to charge fully or to discharge fully.

12. Charging and discharging follow exponential curves.

13. Total series capacitance is less than that of the smallest capacitor in series.

14. Capacitance adds in parallel.

15. Current leads voltage by 90 degrees in a capacitor.

16. Capacitive reactance is the opposition to sinusoidal current and is expressed in ohms.

17. X_C is inversely proportional to frequency and capacitance.

18. The average power in a capacitor is zero; that is, there is no energy loss in an ideal capacitor.

Self-Test

1. Indicate true or false for each of the following statements:
   (a) The plates of a capacitor are conductive.
   (b) The dielectric is the insulating material that separates the plates.
   (c) Constant dc flows through a fully charged capacitor.
   (d) A practical capacitor stores charge indefinitely when it is disconnected from the source.

2. Indicate true or false for each of the following statements:
   (a) There is current through the dielectric of a charging capacitor.
   (b) When a capacitor is connected to a dc voltage source, it will charge to the value of the source voltage.
   (c) You can discharge an ideal capacitor by simply disconnecting it from the voltage source.
   (d) When a capacitor is completely discharged, the voltage across its plates is zero.

3. A 0.006-μF capacitor is charged to 10 V. How many coulombs of charge does it store?

4. A given capacitor stores 50×10^{-6} coulomb of charge when the voltage across its plates is 5 V. Determine its capacitance.

5. A 0.01-μF capacitor stores 5×10^{-8} C. What is the voltage across its plates?

6. Convert 0.005 μF to picofarads.

7. Convert 2000 pF to microfarads.

8. Convert 1500 μF to farads.

9. A 10-μF capacitor is charged to 100 V. How many joules of energy are stored?

10. Calculate the absolute permittivity, ϵ, for ceramic.

11. Calculate the capacitance for the following physical parameters: $A = 0.009$ m^2, $d = 0.0015$ m, and a dielectric of Teflon®.

12. A circuit has $R = 2$ kΩ in series with $C = 0.05$ μF. What is the time constant?

13. The capacitor in an RC circuit is charged to 15 V. If it is allowed to discharge for an interval equal to one time constant, what is the voltage across the capacitor?

14. A series RC circuit with a time constant of 10 ms is connected across a 10-V dc source with a switch. If the capacitor is initially uncharged (0 V), what voltage will it reach at 15 ms after switch closure?

15. Determine the total capacitance in each circuit in Figure 12–49.

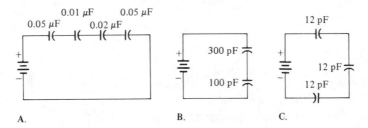

A.   B.   C.

FIGURE 12–49

16. Find the voltage across each capacitor in Figure 12–50.

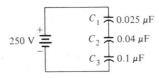

FIGURE 12–50

17. Find C_T in each circuit of Figure 12–51.

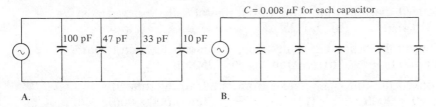

A.

B.

FIGURE 12–51

18. A 2-MHz sinusoidal voltage is applied to a 1-μF capacitor. What is the reactance?

19. The frequency of the source in Figure 12–52 can be adjusted. In order to obtain 50 mA of rms current, to what frequency must the source be adjusted?

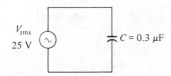

FIGURE 12–52

20. Determine the average power and the reactive power when the circuit in Figure 12–52 is adjusted as required in Problem 19.

21. An ohmmeter shows 1000 Ω when connected across a capacitor that is suspected to be faulty. What is your conclusion?

Problems

Section 12–1

12–1. (a) Find the capacitance when $Q = 50\ \mu$C and $V = 10$ V.
(b) Find the charge when $C = 0.001\ \mu$F and $V = 1$ kV.
(c) Find the voltage when $Q = 2$ mC and $C = 200\ \mu$F.

12–2. The total charge stored by the series capacitors in Figure 12–53 is 10 μC. Determine the voltage across each of the capacitors.

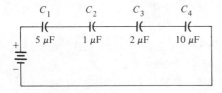

FIGURE 12–53

12–3. Convert the following values from microfarads to picofarads:
(a) 0.1 μF   (b) 0.0025 μF   (c) 5 μF

12–4. Convert the following values from picofarads to microfarads:
(a) 1000 pF   (b) 3500 pF   (c) 250 pF

12–5. Convert the following values from farads to microfarads:
(a) 0.0000001 F   (b) 0.0022 F   (c) 0.0000000015 F

12–6. Calculate the force of repulsion between two electrons 0.001 m apart.

12–7. What size capacitor is capable of storing 10 mJ of energy with 100 volts across its plates?

12–8. Calculate the absolute permittivity, ϵ, for each of the following materials:
(a) air   (b) oil   (c) glass   (d) Teflon®

12–9. A mica capacitor has a plate area of 0.04 m² and a dielectric thickness of 0.008 m. What is its capacitance?

12–10. An air capacitor has 0.1-m square plates. The plates are separated by 0.01 m. Calculate the capacitance.

12–11. At ambient temperature (25°C), a certain capacitor is specified to be 1000 pF. It has a negative temperature coefficient of 200 ppm/°C. What is its capacitance at 75°C?

12–12. A 0.001-μF capacitor has a positive temperature coefficient of 500 ppm/°C. How much change in capacitance will a 25°C increase in temperature cause?

Section 12–2

12–13. In the construction of a mica capacitor, how is the plate area increased?

12–14. What type of capacitor has the highest dielectric constant?

12–15. Show how to connect an electrolytic capacitor between points A and B in Figure 12–54.

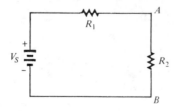

FIGURE 12–54

12–16. Name two types of electrolytic capacitors.

Section 12–3

12–17. Determine the time constant for each of the following series RC combinations:
(a) $R = 100\ \Omega$, $C = 1\ \mu F$ (b) $R = 10\ M\Omega$, $C = 50$ pF
(c) $R = 4.7\ k\Omega$, $C = 0.005\ \mu F$ (d) $R = 1.5\ M\Omega$, $C = 0.01\ \mu F$

12–18. Determine how long it takes the capacitor to reach full charge for each of the following combinations:
   (a) $R = 50\ \Omega$, $C = 50\ \mu F$ **(b)** $R = 3300\ \Omega$, $C = 0.015\ \mu F$
   (c) $R = 22\ k\Omega$, $C = 100\ pF$ **(d)** $R = 5\ M\Omega$, $C = 10\ pF$

12–19. In the circuit of Figure 12–55, the capacitor is initially uncharged. Determine the capacitor voltage at the following times after the switch is closed.
   (a) $10\ \mu s$ **(b)** $20\ \mu s$ **(c)** $30\ \mu s$ **(d)** $40\ \mu s$ **(e)** $50\ \mu s$

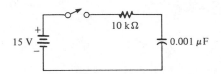

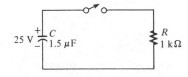

FIGURE 12–55   **FIGURE 12–56**

12–20. In Figure 12–56, the capacitor is charged to 25 V. When the switch is closed, what is the capacitor voltage after the following times?
   (a) 1.5 ms   **(b)** 4.5 ms   **(c)** 6 ms   **(d)** 7.5 ms

12–21. Repeat Problem 12–19 for the following time intervals:
   (a) $2\ \mu s$ **(b)** $5\ \mu s$ **(c)** $15\ \mu s$

12–22. Repeat Problem 12–20 for the following times:
   (a) 0.5 ms   **(b)** 1 ms   **(c)** 2 ms

12–23. Derive the formula for finding the time at any point on an *increasing* exponential voltage curve. Use this formula to find the time at which the voltage in Figure 12–57 reaches 6 V after switch closure.

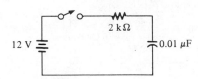

FIGURE 12–57

12–24. How long does it take C to discharge to 3 V in Figure 12–56?

12–25. How long does it take C to charge to 8 V in Figure 12–55?

Section 12–4

12–26. Five 1000-pF capacitors are in series. What is the total capacitance?

12–27. Find the total capacitance for each circuit in Figure 12–58.

12–28. For each circuit in Figure 12–58, determine the voltage across each capacitor.

12–29. Two series capacitors (one 1-μF, the other of unknown value) are charged from a 12-V source. The 1-μF capacitor is charged to 8 V, and the other to 4 V. What is the value of the unknown capacitor?

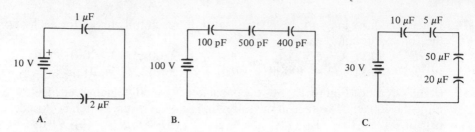

FIGURE 12–58

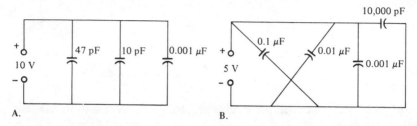

FIGURE 12–59

Section 12–5

12–30. Determine C_T for each circuit in Figure 12–59.

12–31. What is the charge on each capacitor in Figure 12–59?

12–32. Determine C_T for each circuit in Figure 12–60.

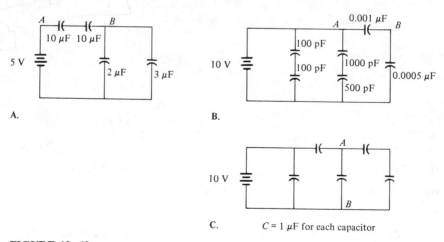

FIGURE 12–60

12–33. What is the voltage between points A and B in each circuit in Figure 12–60?

Section 12–6

12–34. What is the value of the total capacitive reactance in each circuit in Figure 12–61?

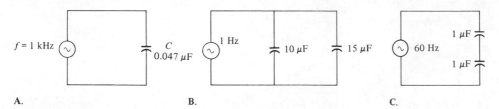

FIGURE 12–61

12–35. In Figure 12–60, each dc voltage source is replaced by a 10-V rms, 2-kHz ac source. Determine the reactance in each case.

12–36. In each circuit of Figure 12–61, what frequency is required to produce an X_C of 100 Ω? An X_C of 1 kΩ?

12–37. A sinusoidal voltage of 20 V rms produces an rms current of 100 mA when connected to a certain capacitor. What is the reactance?

12–38. A 10-kHz voltage is applied to a 0.0047-μF capacitor, and 1 mA of rms current is measured. What is the value of the voltage?

12–39. Determine the average power and the reactive power in Problem 12–38.

Section 12–8

12–40. Write a computer program in BASIC that will compute and tabulate the capacitive reactance for specified values of R and C over a specified range of frequencies in specified increments.

12–41. Develop a program similar to that in Section 12–8 for capacitor discharging.

Advanced Problems

12–42. Determine the time constant for the circuit in Figure 12–62.

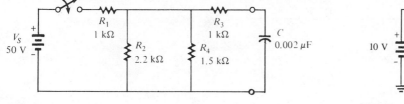

FIGURE 12–62

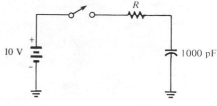

FIGURE 12–63

12–43. In Figure 12–63, the capacitor is initially uncharged. At $t = 10$ μs after the switch is closed, the instantaneous capacitor voltage is 7.2 V. Determine the value of R.

12–44. (a) The capacitor in Figure 12–64 is uncharged when the switch is thrown into position 1. The switch remains in position 1 for 10 ms and is then thrown into position 2, where it remains indefinitely. Sketch the complete wave form for the capacitor voltage.

(b) If the switch is thrown back to position 1 after 5 ms in position 2, and then left in position 1, how would the wave form appear?

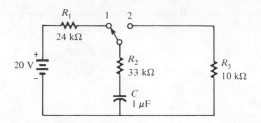

FIGURE 12–64

12–45. Determine the ac voltage across each capacitor and the current in each branch of the circuit in Figure 12–65.

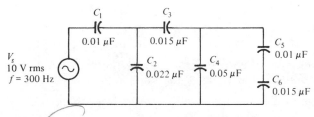

FIGURE 12–65

12–46. Find the value of C_1 in Figure 12–66.

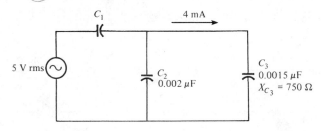

FIGURE 12–66

Answers to Section Reviews

Section 12–1:
1. The ability (capacity) to store electrical charge. **2. (a)** 10^6, **(b)** 10^{12}, **(c)** 10^6.
3. 1500 pF; 0.0000000015 F. **4.** 1.125 μJ. **5. (a)** increase, **(b)** decrease.
6. 10 kV. **7.** 0.425 μF. **8.** 2.01 μF.

Section 12–2:
1. By the dielectric material. **2.** A fixed capacitance cannot be changed; a variable can. **3.** Electrolytic. **4.** Make sure the voltage rating is sufficient. Connect the positive end to the positive side of the circuit.

Section 12–3:
1. 1.2 μs. **2.** 6 μs, 5 V. **3.** 8.6 V, 9.5 V, 9.8 V, 9.9 V. **4.** 37 V. **5.** 4.93 V.
6. 9.43 μs.

Section 12–4:
1. Less. **2.** 62.5 pF. **3.** 0.006 μF. **4.** 20 pF. **5.** 75 V.

Section 12–5:
1. The individual capacitors are added. **2.** By using five 0.01-μF capacitors in parallel. **3.** 98 pF.

Section 12–6:
1. Current leads voltage by 90 degrees. **2.** 637 kΩ. **3.** 796 Hz. **4.** 628 mA.
5. 0. **6.** 0.453 VAR.

Section 12–7:
1. Short its leads. **2.** Initially, the needle jumps to zero; then it slowly moves to the high-resistance end of the scale.

Section 12–8:
1. Line 100; lines 170–220. **2.** Line 100 is the time delay loop; lines 170–220 calculate and print capacitor voltage and percent of full charge.

Inductance is the property of a coil of wire that opposes a change in current. The basis for inductance is the magnetic field that surrounds any conductor when there is current through it. The electrical component designed to have inductance is called an *inductor, coil,* or *choke.* All of these terms have essentially the same meaning.

Magnetic theory is introduced only to the extent necessary to provide a sufficient background for understanding inductance. Also, the basic inductor is introduced, and its characteristics are studied. Various types of inductors are covered in terms of their physical construction and their electrical properties.

The basic behavior of inductors in both dc and ac circuits is studied, and series and parallel combinations are analyzed. Methods for testing inductors are also discussed.

In this chapter, you will learn:

1. The basic properties of magnetic fields.
2. The principles of electromagnetism.
3. How force is exerted on a conductor in a magnetic field.
4. How voltage is induced in an inductor.
5. How inductance is measured.
6. How an inductor stores energy.
7. The definitions of Faraday's law and Lenz's law.
8. How various physical characteristics affect inductance.
9. The classifications of inductors.
10. The definition of time constant, and how to determine the charging and discharging time of an inductor.
11. How to evaluate instantaneous values on a charging or discharging curve.
12. How to obtain a desired inductance value by combining inductors in series.
13. How to obtain a desired inductance value by combining inductors in parallel.
14. How an inductor introduces a phase shift between current and voltage.
15. The definition of inductive reactance, and how to determine its value in any given circuit application.
16. Why an inductor ideally does not dissipate power.
17. The meaning of reactive power in an inductive circuit.

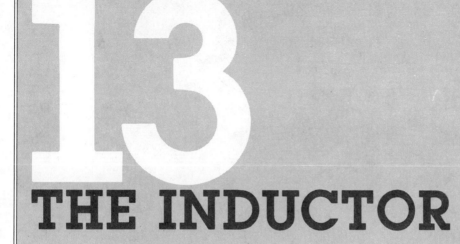

13
THE INDUCTOR

13-1

Magnetic Fields

Magnetic fields are important in the study of inductance. In this section, the basic aspects of magnetic fields are studied as a background for understanding inductance.

A permanent magnet, such as the bar magnet shown in Figure 13–1, has a magnetic field surrounding it. The magnetic field consists of *lines of force* that radiate from the north pole (N) to the south pole (S) and back to the north pole through the magnetic material. For clarity, only a few lines of force are shown in the figure. Imagine, however, that many lines surround the magnet in three dimensions.

Attraction and Repulsion of Magnetic Poles: When *unlike* poles of two permanent magnets are placed close together, an attractive force is produced by the magnetic fields, as indicated in Figure 13–2A. When two *like* poles are brought close together, they repel each other, as shown in Part B of the figure.

Altering a Magnetic Field: When a nonmagnetic material, such as paper, glass, wood, or plastic, is placed in a magnetic field, the lines of force are

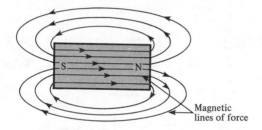

Magnetic
lines of force

FIGURE 13–1 *Magnetic lines of force around a bar magnet.*

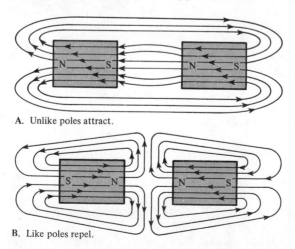

A. Unlike poles attract.

B. Like poles repel.

FIGURE 13–2 *Magnetic attraction and repulsion.*

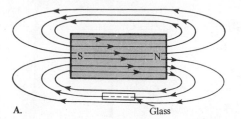
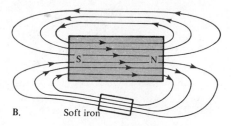

A.   Glass   B.   Soft iron

FIGURE 13–3 *Effect of nonmagnetic and magnetic materials on a magnetic field.*

unaltered, as shown in Figure 13–3A. However, when a *magnetic* material such as iron is placed in the magnetic field, the lines of force tend to change course and pass through the iron rather than through the surrounding air. They do so because the iron provides a magnetic path that is more easily established than that of air. Figure 13–3B illustrates this principle.

Magnetic Flux: The entire group of force lines surrounding a magnet is called the *magnetic flux,* symbolized by ϕ (Greek letter phi). The number of lines of force in a magnetic field determines the value of the flux. The more lines of force, the greater the flux and the stronger the magnetic field.

    The unit of magnetic flux is the *weber* (Wb). One weber equals 10^8 lines. In most practical situations, the weber is too large a unit; thus the microweber (μWb) is more common. One microweber equals 100 lines of magnetic flux.

Magnetic Flux Density: The *flux density* is the amount of flux per unit area in the magnetic field. Its symbol is B, and its unit is the tesla (T). One tesla equals one weber per square meter (Wb/m^2). The following formula expresses the flux density:

$$B = \frac{\phi}{A}$$

(13–1)

where ϕ is the flux and A is the cross-sectional area of the magnetic field.

$4e \times 10^{-7} \ \dfrac{wb}{at. \, m}$

Example 13–1

Find the flux density in a magnetic field in which the flux in 0.1 square meter is 800 μWb.

Solution:

$$B = \frac{\phi}{A} = \frac{800 \ \mu Wb}{0.1 \ m^2} = 8000 \times 10^{-6} \ T$$

$B = M_0 H$ magnetic field intensity

$= tesla$ permeability

$I = \oint H dl$

Gauss

$1 \ wb/m^2 = 10,000 \ gauss$

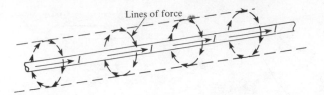

FIGURE 13–4 *Magnetic field around a current-carrying conductor.*

Electromagnetism

Current produces a magnetic field around a conductor, as illustrated in Figure 13–4. This field is called an *electromagnetic field*. The invisible lines of force of the magnetic field form a concentric circular pattern around the conductor and are continuous along its length.

Although the magnetic field cannot be seen, it is capable of producing visible effects. For example, if a current-carrying wire is inserted through a sheet of paper in a perpendicular direction, iron filings placed on the surface of the paper arrange themselves along the magnetic lines of force in concentric rings, as illustrated in Figure 13–5A. Part B of the figure illustrates that the north pole of a compass placed in the electromagnetic field will point in the direction of the lines of force. The field is stronger closer to the conductor and becomes weaker with increasing distance from the conductor.

Direction of the Lines of Force: The direction of the lines of force surrounding the conductor are indicated in Figure 13–6. When the direction of current is left to right, as in Part A, the lines are in a *clockwise* direction. When current is right to left, as in Part B, the lines are in a *counterclockwise* direction.

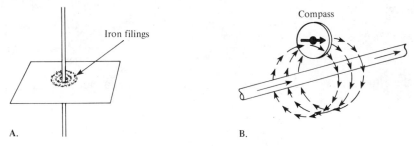

FIGURE 13–5 *Visible effects of an electromagnetic field.*

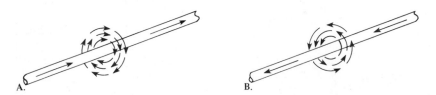

FIGURE 13–6 *Magnetic lines of force around a current-carrying conductor.*

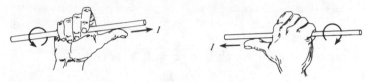

FIGURE 13-7 *Right-hand rule.*

Right-Hand Rule: An aid to remembering the direction of the lines of force is illustrated in Figure 13–7. Imagine that you are grasping the conductor with your right hand, with your thumb pointing in the direction of current. Your fingers indicate the direction of the magnetic lines of force.

Electromagnetic Properties

Several important properties relating to electromagnetic fields are presented in the following discussion.

Magnetomotive Force (mmf): As you have learned, current in a conductor produces a magnetic field. The force that produces the magnetic field is called the *magnetomotive force* (mmf). The unit of mmf, the ampere-turn (At), is established on the basis of the current in a single loop (turn) of wire. The formula for mmf is as follows:

$$F_m = NI \implies V_m = NI \qquad \textbf{(13–2)}$$

where F_m is the magnetomotive force, N is the number of turns of wire, and I is the current.

Reluctance: Reluctance ($\mathcal{R}$) is the opposition to the establishment of a magnetic field in an electromagnetic circuit. It is the ratio of the mmf required to establish a given flux to the amount of flux, and its units are ampere-turns per weber. The formula for reluctance is

$$\mathcal{R} = \frac{F_m}{\phi} \qquad \textbf{(13–3)}$$

This equation is sometimes known as "Ohm's law for magnetic circuits," because the reluctance is analogous to the resistance in electrical circuits.

Permeability: The ease with which a magnetic field can be established in a given material is measured by the *permeability* of that material. The higher the permeability, the more easily a magnetic field can be established.

The symbol of permeability is μ, and the formula is as follows:

$$\mu = \frac{l}{\mathcal{R}A} \qquad \textbf{(13–4)}$$

where $\mathcal{R}$ is the reluctance in ampere-turns per weber, l is the length of the material in meters, and A is the cross-sectional area in square meters. For reference, the permeability of a vacuum is $4\pi \times 10^{-7}$ Wb/At · m.

Example 13–2

There are two amperes of current through a wire with 5 turns.
(a) What is the mmf?
(b) What is the reluctance of the circuit if the flux is 250 μWb?

Solution:

(a) $N = 5$ and $I = 2$ A
$$F_m = NI = (5)(2 \text{ A}) = 10 \text{ At}$$

(b) $\mathcal{R} = \dfrac{F_m}{\phi} = \dfrac{10 \text{ At}}{250 \ \mu\text{Wb}} = 0.04 \times 10^6 \text{ At/Wb}$

Electromagnetic Induction

When a conductor is moved through a magnetic field, a voltage is produced across the conductor. This principle is known as *electromagnetic induction,* and the resulting voltage is an *induced voltage.*

The principle of electromagnetic induction is widely applied in electrical circuits, as you will learn in this chapter and in the study of transformers. The operation of electrical motors and generators is also based on this principle.

Relative Motion: When a wire is moved across a magnetic field, there is a relative motion between the wire and the magnetic field. Likewise, when a magnetic field is moved past a stationary wire, there is also relative motion. In either case, there is an induced voltage in the wire as a result of this motion, as Figure 13–8 indicates.

The amount of the induced voltage depends on the *rate* at which the wire and the magnetic field move with respect to each other. The faster the relative speed, the greater the induced voltage.

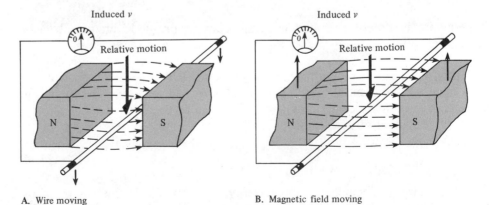

FIGURE 13–8   *Relative motion between a wire and a magnetic field.*

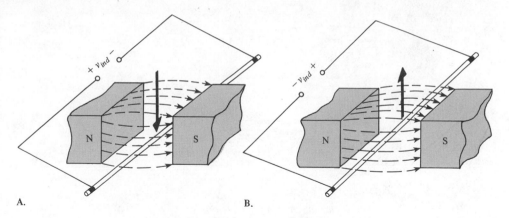

A.   B.

FIGURE 13–9 *Polarity of induced voltage depends on direction of motion.*

Polarity of the Induced Voltage: If the conductor in Figure 13–8 is moved first one way and then another in the magnetic field, a reversal of the polarity of the induced voltage will be observed. As the wire is moved downward, a voltage is induced with the polarity as indicated in Figure 13–9A. As the wire is moved upward, the polarity is as indicated in Part B of the figure.

Induced Current: When a load is connected to the wire in Figure 13–9, the voltage induced by the relative motion in the magnetic field will cause a current in the load, as shown in Figure 13–10. This current is called the *induced current*.

The principle of producing a voltage and a current in a load by moving a conductor across a magnetic field is the basis for electrical generators. The concept of a conductor existing in a moving magnetic field is fundamental to inductance in an electrical circuit.

Forces on a Current-Carrying Conductor in a Magnetic Field: Figure 13–11A shows current outward through a wire in a magnetic field. The electromagnetic

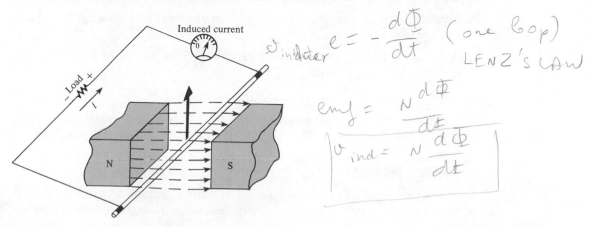

FIGURE 13–10 *Induced current in a load.*

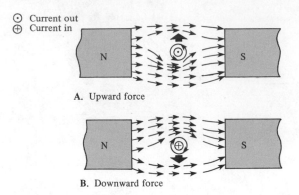

○ Current out
⊕ Current in

A. Upward force

B. Downward force

FIGURE 13–11   *Forces on a current-carrying conductor.*

field set up by the current interacts with the permanent magnetic field; as a result, the permanent lines of force *above* the wire tend to be deflected down under the wire, because they are opposite in direction to the electromagnetic lines of force. Therefore, the flux density above is reduced, and the magnetic field is weakened. The flux density below the conductor is increased, and the magnetic field is strengthened. An upward force on the conductor results, and the conductor tends to move toward the weaker magnetic field.

Figure 13–11B shows the current flowing inward, resulting in a force on the conductor in the downward direction. This principle is the basis for electrical motors.

Review for 13–1

1. The magnetic field surrounding a magnet consists of lines of force that radiate from the _____ pole to the _____ pole.

2. Like poles attract and unlike poles repel (T or F).

3. Determine the flux density in a given magnetic field if there are 500 μWb in an area of 0.000025 m².

4. A magnetic field is created by an electric current in a conductor (T or F).

5. What does *mmf* stand for?

6. Define *reluctance*.

7. When I = 3 A, T = 10 turns, and ϕ = 500 μWb, determine **(a)** the mmf and **(b)** the reluctance.

8. What is the induced voltage across a stationary conductor in a stationary magnetic field?

9. When the speed at which a conductor is moved through a magnetic field is increased, does the induced voltage increase, decrease, or remain the same?

10. When there is current through a conductor in a magnetic field, what happens?

THE BASIC INDUCTOR

13-2

When a length of wire is formed into a coil, as shown in Figure 13–12A, it becomes a basic *inductor*. Current flowing through the coil produces a magnetic field, as illustrated in Figure 13–12B. The magnetic lines of force around each loop (turn) in the coil effectively add to the lines of force around the adjoining loops, forming a strong magnetic field within and around the coil, as shown. The net direction of the total magnetic field creates a north and a south pole, as shown in Part B.

To understand the formation of the total magnetic field in a coil, let's discuss the interaction of the magnetic fields around two adjacent loops. The magnetic lines of force around adjacent loops are deflected into an outer path when the loops are brought close together. This effect occurs because the magnetic lines of force are in *opposing* directions between adjacent loops, as illustrated in Figure 13–13A. The total magnetic field for the two loops is depicted in Part B of the figure. For simplicity, only single lines of force are shown. This effect is additive for many closely adjacent loops in a coil; that is, each additional loop adds to the strength of the electromagnetic field.

Faraday's Law

Michael Faraday discovered the principle of electromagnetic induction in 1831. Faraday found that by moving a magnet through a coil of wire, a voltage was induced across the coil, and that when a complete path was provided, the induced voltage caused an induced current.

The amount of induced voltage is directly proportional to the rate of change of the magnetic field with respect to the coil. This principle is illustrated in Figure 13–14, where a bar magnet is moved through a coil of wire. An induced

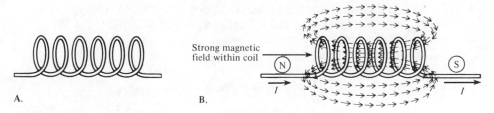

Strong magnetic field within coil

A.   B.

FIGURE 13–12 *Basic inductor and magnetic field.*

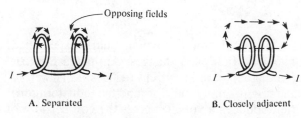

Opposing fields

A. Separated   B. Closely adjacent

FIGURE 13–13 *Interaction of magnetic lines of force in two loops in a coil.*

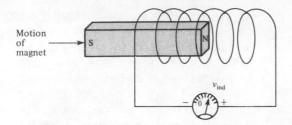

FIGURE 13–14 *Induced voltage created by a changing magnetic field.*

voltage is indicated by the voltmeter connected across the coil. The *faster* the magnet is moved, the *greater* is the induced voltage.

When a wire is formed into a certain number of loops or turns and is exposed to a *changing magnetic field,* a voltage is induced across the coil. The induced voltage is proportional to the *number of turns* of wire in the coil, N, and to the *rate* at which the magnetic field changes. The rate of change of the magnetic field is designated $d\phi/dt$, where ϕ is the *magnetic flux*. $d\phi/dt$ is expressed in webers/second (Wb/s). Faraday's law expresses this relationship in concise form as follows:

$$v_{ind} = N\left(\frac{d\phi}{dt}\right) \qquad\qquad \textbf{(13–5)}$$

This formula states that the induced voltage across a coil is equal to the number of turns (loops) times the rate of flux change.

Example 13–3

Apply Faraday's law to find the induced voltage across a coil with 100 turns that is located in a magnetic field that is changing at a rate of 5 Wb/s.

Solution:

$$v_{ind} = N\left(\frac{d\phi}{dt}\right) = (100 \text{ T})\,(5 \text{ Wb/s}) = 500 \text{ V}$$

Self-Inductance

When there is current through an inductor, a magnetic field is established. When the current changes, the magnetic field also changes. An increase in current expands the magnetic field, and a decrease in current reduces it. Therefore, a changing current produces a changing magnetic field around the inductor (coil). In turn, the changing magnetic field induces a voltage across the coil because of a property called *self-inductance*.

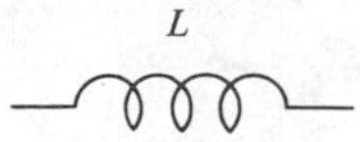

L in henry

(CORE)

L

FIGURE 13–15   *Symbol for inductor.*

Self-inductance is a measure of a coil's ability to establish an induced voltage as a result of a change in its current. Self-inductance is usually referred to as simply *inductance*. Inductance is symbolized by L.

The Unit of Inductance

The *henry,* symbolized by H, is the basic unit of inductance. By definition, the inductance is *one henry* when current through the coil, changing at the rate of *one ampere per second,* induces *one volt* across the coil. In many practical applications, *millihenries* (mH) and microhenries (μH) are the more common units. A common schematic symbol for the inductor is shown in Figure 13–15.

Lenz's Law

$v_{ind} = -\dfrac{d\Phi}{dt}$

The direction of the current induced in a coil is always such that it opposes the change in the magnetic field that produced it. When the current through a coil changes, it causes the magnetic field to change. The changing magnetic field then induces a voltage that opposes the original change in current. You can think of the induced voltage as producing an induced current in a direction opposing the initializing current. The principle of self-inductance is illustrated in Figure 13–16.

Figure 13–16A shows the current I_1 through the coil. As long as the current is a constant value, there is no induced voltage, because the magnetic field surrounding the coil is unchanging. In Part B, the switch is closed to introduce less resistance into the circuit, causing the current to attempt to increase to a new value, $I_1 + I_2$. This change creates an induced voltage in a direction to oppose the attempted increase in current. This is indicated by the polarity marks across the inductor. After a time, the induced voltage dies out, and the coil current reaches a constant value of $I_1 + I_2$, as shown in Part C. In Part D, the switch is opened, and the current tries to decrease back to its original I_1 value. As a result, a voltage and current are induced in a direction to aid the existing current and thus oppose its decrease. After a time, the induced voltage dies out, and the current returns to the I_1 value, as in Part A.

The Induced Voltage Depends on L and di/dt

L is the symbol for the inductance of a coil, and di/dt is the time rate of change of the current. A change in current causes a change in the magnetic field, which, in turn, induces a voltage across the coil, as you know. The induced voltage is directly proportional to L and di/dt, as stated by the following formula:

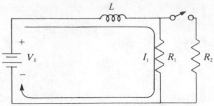

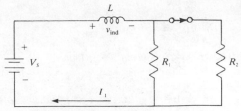

A. Initially, a constant I_1 flows. There is no induced voltage.

B. Switch closure attempts to increase coil current. v_{ind} is in a direction to oppose this *increase*.

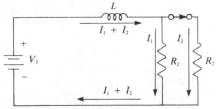

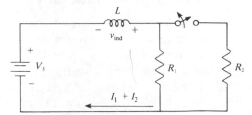

C. After current reaches constant value of $I_1 + I_2$, there is no v_{ind}.

D. Switch opening attempts to decrease coil current. v_{ind} is in a direction to oppose this *decrease*.

FIGURE 13–16

$$v_{ind} = L\left(\frac{di}{dt}\right) \tag{13–6}$$

This equation indicates that the greater the inductance, the greater the induced voltage. Also, the faster the coil current changes (greater di/dt), the greater the induced voltage. Notice the similarity of Equation (13–6) to Equation (12–23): $i = C(dv/dt)$.

Example 13–4

Determine the induced voltage across a 1-henry (1-H) inductor when the current is changing at a rate of 2 A/s.

Solution:

$$v_{ind} = L\left(\frac{di}{dt}\right) = (1\ \text{H})(2\ \text{A/s}) = 2\ \text{V}$$

Energy Storage

An inductor stores energy in the magnetic field created by the current.

The energy stored is expressed as follows:

$$\mathscr{E} = \tfrac{1}{2}LI^2 \tag{13-7}$$

As you can see, the energy stored is proportional to the inductance and the square of the current. When I is in amperes and L is in henries, the energy is in joules.

Inductor Characteristics

The following parameters are important in establishing the inductance of a coil:

Core Material: As discussed earlier, an inductor is basically a coil of wire. The material around which the coil is formed is called the *core*. Coils are wound on either nonmagnetic or magnetic materials. Examples of nonmagnetic materials are air, wood, copper, plastic, and glass. The permeabilities of these materials are the same as for a vacuum. Examples of magnetic materials are iron, nickel, steel, cobalt, or alloys. These materials have permeabilities that are hundreds or thousands of times greater than that of a vacuum and are classified as *ferromagnetic*. A ferromagnetic core provides a better path for the magnetic lines of force and thus permits a stronger magnetic field.

As you have learned, the permeability (μ) of the core material determines how easily a magnetic field can be established. *The inductance is directly proportional to the permeability of the core material.*

Physical Parameters: The number of turns of wire, the length, and the cross-sectional area of the core, as indicated in Figure 13–17, are factors in setting the value of inductance. The inductance is inversely proportional to the length of the core and directly proportional to the cross-sectional area. Also, the inductance is directly related to the number of turns squared. This relationship is as follows:

$$\text{Henry} \longleftarrow L = \frac{N^2 \mu A}{l} \tag{13-8}$$

where L is the inductance in henries, N is the number of turns, μ is the permeability, A is the cross-sectional area in meters squared, and l is the core length in meters.

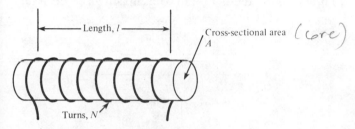

Length, l

Cross-sectional area (core)
A

Turns, N

FIGURE 13–17   *Parameters of an inductor.*

Example 13–5

Determine the inductance of the coil in Figure 13–18. The permeability of the core is 0.25×10^{-3}.

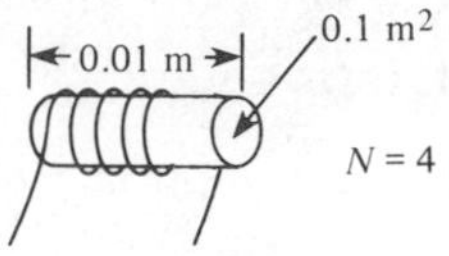

FIGURE 13–18

Solution:

$$L = \frac{N^2 \mu A}{l} = \frac{(4)^2 (0.25 \times 10^{-3})(0.1)}{0.01} = 40 \text{ mH}$$

Winding Resistance

When a coil is made with, for example, insulated copper wire, that wire has a certain *resistance* per unit of length. When many turns of wire are used to construct a coil, the total resistance may be significant. This inherent resistance is called the *dc resistance* or the *winding resistance* (R_W). It effectively appears in series with the inductance of the coil, as shown in Figure 13–19. In many applications, the winding resistance can be ignored and the coil considered as an ideal inductor. In other cases, the resistance must be considered.

Winding Capacitance

When two conductors are placed side by side, there is always some capacitance between them. Thus, when many turns of wire are placed close together in a coil, a certain amount of *stray* capacitance is a natural side effect. In many applications, this stray capacitance is very small and has no significant effect. In other cases, particularly at high frequencies, it may become quite important.

The equivalent circuit for an inductor with both its winding resistance (R_W) and capacitance (C_W) is shown in Figure 13–20. The capacitance effectively acts in parallel.

FIGURE 13–19 *Equivalent circuit diagram for a coil and its winding resistance.*

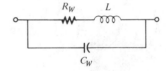

FIGURE 13–20 *Complete equivalent circuit for a nonideal inductor.*

Review for 13-2

1. List the parameters that contribute to the inductance of a coil.

2. The current through a 15-mH inductor is changing at the rate of 500 mA/s. What is the induced voltage?

3. Describe what happens to L when:
   (a) N is increased.
   (b) The core length is increased.
   (c) The cross-sectional area of the core is decreased.
   (d) A ferromagnetic core is replaced by an air core.

4. Explain why inductors inevitably have some winding resistance.

TYPES OF INDUCTORS

13-3

Inductors are made in a variety of shapes and sizes. Basically, they fall into two general categories: *fixed* and *variable*. The standard schematic symbols are shown in Figure 13-21.

Both fixed and variable inductors can be classified according to the type of core material. Three common types are the air core, the iron core, and the ferrite core. Each has a unique symbol, as shown in Figure 13-22.

Adjustable (variable) inductors usually have a screw-type adjustment that moves a sliding core in and out, thus changing the inductance. A wide variety of inductors exists, and some are shown in Figure 13-23.

A. Fixed   B. Variable

FIGURE 13-21 *Symbols for fixed and variable inductors.*

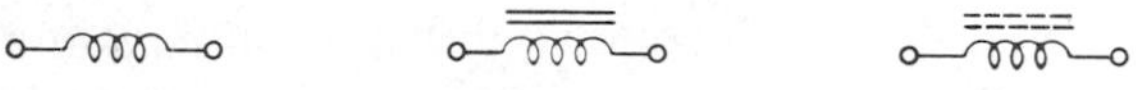

A. Air core   B. Iron core   C. Ferrite core

FIGURE 13-22 *Inductor symbols.*

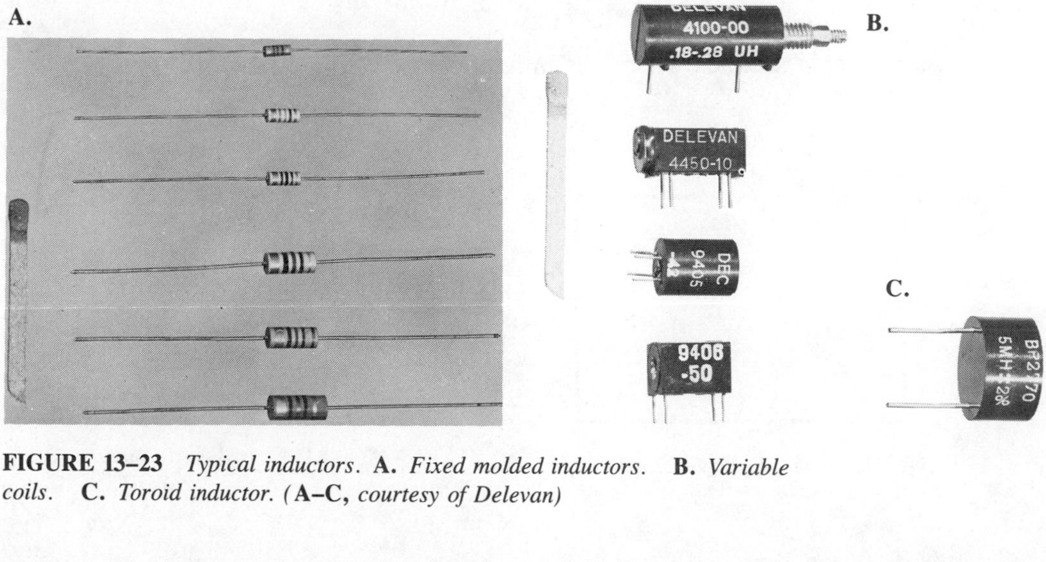

A.

B.

C.

FIGURE 13-23 *Typical inductors.* **A.** *Fixed molded inductors.* **B.** *Variable coils.* **C.** *Toroid inductor.* (**A-C,** *courtesy of Delevan*)

Review for 13–3

1. Name two general categories of inductors.
2. Identify the inductor symbols in Figure 13–24.

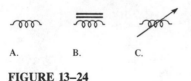

A.   B.   C.

FIGURE 13–24

Inductors act as short under a DC source

INDUCTORS IN dc CIRCUITS

13–4   When constant direct current flows in an inductor, there is no induced voltage. There is, however, a voltage drop due to the winding resistance of the coil. The inductance itself appears as a *short* to dc. Energy is stored in the magnetic field according to the formula previously stated in Equation (13–7), $\mathcal{E} = \frac{1}{2}LI^2$. The only energy loss occurs in the winding resistance ($P = I^2R_W$). This condition is illustrated in Figure 13–25.

Time Constant

Because the inductor's basic action is to *oppose a change in its current,* it follows that *current cannot change instantaneously in an inductor.* A certain time is required for the current to make a change from one value to another. The rate at which the current changes is determined by the *time constant.* The time constant for a series *RL* circuit is

$$\tau = \frac{L}{R} \tag{13–9}$$

where τ is in seconds when L is in henries and R is in ohms.

$$i_L(\bar{0}) = i_L(0^+)$$

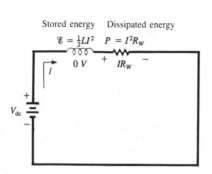

Stored energy   Dissipated energy
$\mathcal{E} = \frac{1}{2}LI^2$ $P = I^2R_W$

FIGURE 13–25   *Energy storage and loss in an inductor in a dc circuit.*

Example 13–6

A series *RL* circuit has a resistance of 1 kΩ and an inductance of 1 mH. What is the time constant?

Solution:

$$\tau = \frac{L}{R} = \frac{1 \text{ mH}}{1 \text{ k}\Omega} = \frac{1 \times 10^{-3} \text{ H}}{1 \times 10^{3} \text{ }\Omega}$$

$$= 1 \times 10^{-6} \text{ s} = 1 \text{ }\mu\text{s}$$

Energizing Current in an Inductor

In a series *RL* circuit, the current will increase to 63% of its full value in one time-constant interval after the switch is closed. This build-up of current is analogous to the build-up of capacitor voltage during the charging in an *RC* circuit; they both follow an exponential curve and reach the approximate percentages of final value as indicated in Table 13–1.

TABLE 13–1 *Percentage of final current after each time-constant interval during current build-up.*

Number of Time Constants	% Final Value
1	63
2	86
3	95
4	98
5	99 (considered 100%)

The change in current over five time-constant intervals is illustrated in Figure 13–26. When the current reaches its final value at approximately 5τ, it ceases to change. At this time, the inductor acts as a short (except for winding resistance) to the constant current. The final value of the current is $V_S/R = 10 \text{ V}/1 \text{ k}\Omega = 10 \text{ mA}$.

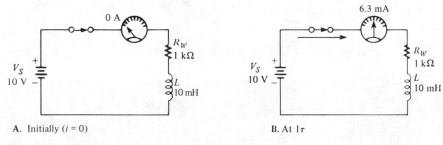

A. Initially ($i = 0$)   B. At 1τ

FIGURE 13–26 *Current build-up in an inductor.*

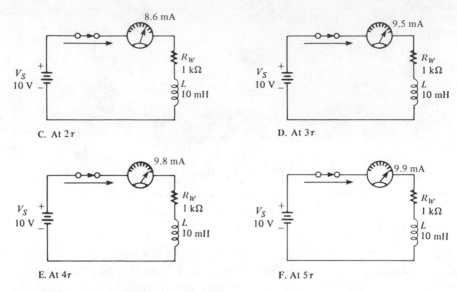

C. At 2τ D. At 3τ

E. At 4τ F. At 5τ

FIGURE 13–26   (continued)

Example 13–7

Calculate the time constant for Figure 13–27. Then determine the current and the time at each time-constant interval, measured from the instant the switch is closed.

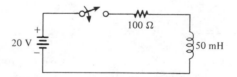

FIGURE 13–27

Solution:

$$I_{\text{final}} = \frac{V_S}{R} = \frac{20 \text{ V}}{100 \text{ }\Omega}$$

$$= 0.2 \text{ A}$$

$$\tau = \frac{L}{R} = \frac{50 \text{ mH}}{100 \text{ }\Omega} = 0.5 \text{ ms}$$

At $1\tau = 0.5$ ms: $i = 0.63(0.2 \text{ A}) = 0.126 \text{ A}$

At $2\tau = 1$ ms: $i = 0.86(0.2 \text{ A}) = 0.172 \text{ A}$

At $3\tau = 1.5$ ms: $i = 0.95(0.2$ A$) = 0.19$ A

At $4\tau = 2$ ms: $i = 0.98(0.2$ A$) = 0.196$ A

At $5\tau = 2.5$ ms: $i = 0.99(0.2$ A$) = 0.198$ A

$$\cong 0.2 \text{ A}$$

Deenergizing Current in an Inductor

Current in an inductor decreases exponentially according to the approximate percentage values in Table 13–2.

TABLE 13–2 *Percentage of initial current after each time-constant interval while current is decreasing.*

Number of Time Constants	% Initial Value
1	37
2	14
3	5
4	2
5	1 (considered 0)

Figure 13–28A shows a constant current of 10 mA through the inductor. If switch 2 is closed at the *same* instant that switch 1 is opened, as shown in Part B, the current decreases to zero in five time constants ($5L/R$) as the inductor deenergizes. This exponential decrease in current is shown in Part C.

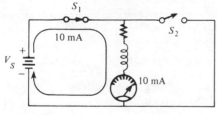

A. Initially

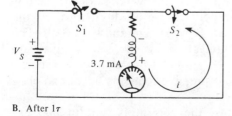

B. After 1τ

C.

FIGURE 13–28 *Deenergizing current in an inductor.*

450

Example 13–8

In Figure 13–29, S_1 is opened at the instant that S_2 is closed.
(a) What is the time constant?
(b) What is the initial coil current at the instant of switching?
(c) What is the coil current at 1τ?
Assume steady state current through the coil prior to switch change.

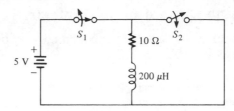

FIGURE 13–29

Solution:

(a) $\tau = L/R = 200\ \mu\text{H}/10\ \Omega = 20\ \mu\text{s}$.
(b) Current cannot change instantaneously in an inductor. Therefore, the current at the instant of the switch change is the same as the steady state current.

$$I = \frac{5\text{ V}}{10\ \Omega} = 0.5\text{ A}$$

(c) At 1τ, the current has decreased to 37% of its initial value:

$$i = 0.37(0.5\text{ A}) = 0.185\text{ A}$$

Induced Voltage in the Series *RL* Circuit

When current changes in an inductor, a voltage is induced. We now examine what happens to the voltages across the resistor and the coil in a series circuit when a change in current occurs.

Look at the circuit in Figure 13–30A. When the switch is open, there is no current, and the resistor voltage and the coil voltage are both zero. At the instant the switch is closed, as indicated in Part B, v_R is zero and v_L is 10 V. The reason for this change is that the induced voltage across the coil is equal and opposite to the applied voltage to prevent the current from changing instantaneously. Therefore, at the instant of switch closure, *L* effectively acts as an *open* with all the applied voltage across it.

During the first five time constants, the current is building up exponentially, and the induced coil voltage is decreasing. The resistor voltage increases with the current, as Part C illustrates.

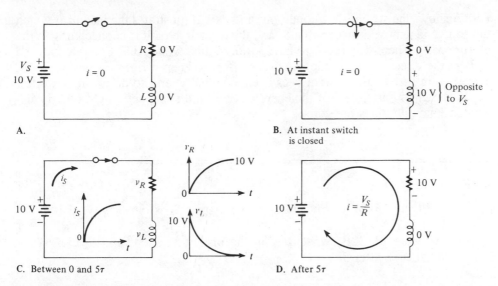

A.

B. At instant switch
is closed

C. Between 0 and 5τ

D. After 5τ

FIGURE 13–30 *Voltage in an RL circuit as the inductor energizes.*

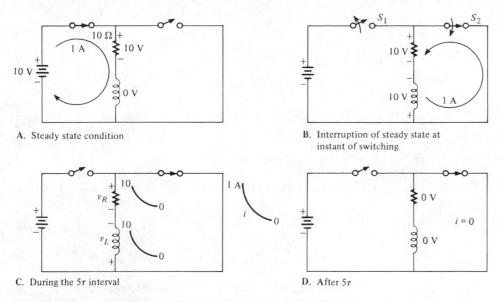

A. Steady state condition

B. Interruption of steady state at
instant of switching

C. During the 5τ interval

D. After 5τ

FIGURE 13–31 *Voltage in an RL circuit as the inductor deenergizes.*

After five time constants have elapsed, the current has reached its final value, V_S/R. At this time, all of the applied voltage is dropped across the resistor and none across the coil. Thus, L effectively acts as a *short* to nonchanging current, as Part D illustrates. Keep in mind that the inductor always reacts to a change in current by creating an induced voltage in order to counteract that change.

Now let us examine the case illustrated in Figure 13–31, where the steady state current is switched out, and the inductor discharges through another path.

Part A shows the steady state condition, and Part B illustrates the instant at which the source is removed by opening S_1 and the discharge path is connected with the closure of S_2. There was 1 A through L prior to this. Notice that 10 V are induced in L in the direction to aid the 1 A in an effort to keep it from changing. Then, as shown in Part C, the current decays exponentially, and so do v_R and v_L. After 5τ, as shown in Part D, all of the energy stored in the magnetic field of L is dissipated, and all values are zero.

Example 13–9 $V t = 25^\checkmark$ $V_L = 0^\checkmark$

(a) In Figure 13–32A, what is v_L at the instant S_1 is closed? What is v_L after 5τ?
                                                                                    100

(b) In Figure 13–32B, what is v_L at the instant S_1 opens and S_2 closes? What is v_L after 5τ? $0 \checkmark$

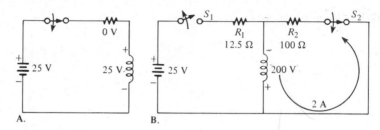

A.   B.

FIGURE 13–32

Solution:

(a) At the instant the switch is closed, all of the voltage is across L. Thus, $v_L = 25$ V, with the polarity as shown. After 5τ, L acts as a short, so $v_L = 0$ V.

(b) With S_1 closed and S_2 open, the steady state current is

$$\frac{25 \text{ V}}{12.5 \text{ }\Omega} = 2 \text{ A}$$

When the switches are thrown, an induced voltage is created across L sufficient to keep this 2-A current for an instant. In this case, it takes $v_L = IR_2 = (2 \text{ A})(100 \text{ }\Omega) = 200$ V. After 5τ, the inductor voltage is zero. These results are indicated in the circuit diagrams.

The Exponential Formulas

The formulas for the exponential current and voltage in an RL circuit are similar to those used in the last chapter for the RC circuit, and the universal exponential curves in Figure 12–26 apply to inductors as well as capacitors. The general formulas for RL circuits are stated as follows:

$$v = V_F + (V_i - V_F)e^{-Rt/L} \qquad \text{(13–10)}$$

$$i = I_F + (I_i - I_F)e^{-Rt/L} \qquad \text{(13–11)}$$

where V_F and I_F are the final values; V_i and I_i are the initial values; and v and i are the instantaneous values of the inductor voltage or current at time t.

Increasing Current: The formula for the special case in which an increasing exponential current curve begins at zero ($I_i = 0$) is

$$i = I_F(1 - e^{-Rt/L}) \qquad \text{(13–12)}$$

Using Equation (13–12), we can calculate the value of the increasing inductor current at any instant of time. The same is true for voltage.

Example 13–10

In Figure 13–33, determine the inductor current 30 μs after the switch is closed.

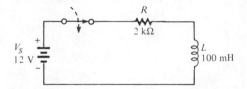

FIGURE 13–33

Solution:

The time constant is $L/R = 100 \text{ mH}/2 \text{ k}\Omega = 50 \ \mu$s. The final current is $V_S/R = 12 \text{ V}/2 \text{ k}\Omega = 6 \text{ mA}$. The initial current is zero. Notice that 30 μs is less than one time constant, so the current will reach less than 63% of its final value in that time.

$$i_L = I_F(1 - e^{-Rt/L}) = 6 \text{ mA}(1 - e^{-0.6})$$
$$= 6 \text{ mA}(1 - 0.549) = 2.71 \text{ mA}$$

Decreasing Current: The formula for the special case in which a decreasing exponential current has a final value of zero is as follows:

$$i = I_i e^{-Rt/L} \qquad \text{(13–13)}$$

This formula can be used to calculate the deenergizing current at any instant, as the following example shows.

Example 13–11

Determine the inductor current in Figure 13–34 at a point in time 2 ms after the switches are thrown (S_1 opened and S_2 closed).

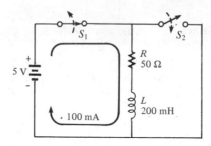

FIGURE 13–34

Solution:

The deenergizing time constant is $L/R = 200 \text{ mH}/50 \ \Omega = 4$ ms. The initial current in the inductor is 100 mA. Notice that 2 ms is less than one time constant, so the current will show a less than 63% decrease. Therefore, the current will be greater than 37% of its initial value at 2 ms after the switches are thrown.

$$i = I_i e^{-Rt/L} = (100 \text{ mA})e^{-0.5} = 60.65 \text{ mA}$$

Review for 13–4

1. A 15-mH inductor with a winding resistance of 10 Ω has a constant direct current of 10 mA through it. What is the voltage drop across the inductor?

2. A 20-V dc source is connected to a series RL circuit with a switch. At the instant of switch closure, what are the values of v_R and v_L?

3. In the same circuit, after a time interval equal to 5τ from switch closure, what are v_R and v_L?

4. In a series RL circuit where $R = 1 \text{ k}\Omega$ and $L = 500 \ \mu\text{H}$, what is the time constant? Determine the current 0.25 μs after a switch connects 10 V across the circuit.

SERIES INDUCTORS

13–5

When inductors are connected in series, as in Figure 13–35, *the total inductance, L_T, is the sum of the individual inductances.* The formula for L_T is expressed in the following equation for the general case of n inductors in series:

$$L_T = L_1 + L_2 + L_3 + \cdots + L_n \qquad (13\text{–}14)$$

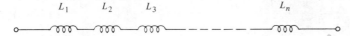

FIGURE 13–35 *Inductors in series.*

Notice that inductance in series is similar to resistance in series.

Example 13–12

Determine the total inductance for each of the series connections in Figure 13–36.

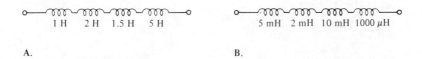

A.   B.

FIGURE 13–36

Solution:

Part A: $L_T = 1\,H + 2\,H + 1.5\,H + 5\,H = 9.5\,H$

Part B: $L_T = 5\,mH + 2\,mH + 10\,mH + 1\,mH = 18\,mH$

Note: $1000\,\mu H = 1\,mH$

Review for 13–5

1. State the rule for combining inductors in series.

2. What is L_T for a series connection of 100 μH, 500 μH, and 2 mH?

3. Five 100-mH coils are connected in series. What is the total inductance?

PARALLEL INDUCTORS

13–6

When inductors are connected in parallel, as in Figure 13–37, *the total inductance is less than the smallest inductance*. The formula for total inductance in parallel is very similar to that for total parallel resistance or total series capacitance:

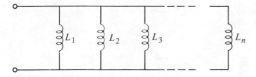

FIGURE 13–37 *Inductors in parallel.*

$$\frac{1}{L_T} = \frac{1}{L_1} + \frac{1}{L_2} + \frac{1}{L_3} + \cdots + \frac{1}{L_n} \qquad (13\text{–}15)$$

This general formula states that the reciprocal of the total inductance is equal to the sum of the reciprocals of the individual inductances. L_T can be found by taking the reciprocal of both sides of Equation (13–15).

$$L_T = \frac{1}{\left(\frac{1}{L_1}\right) + \left(\frac{1}{L_2}\right) + \left(\frac{1}{L_3}\right) + \cdots + \left(\frac{1}{L_n}\right)} \qquad (13\text{–}16)$$

Special Case of Two Parallel Inductors

When only two inductors are in parallel, a special *product over sum* form of Equation (13–15) can be used.

$$L_T = \frac{L_1 L_2}{L_1 + L_2} \qquad (13\text{–}17)$$

Equal-Value Parallel Inductors

This is another special case in which a short-cut formula can be used. This formula is also derived from the general Equation (13–15) and is stated as follows for n equal-value inductors in parallel:

$$L_T = \frac{L}{n} \qquad (13\text{–}18)$$

Example 13–13

Determine L_T in Figure 13–38.

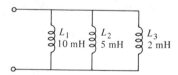

FIGURE 13–38

Solution:

$$\frac{1}{L_T} = \frac{1}{L_1} + \frac{1}{L_2} + \frac{1}{L_3}$$

$$= \frac{1}{10 \text{ mH}} + \frac{1}{5 \text{ mH}} + \frac{1}{2 \text{ mH}}$$

$$L_T = \cfrac{1}{\cfrac{1}{10 \text{ mH}} + \cfrac{1}{5 \text{ mH}} + \cfrac{1}{2 \text{ mH}}}$$

$$= \frac{1}{0.8 \text{ mH}} = 1.25 \text{ mH}$$

Example 13–14

Find L_T for both circuits in Figure 13–39.

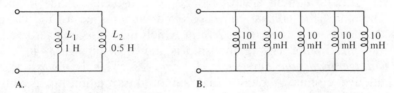

A.

B.

FIGURE 13–39

Solution:

Part A: Using the special formula for two parallel inductors, we obtain

$$L_T = \frac{L_1 L_2}{L_1 + L_2} = \frac{(1 \text{ H})(0.5 \text{ H})}{1.5 \text{ H}} = 0.33 \text{ H}$$

Part B: Using the special formula for equal parallel inductors, we obtain

$$L_T = \frac{L}{n} = \frac{10 \text{ mH}}{5} = 2 \text{ mH}$$

Review for 13–6

1. Compare the total inductance in parallel with the smallest-valued individual inductor.
2. The calculation of total parallel inductance is similar to that for parallel resistance (T or F).
3. Determine L_T for each parallel combination:
   (a) 100 mH, 50 mH, and 10 mH
   (b) 40 μH and 60 μH
   (c) Ten 1-H coils

INDUCTORS IN ac CIRCUITS

13–7 The concept of the derivative was introduced in Chapter 12. The expression for induced voltage in an inductor was stated earlier in Equation (13–6). This formula is $v_{\text{ind}} = L(di/dt)$.

Phase Relationship of Current and Voltage in an Inductor

From the formula for induced voltage, you can see that the faster the current through an inductor changes, the greater the induced voltage will be. For example, if the rate of change of current is zero, the voltage is zero [$v_{\text{ind}} = L(di/dt) = L(0) = 0$ V]. When di/dt is a positive-going maximum, v_{ind} is a positive maximum; when di/dt is a negative-going maximum, v_{ind} is a negative maximum.

A sinusoidal current always induces a sinusoidal voltage in inductive circuits. Therefore, the voltage can be plotted with respect to the current by knowing the points on the current curve at which the voltage is zero and those at which it is maximum. This relationship is shown in Figure 13–40A. Notice that *the voltage leads the current by 90 degrees*. This is always true in a purely inductive circuit. A phasor diagram of this relationship is shown in Figure 13–40B.

Inductive Reactance

Inductive reactance is the opposition to sinusoidal current, expressed in ohms. The symbol for inductive reactance is X_L. To develop a formula for X_L, we use the relationship $v_{\text{ind}} = (di/dt)$ and the curves in Figure 13–41.

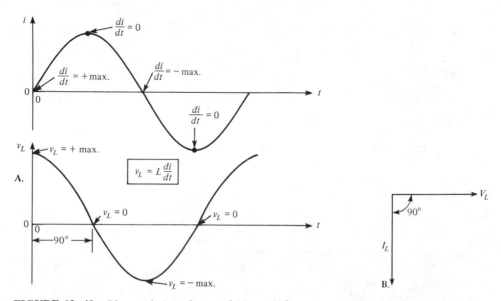

FIGURE 13–40 *Phase relation of v_{ind} and i in an inductor.*

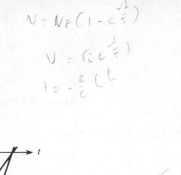

$$V = NF(1 - e^{\frac{-t}{\tau}})$$

$$V = r_i e^{\frac{-t}{\tau}}$$

$$t = -\frac{e}{c}(L$$

Slope A

Slope B

B

A

t

FIGURE 13–41 *Slope indicates rate of change. Sine wave A has a greater rate of change at the zero crossing than B, and thus A has a higher frequency.*

The rate of change of current is directly related to frequency. The faster the current changes, the higher the frequency. For example, you can see that in Figure 13–41, the slope of sine wave A at the zero crossings is greater than that of sine wave B. Recall that the slope of a curve at a point indicates the rate of change at that point. Sine wave A has a higher frequency than sine wave B, as indicated by a greater maximum rate of change (di/dt is greater at the zero crossings).

When frequency increases, di/dt increases, and thus v_{ind} increases. When frequency decreases, di/dt decreases, and thus v_{ind} decreases. The induced voltage is directly dependent on frequency:

$$\uparrow \qquad \uparrow$$
$$v_{ind} = L(di/dt) \qquad \text{and} \qquad v_{ind} = L(di/dt)$$
$$\downarrow \qquad \downarrow$$

An increase in induced voltage means more opposition (X_L is greater). Therefore, X_L *is directly proportional to induced voltage and thus directly proportional to frequency:*

$$X_L \text{ is proportional to } f$$

Now, if di/dt is constant and the inductance is varied, an increase in L produces an increase in v_{ind}, and a decrease in L produces a decrease in v_{ind}, as indicated:

$$\uparrow \qquad \uparrow$$
$$v_{ind} = L(di/dt) \qquad \text{and} \qquad v_{ind} = L(di/dt)$$
$$\downarrow \qquad \downarrow$$

Again, an increase in v_{ind} means more opposition (greater X_L). Therefore, X_L *is directly proportional to induced voltage and thus directly proportional to inductance.*

The inductive reactance is directly proportional to both f and L.

$$X_L \text{ is proportional to } fL$$

The complete formula for X_L is as follows:

$$X_L = 2\pi fL \qquad\qquad \textbf{(13–19)}$$

Notice that 2π appears as a constant factor in the equation. This comes from the relationship of a sine wave to rotational motion, as derived in Appendix F. X_L is in ohms when f is in hertz and L is in henries.

Example 13–15

A sinusoidal voltage is applied to the circuit in Figure 13–42. The frequency is 1 kHz. Determine the inductive reactance.

FIGURE 13–42

Solution:

$$1 \text{ kHz} = 1 \times 10^3 \text{ Hz}$$

$$5 \text{ mH} = 5 \times 10^{-3} \text{ H}$$

$$X_L = 2\pi f L = 2\pi(1 \times 10^3 \text{ Hz})(5 \times 10^{-3} \text{ H}) = 31.4 \text{ } \Omega$$

Ohm's Law as Applied to X_L: Ohm's law applies to inductive reactance as well as to resistance and capacitive reactance. The three equivalent forms are as follows:

$$V = IX_L \tag{13-20}$$

$$I = \frac{V}{X_L} \tag{13-21}$$

$$X_L = \frac{V}{I} \tag{13-22}$$

Current and voltage must be expressed in the same units, such as rms, peak, peak-to-peak, or average values.

Example 13–16

Determine the rms current in Figure 13–43.

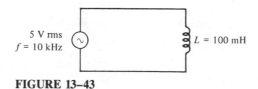

FIGURE 13–43

Solution:

$$10 \text{ kHz} = 10 \times 10^3 \text{ Hz}$$

$$100 \text{ mH} = 100 \times 10^{-3} \text{ H}$$

First calculate X_L:

$$X_L = 2\pi f L = 2\pi(10 \times 10^3 \text{ Hz})(100 \times 10^{-3} \text{ H}) = 6283 \ \Omega$$

Using Ohm's law, we obtain

$$I_{\text{rms}} = \frac{V_{\text{rms}}}{X_L} = \frac{5 \text{ V}}{6283 \ \Omega} = 795.8 \ \mu\text{A}$$

Complex Form of Inductive Reactance: X_L can be treated as a complex quantity, because it introduces a phase angle of 90 degrees between the current and the voltage. Italic X_L represents just the magnitude; boldface $\mathbf{X}_L$ is a phasor quantity representing both magnitude and angle. $\mathbf{X}_L$ causes the current to lag the voltage by 90 degrees; so in polar form, it is assigned a 90-degree angle and is expressed as $X_L \angle 90°$. In rectangular form, it is expressed as jX_L, where the j indicates a 90-degree phase shift.

We utilize these complex forms for circuit analysis in later chapters.

Power in an Inductor

As discussed earlier, an inductor stores energy in its magnetic field when there is current through it. An ideal inductor (assuming no winding resistance) does not dissipate energy; it only stores it. When an ac voltage is applied to an inductor, energy is stored by the inductor during a portion of the cycle; then the stored energy is returned to the source during another portion of the cycle. *There is no net energy loss.* Figure 13–44 shows the power curve that results from one cycle of inductor current and voltage.

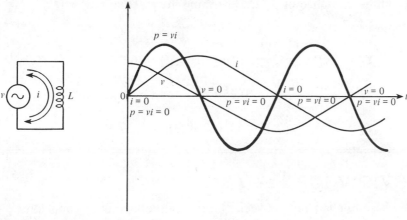

FIGURE 13–44 *Power curve.*

Instantaneous Power (p): *The product of v and i gives instantaneous power, p.* At points where v or i is zero, p is also zero. When both v and i are positive, p is also positive. When either v or i is positive and the other negative, p is negative. When both v and i are negative, p is positive. As you can see in Figure 13–44, the power follows a sinusoidal curve. Positive values of power indicate that energy is stored by the inductor. Negative values of power indicate that energy is returned from the inductor to the source. Note that the power fluctuates at a frequency twice that of the voltage or current as energy is alternately stored and returned to the source.

Average Power (P_avg): Ideally, all of the energy stored by an inductor during the positive portion of the power cycle is returned to the source during the negative portion. *No net energy is consumed in the inductance,* so the average power is zero. Actually, because of winding resistance in a practical inductor, some power is always dissipated.

$$P_{\text{avg}} = (I_{\text{rms}})^2 R_W \qquad (13\text{--}23)$$

Reactive Power (P_r): The rate at which an inductor stores or returns energy is called its *reactive power, P_r.* The reactive power is a nonzero quantity, because at any instant in time, the inductor is actually taking energy from the source or returning energy to it. Reactive power does not represent an energy loss. The following formulas apply:

$$P_r = V_{\text{rms}} I_{\text{rms}} \qquad (13\text{--}24)$$

$$P_r = \frac{V_{\text{rms}}^2}{X_L} \qquad (13\text{--}25)$$

$$P_r = I_{\text{rms}}^2 X_L \qquad (13\text{--}26)$$

Example 13–17

A 10-V rms signal with a frequency of 1 kHz is applied to a 10-mH coil with a winding resistance of 5 Ω. Determine the reactive power (P_r).

Solution:

$$X_L = 2\pi f L = 2\pi(1 \text{ kHz})(10 \text{ mH}) = 62.8 \ \Omega$$

$$I = \frac{V_S}{X_L} = \frac{10 \text{ V}}{63 \ \Omega} = 0.159 \text{ A}$$

$$P_r = I^2 X_L = (0.159 \text{ A})^2 (62.8 \ \Omega) = 1.59 \text{ VAR}$$

Review for 13–7

1. State the phase relationship between current and voltage in an inductor.
2. Calculate X_L for $f = 5$ kHz and $L = 100$ mH.

3. At what frequency is the reactance of a 50-μH inductor equal to 800 Ω?

4. Calculate the rms current in Figure 13–45.

V_{rms} = 1 V
f = 1 MHz

L = 10 μH

FIGURE 13–45

5. A 50-mH inductor is connected to a 12-V rms source. What is the average power? What is the reactive power at a frequency of 1 kHz?

TESTING INDUCTORS

13–8

The most common failure in an inductor is an open. To check for an open, the coil should be removed from the circuit. If there is an open, an ohmmeter check will indicate infinite resistance, as shown in Figure 13–46A.

If the coil is good, the ohmmeter will show the winding resistance. The value of the winding resistance depends on the wire size and length of the coil. It can be anywhere from one ohm to several hundred ohms. Figure 13–46B shows a good reading.

Occasionally, when an inductor is overheated with excessive current, the wire insulation will melt, and some coils will short together. This produces a reduction in the inductance by reducing the effective number of turns and a reduction in winding resistance. This must be tested on an inductance bridge.

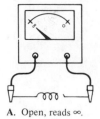

A. Open, reads ∞.

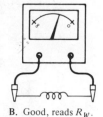

B. Good, reads R_W.

FIGURE 13–46 *Checking a coil by measuring the resistance.*

Review for 13–8

1. When a coil is checked, a reading of infinity on the ohmmeter indicates a partial short (T or F).

2. An ohmmeter check of a good coil will indicate the value of the inductance (T or F).

COMPUTER ANALYSIS

13–9   The program in this section allows you to compute the inductive reactance as a function of frequency. Inputs required are the inductance, the frequency range, and the increments of frequency.

```
10 CLS
20 PRINT "THIS PROGRAM COMPUTES INDUCTIVE REACTANCE AS A"
30 PRINT "FUNCTION OF FREQUENCY.  THE INPUTS REQUIRED ARE"
40 PRINT "INDUCTANCE AND THE FREQUENCY RANGE AND INCREMENTS."
50 FOR T = 1 TO 4000:NEXT
60 CLS
70 INPUT "ENTER L IN HENRIES";L
80 INPUT "LOWEST FREQUENCY IN HERTZ";FL
90 INPUT "HIGHEST FREQUENCY IN HERTZ";FH
100 INPUT "INCREMENTS OF FREQUENCY IN HERTZ";FI
110 CLS
120 PRINT "FREQUENCY (HZ)",,"XL (OHMS)"
130 FOR F=FL TO FH STEP FI
140 XL = 2*3.1416*F*L
150 PRINT F,,XL
160 NEXT
```

Review for 13–9

1. In line 130, explain the purpose of "STEP FI."

2. What is 3.1416 in line 140?

Formulas

$$B = \frac{\phi}{A} \tag{13–1}$$

$$F_m = NI \tag{13–2}$$

$$\mathcal{R} = \frac{F_m}{\phi} \tag{13–3}$$

$$\mu = \frac{1}{\mathcal{R}A} \tag{13–4}$$

$$v_{\text{ind}} = N\left(\frac{d\phi}{dt}\right) \tag{13–5}$$

$$v_{\text{ind}} = L\left(\frac{di}{dt}\right) \tag{13–6}$$

$$\mathcal{E} = \tfrac{1}{2}LI^2 \tag{13–7}$$

$$L = \frac{N^2\mu A}{l} \tag{13–8}$$

$$\tau = \frac{L}{R} \tag{13-9}$$

$$v = V_F + (V_i - V_F)e^{-Rt/L} \tag{13-10}$$

$$i = I_F + (I_i - I_F)e^{-Rt/L} \tag{13-11}$$

$$i = I_F(1 - e^{-Rt/L}) \tag{13-12}$$

$$i = I_i e^{-Rt/L} \tag{13-13}$$

$$L_T = L_1 + L_2 + L_3 + \cdots + L_n \tag{13-14}$$

$$\frac{1}{L_T} = \frac{1}{L_1} + \frac{1}{L_2} + \frac{1}{L_3} + \cdots + \frac{1}{L_n} \tag{13-15}$$

$$L_T = \frac{1}{\left(\dfrac{1}{L_1}\right) + \left(\dfrac{1}{L_2}\right) + \left(\dfrac{1}{L_3}\right) + \cdots + \left(\dfrac{1}{L_n}\right)} \tag{13-16}$$

$$L_T = \frac{L_1 L_2}{L_1 + L_2} \tag{13-17}$$

$$L_T = \frac{L}{n} \tag{13-18}$$

$$X_L = 2\pi f L \tag{13-19}$$

$$V = I X_L \tag{13-20}$$

$$I = \frac{V}{X_L} \tag{13-21}$$

$$X_L = \frac{V}{I} \tag{13-22}$$

$$P_{avg} = (I_{rms})^2 R_W \tag{13-23}$$

$$P_r = V_{rms} I_{rms} \tag{13-24}$$

$$P_r = \frac{(V_{rms})^2}{X_L} \tag{13-25}$$

$$P_r = (I_{rms})^2 X_L \tag{13-26}$$

Summary

1. A magnetic field exists around a conductor in which current is flowing.
2. An inductor can also be referred to as a *coil* or a *choke*.
3. Self-inductance is a measure of a coil's ability to establish an induced voltage as a result of a change in its current.
4. An inductor opposes a change in its own current.

5. Faraday's law states that relative motion between a magnetic field and a coil produces a voltage across the coil.

6. The amount of induced voltage is directly proportional to the inductance and to the rate of change in current.

7. Lenz's law states that the polarity of induced voltage is such that the resulting induced current is in a direction that opposes the change in the magnetic field that produced it.

8. Energy is stored by an inductor in its magnetic field.

9. The henry is the unit of inductance.

10. *One henry* is the amount of inductance when current, changing at the rate of *one ampere per second,* induces *one volt* across the inductor.

11. Inductance is directly proportional to the square of the turns, the permeability, and the cross-sectional area of the core. It is inversely proportional to the length of the core.

12. The permeability of a core material is an indication of the ability of the material to establish a magnetic field.

13. The time constant for a series *RL* circuit is the inductance divided by the resistance.

14. In an *RL* circuit, the voltage and current in an energizing or deenergizing inductor make a 63% change during each time-constant interval.

15. Energizing and deenergizing follow exponential curves.

16. Inductors add in series.

17. Total parallel inductance is less than that of the smallest inductor in parallel.

18. Voltage leads current by 90 degrees in an inductor.

19. Inductive reactance is the opposition to sinusoidal current and is expressed in ohms.

20. X_L is directly proportional to frequency and inductance.

21. The average power in an inductor is zero; that is, there is no energy loss in an ideal inductor.

Self-Test

1. What is the flux density in webers per square meter when the flux in one square millimeter is 1200 μWb?

2. There is 100 mA through a coil with 50 turns. Determine the mmf.

3. An inductor with 50 turns of wire is placed in a magnetic field that is changing at a rate of 25 Wb/s. What is the induced voltage?

4. Convert 1500 μH to millihenries. Convert 20 mH to microhenries.

5. The current through a 100-mH coil is changing at a rate of 200 mA/s. How much voltage is induced across the coil?

6. How many turns are required to produce 30 mH with a coil wound on a cylindrical core having a cross-sectional area of 10×10^{-5} m^2 and a length of 0.05 m? The core has a permeability of 1.2×10^{-6}.

7. A 12-V battery is connected across a coil with a winding resistance of 12 Ω. How much current is there in the coil?

8. How much energy is stored by a 100-mH inductor with a current of 1 A?

9. A circuit has $R = 2$ kΩ in series with $L = 10$ mH. What is the time constant?

10. Three 25-μH coils and two 10-μH coils are connected in series. What is the total inductance?

11. The following coils are connected in parallel: two 1-H, two 0.5-H, and one 2-H. What is L_T?

12. A 2-MHz sinusoidal voltage is applied across a 15-μH inductor. What is the frequency of the resulting current? What is the reactance of the inductor?

13. The frequency of the source in Figure 13–47 is adjustable. To get 50 mA of rms current, to what frequency must you adjust the source?

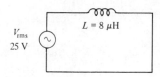

$L = 8\ \mu\text{H}$

V_{rms}
25 V

FIGURE 13–47

14. What inductance is required for an inductor to produce an X_L of 10 MΩ at 50 MHz?

Problems

Section 13–1

13–1. How many webers do one thousand lines of magnetic force equal?

13–2. What is the mmf in 10 turns of wire when there are 25 mA flowing through it?

13–3. How many turns are required to produce an mmf of 1000 At when the current is 2.5 A?

13–4. What mmf is necessary to produce a flux of 500 μWb when the reluctance of a given magnetic material is 3000 At/Wb?

Sections 13–2 and 13–3

13–5. Convert the following to millihenries:
    (a) 1 H   (b) 250 μH   (c) 10 μH   (d) 0.0005 H

13–6. Convert the following to microhenries:
    (a) 300 mH   (b) 0.08 H   (c) 5 mH   (d) 0.00045 mH

468

13–7. What is the voltage across a coil when $di/dt = 10 \text{ mA}/\mu s$ and $L = 5 \ \mu H$?

13–8. Fifty volts are induced across a 25-mH coil. At what rate is the current changing?

13–9. A core has the following parameters: $A = 0.005 \text{ m}^2$, $l = 0.02 \text{ m}$, and $\mu = 50 \times 10^{-6}$. How many turns are needed to produce 100 mH?

Section 13–4

13–10. Determine the time constant for each of the following series RL combinations:
(a) $R = 100 \ \Omega$, $L = 100 \ \mu H$ **(b)** $R = 4.7 \text{ k}\Omega$, $L = 10 \text{ mH}$
(c) $R = 1.5 \text{ M}\Omega$, $L = 3 \text{ H}$

13–11. In a series RL circuit, determine how long it takes the current to build up to its full value for each of the following:
(a) $R = 50 \ \Omega$, $L = 50 \ \mu H$ **(b)** $R = 3300 \ \Omega$, $L = 15 \text{ mH}$
(c) $R = 22 \text{ k}\Omega$, $L = 100 \text{ mH}$

13–12. In the circuit of Figure 13–48, there is initially no current. Determine the inductor voltage at the following times after the switch is closed:
(a) 10 μs   **(b)** 20 μs   **(c)** 30 μs   **(d)** 40 μs   **(e)** 50 μs

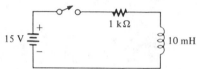

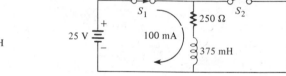

FIGURE 13–48   **FIGURE 13–49**

13–13. In Figure 13–49, there are 100 mA through the coil. When S_1 is opened and S_2 simultaneously closed, find the inductor voltage at the following times:
(a) Initially   **(b)** 1.5 ms   **(c)** 4.5 ms   **(d)** 6 ms

13–14. Repeat Problem 13–12 for the following times:
(a) 2 μs   **(b)** 5 μs   **(c)** 15 μs

13–15. Repeat Problem 13–13 for the following times:
(a) 0.5 ms   **(b)** 1 ms   **(c)** 2 ms

13–16. In Figure 13–48, at what time after switch closure does the inductor voltage reach 5 V?

13–17. What is the polarity of the induced voltage in Figure 13–50 when the switch is closed? What is the final value of current if $R_W = 10 \ \Omega$?

Section 13–5

13–18. Five inductors are connected in series. The lowest value is 5 μH. If the value of each inductor is twice that of the preceding one, and if the inductors are connected in order of ascending values, what is the total inductance?

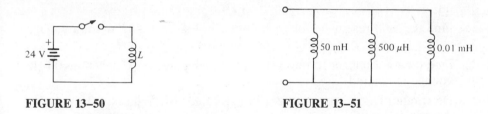

FIGURE 13–50   **FIGURE 13–51**

13–19. Suppose that you require a total inductance of 50 mH. You have available a 10-mH coil and a 22-mH coil. How much additional inductance do you need?

13–20. Determine the total inductance in Figure 13–51.

Section 13–6

13–21. Determine the total parallel inductance for the following coils in parallel: 75 μH, 50 μH, 25 μH, and 15 μH.

13–22. You have a 12-mH inductor, and it is your smallest value. You need an inductance of 8 mH. What value can you use in parallel with the 12-mH to obtain 8 mH?

13–23. Determine the total inductance of each circuit in Figure 13–52.

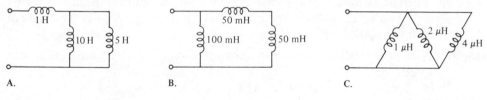

A.   B.   C.

FIGURE 13–52

13–24. Determine the total inductance of each circuit in Figure 13–53.

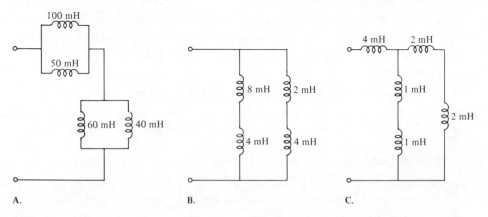

A.   B.   C.

FIGURE 13–53

Section 13–7

13–25. Find the total reactance for each circuit in Figure 13–52 when a voltage with a frequency of 5 kHz is applied across the terminals.

13–26. Find the total reactance for each circuit in Figure 13–53 when a 400-Hz voltage is applied.

13–27. Determine the total rms current in Figure 13–54. What are the currents through L_2 and L_3?

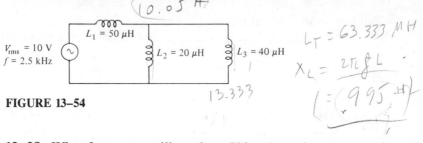

FIGURE 13–54

13–28. What frequency will produce 500-mA total rms current in each circuit of Figure 13–53 with an rms input voltage of 10 V?

13–29. Determine the reactive power in Figure 13–54.

Section 13–8

13–30. A certain coil that is supposed to have a 5-Ω winding resistance is measured with an ohmmeter. The meter indicates 2.8 Ω. What is the problem with the coil?

13–31. What is the indication corresponding to each of the following failures in a coil?
(a) open   **(b)** shorted   **(c)** some windings shorted

Section 13–9

13–32. Write a program to compute and display the inductive current at any value of time after a switch is closed in a series *RL* circuit. The values of source voltage, resistance, and inductance are to be specified as input variables.

13–33. Develop a program to determine the energy stored by an inductor and the power dissipated in the winding resistance for specified values of inductance, winding resistance, and direct current.

Advanced Problems

13–34. Determine the time constant for the circuit in Figure 13–55.

13–35. Find the inductor current at 10 μs after the switch is thrown from position 1 to position 2 in Figure 13–56. For simplicity, assume that the switch makes contact with position 2 at the same instant it breaks contact with position 1.

13–36. In Figure 13–57, switch 1 is opened and switch 2 is closed at the same instant (t_0). What is the instantaneous voltage across R_2 at t_0?

13–37. Determine the value of I_{L2} in Figure 13–58.

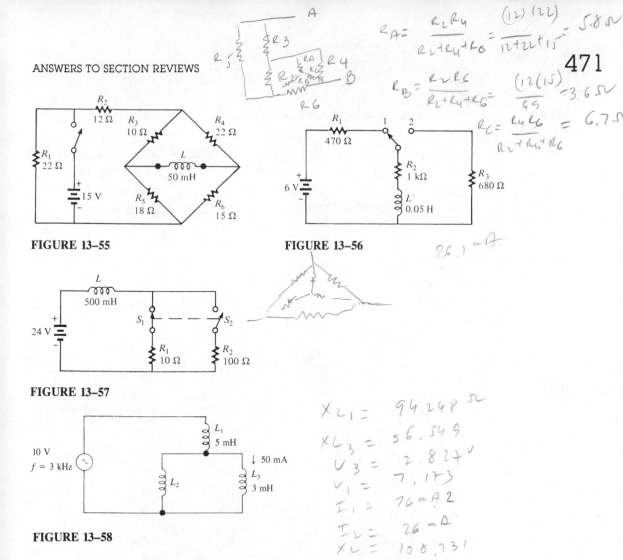

FIGURE 13–55

FIGURE 13–56

FIGURE 13–57

FIGURE 13–58

Answers to Section Reviews

Section 13–1:
1. North, south. **2.** F. **3.** 20 Wb/m². **4.** T. **5.** Magnetomotive force. **6.** The opposition to the establishment of a magnetic field in magnetic material. **7.** 30 At, 6×10^4 At/Wb. **8.** Zero. **9.** Increase. **10.** A force on the conductor is created.

Section 13–2:
1. Turns, permeability, cross-sectional area, and length. **2.** 7.5 mV. **3.** (a) L increases, (b) L decreases, (c) L decreases, (d) L decreases. **4.** All wire has some resistance, and since inductors are made from turns of wire, there is always resistance.

Section 13–3:
1. Fixed and variable. **2.** Air core, iron core, variable.

Section 13–4:
1. 0.1 V. **2.** $v_R = 0$ V, $v_L = 20$ V. **3.** $v_R = 20$ V, $v_L = 0$ V. **4.** 0.5 μs, 3.93 mA.

Section 13–5:
1. Inductances are added in series. **2.** 2600 μH. **3.** 500 mH.

Section 13–6:
1. The total parallel inductance is smaller than that of the smallest individual inductor in parallel. **2.** T. **3.** **(a)** 7.69 mH, **(b)** 24 μH, **(c)** 0.1 H.

Section 13–7:
1. Voltage leads current by 90 degrees. **2.** 3.14 kΩ. **3.** 2.55 MHz. **4.** 15.9 mA. **5.** 0, 458.4 mVAR.

Section 13–8:
1. F. **2.** F.

Section 13–9:
1. To set the increments of frequency between the lowest and the highest. **2.** π.

An *RC* circuit contains both *resistance* and *capacitance*. It is one of the basic types of reactive circuits that will be studied. In this chapter, basic series and parallel *RC* circuits and their responses to sinusoidal ac voltages are covered. Series-parallel combinations are also analyzed. Power in *RC* circuits is studied, and basic applications are introduced. Pulse response of *RC* circuits is covered, and troubleshooting and computer analysis are introduced at the end of the chapter. Applications of the *RC* circuit include filters, amplifier coupling, oscillators, and wave-shaping circuits.

In this chapter, you will learn:
1. How to calculate the impedance of a series *RC* circuit.
2. How to analyze a series *RC* circuit in terms of phase angle, current, and voltages using Ohm's law and Kirchhoff's voltage law.
3. How to determine the effects of frequency on a series *RC* circuit.
4. How to calculate the impedance of a parallel *RC* circuit.
5. How to use the concepts of conductance, susceptance, and admittance in the analysis of parallel *RC* circuits.
6. How to analyze parallel *RC* circuits in terms of phase angle, currents, and voltage using Ohm's law and Kirchhoff's current law.
7. How to convert a parallel *RC* circuit to an equivalent series form.
8. How to analyze circuits with combinations of series and parallel elements.
9. How to determine the average power, reactive power, apparent power, and the power factor in *RC* circuits.
10. How to analyze *RC* lead and lag networks and basic *RC* filters.
11. How to troubleshoot *RC* circuits for basic types of component failures.

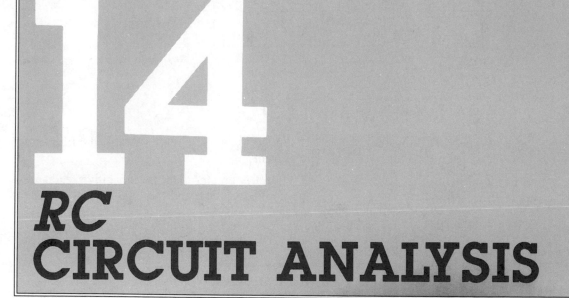

14

RC CIRCUIT ANALYSIS

14–1

When a sinusoidal voltage is applied to an *RC* circuit, *each resulting voltage drop and the current in the circuit is also sinusoidal with the same frequency as the applied voltage.* As shown in Figure 14–1, the resistor voltage, the capacitor voltage, and the current are all sine waves with the frequency of the source.

Phase shifts are introduced because of the capacitance. As you will learn, the resistor voltage and current lead the source voltage, and the capacitor voltage lags the source voltage. The phase angle between the current and the capacitor voltage is always 90 degrees. These generalized phase relationships are indicated in the figure.

The amplitudes and the phase relationships of the voltages and current depend on the ohmic values of the resistance and the capacitive reactance. When a circuit is purely resistive, the phase angle between the applied (source) voltage and the total current is zero. When a circuit is purely capacitive, the phase angle between the applied voltage and the total current is 90 degrees, with the current leading the voltage. When there is a combination of both resistance and capacitive reactance in a circuit, the phase angle between the applied voltage and the total current is somewhere between zero and 90 degrees, depending on the relative values of the resistance and the reactance.

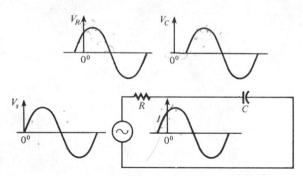

FIGURE 14–1 *Illustration of sinusoidal response with general phase relationships indicated. I and V_R are in phase. I leads V_C by 90 degrees.*

Review for 14–1

1. A 60-Hz sinusoidal voltage is applied to an *RC* circuit. What is the frequency of the capacitor voltage? The current?

2. When the resistance in an *RC* circuit is greater than the capacitive reactance, is the phase angle between the applied voltage and the total current closer to zero or to 90 degrees?

IMPEDANCE OF A SERIES *RC* CIRCUIT

14–2

Impedance is the total opposition to sinusoidal current expressed in units of ohms. In a purely resistive circuit, the impedance is simply equal to the total re-

476

$$V_C = (X_C \angle -90°)(I_C \angle \theta)$$

$$V_C = X_C I_C \angle \theta - 90°$$

sistance. In a purely capacitive circuit, the impedance is equal to the total capacitive reactance. The impedance of a series *RC* circuit is determined by both the resistance and the capacitive reactance.

Recall from Chapter 12 that capacitive reactance is expressed as a complex number in rectangular form as

$$\mathbf{X}_C = -jX_C \tag{14-1}$$

where boldface $\mathbf{X}_C$ designates a phasor quantity (representing both magnitude and angle) and X_C is just the magnitude.

In the series *RC* circuit of Figure 14-2, the total impedance is the phasor sum of R and $-jX_C$ and is expressed as

$$\mathbf{Z} = R - jX_C \tag{14-2}$$

The Impedance Triangle

In ac analysis, both R and X_C are treated as phasor quantities, as shown in the phasor diagram of Figure 14-3A, with X_C appearing at a -90-degree angle with respect to R. This relationship comes from the fact that the capacitor voltage in a series *RC* circuit lags the current, and thus the resistor voltage, by 90 degrees. Since $\mathbf{Z}$ is the phasor sum of R and $-jX_C$, its phasor representation is shown in Figure 14-3B. A repositioning of the phasors, as shown in Part C, forms a right triangle. This is called the *impedance triangle*. The length of each phasor represents the *magnitude* in ohms, and the angle θ is the phase angle of the *RC* circuit and represents the phase difference between the applied voltage and the current.

From right-angle trigonometry, the magnitude (length) of the impedance can be expressed in terms of the resistance and reactance as

$$Z = \sqrt{R^2 + X_C^2} \tag{14-3}$$

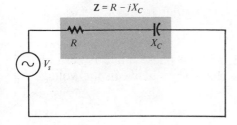

FIGURE 14-2 *Series RC circuit.*

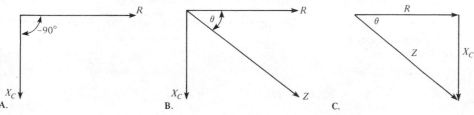

FIGURE 14-3 *Development of the impedance triangle for a series RC circuit.*

478

This is the *magnitude* of **Z** and is expressed in ohms.

The phase angle, θ, is expressed as

$$\theta = -\arctan\left(\frac{X_C}{R}\right) \tag{14-4}$$

Arctan can be found on a calculator by pressing INV, then TAN. Combining the magnitude and the angle, the impedance can be expressed in polar form as

$$\mathbf{Z} = \sqrt{R^2 + X_C^2} \, \angle -\arctan\left(\frac{X_C}{R}\right) \tag{14-5}$$

Example 14–1

For each circuit in Figure 14–4, express the impedance in both rectangular form and polar form.

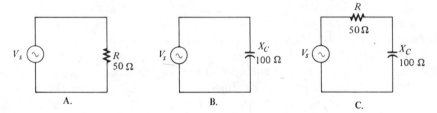

FIGURE 14–4

Solution:

Circuit A:

$\mathbf{Z} = R - j0 = R = 50 \ \Omega$ in rectangular form ($X_C = 0$)

$\mathbf{Z} = R \angle 0° = 50\angle 0° \ \Omega$ in polar form

The impedance is simply the resistance, and the phase angle is zero because pure resistance does not cause a phase shift between the voltage and current.

Circuit B:

$\mathbf{Z} = 0 - jX_C = -j100 \ \Omega$ in rectangular form ($R = 0$)

$\mathbf{Z} = X_C \angle -90° = 100\angle -90° \ \Omega$ in polar form

The impedance is simply the capacitive reactance, and the phase angle is -90 degrees because the capacitance causes the current to lead the voltage by 90 degrees.

Circuit C:

$\mathbf{Z} = R - jX_C = 50 \ \Omega - j100 \ \Omega$ in rectangular form

$$\mathbf{Z} = \sqrt{R^2 + X_C^2} \angle -\arctan\left(\frac{X_C}{R}\right)$$

$$= \sqrt{(50\ \Omega)^2 + (100\ \Omega)^2} \angle -\arctan\left(\frac{100\ \Omega}{50\ \Omega}\right) = 111.8 \angle -63.4°\ \Omega$$

The impedance is the phasor sum of the resistance and the capacitive reactance. The phase angle is fixed by the relative values of X_C and R.

Review for 14–2

1. The impedance of a certain *RC* circuit is 150 Ω − *j*220 Ω. What is the value of the resistance? The capacitive reactance?

2. A series *RC* circuit has a total resistance of 33 kΩ and a capacitive reactance of 50 kΩ. Write the expression for the impedance in rectangular form.

3. For the circuit in Problem 2, what is the magnitude of the impedance? What is the phase angle?

ANALYSIS OF SERIES *RC* CIRCUITS

In the previous section, you learned how to express the impedance of a series *RC* circuit. In this section, Ohm's law and Kirchhoff's voltage law are utilized in the analysis of *RC* circuits.

14–3

Ohm's Law

The application of Ohm's law to series *RC* circuits involves the use of the phasor quantities of **Z**, **V**, and **I**. Keep in mind that the use of boldface letters indicates that both magnitude and angle are included. The three equivalent forms of Ohm's law are as follows:

$$\mathbf{V} = \mathbf{IZ} \tag{14–6}$$

$$\mathbf{I} = \frac{\mathbf{V}}{\mathbf{Z}} \tag{14–7}$$

$$\mathbf{Z} = \frac{\mathbf{V}}{\mathbf{I}} \tag{14–8}$$

From your knowledge of phasor algebra, you should recall that multiplication and division are most easily accomplished with the *polar* forms. Since Ohm's law calculations involve multiplications and divisions, the voltage, current, and impedance should be expressed in polar form, as the next examples show.

Example 14–2

If the current in Figure 14–5 is expressed in polar form as $\mathbf{I} = 0.2\angle 0°$ mA, determine the source voltage, and express it in polar form.

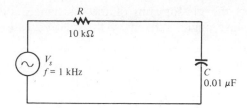

FIGURE 14–5

Solution:

The capacitive reactance is

$$X_C = \frac{1}{2\pi f C} = \frac{1}{2\pi(1000 \text{ Hz})(0.01 \ \mu\text{F})} = 15.9 \text{ k}\Omega$$

The impedance is

$$\mathbf{Z} = R - jX_C = 10 \text{ k}\Omega - j15.9 \text{ k}\Omega$$

Converting to polar form:

$$\mathbf{Z} = \sqrt{(10 \text{ k}\Omega)^2 + (15.9 \text{ k}\Omega)^2} \angle -\arctan\left(\frac{15.9 \text{ k}\Omega}{10 \text{ k}\Omega}\right)$$

$$= 18.78\angle -57.83° \ \Omega$$

Applying Ohm's law:

$$\mathbf{V}_s = \mathbf{IZ} = (0.2\angle 0° \text{ mA})(18.78\angle -57.83° \text{ k}\Omega) = 3.76\angle -57.83° \text{ V}$$

The magnitude of the source voltage is 3.76 V at an angle of $-57.83°$ with respect to the current; that is, the voltage lags the current by 57.83°.

Example 14–3

Determine the current in the circuit of Figure 14–6.

Solution:

$$X_C = \frac{1}{2\pi f C} = \frac{1}{2\pi(1.5 \text{ kHz})(0.02 \ \mu\text{F})} = 5.3 \text{ k}\Omega$$

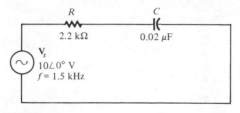

FIGURE 14–6

The total impedance is

$$\mathbf{Z} = R - jX_C = 2.2 \text{ k}\Omega - j5.3 \text{ k}\Omega$$

Converting to polar form:

$$\mathbf{Z} = \sqrt{(2.2 \text{ k}\Omega)^2 + (5.3 \text{ k}\Omega)^2} \angle -\arctan\left(\frac{5.3 \text{ k}\Omega}{2.2 \text{ k}\Omega}\right)$$

$$= 5.74 \angle -67.46° \text{ k}\Omega$$

Applying Ohm's law:

$$\mathbf{I} = \frac{\mathbf{V}}{\mathbf{Z}} = \frac{10\angle 0° \text{ V}}{5.74\angle -67.46° \text{ k}\Omega} = 1.74\angle 67.46° \text{ mA}$$

The magnitude of the current is 1.74 mA. The positive phase angle of 67.46 degrees indicates that the current leads the voltage by that amount.

Relationships of the Current and Voltages in a Series *RC* Circuit

In a series circuit, the current is the same through both the resistor and the capacitor. Thus, *the resistor voltage is in phase with the current, and the capacitor voltage lags the current by 90 degrees.* Therefore, there is a phase difference of 90 degrees between the resistor voltage, V_R, and the capacitor voltage, V_C, as shown in the wave form diagram of Figure 14–7.

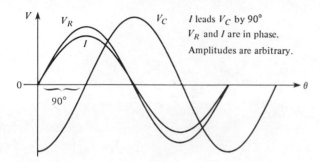

FIGURE 14–7 *Phase relation of voltages and current in a series RC circuit.*

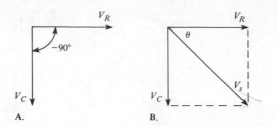

FIGURE 14–8 *Voltage phasor diagram for a series RC circuit.*

We know from Kirchhoff's voltage law that the sum of the voltage drops must equal the applied voltage. However, since V_R and V_C are not in phase with each other, they must be added as phasor quantities, with V_C lagging V_R by 90 degrees, as shown in Figure 14–8A. As shown in Part B, $\mathbf{V}_s$ is the phasor sum of $\mathbf{V}_R$ and $\mathbf{V}_C$, as expressed in the following equation:

$$\mathbf{V}_s = V_R - jV_C \qquad (14\text{–}9)$$

This equation can be expressed in polar form as

$$\mathbf{V}_s = \sqrt{V_R^2 + V_C^2}\angle -\arctan\!\left(\frac{V_C}{V_R}\right) \qquad (14\text{–}10)$$

where the magnitude of the source voltage is

$$V_s = \sqrt{V_R^2 + V_C^2} \qquad (14\text{–}11)$$

and the phase angle between the resistor voltage and the source voltage is

$$\theta = -\arctan\!\left(\frac{V_C}{V_R}\right) \qquad (14\text{–}12)$$

Since the resistor voltage and the current are in phase, θ also represents the phase angle between the source voltage and the current. Figure 14–9 shows a complete voltage and current phasor diagram representing the wave form diagram of Figure 14–7.

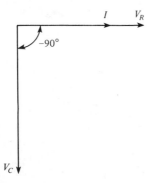

FIGURE 14–9 *Voltage and current phasor diagram for the wave forms in Figure 14–7.*

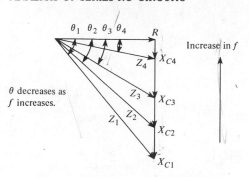

FIGURE 14–10 *Effect of frequency on phase angle.*

Variation of Impedance and Phase Angle with Frequency

The impedance triangle is useful in visualizing how the frequency of the applied sinusoidal voltage affects the *RC* circuit. As you know, capacitive reactance varies *inversely* with frequency. Since $Z = \sqrt{R^2 + X_C^2}$, you can see that when X_C increases, the magnitude of the total impedance also increases; and when X_C decreases, the magnitude of the total impedance also decreases. Therefore, *Z is inversely dependent on frequency.*

The phase angle θ also varies *inversely* with frequency, because $\theta = -\arctan(X_C/R)$. As X_C increases, so does θ and vice versa.

Figure 14–10 uses the impedance triangle to illustrate the variations in X_C, Z, and θ as the frequency changes. Of course, R remains constant. The key point is that *because X_C varies inversely as the frequency, so also do the magnitude of the total impedance and the phase angle.* Example 14–4 illustrates this.

Example 14–4

For the series *RC* circuit in Figure 14–11, determine the magnitude of the total impedance and phase angle for each of the following values of input frequency: 10 kHz, 20 kHz, and 30 kHz.

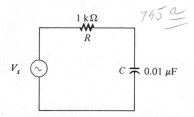

FIGURE 14–11

Solution:

For $f = 10$ kHz:

Example 14–4 (continued)

$$X_C = \frac{1}{2\pi f C} = \frac{1}{2\pi (10 \text{ kHz})(0.01 \ \mu\text{F})} = 1592 \ \Omega$$

$$\mathbf{Z} = \sqrt{R^2 + X_C^2} \ \angle -\arctan\left(\frac{X_C}{R}\right)$$

$$= \sqrt{(1000 \ \Omega)^2 + (1592 \ \Omega)^2} \ \angle -\arctan\left(\frac{1592}{1000}\right)$$

$$= 1880 \angle -57.87° \ \Omega$$

For $f = 20$ kHz:

$$X_C = \frac{1}{2\pi (20 \text{ kHz})(0.01 \ \mu\text{F})} = 796 \ \Omega$$

$$\mathbf{Z} = \sqrt{(1000 \ \Omega)^2 + (796 \ \Omega)^2} \ \angle -\arctan\left(\frac{796}{1000}\right)$$

$$= 1278 \angle -38.52° \ \Omega$$

For $f = 30$ kHz:

$$X_C = \frac{1}{2\pi (30 \text{ kHz})(0.01 \ \mu\text{F})} = 531 \ \Omega$$

$$\mathbf{Z} = \sqrt{(1000 \ \Omega)^2 + (531 \ \Omega)^2} \ \angle -\arctan\left(\frac{531}{1000}\right)$$

$$= 1132 \angle -27.97° \ \Omega$$

Notice that as the frequency increases, X_C, Z, and θ decrease.

Review for 14–3

1. In a certain series *RC* circuit, $V_R = 4$ V, and $V_C = 6$ V. What is the magnitude of the total voltage?

2. In Problem 1, what is the phase angle between the total voltage and the current?

3. What is the phase difference between the capacitor voltage and the resistor voltage in a series *RC* circuit?

4. When the frequency of the applied voltage in a series *RC* circuit is increased, what happens to the capacitive reactance? What happens to the magnitude of the total impedance? What happens to the phase angle?

IMPEDANCE OF A PARALLEL *RC* CIRCUIT

14–4

A basic parallel *RC* circuit is shown in Figure 14–12. The expression for the total impedance is developed as follows using the rules of phasor algebra.

$$\mathbf{Z} = \frac{(R \angle 0°)(X_C \angle -90°)}{R - jX_C}$$

By multiplying the magnitudes, adding the angles in the numerator, and converting the denominator to polar form, we get

$$\mathbf{Z} = \frac{RX_C \angle (0° - 90°)}{\sqrt{R^2 + X_C^2} \angle -\arctan\left(\dfrac{X_C}{R}\right)}$$

Now, dividing the magnitude expression in the numerator by that in the denominator, and by subtracting the angle in the denominator from that in the numerator, we get

$$\mathbf{Z} = \left(\frac{RX_C}{\sqrt{R^2 + X_C^2}}\right) \angle \left(-90° + \arctan\left(\frac{X_C}{R}\right)\right) \tag{14–13}$$

Equation (14–13) is the expression for the total parallel impedance, where the magnitude is

$$Z = \frac{RX_C}{\sqrt{R^2 + X_C^2}} \tag{14–14}$$

and the phase angle between the applied voltage and the total current is

$$\theta = -90° + \arctan\left(\frac{X_C}{R}\right) \tag{14–15}$$

Equivalently, this expression can be written as

$$\theta = -\arctan\left(\frac{R}{X_C}\right)$$

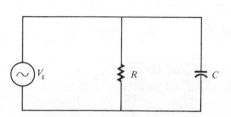

FIGURE 14–12 *Parallel RC circuit.*

Example 14–5

For each circuit in Figure 14–13, determine the magnitude of the total impedance and the phase angle.

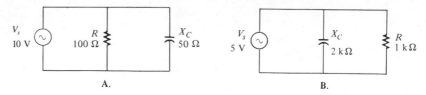

FIGURE 14–13

Solution:

Circuit A:

$$\mathbf{Z} = \left(\frac{RX_c}{\sqrt{R^2 + X_C^2}}\right)\angle\left(-90° + \arctan\left(\frac{X_C}{R}\right)\right)$$

$$= \left[\frac{(100\ \Omega)(50\ \Omega)}{\sqrt{(100\ \Omega)^2 + (50\ \Omega)^2}}\right]\angle\left(-90° + \arctan\left(\frac{50}{100}\right)\right)$$

$$= 44.72\angle-63.43°\ \Omega$$

Thus, $Z = 44.72\ \Omega$, and $\theta = -63.43°$.

Circuit B:

$$\mathbf{Z} = \left[\frac{(1\ k\Omega)(2\ k\Omega)}{\sqrt{(1\ k\Omega)^2 + (2\ k\Omega)^2}}\right]\angle\left(-90° + \arctan\left(\frac{2000}{1000}\right)\right)$$

$$= 894.4\angle-26.57°\ \Omega$$

Thus, $Z = 894.4\ \Omega$, and $\theta = -26.57°$.

Conductance, Susceptance, and Admittance

Recall that *conductance* is the reciprocal of resistance and is expressed as

$$G = \frac{1}{R} \tag{14–16}$$

Two new terms are now introduced for use in parallel *RC* circuits. *Capacitive susceptance* (B_C) is the reciprocal of capacitive reactance and is expressed as

$$B_C = \frac{1}{X_C} \tag{14–17}$$

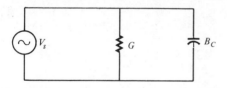

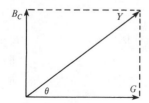

FIGURE 14-14 *Admittance in a parallel RC circuit.*

Admittance (Y) is the reciprocal of impedance and is expressed as

$$Y = \frac{1}{Z}$$ (14-18)

The unit of each of these terms is, of course, the *siemen,* which is the reciprocal of the ohm.

In working with parallel circuits, it is often easier to use G, B_C, and Y rather than R, X_C, and Z, as we now discuss. In a parallel *RC* circuit, as shown in Figure 14-14, the total admittance is simply the phasor sum of the conductance and the susceptance.

$$\mathbf{Y} = G + jB_C$$ (14-19)

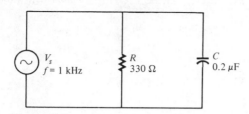

Example 14-6

Determine the admittance in Figure 14-15.

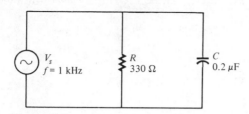

FIGURE 14-15

Solution:

$R = 470 \ \Omega$; thus $G = 1/R = 1/330 \ \Omega = 0.003$ S.

$$X_C = \frac{1}{2\pi(1000 \ \text{Hz})(0.2 \ \mu\text{F})} = 796 \ \Omega$$

Example 14–6 (continued)

$$B_C = \frac{1}{X_C} = \frac{1}{796 \ \Omega} = 0.00126 \ \text{S}$$

$$\mathbf{Y} = G + jB_C = 0.003 \ \text{S} + j0.00126 \ \text{S}$$

Y can be expressed in polar form as follows:

$$\mathbf{Y} = \sqrt{(0.003 \ \text{S})^2 + (0.00126 \ \text{S})^2} \angle \arctan\left(\frac{0.00126}{0.003}\right)$$

$$= 0.00325 \angle 22.78° \ \text{S}$$

Now, this can be converted to impedance.

$$\mathbf{Z} = \frac{1}{Y} = \frac{1}{(0.00325 \angle 22.78° \ \text{S})} = 307.69 \angle -22.78° \ \Omega$$

Review for 14–4

1. Define *conductance, capacitive susceptance,* and *admittance.*
2. If $Z = 100 \ \Omega$, what is the value of Y?
3. In a certain parallel *RC* circuit, $R = 50 \ \Omega$ and $X_C = 75 \ \Omega$. Determine **Y**.
4. In Problem 3, what is the magnitude of **Y**, and what is the phase angle between the total current and the applied voltage?

ANALYSIS OF PARALLEL *RC* CIRCUITS

14–5   For convenience in the analysis of parallel circuits, the Ohm's law formulas using impedance, previously stated, can be rewritten for admittance using the relation $Y = 1/Z$.

$$V = \frac{I}{Y} \qquad\qquad (14\text{–}20)$$

$$I = VY \qquad\qquad (14\text{–}21)$$

$$Y = \frac{I}{V} \qquad\qquad (14\text{–}22)$$

Example 14–7

Determine the total current and phase angle in Figure 14–16.

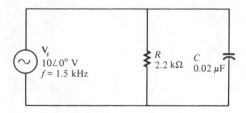

FIGURE 14–16

Solution:

$$X_C = \frac{1}{2\pi(1.5 \text{ kHz})(0.02 \text{ }\mu\text{F})} = 5.3 \text{ k}\Omega$$

The susceptance is

$$B_C = \frac{1}{X_C} = \frac{1}{5.3 \text{ k}\Omega} = 0.189 \text{ mS}$$

The conductance is

$$G = \frac{1}{R} = \frac{1}{2.2 \text{ k}\Omega} = 0.455 \text{ mS}$$

The total admittance is

$$\mathbf{Y} = G + jB_C = 0.455 \text{ mS} + j0.189 \text{ mS}$$

Converting to polar form:

$$\mathbf{Y} = \sqrt{(0.455 \text{ mS})^2 + (0.189 \text{ mS})^2} \angle \arctan\left(\frac{0.189}{0.455}\right)$$

$$= 0.493\angle22.56° \text{ mS}$$

Applying Ohm's law:

$$\mathbf{I}_T = \mathbf{VY} = (10\angle0° \text{ V})(0.493\angle22.56° \text{ mS}) = 4.93\angle22.56° \text{ mA}$$

The magnitude of the total current is 4.93 mA, and it leads the applied voltage by 22.56 degrees.

Relationships of the Currents and Voltages in a Parallel *RC* Circuit

Figure 14–17A shows all the currents and voltages in a basic parallel *RC* circuit. As you can see, the applied voltage, V_s, appears across both the resistive and the capacitive branches, so V_s, V_R, and V_C are all in phase and of the same magnitude. The total current, I_T, divides at the junction into the two branch currents, I_R and I_C.

490

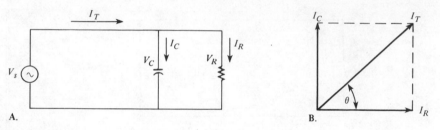

FIGURE 14–17 *Currents and voltages in a parallel RC circuit.*

The current through the resistor is in phase with the voltage. The current through the capacitor leads the voltage, and thus the resistive current, by 90 degrees. By Kirchhoff's current law, the total current is the phasor sum of the two branch currents, as shown by the phasor diagram in Figure 14–17B. The total current is expressed as

$$\mathbf{I}_T = I_R + jI_C \tag{14-23}$$

This equation can be expressed in polar form as

$$\mathbf{I}_T = \sqrt{I_R^2 + I_C^2}\ \angle\arctan\!\left(\frac{I_C}{I_R}\right) \tag{14-24}$$

where the magnitude of the total current is

$$I_T = \sqrt{I_R^2 + I_C^2} \tag{14-25}$$

and the phase angle between the resistor current and the total current is

$$\theta = \arctan\!\left(\frac{I_C}{I_R}\right) \tag{14-26}$$

Since the resistor current and the applied voltage are in phase, θ also represents the phase angle between the total current and the applied voltage. Figure 14–18 shows a complete current and voltage phasor diagram.

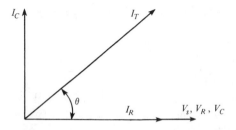

FIGURE 14–18 *Current and voltage phasor diagram for a parallel RC circuit (amplitudes are arbitrary).*

Example 14–8

Determine the value of each current in Figure 14–19, and describe the phase relationship of each with the applied voltage.

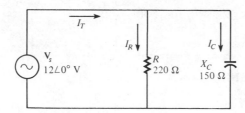

FIGURE 14–19

Solution:

$$\mathbf{I}_R = \frac{\mathbf{V}_s}{\mathbf{R}} = \frac{12\angle 0° \text{ V}}{220\angle 0° \text{ }\Omega} = 54.55\angle 0° \text{ mA}$$

$$\mathbf{I}_C = \frac{\mathbf{V}_s}{\mathbf{X}_C} = \frac{12\angle 0° \text{ V}}{150\angle -90° \text{ }\Omega} = 80\angle 90° \text{ mA}$$

$$\mathbf{I}_T = I_R + jI_C = 54.55 \text{ mA} + j80 \text{ mA}$$

Converting I_T to polar form:

$$\mathbf{I}_T = \sqrt{(54.55 \text{ mA})^2 + (80 \text{ mA})^2} \angle \arctan\left(\frac{80}{54.55}\right)$$

$$= 96.83\angle 55.71° \text{ mA}$$

As the results show, the resistor current is 54.55 mA and is in phase with the voltage. The capacitor current is 80 mA and leads the voltage by 90 degrees. The total current is 96.83 mA and leads the voltage by 55.71 degrees.

Conversion from Parallel to Series Form

For every parallel *RC* circuit, there is an *equivalent* series *RC* circuit. Two circuits are equivalent when they both present an equal impedance at their terminals; that is, the magnitude of impedance and the phase angle are identical.

To obtain the equivalent series circuit from a given parallel circuit, express the total impedance of the parallel circuit in *rectangular* form. From this, the values of *R* and X_C are obtained. An example best illustrates this approach.

Example 14–9

Convert the parallel circuit in Figure 14–20 to a series form.

Example 14–9 (continued)

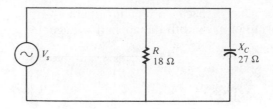

FIGURE 14–20

Solution:

First, find the impedance of the parallel circuit.

$$G = \frac{1}{R} = \frac{1}{18\ \Omega} = 0.055\ \text{S}$$

$$B_C = \frac{1}{X_C} = \frac{1}{27\ \Omega} = 0.037\ \text{S}$$

$$\mathbf{Y} = G + jB_C = 0.055\ \text{S} + j0.037\ \text{S}$$

Converting to polar form:

$$\mathbf{Y} = \sqrt{(0.055\ \text{S})^2 + (0.037\ \text{S})^2}\ \angle \arctan\left(\frac{0.037}{0.055}\right) = 0.066\angle 33.93°\ \text{S}$$

The total impedance is

$$\mathbf{Z} = \frac{1}{\mathbf{Y}} = \frac{1}{0.066\angle 33.93°\ \text{S}} = 15.15\angle -33.93°\ \Omega$$

Converting to rectangular form:

$$\mathbf{Z} = 15.15\cos(-33.93) + j15.15\sin(-33.93)$$
$$= 12.57\ \Omega - j8.46\ \Omega$$

The equivalent series *RC* circuit is a 12.57-Ω resistor in series with a capacitive reactance of 8.46 Ω. This is shown in Figure 14–21.

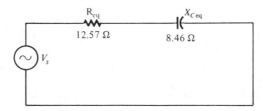

FIGURE 14–21

Review for 14-5

1. The admittance of an RC circuit is 0.0035 S, and the applied voltage is 6 V. What is the total current?

2. In a certain parallel RC circuit, the resistor current is 10 mA, and the capacitor current is 15 mA. Determine the magnitude and phase angle of the total current. This phase angle is measured with respect to what?

3. What is the phase angle between the capacitor current and the applied voltage in a parallel RC circuit?

SERIES-PARALLEL ANALYSIS

In this section, we use the concepts studied in the previous sections to analyze circuits with combinations of both series and parallel R and C elements. Examples are used to demonstrate the procedures.

14-6

Example 14-10

In the circuit of Figure 14-22, determine the following:
(a) Total impedance
(b) Total current
(c) Phase angle by which I_T leads V_s.

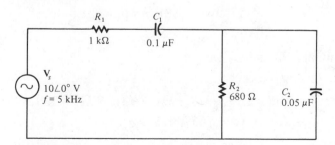

FIGURE 14-22

Solution:

(a) First, calculate the magnitudes of capacitive reactance:

$$X_{C1} = \frac{1}{2\pi(5 \text{ kHz})(0.1 \ \mu\text{F})} = 318.3 \ \Omega$$

$$X_{C2} = \frac{1}{2\pi(5 \text{ kHz})(0.05 \ \mu\text{F})} = 636.6 \ \Omega$$

One approach is to find the impedance of the series portion and the impedance of the parallel portion and combine them to get the total impedance.

Example 14–10 (continued)

R_1 and C_1 are in series. R_2 and C_2 are in parallel. The series portion and the parallel portion are in series with each other.

$$\mathbf{Z}_1 = R_1 - jX_{C1} = 1 \text{ k}\Omega - j318.3 \text{ }\Omega$$

$$G_2 = \frac{1}{R_2} = \frac{1}{680 \text{ }\Omega} = 0.00147 \text{ S}$$

$$B_{C2} = \frac{1}{X_{C2}} = \frac{1}{636.6 \text{ }\Omega} = 0.00157 \text{ S}$$

$$\mathbf{Y}_2 = G_2 + jB_{C2} = 0.00147 \text{ S} + j0.00157 \text{ S}$$

Converting to polar form:

$$\mathbf{Y}_2 = \sqrt{(0.00147 \text{ S})^2 + (0.00157 \text{ S})^2} \angle \arctan\left(\frac{0.00157}{0.00147}\right)$$

$$= 0.00215\angle 46.88° \text{ S}$$

Then

$$\mathbf{Z}_2 = \frac{1}{\mathbf{Y}_2} = \frac{1}{0.00215\angle 46.88° \text{ S}} = 465.12\angle -46.88° \text{ }\Omega$$

Converting to rectangular form:

$$\mathbf{Z}_2 = 465.12 \cos(46.88) - j465.12 \sin(46.88)$$
$$= 317.9 \text{ }\Omega - j339.5 \text{ }\Omega$$

Combining $\mathbf{Z}_1$ and $\mathbf{Z}_2$ we get

$$\mathbf{Z}_T = \mathbf{Z}_1 + \mathbf{Z}_2$$
$$= (1000 \text{ }\Omega - j318.3 \text{ }\Omega) + (317.9 \text{ }\Omega - 339.5 \text{ }\Omega)$$
$$= 1317.9 \text{ }\Omega - j657.8 \text{ }\Omega$$

Expressing $\mathbf{Z}_T$ in polar form:

$$\mathbf{Z}_T = \sqrt{(1317.9 \text{ }\Omega)^2 + (657.8 \text{ }\Omega)^2} \angle -\arctan\left(\frac{657.8}{1317.9}\right)$$

$$= 1472.9\angle -26.53° \text{ }\Omega$$

(b) The total current can be found with Ohm's law.

$$\mathbf{I}_T = \frac{\mathbf{V}_s}{\mathbf{Z}_T} = \frac{10\angle 0° \text{ V}}{1472.9\angle -26.53° \text{ }\Omega}$$

$$= 6.79\angle 26.53° \text{ mA}$$

(c) The total current leads the applied voltage by 26.53 degrees.

Example 14–11

Determine all currents in Figure 14–23. Sketch a current phasor diagram.

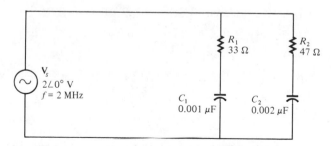

FIGURE 14–23

Solution:

First, calculate X_{C1} and X_{C2}:

$$X_{C1} = \frac{1}{2\pi(2 \text{ MHz})(0.001 \ \mu\text{F})} = 79.58 \ \Omega$$

$$X_{C2} = \frac{1}{2\pi(2 \text{ MHz})(0.002 \ \mu\text{F})} = 39.79 \ \Omega$$

Now, we determine the impedance of each of the two parallel branches.

$$\mathbf{Z}_1 = R_1 - jX_{C1} = 33 \ \Omega - j79.58 \ \Omega$$
$$\mathbf{Z}_2 = R_2 - jX_{C2} = 47 \ \Omega - j39.79 \ \Omega$$

Convert these to polar form:

$$\mathbf{Z}_1 = \sqrt{(33 \ \Omega)^2 + (79.58 \ \Omega)^2} \ \angle -\arctan\left(\frac{79.58}{33}\right) = 86.15\angle-67.48° \ \Omega$$

$$\mathbf{Z}_2 = \sqrt{(47 \ \Omega)^2 + (39.78 \ \Omega)^2} \ \angle -\arctan\left(\frac{39.79}{47}\right) = 61.57\angle-40.25° \ \Omega$$

Calculate each branch current:

$$\mathbf{I}_1 = \frac{\mathbf{V}_s}{\mathbf{Z}_1} = \frac{2\angle 0° \text{ V}}{86.15\angle-67.48° \ \Omega} = 23.22\angle67.48° \text{ mA}$$

$$\mathbf{I}_2 = \frac{\mathbf{V}_s}{\mathbf{Z}_2} = \frac{2\angle 0° \text{ V}}{61.57\angle-40.25° \ \Omega} = 32.48\angle40.25° \text{ mA}$$

To get the total current, we must express each branch current in rectangular form so that they can be added.

Example 14–11 (continued)

$$\mathbf{I}_1 = 23.22 \cos(67.48) + j23.22 \sin(67.48)$$
$$= 8.89 \text{ mA} + j21.45 \text{ mA}$$

$$\mathbf{I}_2 = 32.48 \cos(40.25) + j32.48 \sin(40.25)$$
$$= 24.79 \text{ mA} + j20.99 \text{ mA}$$

$$\mathbf{I}_T = \mathbf{I}_1 + \mathbf{I}_2$$
$$= (8.89 \text{ mA} + j21.45 \text{ mA}) + (24.79 \text{ mA} + j20.99 \text{ mA})$$
$$= 33.68 \text{ mA} + j42.44 \text{ mA}$$

Converting to polar form:

$$\mathbf{I}_T = \sqrt{(33.68 \text{ mA})^2 + (42.44 \text{ mA})^2} \angle \arctan\left(\frac{42.44}{33.68}\right)$$

$$= 54.18 \angle 51.56° \text{ mA}$$

The current phasor diagram is shown in Figure 14–24.

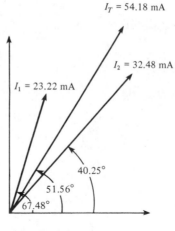

FIGURE 14–24

The two previous examples should give you a feel for how to approach the analysis of complex *RC* networks. Further problem work will sharpen your skills.

Review for 14–6

1. What is the equivalent series *RC* circuit for the series-parallel circuit in Figure 14–22?

2. What is the total impedance of the circuit in Figure 14–23?

In a purely resistive ac circuit, all of the energy delivered by the source is dissipated in the form of heat by the resistance. In a purely capacitive ac circuit, all of the energy delivered by the source is stored by the capacitor during a portion of the voltage cycle and then returned to the source during another portion of the cycle so that there is no net energy loss. What happens when both resistance and capacitance exist in a circuit? *Some of the energy is alternately stored and returned by the capacitance, and some is dissipated by the resistance.* The amount of energy loss is determined by the relative values of the resistance and the capacitive reactance.

It is reasonable to assume that when the resistance is greater than the reactance, more of the total energy delivered by the source is dissipated by the resistance than is stored by the capacitance. Likewise, when the reactance is greater than the resistance, more of the total energy is stored and returned than is lost.

The power in a resistor, sometimes called *average power* (P_{avg}), and the power in a capacitor, called *reactive power* (P_r), were developed in previous chapters and are restated here. The unit of average power is the *watt,* and the unit of reactive power is the VAR (volt-ampere reactive).

$$P_{avg} = I^2 R \qquad (14\text{–}27)$$

$$P_r = I^2 X_C \qquad (14\text{–}28)$$

The Power Triangle

The generalized impedance phasor diagram is shown in Figure 14–25A. A phasor relationship for the powers can also be represented by a similar diagram, because the respective magnitudes of the powers, P_{avg} and P_r, differ from R and X_C by a factor of I^2. This is shown in Figure 14–25B.

The resultant power phasor, $I^2 Z$, represents the *apparent power, P_a*. At any instant in time, P_a is the total power that *appears* to be transferred between the source and the *RC* circuit. The unit of apparent power is the *volt-ampere,* VA. The expression for apparent power is

$$P_a = I^2 Z \qquad (14\text{–}29)$$

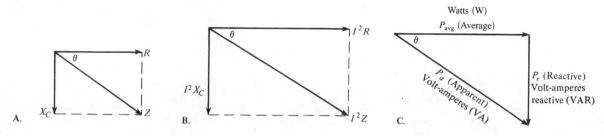

FIGURE 14–25 *Development of the power triangle for an RC circuit.*

The power phasor diagram in Figure 14–25B can be rearranged in the form of a right triangle, as shown in Figure 14–25C. This is called the *power triangle*. Using the rules of trigonometry, P_{avg} can be expressed as

$$P_{avg} = P_a \cos \theta \qquad (14\text{–}30)$$

Since P_a equals $I^2 Z$ or VI, the equation for the average power loss in an *RC* circuit can be written as

$$P_{avg} = VI \cos \theta \qquad (14\text{–}31)$$

where V is the applied voltage and I is the total current.

For the case of a purely resistive circuit, $\theta = 0°$ and $\cos 0° = 1$, so P_{avg} equals VI. For the case of a purely capacitive circuit, $\theta = 90°$ and $\cos 90° = 0$, so P_{avg} is zero. As you already know, there is no power loss in an ideal capacitor.

Power Factor

The term $\cos \theta$ is called the *power factor* and is stated as follows:

$$PF = \cos \theta \qquad (14\text{–}32)$$

As the phase angle between applied voltage and total current increases, the power factor decreases, indicating an increasingly reactive circuit. The smaller the power factor, the smaller the power dissipation.

Example 14–12

Determine the power factor and the average power in the circuit of Figure 14–26.

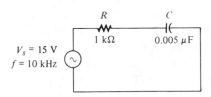

FIGURE 14–26

Solution:

$$X_C = \frac{1}{2\pi f C} = \frac{1}{2\pi (10 \text{ kHz})\,(0.005 \ \mu\text{F})} = 3183 \ \Omega$$

$$\mathbf{Z} = R - jX_C = 1 \text{ k}\Omega - j3183 \ \Omega$$

$$= \sqrt{(1000 \ \Omega)^2 + (3183 \ \Omega)^2} \ \angle -\arctan\!\left(\frac{3183}{1000}\right) = 3336.4 \angle -72.56°$$

The angle associated with the impedance is θ, the angle between the applied voltage and the total current; therefore

$$PF = \cos \theta = \cos(-72.56°) = 0.2997$$

$$I = \frac{V_s}{Z} = \frac{15 \text{ V}}{3336.4 \text{ Ω}} = 4.496 \text{ mA}$$

The average power is

$$P_{avg} = V_s I \cos \theta = (15 \text{ V})(4.496 \text{ mA})(0.2997) = 20.21 \text{ mW}$$

Example 14–13

For the circuit in Figure 14–27, find the average power, the reactive power, and the apparent power. X_C has been determined to be 2 kΩ.

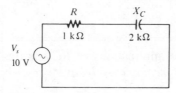

FIGURE 14–27

Solution:

We first find the total impedance so that the current can be calculated.

$$\mathbf{Z} = R - jX_C = 1 \text{ kΩ} - j2 \text{ kΩ}$$

$$= \sqrt{(1000 \text{ Ω})^2 + (2000 \text{ Ω})^2} \angle -\arctan\left(\frac{2000}{1000}\right)$$

$$= 2236 \angle -63.44° \text{ Ω}$$

$$I = \frac{V_s}{Z} = \frac{10 \text{ V}}{2236 \text{ Ω}} = 4.47 \text{ mA}$$

The phase angle, θ, is indicated in the polar expression for impedance.

$$\theta = -63.44°$$

The average power is

$$P_{avg} = V_s I \cos \theta = (10 \text{ V})(4.47 \text{ mA}) \cos(-63.44°) = 19.99 \text{ mW}$$

(The same result is realized using the formula $P_{avg} = I^2 R$.) The reactive power is

$$P_r = I^2 X_C = (4.47 \text{ mA})^2 (2 \text{ kΩ}) = 39.96 \text{ mVAR}$$

The apparent power is

$$P_a = I^2 Z = (4.47 \text{ mA})^2 (2236 \text{ Ω}) = 44.68 \text{ mVA}$$

Example 14–13 (continued)

The apparent power is also the phasor sum of P_{avg} and P_r.
$$P_a = \sqrt{P_{avg}^2 + P_r^2} = 44.68 \text{ mVA}$$

Review for 14–7

1. To which component in an *RC* circuit is the energy loss due?
2. The phase angle, θ, is 45 degrees. What is the power factor?
3. A certain series *RC* circuit has the following parameter values: $R = 300 \ \Omega$, $X_C = 460 \ \Omega$, and $I = 2$ A. Determine the average power, the reactive power, and the apparent power.

BASIC APPLICATIONS

14–8

In this section, two basic applications of *RC* circuits are discussed. These are *phase shift networks* and *frequency-selective networks (filters)*.

The *RC* Lag Network

The first type of phase shift network that we cover causes the output to lag the input by a specified amount. Figure 14–28A shows a series *RC* circuit with the *output voltage taken across the capacitor*. The source voltage is the *input*, V_{in}. As you know, θ, the phase angle between the current and the input voltage, is also the phase angle between the resistor voltage and the input voltage, because V_R and I are in phase with each other.

Since V_C *lags* V_R by 90 degrees, the phase angle between the capacitor voltage and the input voltage is the difference between -90 degrees and θ, as shown in Figure 14–28B. The capacitor voltage is the *output*, and it lags the input, thus creating a basic *lag* network.

When the input and output wave forms of the lag network are displayed on an oscilloscope, a relationship similar to that in Figure 14–29 is observed. The amount of phase difference between the input and the output is dependent on the

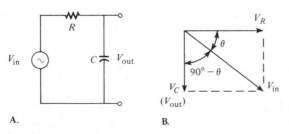

A.   B.

FIGURE 14–28   *RC lag network.*

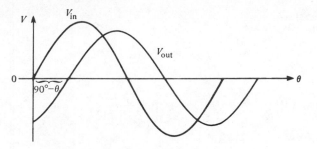

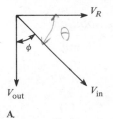

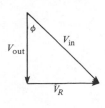

A.   B.

FIGURE 14–29 *Input and output voltages of an RC lag network.*   **FIGURE 14–30**

relative sizes of the capacitive reactance and the resistance, as is the magnitude of the output voltage.

Phase Difference between Input and Output: As already established, θ is the phase angle between I and V_{in}. The angle between V_{out} and V_{in} is designated ϕ (phi) and is developed as follows. The polar expressions for the input voltage and the current are $V_{in}\angle 0°$ and $I\angle\theta$, respectively. The output voltage is

$$\mathbf{V}_{out} = (I\angle\theta)(X_C\angle -90°) = IX_C\angle(-90° + \theta)$$

This shows that the output voltage is at an angle of $-90° + \theta$ with respect to the input voltage. Since $\theta = -\arctan(X_C/R)$, the angle between the input and output is

$$= -\tan^{-1}(R/x_C) = -\tan^{-1}\left(\frac{V_R}{V_{out}}\right)$$

$$\phi = -90° + \arctan\left(\frac{X_C}{R}\right) \qquad\qquad \textbf{(14–33)}$$

This angle is always negative, indicating that the output voltage *lags* the input voltage, as shown in Figure 14–30.

Example 14–14

Determine the amount of phase lag from input to output in each lag network in Figure 14–31.

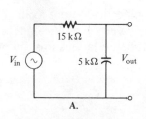

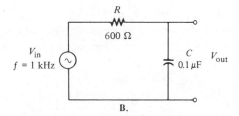

A.   B.

FIGURE 14–31

Example 14–14 (continued)

Solution:

Circuit A:

$$\phi = -90° + \arctan\left(\frac{X_C}{R}\right) = -90° + \arctan\left(\frac{5 \text{ k}\Omega}{15 \text{ k}\Omega}\right)$$

$$= -90° + 18.44° = -71.56°$$

The output lags the input by 71.56°.
  Circuit B:

$$X_C = \frac{1}{2\pi f C} = \frac{1}{2\pi(1 \text{ kHz})(0.1 \text{ }\mu F)} = 1592 \text{ }\Omega$$

$$\phi = -90° + \arctan\left(\frac{X_C}{R}\right) = -90° + \arctan\left(\frac{1592 \text{ }\Omega}{600 \text{ }\Omega}\right) = -20.65°$$

The output lags the input by 20.65°.

Magnitude of the Output Voltage: To evaluate the output voltage in terms of its magnitude, visualize the *RC* lag network as a voltage divider. A portion of the total input voltage is dropped across *R* and a portion across *C*. Since the output voltage is V_C, it can be calculated as

$$V_{\text{out}} = \left(\frac{X_C}{\sqrt{R^2 + X_C^2}}\right) V_{\text{in}} \qquad\qquad \textbf{(14–34)}$$

Or it can be calculated using Ohm's law as

$$V_{\text{out}} = I X_C \qquad\qquad \textbf{(14–35)}$$

The total phasor expression for the output voltage of a lag network is

$$\mathbf{V}_{\text{out}} = V_{\text{out}} \angle \phi \qquad\qquad \textbf{(14–36)}$$

Example 14–15

For the lag network in Figure 14–31B, determine the output voltage in phasor form when the input voltage has an rms value of 10 V. Sketch the input and output wave forms showing the proper relationships. ϕ was found in Example 14–14.

Solution:

$$\mathbf{V}_{\text{out}} = \left(\frac{X_C}{\sqrt{R^2 + X_C^2}}\right) V_{\text{in}} \angle \phi$$

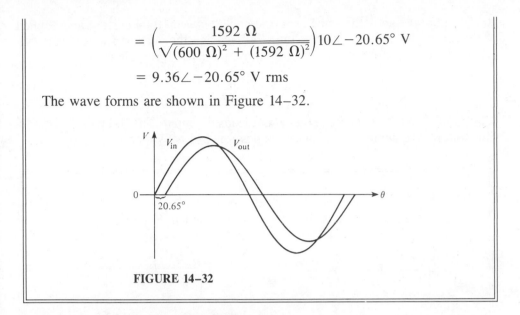

$$= \left(\frac{1592\ \Omega}{\sqrt{(600\ \Omega)^2 + (1592\ \Omega)^2}}\right) 10 \angle -20.65° \text{ V}$$

$$= 9.36 \angle -20.65° \text{ V rms}$$

The wave forms are shown in Figure 14–32.

FIGURE 14–32

The *RC* Lead Network

The second basic type of phase shift network is the *RC* lead network. When the output of a series *RC* circuit is taken across the resistor rather than across the capacitor, as shown in Figure 14–33A, it becomes a *lead* network. Lead networks cause the phase of the output voltage to *lead* the input by a specified amount.

Phase Difference between Input and Output: In a series *RC* circuit, the current leads the input voltage. Also, as you know, the resistor voltage is in phase with the current. Since the output voltage is taken across the resistor, the output *leads* the input, as indicated by the phasor diagram in Figure 14–33B. The wave form diagrams are shown in Part C.

As in the lag network, the amount of phase difference between the input and output and also the magnitude of the output voltage in the lead network is dependent on the relative values of the resistance and the capacitive reactance. When the input voltage is assigned a reference angle of 0 degrees, the angle of

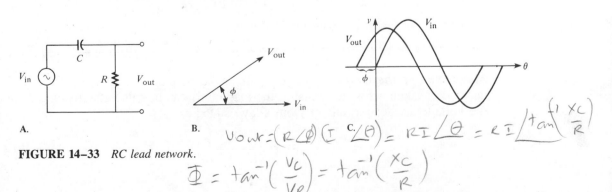

A.

B.

FIGURE 14–33   *RC lead network.*

$$V_{out} = (R \angle \phi)(I \angle \theta) = RI \angle \theta = RI \angle \tan^{-1}\left(\frac{X_C}{R}\right)$$

$$\phi = \tan^{-1}\left(\frac{V_C}{V_R}\right) = \tan^{-1}\left(\frac{X_C}{R}\right)$$

the output voltage is the same as θ (the angle between total current and applied voltage), because the resistor voltage (output) and the current are in phase with each other. Therefore, since $\phi = \theta$ in this case, the expression is

$$\phi = \arctan\left(\frac{X_C}{R}\right) \tag{14–37}$$

This angle is positive, because the output leads the input. The following example illustrates the computation of phase angles for lead networks.

Example 14–16

Calculate the output phase angle for each circuit in Figure 14–34.

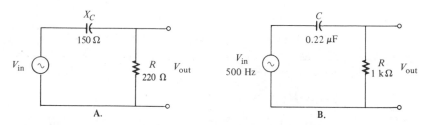

FIGURE 14–34

Solution:

Circuit A:

$$\phi = \arctan\left(\frac{X_C}{R}\right) = \arctan\left(\frac{150\ \Omega}{220\ \Omega}\right) = 34.29°$$

The output leads the input by 34.29°.
   Circuit B:

$$X_C = \frac{1}{2\pi f C} = \frac{1}{2\pi (500\ \text{Hz})(0.22\ \mu\text{F})} \doteq 1446.86\ \Omega$$

$$\phi = \arctan\left(\frac{X_C}{R}\right) = \arctan\left(\frac{1446.86\ \Omega}{1000\ \Omega}\right) = 55.35°$$

The output leads the input by 55.35°.

Magnitude of the Output Voltage: Since the output voltage of an *RC* lead network is taken across the resistor, the magnitude can be calculated using either the voltage divider formula or Ohm's law, stated as

$$V_{out} = \left(\frac{R}{\sqrt{R^2 + X_C^2}}\right) V_{in} \tag{14–38}$$

$$V_{out} = IR \tag{14-39}$$

The expression for the output voltage in phasor form is

$$\mathbf{V}_{out} = V_{out} \angle \phi \tag{14-40}$$

Example 14–17

The input voltage in Figure 14–34B has an rms value of 10 V. Determine the phasor expression for the output voltage. Sketch the wave form relationships for the input and output voltages showing peak values.

Solution:

The phase angle was found to be 55.35 degrees in Example 14–16.

$$\mathbf{V}_{out} = \left(\frac{R}{\sqrt{R^2 + X_C^2}} \right) (V_{in} \angle \phi)$$

$$= \left(\frac{1000 \ \Omega}{1759 \ \Omega} \right) (10 \angle 55.35° \ V) = 5.69 \angle 55.35° \ V \ rms$$

The peak value of the input voltage is 1.414(10 V) = 14.14 V. The peak value of the output voltage is 1.414(5.69 V) = 8.05 V. The wave forms are shown in Figure 14–35.

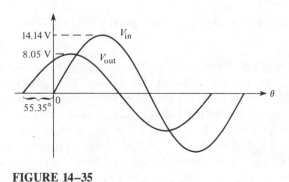

FIGURE 14–35

The *RC* Circuit as a Filter

Filters are frequency-selective circuits that permit signals of certain frequencies to pass from the input to the output while blocking all others. That is, all frequencies but the selected ones are *filtered* out.

Series *RC* circuits exhibit a frequency-selective characteristic and therefore act as basic filters. There are two types. The first one that we examine, called a *low-pass filter,* is realized by taking the output across the capacitor, just as in a lag network. The second type, called a *high-pass filter,* is implemented by taking the output across the resistor, as in a lead network.

506

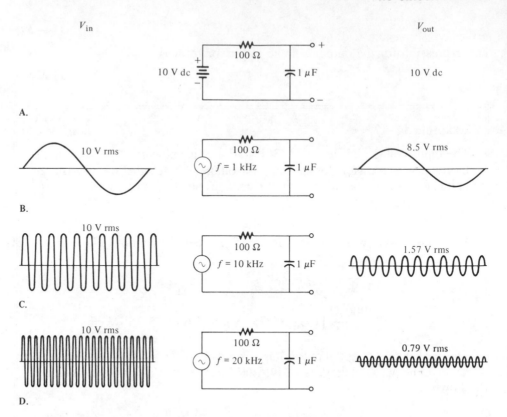

FIGURE 14–36 *Low-pass filter action. (Phase shifts are not indicated.)*

Low-Pass Filter: You have already seen what happens to the output magnitude and phase angle in the lag network. In terms of its filtering action, we are interested primarily in the variation of the output magnitude with frequency.

Figure 14–36 shows the filtering action of a series *RC* circuit using specific values for illustration. In Part A of the figure, the input is *zero* frequency (dc). Since the capacitor blocks constant direct current, the output voltage equals the full value of the input voltage, because there is no voltage dropped across *R*. Therefore, the circuit passes all of the input voltage to the output (10 V in, 10 V out).

In Figure 14–36B, the frequency of the input voltage has been increased to 1 kHz, causing the capacitive reactance to *decrease* to 159 Ω. For an input voltage of 10 V rms, the output voltage is approximately 8.5 V rms, which can be calculated using the voltage divider approach or Ohm's law.

In Figure 14–36C, the input frequency has been increased to 10 kHz, causing the capacitive reactance to decrease further to 15.9 Ω. For a constant input voltage of 10 V rms, the output voltage is now 1.57 V rms.

As the input frequency is increased further, the output voltage continues to decrease and approaches zero as the frequency becomes very high, as shown in Figure 14–36D. A description of the circuit action is as follows: As the frequency of the input increases, the capacitive reactance decreases. Because the resistance is constant and the capacitive reactance decreases, the voltage across the capacitor (output voltage) also decreases according to the voltage divider principle. The input frequency can be increased until it reaches a value at which the reactance is so small compared to the resistance that the output voltage can be neglected, because it is very small compared to the input voltage. At this value of frequency, the circuit is essentially completely blocking the input signal.

As shown in Figure 14–36, the circuit passes dc (zero frequency) completely. As the frequency of the input increases, less of the input voltage is passed through to the output; that is, the output voltage decreases as the frequency increases. It is apparent that the lower frequencies pass through the circuit much better than the higher frequencies. This *RC* circuit is therefore a very basic form of *low-pass filter.*

Figure 14–37 shows a graph of output voltage magnitude versus frequency for the low-pass filter. This graph, called a *response curve,* indicates that the output decreases as the frequency increases.

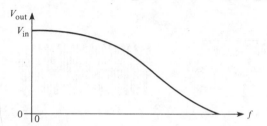

FIGURE 14–37 *Frequency response curve for a low-pass filter.*

High-Pass Filter: Next, refer to Figure 14–38A, where the output is taken across the resistor, just as in a lead network. When the input voltage is dc (zero frequency), the output is zero volts, because the capacitor blocks direct current; therefore no voltage is developed across *R*.

In Figure 14–38B, the frequency of the input signal has been increased to 100 Hz with an rms value of 10 V. The output voltage is 0.63 V rms. Thus, only a small percentage of the input voltage appears on the output at this frequency.

In Figure 14–38C, the input frequency is increased further to 1 kHz, causing more voltage to be developed across the resistor because of the further decrease in the capacitive reactance. The output voltage at this frequency is 5.32 V rms. As you can see, the output voltage increases as the frequency increases. A value of frequency is reached at which the reactance is negligible compared to the resistance, and most of the input voltage appears across the resistor, as shown in Figure 14–38D.

As illustrated, this circuit tends to prevent lower frequencies from appearing on the output but allows higher frequencies to pass through from input to output. Therefore, this *RC* circuit is a very basic form of *high-pass filter.*

Figure 14–39 shows a plot of output voltage magnitude versus frequency for the high-pass filter. This is a response curve; it shows that the output increases

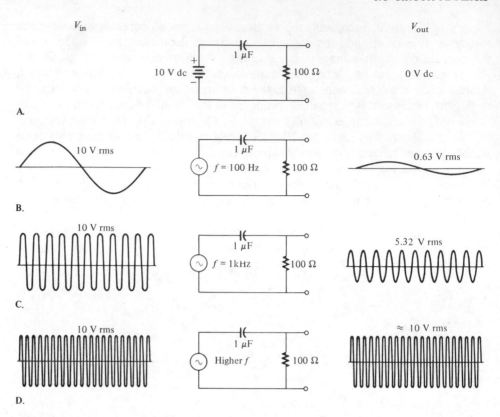

FIGURE 14–38   *High-pass filter action. (Phase shifts are not indicated.)*

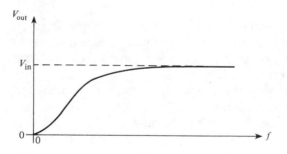

FIGURE 14–39   *Frequency response curve for a high-pass filter.*

as the frequency increases and then levels off, approaching the value of the input voltage.

RC Coupling: A common application of the *RC* high-pass filter is in amplifier coupling. The *RC* circuit as a coupling network is shown in Figure 14–40A. Its purpose is to completely pass or *couple* the ac input signal to the output and block any dc voltage. The capacitance is chosen large enough so that X_C is practically a *short to ac*. The capacitor appears as an *open to dc*, as indicated in Figure 14–40B and C.

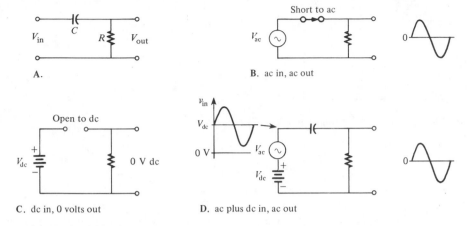

A.

B. ac in, ac out

C. dc in, 0 volts out

D. ac plus dc in, ac out

FIGURE 14–40 *RC coupling.*

When both ac and dc voltages are superimposed, only the ac is coupled to the output, as shown in Part D. This situation occurs in transistor amplifiers where a dc bias voltage and an ac signal voltage are superimposed. The signal is passed to the next amplifier stage, but the dc is blocked to prevent it from altering the operation of the next stage.

Review for 14–8

1. A certain *RC* lag network consists of a 4.7-kΩ resistor and a 0.022-μF capacitor. Determine the phase shift between input and output at a frequency of 3 kHz.

2. An *RC* lead network has the same component values as the lag network in Problem 1. What is the magnitude of the output voltage at 3 kHz when the input is 10 V rms?

3. When an *RC* circuit is used as a low-pass filter, across which component is the output taken?

4. In a coupling network, is the criterion for effective operation $X_C \ll R$, $X_C = R$, or $X_C \gg R$?

PULSE RESPONSE OF *RC* CIRCUITS

14–9

Figure 14–41 shows a series *RC* circuit with an ideal pulse input. We now investigate how this circuit responds in terms of the current, resistor voltage, and capacitor voltage during and after the application of the pulse.

For an ideal pulse, both edges are considered to occur instantaneously. Two basic rules of capacitor behavior aid in analyzing *RC* circuit responses to pulse inputs: (1) the capacitor appears as a *short* to an instantaneous change in current and as an open to dc; (2) the voltage across the capacitor cannot change instantaneously—it can change only exponentially.

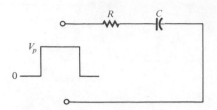

FIGURE 14–41 *Series RC circuit with an ideal pulse input.*

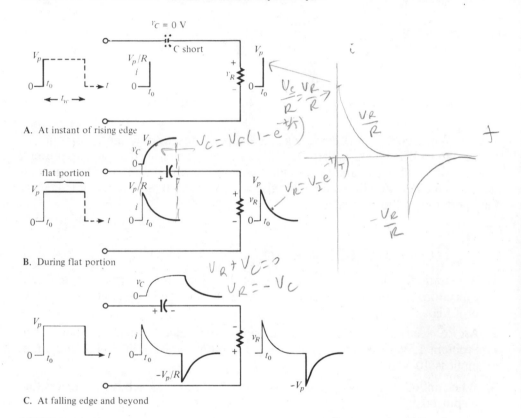

A. At instant of rising edge

B. During flat portion

C. At falling edge and beyond

FIGURE 14–42 *Pulse response for condition where the capacitor fully charges ($t_w \geqq 5RC$).*

The following is a description of the ideal basic circuit response, which is illustrated in Figure 14–42. At the instant of the *rising edge* (t_0) of the pulse, the capacitor appears as a short; therefore, the current is

$$i = \frac{V_p}{R}$$

where V_p is the pulse amplitude. The instantaneous resistor voltage is

$$v_R = iR = V_p$$

Since the voltage across the capacitor cannot change instantaneously, it is initially zero.

$$v_C = 0 \text{ V}$$

During the *flat portion* of the pulse, the capacitor begins to charge exponentially toward V_p; as a result, the current decreases exponentially toward zero. According to Kirchhoff's voltage law, the sum of the voltage drops at any instant must equal the applied voltage. Therefore, as the capacitor voltage builds up, the resistor voltage must decrease correspondingly such that $v_C + v_R = V_p$. The decrease in resistor voltage results in an exponential decrease in current.

If the pulse width of the input is sufficiently long, the capacitor will *fully charge* to V_p; if not, the capacitor only *partially charges*. Each of these two conditions is now discussed.

When the Capacitor Fully Charges during the Pulse

For the capacitor to fully charge, the pulse width, t_w, must be equal to or greater than five time constants.

$$t_w \geqq 5RC$$

In this situation, as illustrated in Figure 14–42, v_C reaches the value of the pulse amplitude, and both i and v_R decrease to zero at or before the end of the pulse. At the instant of the falling edge, the charged capacitor appears as a voltage source equal to $-V_p$. This produces an instantaneous current of $-V_p/R$ back through R and the input pulse source. The negative sign indicates a direction opposite to that of the current on the rising edge and during the pulse (the pulse source is ideally considered to be a short when the input is zero). This instantaneous current produces an instantaneous negative voltage across R equal to $-V_p$. After the instant of the falling edge, both the current and the resistor voltage go to approximately zero exponentially in five time constants as the capacitor discharges, as shown in the figure.

Example 14–18

A pulse is applied to the *RC* circuit as shown in Figure 14–43. Determine the complete wave shape and values for v_C, v_R, and i.

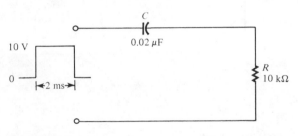

FIGURE 14–43

512

RC CIRCUIT ANALYSIS

Example 14–18 (continued)

Solution:

The circuit time constant is

$$\tau = RC = (10 \text{ k}\Omega)(0.02 \text{ }\mu\text{F}) = 200 \text{ }\mu\text{s}$$

Since 5τ is less than t_w, the capacitor will fully charge to 10 V and remain there until the end of the pulse.

At the rising edge of the pulse:

$$v_C = 0 \text{ V}$$

$$i = \frac{V_p}{R} = \frac{10 \text{ V}}{10 \text{ k}\Omega} = 1 \text{ mA}$$

$$v_R = iR = (1 \text{ mA})(10 \text{ k}\Omega) = 10 \text{ V}$$

Since the capacitor is initially a short, all of the input voltage appears across R.

During the pulse:

    v_C charges exponentially to 10 V in 5(200 μs) = 1 ms.
    i decreases exponentially to zero in 1 ms.
    v_R decreases exponentially to zero in 1 ms.

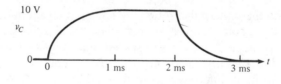

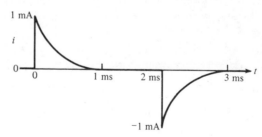

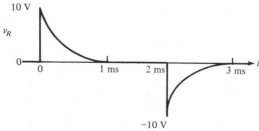

FIGURE 14–44

At the falling edge of the pulse:

$$v_C = 10 \text{ V}$$

$$i = \frac{-v_C}{R} = \frac{-10 \text{ V}}{10 \text{ k}\Omega} = -1 \text{ mA}$$

$$v_R = iR = (-1 \text{ mA})(10 \text{ k}\Omega) = -10 \text{ V}$$

After the pulse:
  v_C discharges exponentially to zero in 1 ms.
  i decreases exponentially to zero in 1 ms.
  v_R decreases exponentially to zero in 1 ms.
The wave forms are shown in Figure 14–44.

When the Capacitor Partially Charges

When the pulse width of the input is less than five time constants, the capacitor does not have time to charge fully. The amount that it charges depends on the duration of the input pulse compared to five time constants and can be determined from the following exponential formula (restated from Chapter 12).

$$v_C = V_F(1 - e^{-t_w/RC}) \qquad \textbf{(14–41)}$$

where V_F is the final voltage toward which the capacitor is charging.

The response of the circuit to the instantaneous rising edge is the same as described for the previous case. The difference comes during the pulse and at the falling edge, as Figure 14–45 shows. During the pulse, the capacitor reaches a value less than V_F, which depends on the length of the capacitor's charge time as controlled by the pulse width of the input. The resistor voltage does not reach zero by the end of the pulse, because the capacitor is still charging and there is still current through R.

At the instant of the falling edge, the capacitor voltage does not change. Thus, since the input has suddenly gone to zero, the resistor voltage instantaneously goes to a negative value equal to the capacitor voltage. This happens in accordance with Kirchhoff's voltage law, which states that the sum of v_C and v_R must equal the input, which is zero.

$$v_C + v_R = 0$$

$$v_R = -v_C$$

The current also reverses direction when the input jumps to zero, and both i and v_R then exponentially go to zero as the capacitor discharges over a 5τ interval, as shown and as determined by the following equation (restated from Chapter 12).

$$v_C = V_i e^{-t_w/RC} \qquad \textbf{(14–42)}$$

where V_i is the initial voltage across the capacitor.

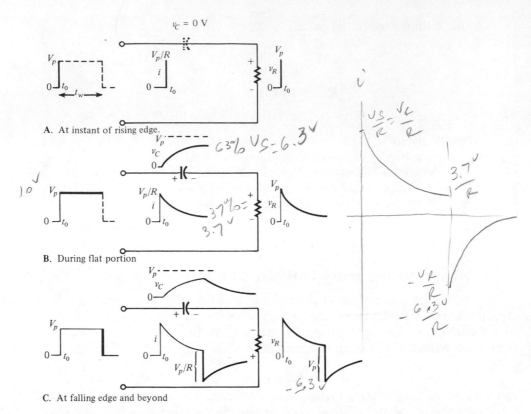

A. At instant of rising edge.

B. During flat portion

C. At falling edge and beyond

FIGURE 14–45 *Pulse response for condition where the capacitor does not fully charge* $(t_w < 5RC).$

Example 14–19

Determine the pulse response in Figure 14–46. Sketch the wave forms for v_C, V_R, and i.

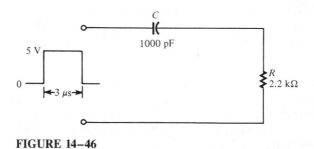

FIGURE 14–46

Solution:

The time constant is

$$\tau = RC = (2.2 \text{ k}\Omega)(1000 \text{ pF}) = 2.2 \ \mu\text{s}$$

Since $5\tau > t_w$, the capacitor does not fully charge during the input pulse. At the rising edge of the pulse:

$$v_C = 0 \text{ V}$$

$$i = \frac{V_p}{R} = \frac{5 \text{ V}}{2.2 \text{ k}\Omega} = 2.27 \text{ mA}$$

$$v_R = iR = (2.27 \text{ mA})(2.2 \text{ k}\Omega) = 5 \text{ V}$$

During the pulse: The capacitor charges exponentially to

$$v_C = V_F(1 - e^{-t_w/RC}) = 5 \text{ V}(1 - e^{-3 \text{ }\mu s/2.2 \text{ }\mu s}) = 3.72 \text{ V}$$

where $V_F = V_p$, the amplitude of the input pulse. The resistor voltage decreases exponentially to

$$v_R = V_p - v_C = 5 \text{ V} - 3.72 \text{ V} = 1.28 \text{ V}$$

The current decreases exponentially to

$$i = \frac{v_R}{R} = \frac{1.28 \text{ V}}{2.2 \text{ k}\Omega} = 0.582 \text{ mA}$$

At the falling edge of the pulse:

$$v_C = 3.72 \text{ V}$$

$$v_R = -3.72 \text{ V}$$

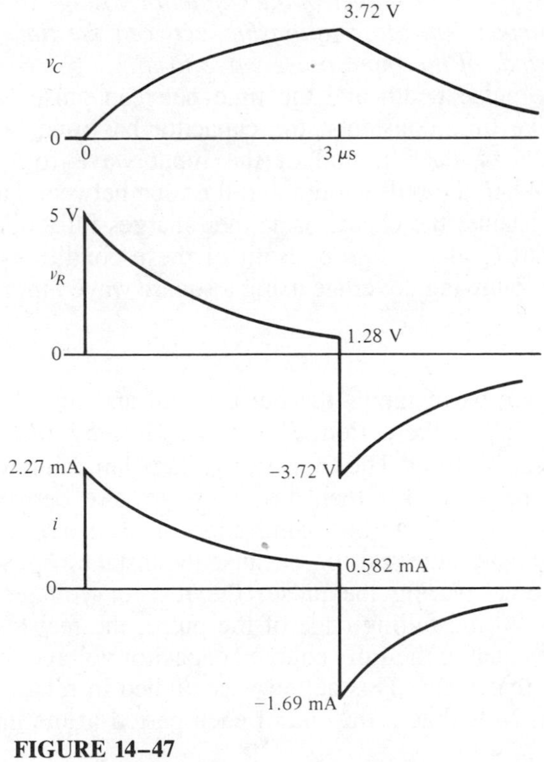

FIGURE 14-47

Example 14–19 (continued)

$$i = \frac{v_R}{R} = \frac{-3.72 \text{ V}}{2.2 \text{ k}\Omega} = -1.69 \text{ mA}$$

The wave forms are shown in Figure 14–47.

Review for 14–9

1. A 10-ms-wide pulse is applied to a series *RC* circuit in which the time constant is 1.5 ms. Does the capacitor fully charge?

2. A 12-V, positive-going, 500-μs pulse is applied to a circuit with $R = 1$ kΩ and $C = 0.05$ μF. Determine v_C, v_R, and i at the rising edge of the pulse.

3. In Problem 2, what are v_C, v_R, and i at the falling edge?

RESPONSE OF *RC* CIRCUITS TO PERIODIC PULSE WAVE FORMS

14–10 In the last section, you were introduced to the response of *RC* circuits to single pulses. Now, those concepts are extended to repetitive pulses.

When a periodic pulse wave form is applied to an *RC* circuit, as shown in Figure 14–48A, *the wave shapes of the capacitor voltage, the resistor voltage, and the current depend on the relationship between the time constant and the frequency (or period) of the input pulse wave form.*

When the pulse width and the time between pulses are each equal to or greater than five time constants, the capacitor has time to charge fully and discharge fully during each period of the input wave form, as indicated in Figure 14–48B. When the pulse width and the time between pulses are each less than five time constants, the capacitor neither charges fully nor discharges fully, as indicated in Part C of the figure. Both of these conditions are examined in more detail in the following coverage using a square wave input to the *RC* circuit.

$T/2 \geq 5\tau$

For a square wave input, the pulse width and time between pulses are each equal to one-half of the period, T. When $T/2 \geq 5\tau$, the capacitor has time to charge and discharge fully. The capacitor voltage builds up to the amplitude of the input during the pulse and then discharges to zero between the pulses, as indicated in Figure 14–49. At the rising edge of each pulse, all of the source voltage is dropped across the resistor, because the instantaneous capacitor voltage is zero at that point. During the pulse, the resistor voltage decreases as the capacitor charges. At the falling edge of the pulse, the resistor voltage reverses polarity to a value equal to the fully charged capacitor voltage, because the source voltage is zero at that point. This action was studied in relation to single pulses in the last section. Note that at the end of each period of the input, the capacitor

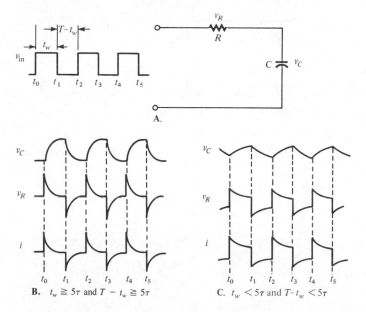

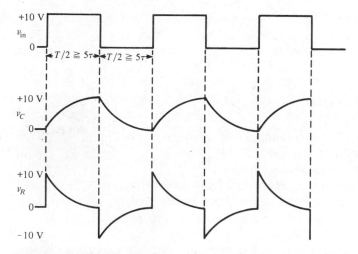

FIGURE 14–48 *Generalized response of an RC circuit to a periodic pulse wave form for two cases:* $t_w \gtrsim 5\tau$, $T - t_w \gtrsim 5\tau$; *and* $t_w < 5\tau$, $T - t_w < 5\tau$.

FIGURE 14–49 *Response to square wave for* $T/2 \gtrsim 5\tau$.

has completely discharged and, therefore, the capacitor and resistor voltage wave forms are identical on each successive period, as indicated in the figure.

$T/2 < 5\tau$

To illustrate this condition, let's take an example of an *RC* circuit with a time constant equal to one-half the period of the input square wave ($\tau = T/2$), as shown in Figure 14–50. This choice simplifies the analysis and demonstrates the

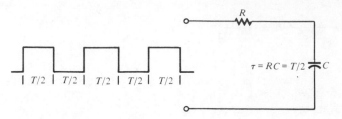

FIGURE 14–50   *RC circuit with square wave input with $\tau = T/2$.*

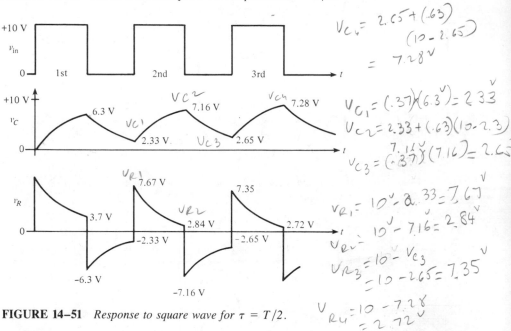

$$V_{C_4} = 2.65 + (.63)$$
$$(10 - 2.65)$$
$$= 7.28\,V$$

$$V_{C_1} = (.37)(6.3\,V) = 2.33$$
$$V_{C_2} = 2.33 + (.63)(10 - 2.3)$$
$$= 7.16\,V$$
$$V_{C_3} = (.37)(7.16) = 2.65$$

$$V_{R_1} = 10\,V - 2.33 = 7.67\,V$$
$$V_{R_2} = 10\,V - 7.16 = 2.84\,V$$
$$V_{R_3} = 10\,V - V_{C_3}$$
$$= 10 - 2.65 = 7.35\,V$$
$$V_{R_4} = 10 - 7.28$$
$$= 2.72\,V$$

FIGURE 14–51   *Response to square wave for $\tau = T/2$.*

basic action of the series *RC* circuit under this condition. At this point, we do not care what the exact time-constant value is, because we know that an *RC* circuit charges 63% during one time-constant interval.

It is assumed that the capacitor in Figure 14–50 is initially uncharged. We now examine the capacitor voltage and the resistor voltage on a pulse-by-pulse basis. The results of this analysis are shown in Figure 14–51.

First Pulse: During the first pulse, the capacitor charges exponentially and reaches 6.3 V (63% of 10 V). On the rising edge, the resistor voltage instantaneously jumps to $+10$ V. Then, as the capacitor charges to 6.3 V, the resistor voltage decreases to 3.7 V. On the falling edge, the resistor voltage makes a negative-going transition to -6.3 V $(-10\text{ V} + 3.7\text{ V} = -6.3\text{ V})$.

Between First and Second Pulses: The capacitor discharges, and its voltage decreases to 37% of its voltage at the beginning of this interval: $(0.37)(6.3\text{ V}) = 2.33$ V. The resistor voltage, which starts at -6.3 V, must increase to -2.33 V, because at the *instant prior to the next pulse,* the input voltage is zero. Therefore,

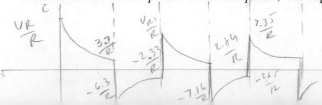

the sum of v_C and v_R must be zero ($+2.33$ V $-$ 2.33 V $= 0$). Remember that $v_C + v_R = v_s$ at all times.

Second Pulse: The capacitor voltage begins at 2.33 V and increases 63% of the way to 10 V. This calculation is as follows: The total charging range is 10 V $-$ 2.33 V $=$ 7.67 V. The capacitor voltage will increase an additional 63% of 7.67 V, which is 4.83 V. Thus, at the end of the second pulse, the capacitor voltage is 2.33 V $+$ 4.83 V $=$ 7.16 V. Notice that the *average* is building up. At the rising edge, the resistor voltage makes an instantaneous, positive-going, 10-V transition from -2.33 V to 7.67 V and then decreases exponentially to $(0.37)(7.67$ V$) = 2.84$ V. At the falling edge, the resistor voltage instantaneously makes a negative-going transition from $+2.84$ V to -7.16 V, as indicated.

Between Second and Third Pulses: The capacitor discharges during this time to 37% of its voltage at the end of the second pulse: $(0.37)(7.16$ V$) = 2.65$ V. The resistor voltage starts at -7.16 V and increases to -2.65 V, because the capacitor voltage and the resistor voltage must add up to zero at the instant prior to the third pulse (the input is zero at this point).

Third Pulse: At the start of the third pulse, the capacitor voltage begins at 2.65 V and then increases 63% of the way from 2.65 V to 10 V: $(0.63)(10$ V $-$ 2.65 V$) = 4.63$ V. Therefore, the capacitor voltage at the end of the third pulse is 2.65 V $+$ 4.63 V $=$ 7.28 V. On the rising edge, the resistor voltage makes an instantaneous 10-V transition from -2.65 V to $+7.35$ V. Then, as the capacitor charges, the resistor voltage drops to $(0.37)(7.35$ V$) = 2.72$ V. On the falling edge, the resistor voltage instantaneously goes from $+2.72$ V down to -7.28 V.

After the third pulse, five time constants have elapsed (2.5 periods of the input in this case), and both the capacitor voltage and the resistor voltage are close to their *steady state*. As you have seen, the capacitor voltage gradually built up and then leveled off. It takes approximately 5τ for the capacitor voltage to build up to a constant *average* value equal to the average of the input voltage. This is called the steady state and is illustrated in Figure 14–52, which shows the approximate values.

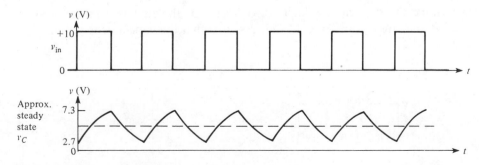

FIGURE 14–52   *Output reaches steady state after 5τ.*

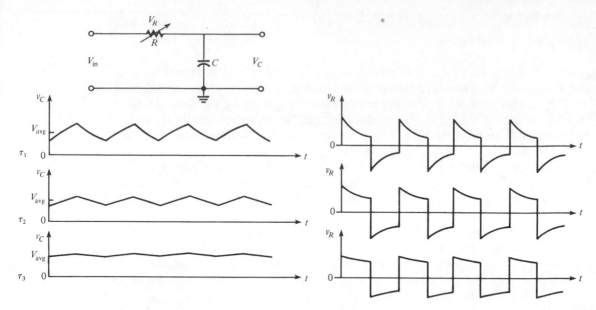

FIGURE 14–53 *Response to longer time constants ($\tau_3 > \tau_2 > \tau_1$).*

Increase in Time Constant

What happens to the voltages when the time constant of the circuit is increased, for example, with a variable resistor, as in Figure 14–53? As the time constant is increased, the capacitor charges less during a pulse and discharges less between pulses. The result is a smaller fluctuation in the capacitor voltage and a resistor voltage that appears more like the input, as illustrated in the figure.

As the time constant is made very large compared to the pulse width of the input, the capacitor voltage approaches a constant dc value equal to the average value of the input, and the resistor voltage approaches the wave shape of the input but with an average value of zero, as indicated.

Example 14–20

Determine the capacitor voltage wave form for the first two pulses applied to the circuit in Figure 14–54. Assume that the capacitor is initially uncharged.

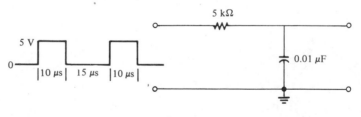

FIGURE 14–54

Solution:

First calculate the circuit time constant:

$$\tau = RC = (5 \text{ k}\Omega)(0.01 \text{ } \mu F) = 50 \text{ } \mu s$$

Obviously, the time constant is much longer than the input pulse width or the interval between pulses (notice that the input is not a square wave). As a result, the fixed percentages from the time-constant tables cannot be used. The exponential formulas must be applied; thus, the analysis is relatively difficult. Follow the solution carefully.

Calculation for first pulse: Use Equation (14–41), because C is charging. Note that $V_F = V_p = 5$ V, and t_w is the pulse width of 10 μs. Therefore,

$$v_C = V_F(1 - e^{-t_w/RC}) = (5 \text{ V})(1 - e^{-10 \mu s/50 \mu s})$$
$$= (5 \text{ V})(1 - 0.819) = 0.906 \text{ V}$$

This result is plotted in Figure 14–55A.

Calculation for interval between first and second pulses: Use Equation (14–42), because C is discharging. Note that V_i is 0.906 V, because C begins to discharge from this value at the end of the first pulse. The discharge time is 15 μs. Therefore,

$$v_C = V_i e^{-t_w/RC} = 0.906 e^{-15 \mu s/50 \mu s}$$
$$= (0.906)(0.741) = 0.671 \text{ V}$$

This result is shown in Figure 14–55B.

Calculation for second pulse: At the beginning of the second pulse, the capacitor voltage is 0.671 V. During the second pulse, the capacitor will again charge. In this case, it does not begin at zero volts. It already has 0.671 V from the previous charge-and-discharge cycle. To handle this situation, we must use the general formula of Equation (12–9) in the following form:

$$v_C = V_F + (V_i - V_F)e^{-t_w/RC}$$

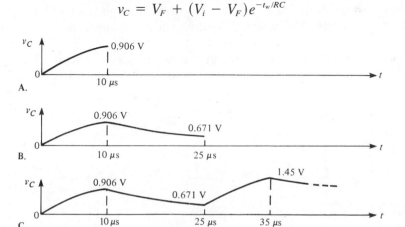

FIGURE 14–55

Example 14–20 (continued)

Using this equation, we can calculate the voltage across the capacitor at the end of the second pulse, as follows:

$$v_C = 5 \text{ V} + (0.671 \text{ V} - 5 \text{ V})e^{-10\,\mu s/50\,\mu s}$$
$$= 5 \text{ V} + (-4.33 \text{ V})(0.819)$$
$$= 5 \text{ V} - 3.55 \text{ V} = 1.45 \text{ V}$$

This result is shown in Figure 14–55C.

Notice that the capacitor wave form builds up on successive input pulses. After approximately 5τ, it reaches its steady state and fluctuates between a constant maximum and constant minimum, with an average equal to the average value of the input. We could demonstrate this pattern by carrying the analysis in this example further.

Relationship of Time Response to Frequency Response

There is a definite relationship between time response and frequency response: *The fast rising and falling edges of a pulse wave form contain the higher-frequency components in that wave form. The flat portions of the pulses represent the slow changes or lower-frequency components. The average value of the pulse wave form is its dc component.* These relationships are indicated in Figure 14–56.

Low-Pass Filter Action: When the output of a series *RC* circuit is taken across the capacitor, the edges of a pulse tend to be "rounded off" exponentially. This rounding off occurs to varying degrees, depending on the relationship of the time constant to the pulse width and period. The rounding off of the edges indicates that the circuit tends to reduce the higher-frequency components of the pulse wave form, as illustrated in Figure 14–57. In pulse applications, the *RC* circuit with the output across the capacitor is often called an *integrator*.

High-Pass Filter Action: When the output of a series *RC* circuit is taken across the resistor, *tilt* is introduced to the flat portion of the pulses; that is, the lower-

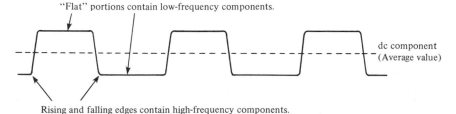

FIGURE 14–56   *General relationship of a pulse wave form to frequency content.*

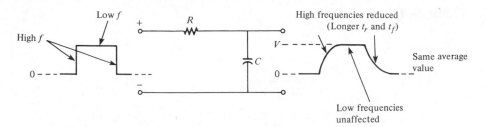

FIGURE 14–57 *Time and frequency response relationship in an integrator (one pulse in a repetitive wave form shown).*

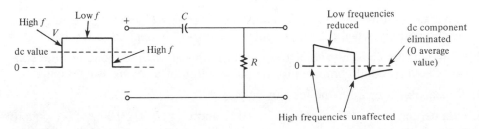

FIGURE 14–58 *Time and frequency response relationship in a differentiator (one pulse in a repetitive wave form shown).*

frequency components are reduced. Also, the dc component of the input is completely eliminated. This action is illustrated in Figure 14–58. In pulse applications, the *RC* circuit with the output across the resistor is often called a *differentiator*.

Formula Relating Rise Time and Fall Time to Frequency: It can be shown that the fast transitions of a pulse (rise and fall times) are related to the *highest-frequency* component, f_h, in that pulse by the following formula:

$$t_r = \frac{0.35}{f_h} \tag{14-43}$$

This formula also applies to fall time, and the fastest transition determines the highest frequency in the pulse wave form. Equation (14–43) can be rearranged to give the highest frequency as follows:

$$f_h = \frac{0.35}{t_r} \tag{14-44}$$

Example 14–21

What is the highest frequency contained in a pulse that has rise and fall times equal to 10 nanoseconds (10 ns)?

Example 14–21 (continued)

Solution:

$$f_h = \frac{0.35}{t_r} = \frac{0.35}{10 \times 10^{-9} \text{ s}} = 0.035 \times 10^9 \text{ Hz}$$

$$= 35 \times 10^6 \text{ Hz} = 35 \text{ MHz}$$

Review for 14–10

1. What conditions allow the capacitor in a series *RC* circuit to charge and discharge fully when a periodic pulse wave form is applied to the input?

2. How will the capacitor voltage wave form appear when the circuit time constant is very small compared to the pulse width of a square wave input?

3. Repeat Problem 2 for the resistor voltage.

4. In a certain series *RC* circuit, $R = 1 \text{ k}\Omega$ and $C = 0.022 \text{ }\mu\text{F}$. A 12-V square wave with a period of 44 μs is applied. To what voltage will the capacitor charge on the first pulse if it is initially uncharged?

TROUBLESHOOTING

14–11   In this section, we consider the effects that typical component failures or degradation have on the response of basic *RC* circuits. The cases to be considered are open resistor, open capacitor, shorted capacitor, and excessive leakage in a capacitor.

Effects of an Open Resistor

It is very easy to see how an open resistor affects the operation of a basic series *RC* circuit, as shown in Figure 14–59. Obviously, there is no path for current, so the capacitor voltage remains at zero; thus, the total voltage appears across the open resistor.

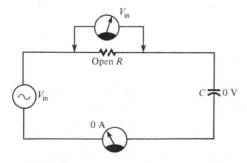

FIGURE 14–59 *Effect of an open resistor.*

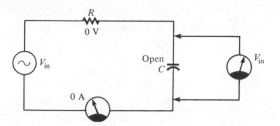

FIGURE 14–60 *Effect of an open capacitor.*

Effects of an Open Capacitor

When the capacitor is open, there is no current; thus, the resistor voltage remains at zero. The total input voltage is across the open capacitor, as shown in Figure 14–60.

Effects of a Shorted Capacitor

When a capacitor shorts out, the voltage across it is zero, the current equals V_{in}/R, and the total voltage appears across the resistor, as shown in Figure 14–61.

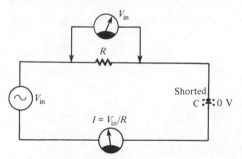

FIGURE 14–61 *Effect of a shorted capacitor.*

Effects of a Leaky Capacitor

When a capacitor exhibits a high leakage current, the leakage resistance effectively appears in parallel with the capacitor, as shown in Figure 14–62A. When the leakage resistance is comparable in value to the circuit resistance, R, the circuit response is drastically affected. The circuit looking from the capacitor toward the source can be Thevenized, as shown in Figure 14–62B. The Thevenin equivalent resistance is R in parallel with R_{leak} (the source appears as a short), and the Thevenin equivalent voltage is determined by the voltage divider action of R and R_{leak}. The Thevenin equivalent circuit is shown in Figure 14–62C.

$$R_{th} = R \| R_{leak}$$

$$R_{th} = \frac{RR_{leak}}{R + R_{leak}} \qquad \textbf{(14–45)}$$

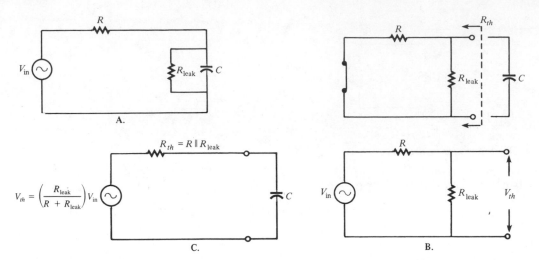

FIGURE 14–62 *Effects of a leaky capacitor.*

$$V_{th} = \frac{R_{leak} V_{in}}{R + R_{leak}} \qquad (14\text{–}46)$$

As you can see, the voltage to which the capacitor will charge is reduced since $V_{th} < V_{in}$. Also, the circuit time constant is reduced, and the current is increased.

Example 14–22

Assume that the capacitor in Figure 14–63 is degraded to a point where its leakage resistance is 10 kΩ. Determine the phase shift from input to output and the output voltage under the degraded condition.

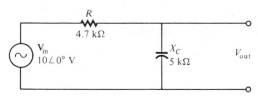

FIGURE 14–63

Solution:

The effective circuit resistance is

$$R_{th} = \frac{R R_{leak}}{R + R_{leak}} = \frac{(4.7 \text{ k}\Omega)(10 \text{ k}\Omega)}{14.7 \text{ k}\Omega} = 3.2 \text{ k}\Omega$$

The phase shift is

$$\phi = -90° + \arctan\left(\frac{X_C}{R_{th}}\right) = -90° + \arctan\left(\frac{5\ k\Omega}{3.2\ k\Omega}\right) = -32.62°$$

$$V_{out} = \left(\frac{X_C}{\sqrt{R_{th}^2 + X_C^2}}\right)(V_{th})$$

$$= \left(\frac{5\ k\Omega}{\sqrt{(3.2\ k\Omega)^2 + (5\ k\Omega)^2}}\right)(6.8\ V) = 5.73\ V$$

Review for 14-11

1. Describe the effect of a leaky capacitor on the response of an *RC* circuit.
2. In a series *RC* circuit, all of the applied voltage appears across an open capacitor (T or F).

COMPUTER ANALYSIS

The program listed here provides for the computation of phase shift and output voltage as functions of frequency for an *RC* lag network. Inputs required are the component values, the upper and lower frequency limits, and the frequency steps.

14-12

```
10 CLS
20 PRINT "THIS PROGRAM COMPUTES THE PHASE SHIFT FROM INPUT TO"
30 PRINT "OUTPUT AND THE NORMALIZED OUTPUT VOLTAGE MAGNITUDE"
40 PRINT "AS FUNCTIONS OF FREQUENCY FOR AN RC LAG NETWORK."
50 PRINT:PRINT:PRINT
60 INPUT "TO CONTINUE PRESS 'ENTER'";X:CLS
70 INPUT "THE VALUE OF R IN OHMS";R
80 INPUT "THE VALUE OF C IN FARADS";C
90 INPUT "THE LOWEST NONZERO  FREQUENCY IN HERTZ";FL
100 INPUT "THE HIGHEST FREQUENCY IN HERTZ";FH
110 INPUT "THE FREQUENCY INCREMENTS IN HERTZ";FI
120 CLS
130 PRINT "FREQUENCY(HZ)","PHASE SHIFT","VOUT"
140 FOR F=FL TO FH STEP FI
150 XC = 1/(2*3.1416*F*C)
160 PHI=-90+ATN(XC/R)*57.3
170 VO=XC/(SQR(R*R+XC*XC))
180 PRINT F,PHI,VO
190 NEXT
```

Review for 14-12

1. What is the purpose of the "STEP FI" portion of line 140?
2. Explain each line in the FOR/NEXT loop.

Formulas

Series RC Circuits:

$$\mathbf{X}_C = -jX_C \qquad (14\text{--}1)$$

$$\mathbf{Z} = R - jX_C \qquad (14\text{--}2)$$

$$Z = \sqrt{R^2 + X_C^2} \qquad (14\text{--}3)$$

$$\theta = -\arctan\left(\frac{X_C}{R}\right) \qquad (14\text{--}4)$$

$$\mathbf{Z} = \sqrt{R^2 + X_C^2} \angle -\arctan\left(\frac{X_C}{R}\right) \qquad (14\text{--}5)$$

$$\mathbf{V} = \mathbf{IZ} \qquad (14\text{--}6)$$

$$\mathbf{I} = \frac{\mathbf{V}}{\mathbf{Z}} \qquad (14\text{--}7)$$

$$\mathbf{Z} = \frac{\mathbf{V}}{\mathbf{I}} \qquad (14\text{--}8)$$

$$\mathbf{V}_s = V_R - jV_C \qquad (14\text{--}9)$$

$$\mathbf{V}_s = \sqrt{V_R^2 + V_C^2} \angle -\arctan\left(\frac{V_C}{V_R}\right) \qquad (14\text{--}10)$$

$$V_s = \sqrt{V_R^2 + V_C^2} \qquad (14\text{--}11)$$

$$\theta = -\arctan\left(\frac{V_C}{V_R}\right) \qquad (14\text{--}12)$$

Parallel RC Circuits:

$$\mathbf{Z} = \left(\frac{RX_C}{\sqrt{R^2 + X_C^2}}\right) \angle \left(-90° + \arctan\left(\frac{X_C}{R}\right)\right) \qquad (14\text{--}13)$$

$$Z = \frac{RX_C}{\sqrt{R^2 + X_C^2}} \qquad (14\text{--}14)$$

$$\theta = -90° + \arctan\left(\frac{X_C}{R}\right) \qquad (14\text{--}15)$$

$$G = \frac{1}{R} \qquad (14\text{--}16)$$

$$B_C = \frac{1}{X_C} \qquad (14\text{--}17)$$

$$Y = \frac{1}{Z} \qquad (14\text{--}18)$$

$$\mathbf{Y} = G + jB_C \qquad (14\text{--}19)$$

$$\mathbf{V} = \frac{\mathbf{I}}{\mathbf{Y}} \qquad (14\text{--}20)$$

$$\mathbf{I} = \mathbf{VY} \qquad (14\text{--}21)$$

$$\mathbf{Y} = \frac{\mathbf{I}}{\mathbf{V}} \qquad (14\text{--}22)$$

$$\mathbf{I}_T = I_R + jI_C \qquad (14\text{--}23)$$

$$\mathbf{I}_T = \sqrt{I_R^2 + I_C^2}\,\angle\arctan\!\left(\frac{I_C}{I_R}\right) \qquad (14\text{--}24)$$

$$I_T = \sqrt{I_R^2 + I_C^2} \qquad (14\text{--}25)$$

$$\theta = \arctan\!\left(\frac{I_C}{I_R}\right) \qquad (14\text{--}26)$$

Power in RC Circuits:

$$P_{\text{avg}} = I^2 R \qquad (14\text{--}27)$$

$$P_r = I^2 X_C \qquad (14\text{--}28)$$

$$P_a = I^2 Z \qquad (14\text{--}29)$$

$$P_a = I^2 Z \qquad (14\text{--}29)$$

$$P_{\text{avg}} = P_a \cos\theta \qquad (14\text{--}30)$$

$$PF = \cos\theta \qquad (14\text{--}32)$$

Lag Network:

$$\phi = -90° + \arctan\!\left(\frac{X_C}{R}\right) \qquad (14\text{--}33)$$

$$V_{\text{out}} = \left(\frac{X_C}{\sqrt{R^2 + X_C^2}}\right) V_{\text{in}} \qquad (14\text{--}34)$$

$$V_{\text{out}} = IX_C \qquad (14\text{--}35)$$

$$\mathbf{V}_{\text{out}} = V_{\text{out}}\angle\phi \qquad (14\text{--}36)$$

Lead Network:

$$\phi = \arctan\!\left(\frac{X_C}{R}\right) \qquad (14\text{--}37)$$

$$V_{\text{out}} = \left(\frac{R}{\sqrt{R^2 + X_C^2}}\right) V_{\text{in}} \qquad (14\text{--}38)$$

$$V_{\text{out}} = IR \qquad (14\text{--}39)$$

$$\mathbf{V}_{out} = V_{out} \angle \phi \tag{14-40}$$

Pulse Response:

$$v_C = V_F (1 - e^{-t_w/RC}) \tag{14-41}$$

$$v_C = V_i e^{-t_w/RC} \tag{14-42}$$

$$t_r = \frac{0.35}{f_h} \tag{14-43}$$

$$f_h = \frac{0.35}{t_r} \tag{14-44}$$

Troubleshooting:

$$R_{th} = \frac{R R_{leak}}{R + R_{leak}} \tag{14-45}$$

$$V_{th} = \frac{R_{leak} V_{in}}{R + R_{leak}} \tag{14-46}$$

Summary

1. A sinusoidal voltage applied to an *RC* circuit produces a sinusoidal current.
2. Current leads voltage in an *RC* circuit.
3. Impedance is the total opposition to sinusoidal current.
4. The unit of impedance is the ohm.
5. In a series *RC* circuit, the total impedance is the phasor sum of the resistance and the capacitive reactance.
6. The resistor voltage is always in phase with the current.
7. The capacitor voltage always lags the current by 90 degrees.
8. The phase angle between applied voltage and current is dependent on the relative values of R and X_C.
9. When $X_C = R$, the phase angle is 45 degrees.
10. The term *arctan* means "the angle whose tangent is."
11. The average power in a capacitor is zero.
12. When a sinusoidal source drives an *RC* circuit, part of the total power delivered by the source is resistive power (average power), and part of it is reactive power.
13. Reactive power is the rate at which energy is stored by a capacitor.
14. The total power being transferred between the source and the circuit is the combination of average power and reactive power and is called the apparent power; its unit is the volt-ampere (VA).

15. The power factor is the cosine of the phase angle.

16. In an *RC* lag network, the output voltage is across the capacitor and lags the input voltage.

17. In an *RC* lead network, the output voltage is across the resistor and leads the input voltage.

18. A low-pass filter passes lower frequencies and rejects higher frequencies.

19. A high-pass filter passes higher frequencies and rejects lower frequencies.

20. The pulse width must be equal to at least five time constants for a capacitor to reach full charge in an *RC* circuit.

21. In an *RC* circuit, the current and voltage during charge and discharge follow exponential curves.

Self-Test

1. When a sinusoidal voltage is applied to an *RC* circuit, what is the wave shape of the voltage appearing across the capacitor?

2. In a certain series *RC* circuit, $R = 2.5$ kΩ and $C = 0.005$ μF. Express the total impedance in rectangular form when $f = 5$ kHz. In polar form.

3. In Problem 2, if the frequency is doubled, what is the magnitude of the total impedance? What is the phase angle?

4. In Figure 14–64, at what frequency does a 45-degree phase angle between the source voltage and the current occur?

5. In a given series *RC* circuit, the capacitance is 0.01 μF. At 5 kHz, what value of resistance will produce a 60-degree phase angle?

6. In the circuit of Figure 14–64, what is the phase difference between the source voltage and the resistor voltage for $f = 100$ kHz?

7. In a given series *RC* circuit, $R = 500$ Ω and $X_C = 1$ kΩ. What is the phase angle?

8. A series *RC* circuit has a resistance that is twice the capacitive reactance at a given frequency. By how many degrees does the voltage lag the current?

9. Determine the impedance in rectangular form for each circuit in Figure 14–65.

10. Determine the current and express it in polar form for each circuit in Figure 14–65, when $V_s = 10\angle 0°$ V.

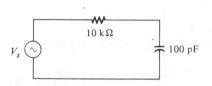

FIGURE 14–64

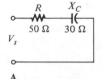

FIGURE 14–65

11. In a certain series *RC* circuit, the resistor voltage is 6 V, and the capacitor voltage is 4.5 V. What is the magnitude of the total voltage applied to the circuit?

12. Determine the impedance and express it in polar form for each circuit in Figure 14–66.

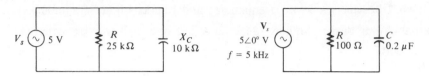

FIGURE 14–66

13. Determine the total current and each branch current in Figure 14–66 when $V_s = 5\angle 0°$ V.

14. What is the phase angle between the applied voltage and the total current in Figure 14–66?

15. For the lag network in Figure 14–67, determine the angle by which the output voltage lags the input voltage (phase shift).

16. What is the magnitude of the output voltage in Figure 14–67?

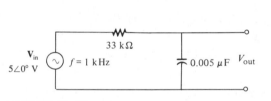

FIGURE 14–67

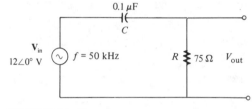

FIGURE 14–68

17. Determine the phase shift from input to output for the lead network in Figure 14–68.

18. What is the magnitude of V_{out} in Figure 14–68?

19. For Figure 14–67, is the output voltage greater at 2 kHz or at 200 Hz?

20. For Figure 14–68, is the output voltage greater at 5 kHz or at 10 Hz?

21. Calculate the average power, the reactive power, and the apparent power for the circuits of Figures 14–67 and 14–68.

22. Calculate the power factors in Problem 21.

23. A 5-V pulse with a width of 5 μs is applied to an *RC* circuit with $\tau = 10$ μs. If the capacitor is initially uncharged, to how many volts does it charge by the end of the pulse?

24. Determine the output voltage for the circuit in Figure 14–69 for the single input pulse shown.

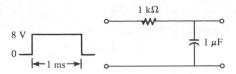

FIGURE 14–69

25. Assume that a square wave with the same pulse width and amplitude as the single pulse in Problem 24 is applied to the circuit in Figure 14–69. Determine the capacitor voltage for two pulses.

Problems

Section 14–1

14–1. An 8-kHz sinusoidal voltage is applied to a series *RC* circuit. What is the frequency of the voltage across the resistor? The capacitor?

14–2. What is the wave shape of the current in the circuit of Problem 14–1?

Section 14–2

14–3. Express the total impedance of each circuit in Figure 14–70 in both polar and rectangular forms.

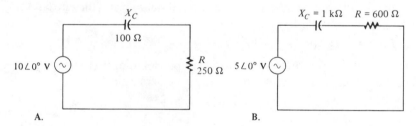

FIGURE 14–70

14–4. Determine the impedance magnitude and phase angle in each circuit in Figure 14–71.

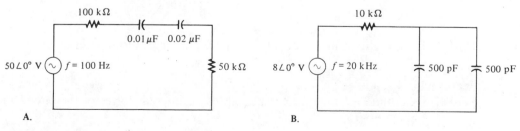

FIGURE 14–71

14–5. For the circuit of Figure 14–72, determine the impedance expressed in rectangular form for each of the following frequencies.
(a) 100 Hz **(b)** 500 Hz **(c)** 1 kHz **(d)** 2.5 kHz

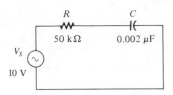

FIGURE 14–72

14–6. Repeat Problem 14–5 for $C = 0.005$ μF.

14–7. Determine the values of R and X_C in a series RC circuit for the following values of total impedance.
(a) $\mathbf{Z} = 30$ $\Omega - j50$ Ω **(b)** $\mathbf{Z} = 300\angle-25°$ Ω
(c) $\mathbf{Z} = 1.8\angle-67.2°$ $k\Omega$ **(d)** $\mathbf{Z} = 789\angle-45°$ Ω

Section 14–3

14–8. Express the current in polar form for each circuit of Figure 14–70.

14–9. Calculate the total current in each circuit of Figure 14–71, and express in polar form.

14–10. Determine the phase angle between the applied voltage and the current for each circuit in Figure 14–71.

14–11. Repeat Problem 14–10 for the circuit in Figure 14–72, using $f = 5$ kHz.

14–12. For the circuit in Figure 14–73, draw the phasor diagram showing all voltages and the total current. Indicate the phase angles.

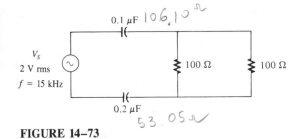

0.1 μF 106.10 Ω

53.05 Ω

FIGURE 14–73

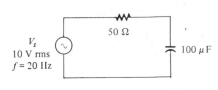

FIGURE 14–74

14–13. For the circuit in Figure 14–74, determine the following in polar form.
(a) $\mathbf{Z}$ **(b)** $\mathbf{I}_T$ **(c)** $\mathbf{V}_R$ **(d)** $\mathbf{V}_C$

Section 14–4

14–14. Determine the impedance and express it in polar form for the circuit in Figure 14–75.

14–15. Determine the impedance magnitude and phase angle in Figure 14–76.

FIGURE 14-75

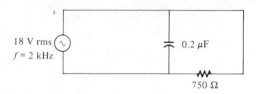

FIGURE 14-76

14-16. Repeat Problem 14-15 for the following frequencies:
(a) 1.5 kHz   (b) 3 kHz   (c) 5 kHz   (d) 10 kHz

Section 14-5

14-17. For the circuit in Figure 14-77, find all the currents and voltages in polar form.

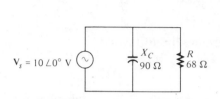

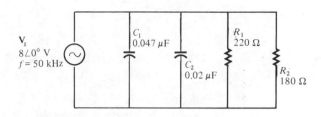

FIGURE 14-77

FIGURE 14-78

14-18. For the parallel circuit in Figure 14-78, find the magnitude of each branch current and the total current. What is the phase angle between the applied voltage and the total current?

14-19. For the circuit in Figure 14-79, determine the following:
(a) $\mathbf{Z}$ (b) $\mathbf{I}_R$ (c) $\mathbf{I}_C$ (d) $\mathbf{I}_T$ (e) θ

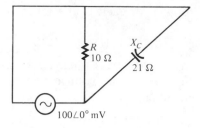

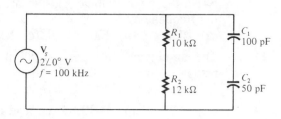

FIGURE 14-79

FIGURE 14-80

14-20. Repeat Problem 14-19 for $R = 5\ k\Omega$, $C = 0.05\ \mu F$, and $f = 500$ Hz.

14-21. Convert the circuit in Figure 14-80 to an equivalent series form.

Section 14-6

14-22. Determine the voltages in polar form across each element in Figure 14-81. Sketch the voltage phasor diagram.

14–23. Is the circuit in Figure 14–81 predominantly resistive or predominantly capacitive?

14–24. Find the current through each branch and the total current in Figure 14–81. Express the currents in polar form. Sketch the current phasor diagram.

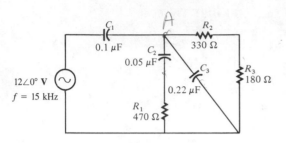

FIGURE 14–81

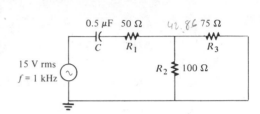

FIGURE 14–82

14–25. For the circuit in Figure 14–82, determine the following:
(a) I_T (b) θ (c) V_{R1} (d) V_{R2} (e) V_{R3} (f) V_C

Section 14–7

14–26. In a certain series *RC* circuit, the average power is 2 W, and the reactive power is 3.5 VAR. Determine the apparent power.

14–27. In Figure 14–74, what is the average power and the reactive power?

14–28. What is the power factor for the circuit of Figure 14–80?

14–29. Determine P_{avg}, P_r, P_a, and *PF* for the circuit in Figure 14–82. Sketch the power triangle.

Section 14–8

14–30. For the lag network in Figure 14–83, determine the phase shift between the input voltage and the output voltage for each of the following frequencies:
(a) 1 Hz   (b) 100 Hz   (c) 1 kHz   (d) 10 kHz

14–31. The lag network in Figure 14–83 also acts as a low-pass filter. Draw a response curve for this circuit by plotting the output voltage versus frequency for 0 Hz to 10 kHz in 1-kHz increments.

14–32. Repeat Problem 14–30 for the lead network in Figure 14–84.

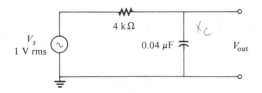

FIGURE 14–83

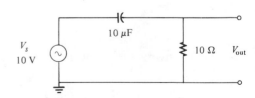

FIGURE 14–84

14–33. Plot the frequency response curve for the lead network in Figure 14–84 for a frequency range of 0 Hz to 10 kHz in 1-kHz increments.

14–34. Draw the voltage phasor diagram for each circuit in Figures 14–83 and 14–84 for a frequency of 5 kHz with $V_s = 1$ V rms.

14–35. What value of coupling capacitor is required in Figure 14–85 so that the signal voltage at the input of amplifier 2 is at least 70.7% of the signal voltage at the output of amplifier 1 when the frequency is 20 Hz? Neglect the input resistance of the amplifier.

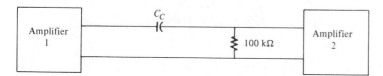

FIGURE 14–85

Section 14–9

14–36. Sketch the output with respect to the input for the circuit in Figure 14–86.

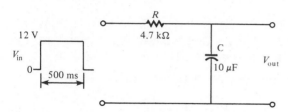

FIGURE 14–86

14–37. For each circuit in Figure 14–87:
    (a) What is τ?   **(b)** Sketch the output voltage.

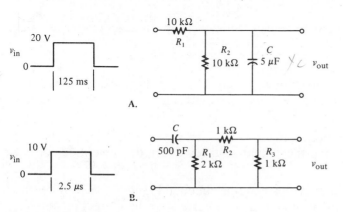

FIGURE 14–87

Since τ > tw C will charge & discharge a negligible amount once it reaches its average charge of 15 volts (average value of Vin) therefore the output is essentially a constant 15V

Section 14–10

14–38. Determine the steady state output voltage in Figure 14–88.

Vout (steady state) = 15 Vdc

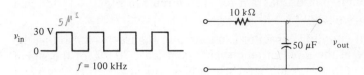

FIGURE 14–88

14–39. Sketch the output voltage in Figure 14–89 showing all voltages.

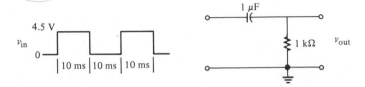

FIGURE 14–89

14–40. A 10-V, 5-kHz pulse wave form with a duty cycle of 50% is applied to a series RC circuit with $\tau = 50$ μs. Graph the resistor voltage for the three initial pulses. C is initially uncharged.

14–41. Sketch the steady state output voltage in Figure 14–90.

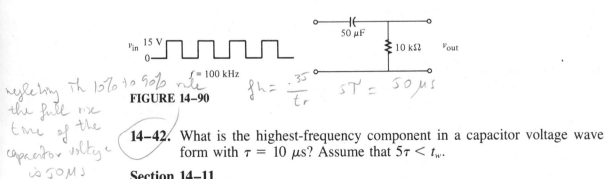

FIGURE 14–90

$f_h = \dfrac{.35}{t_r}$ $5T = 50\,\mu s$

neglecting the 10% to 90% rule the full rise time of the capacitor voltage is 50μs

$f_h = \dfrac{.35}{50\mu s} = 7\,kHz$

14–42. What is the highest-frequency component in a capacitor voltage wave form with $\tau = 10$ μs? Assume that $5\tau < t_w$.

Section 14–11

14–43. Assume that the capacitor in Figure 14–86 is excessively leaky. Show how this degradation affects the output voltage assuming that the leakage resistance is 5 kΩ.

14–44. Each of the capacitors in Figure 14–87 has developed a leakage resistance of 2 kΩ. Determine the output voltages under this condition for each circuit.

14–45. Determine the output voltage for the circuit in Figure 14–87A for each of the following failure modes, and compare it to the correct output:
(a) R_1 open (b) R_2 open (c) C open (d) C shorted

$\tau = (10k)(5\mu F)\ 50\,ms$

$V_c = 20V\,(1 - e^{-ns/50}) = 18.36V$

C partially charged to 18.36V

14–46. Determine the output voltage for the circuit in Figure 14–87B for each of the following failure modes, and compare it to the correct output:
(a) C open   (b) C shorted   (c) R_1 open
(d) R_2 open   (e) R_3 open

Section 14–12

14–47. Modify the program in Section 14–12 to compute and tabulate current in addition to phase shift and output voltage.

14–48. Develop a computer program similar to the program in Section 14–12 for an RC lead network.

Advanced Problems

14–49. Determine the series element or elements that must be installed in the block of Figure 14–91 to meet the following requirements:
(a) $P_{avg} = 400$ W   (b) Leading power factor (I_T leads V_s)

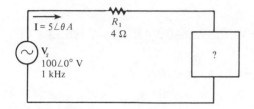

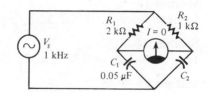

FIGURE 14–91   **FIGURE 14–92**

14–50. A certain semiconductor trigger device requires 12 V on its input to turn on. Design an RC circuit in which a rotary switch permits selected delay times of 1 s, 10 s, 50 s, and 100 s from the time a push-button switch is closed until the device is triggered on. All circuit component values must be specified.

14–51. Determine the value of C_2 in Figure 14–92.

Answers to Section Reviews

Section 14–1:
1. 60 Hz, 60 Hz. **2.** Closer to 0 degrees.

Section 14–2:
1. $R = 150$ Ω, $X_C = 220$ Ω. **2.** $Z = 33$ kΩ $- j50$ kΩ. **3.** 59.9 Ω, $-56.58°$.

Section 14–3:
1. 7.2 V. **2.** $-56.3°$. **3.** $90°$. **4.** X_C decreases, Z decreases, θ decreases.

Section 14–4:
1. Conductance is the reciprocal of resistance, capacitive susceptance is the reciprocal of capacitive reactance, and admittance is the reciprocal of impedance.
2. $Y = 0.01$ S. **3.** $Y = 0.024\angle 33.6°$ S. **4.** 0.024 S, $33.6°$.

Section 14–5:
1. 21 mA. **2.** 18 mA, 56.3°, applied voltage. **3.** 90°.

Section 14–6:
1. See Figure 14–93. **2.** $36.91\angle-51.56°$ Ω.

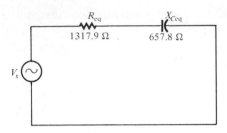

R_{eq} X_{Ceq}
1317.9 Ω 657.8 Ω
V_s

FIGURE 14–93

Section 14–7:
1. Resistance. **2.** 0.707. **3.** $P_{avg} = 1200$ W, $P_r = 1840$ VAR, $P_a = 2196.73$ VA.

Section 14–8:
1. $-62.84°$. **2.** 8.9 V rms. **3.** Capacitor. **4.** $X_C \ll R$.

Section 14–9:
1. Yes. **2.** $v_C = 0$ V, $v_R = 12$ V, $i = 12$ mA. **3.** $v_C = 12$ V, $v_R = -12$ V, $i = -12$ mA.

Section 14–10:
1. $t_w \geq 5\tau$ and $T - t_w \geq 5\tau$. **2.** Similar to the input. **3.** Short-duration positive spikes on positive-going edges and short-duration negative spikes on negative-going edges. **4.** 7.56 V.

Section 14–11:
1. The leakage resistance acts in parallel with C, which alters the circuit time constant. **2.** T.

Section 14–12:
1. To advance the frequency value by *FI* each pass through the loop. **2.** Line 150 calculates X_C; line 160 calculates ϕ; line 170 calculates V_{out}; and line 180 prints frequency, phase angle, and output voltage.

An *RL* circuit contains both *resistance* and *inductance*. It is one of the basic types of reactive circuits to be studied. *RC* circuits were studied in the last chapter. Basic series and parallel *RL* circuits and their response to sinusoidal ac voltages are covered in this chapter. Series-parallel combinations are also analyzed. Power in *RL* circuits is studied, and basic applications are introduced. Pulse response of *RL* circuits is covered, and troubleshooting and computer analysis are introduced. The coverage in this chapter is basically parallel with the coverage of *RC* circuits in Chapter 14. This approach allows you to recognize the similarities and understand the differences in the analysis and response of *RC* and *RL* circuits. In this chapter, you will learn:

1. How to calculate the impedance of a series *RL* circuit.
2. How to analyze a series *RL* circuit in terms of phase angle, current, and voltages using Ohm's law and Kirchhoff's voltage law.
3. How to determine the effects of frequency on a series *RL* circuit.
4. How to calculate the impedance of a parallel *RL* circuit.
5. How to use the concepts of conductance, susceptance, and admittance in the analysis of parallel *RL* circuits.
6. How to analyze parallel *RL* circuits in terms of phase angle, currents, and voltage using Ohm's law and Kirchhoff's current law.
7. How to convert a parallel *RL* circuit to an equivalent series form.
8. How to analyze circuits with combinations of series and parallel elements.
9. How to determine the average power, reactive power, apparent power, and the power factor in *RL* circuits.
10. How to analyze *RL* lead and lag networks.
11. How to analyze basic *RL* filters.

15

RL CIRCUIT ANALYSIS

15–1 As with the *RC* circuit, all currents and voltages in an *RL* circuit are sinusoidal when the input is sinusoidal. Phase shifts are introduced because of the inductance. As you will learn, the resistor voltage and current are in phase but lag the source voltage, and the inductor voltage leads the source voltage. The phase angle between the current and the inductor voltage is always 90 degrees. These generalized phase relationships are indicated in Figure 15–1. Notice that they are opposite from those of the *RC* circuit, as discussed in the last chapter.

The amplitudes and the phase relationships of the voltages and current depend on the ohmic values of the resistance and the inductive reactance. When a circuit is purely inductive, the phase angle between the applied voltage and the total current is 90 degrees, with the current *lagging* the voltage. When there is a combination of both resistance and inductive reactance in a circuit, the phase angle is somewhere between zero and 90 degrees depending on the relative values of R and X_L.

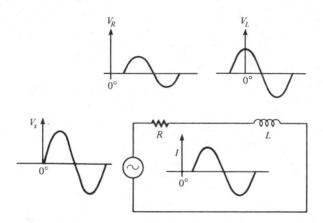

FIGURE 15–1 *Illustration of sinusoidal response with general phase relationships indicated. I and V_R are in phase. I lags V_L by 90°.*

Review for 15–1

1. A 1-kHz sinusoidal voltage is applied to an *RL* circuit. What is the frequency of the resulting current?

2. When the resistance in an *RL* circuit is greater than the inductive reactance, do you think that the phase angle between the applied voltage and the total current is closer to zero or to 90 degrees?

IMPEDANCE OF SERIES *RL* CIRCUITS

15–2 As you know, impedance is the total opposition to sinusoidal current in a circuit and is expressed in ohms. The impedance of a series *RL* circuit is determined by the resistance and the inductive reactance.

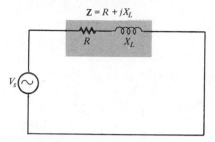

$$\mathbf{Z} = R + jX_L$$

FIGURE 15–2 *Series RL circuit.*

Recall from Chapter 13 that inductive reactance is expressed as a phasor quantity in rectangular form as

$$\mathbf{X}_L = jX_L \qquad (15–1)$$

In the series *RL* circuit of Figure 15–2, the total impedance is the phasor sum of R and jX_L and is expressed as

$$\mathbf{Z} = R + jX_L \qquad (15–2)$$

The Impedance Triangle

In ac analysis, both R and X_L are treated as phasor quantities, as shown in the phasor diagram of Figure 15–3A, with X_L appearing at a +90-degree angle with respect to R. This relationship comes from the fact that the inductor voltage leads the current, and thus the resistor voltage, by 90 degrees. Since $\mathbf{Z}$ is the phasor sum of R and jX_L, its phasor representation is shown in Figure 15–3B. A repositioning of the phasors, as shown in Part C, forms a right triangle. This is called the *impedance triangle*. The length of each phasor represents the magnitude of the quantity, and θ is the phase angle between the applied voltage and the current in the *RL* circuit.

The impedance magnitude of the series *RL* circuit can be expressed in terms of the resistance and reactance as

$$Z = \sqrt{R^2 + X_L^2} \qquad (15–3)$$

This is the magnitude of the impedance and is expressed in ohms.

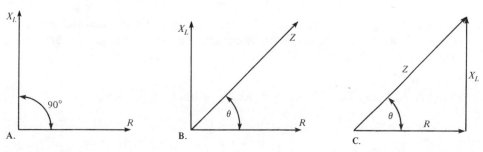

FIGURE 15–3 *Development of the impedance triangle for a series RL circuit.*

The phase angle, θ, is expressed as

$$\theta = \arctan\left(\frac{X_L}{R}\right) \qquad\qquad (15\text{–}4)$$

Combining the magnitude and the angle, the impedance can be expressed in polar form as

$$\mathbf{Z} = \sqrt{R^2 + X_L^2}\,\angle\arctan\left(\frac{X_L}{R}\right) \qquad\qquad (15\text{–}5)$$

Example 15–1

For each circuit in Figure 15–4, express the impedance in both rectangular and polar forms.

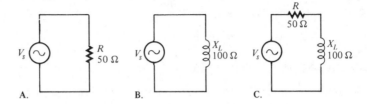

FIGURE 15–4

Solution:

Circuit A:

$$\mathbf{Z} = R + j0 = R = 50\ \Omega \quad \text{in rectangular form } (X_L = 0)$$

$$\mathbf{Z} = R\angle 0° = 50\angle 0°\ \Omega \quad \text{in polar form}$$

The impedance is simply equal to the resistance, and the phase angle is zero because pure resistance does not introduce a phase shift.

      Circuit B:

$$\mathbf{Z} = 0 + jX_L = j100\ \Omega \quad \text{in rectangular form } (R = 0)$$

$$\mathbf{Z} = X_L\angle 90° = 100\angle 90°\ \Omega \quad \text{in polar form}$$

The impedance equals the inductive reactance in this case, and the phase angle is +90 degrees because the inductance causes the current to lag the voltage by 90 degrees.

      Circuit C:

$$\mathbf{Z} = R + jX_L = 50\ \Omega + j100\ \Omega \quad \text{in rectangular form}$$

$$\mathbf{Z} = \sqrt{R^2 + X_L^2}\,\angle\arctan\left(\frac{X_L}{R}\right)$$

$$= \sqrt{(50 \ \Omega)^2 + (100 \ \Omega)^2} \angle \arctan\left(\frac{100}{50}\right) = 111.8 \angle 63.4° \ \Omega$$

The impedance is the phasor sum of the resistance and the inductive reactance. The phase angle is fixed by the relative values of X_L and R.

Review for 15–2

1. The impedance of a certain *RL* circuit is 150 Ω + j220 Ω. What is the value of the resistance? The inductive reactance?

2. A series *RL* circuit has a total resistance of 33 kΩ and an inductive reactance of 50 kΩ. Write the expression for the impedance in rectangular form. Convert the impedance to polar form.

ANALYSIS OF SERIES *RL* CIRCUITS

Ohm's Law

15–3

The application of Ohm's law to series *RL* circuits involves the use of the phasor quantities of **Z**, **V**, and **I**. The three equivalent forms of Ohm's law were stated in Chapter 14 for *RC* circuits. They apply also to *RL* circuits and are restated here for convenience: **V** = **IZ**, **I** = **V/Z**, and **Z** = **V/I**.

Recall that since Ohm's law calculations involve multiplication and division operations, the voltage, current, and impedance should be expressed in polar form.

Example 15–2

The current in Figure 15–5 is expressed as **I** = 0.2∠0° mA. Determine the source voltage and express it also in polar form.

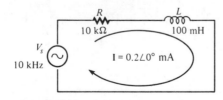

FIGURE 15–5

Solution:

The inductive reactance is

$$X_L = 2\pi fL = 2\pi(10 \text{ kHz})(100 \text{ mH}) = 6.28 \text{ k}\Omega$$

Example 15–2 (continued)

The impedance is

$$\mathbf{Z} = R + jX_L = 10 \text{ k}\Omega + j6.28 \text{ k}\Omega$$

Converting to polar form:

$$\mathbf{Z} = \sqrt{(10 \text{ k}\Omega)^2 + (6.28 \text{ k}\Omega)^2} \angle \arctan\left(\frac{6.28}{10}\right)$$

$$= 11.81 \angle 32.13° \text{ k}\Omega$$

Applying Ohm's law:

$$\mathbf{V}_s = \mathbf{IZ} = (0.2 \angle 0° \text{ mA})(11.81 \angle 32.13° \text{ k}\Omega) = 2.36 \angle 32.13° \text{ V}$$

The magnitude of the source voltage is 2.36 V at an angle of 32.13° with respect to the current; that is, the voltage leads the current by 32.13°. (Note that lower-case subscript denotes an ac quantity.)

Relationships of the Current and Voltages in a Series *RL* Circuit

In a series *RL* circuit, the current is the same through both the resistor and the inductor. Thus, the resistor voltage is in phase with the current, and the inductor voltage leads the current by 90 degrees. Therefore, there is a phase difference of 90 degrees between the resistor voltage, V_R, and the inductor voltage, V_L, as shown in the wave-form diagram of Figure 15–6.

From Kirchhoff's voltage law, the sum of the voltage drops must equal the applied voltage. However, since V_R and V_L are not in phase with each other, they must be added as phasor quantities with V_L leading V_R by 90 degrees, as shown in Figure 15–7A. As shown in Part B, $\mathbf{V}_s$ is the phasor sum of V_R and V_L.

$$\mathbf{V}_s = V_R + jV_L \qquad (15\text{–}6)$$

This equation can be expressed in polar form as

$$\mathbf{V}_s = \sqrt{V_R^2 + V_L^2} \angle \arctan\left(\frac{V_L}{V_R}\right) \qquad (15\text{–}7)$$

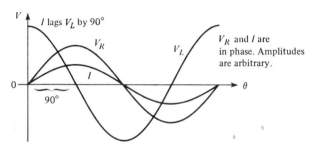

FIGURE 15–6 *Phase relation of voltages and current in a series RL circuit.*

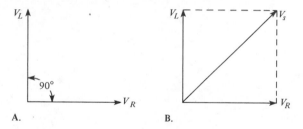

FIGURE 15–7 *Voltage phasor diagram for a series RL circuit.*

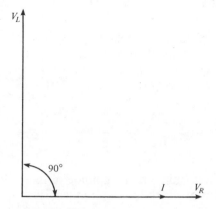

FIGURE 15–8 *Voltage and current phasor diagram for the wave forms in Figure 15–6.*

where the magnitude of the source voltage is

$$V_s = \sqrt{V_R^2 + V_L^2} \qquad \text{(15–8)}$$

and the phase angle between the resistor voltage and the source voltage is

$$\theta = \arctan\left(\frac{V_L}{V_R}\right) \qquad \text{(15–9)}$$

θ is also the phase angle between the source voltage and the current. Figure 15–8 shows a voltage and current phasor diagram that represents the wave-form diagram of Figure 15–6.

Variation of Impedance and Phase Angle with Frequency

The impedance triangle is useful in visualizing how the frequency of the applied voltage affects the *RL* circuit response. As you know, inductive reactance varies *directly* with frequency. When X_L increases, the magnitude of the total impedance also increases; and when X_L decreases, the magnitude of the total impedance decreases. Thus, *Z is directly dependent on frequency.*

The phase angle θ also varies *directly* with frequency, because $\theta = \arctan(X_L/R)$. As X_L increases with frequency, so does θ, and vice versa.

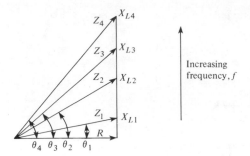

FIGURE 15–9 *Effect of frequency on phase angle.*

The impedance triangle is used in Figure 15–9 to illustrate the variations in X_L, Z, and θ as the frequency changes. Of course, R remains constant. The main point is that *because X_L varies directly as the frequency, so also do the magnitude of the total impedance and the phase angle.* Example 15–3 illustrates this.

Example 15–3

For the series *RL* circuit in Figure 15–10, determine the magnitude of the total impedance and the phase angle for each of the following frequencies: 10 kHz, 20 kHz, and 30 kHz.

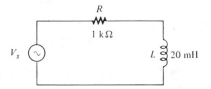

FIGURE 15–10

Solution:

For $f = 10$ kHz:
$$X_L = 2\pi fL = 2\pi(10 \text{ kHz})(20 \text{ mH}) = 1.26 \text{ k}\Omega$$

$$\mathbf{Z} = \sqrt{(1 \text{ k}\Omega)^2 + (1.26 \text{ k}\Omega)^2} \angle \arctan\left(\frac{1.26}{1}\right)$$

$$= 1.61\angle 51.56° \text{ k}\Omega$$

For $f = 20$ kHz:
$$X_L = 2\pi(20 \text{ kHz})(20 \text{ mH}) = 2.52 \text{ k}\Omega$$

$$\mathbf{Z} = \sqrt{(1 \text{ k}\Omega)^2 + (2.52 \text{ k}\Omega)^2} \angle \arctan\left(\frac{2.52}{1}\right)$$

$$= 2.71\angle 68.36° \text{ k}\Omega$$

For $f = 30$ kHz:

$$X_L = 2\pi(30 \text{ kHz})(20 \text{ mH}) = 3.77 \text{ k}\Omega$$

$$\mathbf{Z} = \sqrt{(1 \text{ k}\Omega)^2 + (3.77 \text{ k}\Omega)^2} \angle\arctan\left(\frac{3.77}{1}\right)$$

$$= 3.9\angle75.14° \text{ k}\Omega$$

Notice that as the frequency increases, X_L, Z, and θ also increase.

Review for 15–3

1. In a certain series *RL* circuit, $V_R = 2$ V and $V_L = 3$ V. What is the magnitude of the total voltage?

2. In Problem 1, what is the phase angle between the total voltage and the current?

3. When the frequency of the applied voltage in a series *RL* circuit is increased, what happens to the inductive reactance? What happens to the magnitude of the total impedance? What happens to the phase angle?

IMPEDANCE OF PARALLEL *RL* CIRCUITS

15–4

A basic parallel *RL* circuit is shown in Figure 15–11. The expression for the total impedance is developed as follows.

$$\mathbf{Z} = \frac{(R\angle0°)(X_L\angle90°)}{R + jX_L}$$

$$= \frac{RX_L\angle(0° + 90°)}{\sqrt{R^2 + X_L^2}\angle\arctan\left(\frac{X_L}{R}\right)}$$

$$\mathbf{Z} = \left(\frac{RX_L}{\sqrt{R^2 + X_L^2}}\right)\angle\left(90° - \arctan\left(\frac{X_L}{R}\right)\right) \qquad (15\text{–}10)$$

Equation (15–10) is the expression for the total parallel impedance where the magnitude is

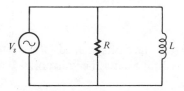

FIGURE 15–11 *Parallel RL circuit.*

$$Z = \frac{RX_L}{\sqrt{R^2 + X_L^2}} \qquad (15\text{--}11)$$

and the phase angle between the applied voltage and the total current is

$$\theta = 90° - \arctan\left(\frac{X_L}{R}\right) \qquad (15\text{--}12)$$

This can also be expressed equivalently as $\theta = \arctan(R/X_L)$.

Example 15–4

For each circuit in Figure 15–12, determine the magnitude of the total impedance and the phase angle.

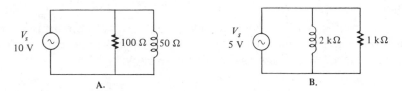

FIGURE 15–12

Solution:

Circuit *A*:

$$\mathbf{Z} = \left(\frac{RX_L}{\sqrt{R^2 + X_L^2}}\right)\angle\left(90° - \arctan\left(\frac{X_L}{R}\right)\right)$$

$$= \left[\frac{(100\ \Omega)(50\ \Omega)}{\sqrt{(100\ \Omega)^2 + (50\ \Omega)^2}}\right]\angle\left(90° - \arctan\left(\frac{50}{100}\right)\right)$$

$$= 44.72\angle63.43°\ \Omega$$

Thus, $Z = 44.72\ \Omega$ and $\theta = 63.43°$.

Circuit *B*:

$$\mathbf{Z} = \left[\frac{(1\ k\Omega)(2\ k\Omega)}{\sqrt{(1\ k\Omega)^2 + (2\ k\Omega)^2}}\right]\angle\left(90° - \arctan\left(\frac{2\ k\Omega}{1\ k\Omega}\right)\right)$$

$$= 894.4\angle26.57°\ \Omega$$

Thus, $Z = 894.4\ \Omega$ and $\theta = 26.57°$.

Notice that the positive angle indicates that the voltage leads the current, as opposed to the *RC* case where the voltage lags the current.

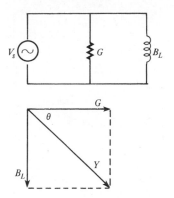

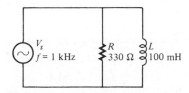

FIGURE 15–13 *Admittance in a parallel RL circuit.*

Susceptance and Admittance

As you know from the previous chapter, conductance is the reciprocal of resistance, susceptance is the reciprocal of reactance, and admittance is the reciprocal of impedance.

For parallel *RL* circuits, inductive susceptance is expressed as

$$\mathbf{B}_L = \frac{1}{X_L \angle 90°} \tag{15–13}$$

and the admittance is

$$\mathbf{Y} = \frac{1}{Z \angle \theta} \tag{15–14}$$

As with the *RC* circuit, the unit for G, B_L, and Y is the siemen (S). In the basic parallel *RL* circuit shown in Figure 15–13, the total admittance is the phasor sum of the conductance, and the susceptance is expressed as follows.

$$\mathbf{Y} = G - jB_L \tag{15–15}$$

Example 15–5

Determine the admittance in Figure 15–14.

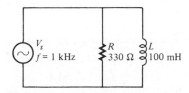

FIGURE 15–14

Example 15–5 (continued)

Solution:

$R = 330 \ \Omega$, thus

$$G = \frac{1}{R} = \frac{1}{330 \ \Omega} = 0.003 \ \text{S}$$

$$X_L = 2\pi(1000 \ \text{Hz})(100 \ \text{mH}) = 628.3 \ \Omega$$

$$B_L = \frac{1}{X_L} = 0.00159 \ \text{S}$$

$$\mathbf{Y} = G - jB_L = 0.003 \ \text{S} - j0.00159 \ \text{S}$$

Y can be expressed in polar form as

$$\mathbf{Y} = \sqrt{(0.003 \ \text{S})^2 + (0.00159 \ \text{S})^2} \angle -\arctan\left(\frac{0.00159}{0.003}\right)$$

$$= 0.0034 \angle -27.92° \ \text{S}$$

Converting to impedance, we get

$$\mathbf{Z} = \frac{1}{Y} = \frac{1}{0.0034 \angle -27.92° \ \text{S}} = 294.12 \angle 27.92° \ \Omega$$

Again, the positive phase angle indicates that the voltage leads the current.

Review for 15–4

1. If $Z = 500 \ \Omega$, what is the value of Y?
2. In a certain parallel *RL* circuit, $R = 50 \ \Omega$ and $X_L = 75 \ \Omega$. Determine the admittance.
3. In the circuit of Problem 2, does the total current lead or lag the applied voltage and by what phase angle?

ANALYSIS OF PARALLEL *RL* CIRCUITS

15–5   The following example applies Ohm's law to the analysis of a parallel *RL* circuit.

Example 15–6

Determine the total current and the phase angle in the circuit of Figure 15–15.

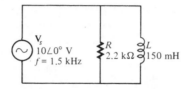

FIGURE 15–15

Solution:

$$X_L = 2\pi(1.5 \text{ kHz})(150 \text{ mH}) = 1.41 \text{ k}\Omega$$

The susceptance magnitude is

$$B_L = \frac{1}{X_L} = \frac{1}{1.41 \text{ k}\Omega} = 0.709 \text{ mS}$$

The conductance magnitude is

$$G = \frac{1}{R} = \frac{1}{2.2 \text{ k}\Omega} = 0.455 \text{ mS}$$

The total admittance is

$$\mathbf{Y} = G - jB_L = 0.455 \text{ mS} - j0.709 \text{ mS}$$

Converting to polar form:

$$\mathbf{Y} = \sqrt{(0.455 \text{ mS})^2 + (0.709 \text{ mS})^2} \angle -\arctan\left(\frac{0.709}{0.455}\right)$$

$$= 0.842 \angle -57.31° \text{ mS}$$

Applying Ohm's law:

$$\mathbf{I}_T = \mathbf{VY} = (10\angle 0° \text{ V})(0.842\angle -57.31° \text{ mS}) = 8.42\angle -57.31° \text{ mA}$$

The magnitude of the total current is 8.42 mA, and it lags the applied voltage by 57.31 degrees, as indicated by the negative angle associated with it.

Relationships of the Currents and Voltages in a Parallel *RL* Circuit

Figure 15–16A shows all the currents and voltages in a basic parallel *RL* circuit. As you can see, the applied voltage, V_s, appears across both the resistive and the inductive branches, so V_s, V_R, and V_L are all in phase and of the same magnitude. The total current, I_T, divides at the junction into the two branch currents, I_R and I_L.

The current through the resistor is in phase with the voltage. The current through the inductor lags the voltage and the resistor current by 90 degrees. By

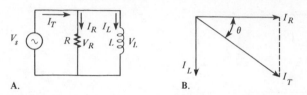

FIGURE 15–16   *Currents and voltages in a parallel RL circuit.*

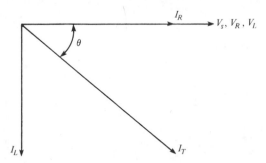

FIGURE 15–17   *Current and voltage phasor diagram for a parallel RL circuit (amplitudes are arbitrary).*

Kirchhoff's current law, the total current is the phasor sum of the two branch currents, as shown by the phasor diagram in Figure 15–16B. The total current is expressed as

$$\mathbf{I}_T = I_R - jI_L \tag{15–16}$$

This equation can be expressed in polar form as

$$\mathbf{I}_T = \sqrt{I_R^2 + I_L^2}\,\angle{-\arctan\!\left(\frac{I_L}{I_R}\right)} \tag{15–17}$$

where the magnitude of the total current is

$$I_T = \sqrt{I_R^2 + I_L^2} \tag{15–18}$$

and the phase angle between the resistor current and the total current is

$$\theta = -\arctan\!\left(\frac{I_L}{I_R}\right) \tag{15–19}$$

Since the resistor current and the applied voltage are in phase, θ also represents the phase angle between the total current and the applied voltage. Figure 15–17 shows a complete current and voltage phasor diagram.

Example 15–7

Determine the value of each current in Figure 15–18, and describe the phase relationship of each with the applied voltage.

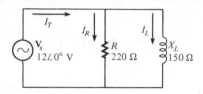

FIGURE 15–18

Solution:

$$I_R = \frac{V_s}{R} = \frac{12\angle 0° \text{ V}}{220\angle 0° \text{ }\Omega} = 54.55\angle 0° \text{ mA}$$

$$I_L = \frac{V_s}{X_L} = \frac{12\angle 0° \text{ V}}{150\angle 90° \text{ }\Omega} = 80\angle -90° \text{ mA}$$

$$I_T = I_R - jI_L = 54.55 \text{ mA} - j80 \text{ mA}$$

Converting I_T to polar form:

$$I_T = \sqrt{(54.55 \text{ mA})^2 + (80 \text{ mA})^2} \angle -\arctan\left(\frac{80}{54.55}\right)$$

$$= 96.83\angle -55.71° \text{ mA}$$

As the results show, the resistor current is 54.55 mA and is in phase with the applied voltage. The inductor current is 80 mA and lags the applied voltage by 90 degrees. The total current is 96.83 mA and lags the voltage by 55.71 degrees.

Review for 15–5

1. The admittance of an *RL* circuit is 0.004 S, and the applied voltage is 8 V. What is the total current?

2. In a certain parallel *RL* circuit, the resistor current is 12 mA, and the inductor current is 20 mA. Determine the magnitude and phase angle of the total current. This phase angle is measured with respect to what?

3. What is the phase angle between the inductor current and the applied voltage in a parallel *RL* circuit?

SERIES-PARALLEL ANALYSIS

15–6

In this section, *RL* circuits with combinations of both series and parallel *R* and *L* elements are analyzed. Examples are used to illustrate the procedures.

Example 15–8

In the circuit of Figure 15–19, determine the following values:
(a) **Z**
(b) $\mathbf{I}_T$
(c) θ

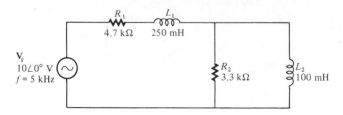

FIGURE 15–19

Solution:

(a) First, the inductive reactances are calculated:

$$X_{L1} = 2\pi(5 \text{ kHz})(250 \text{ mH}) = 7.85 \text{ k}\Omega$$

$$X_{L2} = 2\pi(5 \text{ kHz})(100 \text{ mH}) = 3.14 \text{ k}\Omega$$

R_1 and L_1 are in series with each other. R_2 and L_2 are in parallel. The series portion and the parallel portion are in series with each other.

$$\mathbf{Z}_1 = R_1 + jX_{L1} = 4.7 \text{ k}\Omega + j7.85 \text{ k}\Omega$$

$$G_2 = \frac{1}{R_2} = \frac{1}{3.3 \text{ k}\Omega} = 303 \ \mu\text{S}$$

$$B_{L2} = \frac{1}{X_{L2}} = \frac{1}{3.14 \text{ k}\Omega} = 318.5 \ \mu\text{S}$$

$$\mathbf{Y}_2 = G_2 - jB_{L_y} = 303 \ \mu\text{S} - j318.5 \ \mu\text{S}$$

Converting to polar form:

$$\mathbf{Y}_2 = \sqrt{(303 \ \mu\text{S})^2 + (318.5 \ \mu\text{S})^2} \angle -\arctan\left(\frac{318.5}{303}\right)$$

$$= 439.6\angle -46.43° \ \mu\text{S}$$

Then

$$\mathbf{Z}_2 = \frac{1}{\mathbf{Y}_2} = \frac{1}{439.6\angle -46.43° \ \mu\text{S}} = 2.28\angle 46.43° \text{ k}\Omega$$

Converting to rectangular form:

$$\mathbf{Z}_2 = (2.28 \text{ k}\Omega) \cos(46.43°) + j(2.28 \text{ k}\Omega) \sin(46.43°)$$

$$= 1.57 \text{ k}\Omega + j1.65 \text{ k}\Omega$$

Combining $\mathbf{Z}_1$ and $\mathbf{Z}_2$, we get

$$\mathbf{Z}_T = \mathbf{Z}_1 + \mathbf{Z}_2$$
$$= (4.7 \text{ k}\Omega + j7.85 \text{ k}\Omega) + (1.57 \text{ k}\Omega + j1.65 \text{ k}\Omega)$$
$$= 6.27 \text{ k}\Omega + j9.5 \text{ k}\Omega$$

Expressing $\mathbf{Z}_T$ in polar form:

$$\mathbf{Z}_T = \sqrt{(6.27 \text{ k}\Omega)^2 + (9.5 \text{ k}\Omega)^2} \angle \arctan\left(\frac{9.5}{6.27}\right)$$

$$= 11.38 \angle 56.58° \text{ k}\Omega$$

(b) The total current is found using Ohm's law:

$$\mathbf{I}_T = \frac{\mathbf{V}_s}{\mathbf{Z}_T} = \frac{10 \angle 0° \text{ V}}{11.38 \angle 56.58° \text{ k}\Omega}$$

$$= 0.879 \angle -56.58° \text{ mA}$$

(c) The total current lags the applied voltage by 56.58 degrees.

Example 15–9

Determine the voltage across each element in Figure 15–20. Sketch a voltage phasor diagram.

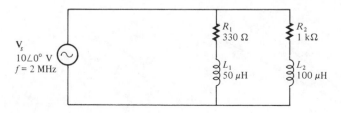

FIGURE 15–20

Solution:

First calculate X_{L1} and X_{L2}:

$$X_{L1} = 2\pi f L_1 = 2\pi(2 \text{ MHz})(50 \text{ }\mu\text{H}) = 628.3 \text{ }\Omega$$

$$X_{L2} = 2\pi f L_2 = 2\pi(2 \text{ MHz})(100 \text{ }\mu\text{H}) = 1.257 \text{ k}\Omega$$

Now, determine the impedance of each branch:

$$\mathbf{Z}_1 = R_1 + jX_{L1} = 330 \text{ }\Omega + j628.3 \text{ }\Omega$$

$$\mathbf{Z}_2 = R_2 + jX_{L2} = 1 \text{ k}\Omega + j1.257 \text{ k}\Omega$$

Example 15–9 (continued)

Converting to polar form:

$$\mathbf{Z}_1 = \sqrt{(330\ \Omega)^2 + (628.3\ \Omega)^2}\ \angle\arctan\left(\frac{628.3}{330}\right)$$

$$= 709.69\angle 62.29°\ \Omega$$

$$\mathbf{Z}_2 = \sqrt{(1\ k\Omega)^2 + (1.257\ k\Omega)^2}\ \angle\arctan\left(\frac{1.257}{1}\right)$$

$$= 1.606\angle 51.5°\ k\Omega$$

Calculate each branch current:

$$\mathbf{I}_1 = \frac{\mathbf{V}_s}{\mathbf{Z}_1} = \frac{10\angle 0°\ V}{709.69\angle 62.29°\ \Omega} = 14.09\angle -62.29°\ mA$$

$$\mathbf{I}_2 = \frac{\mathbf{V}_s}{\mathbf{Z}_2} = \frac{10\angle 0°\ V}{1.606\angle 51.5°\ k\Omega} = 6.23\angle -51.5°\ mA$$

Now, Ohm's law is used to get the voltage across each element:

$$\mathbf{V}_{R1} = \mathbf{I}_1\mathbf{R}_1 = (14.09\angle -62.2°\ mA)(330\angle 0°\ \Omega) = 4.65\angle -62.2°\ V$$

$$\mathbf{V}_{L1} = \mathbf{I}_1\mathbf{X}_{L1} = (14.09\angle -62.2°\ mA)(628.3\angle 90°\ \Omega) = 8.85\angle 27.8°\ V$$

$$\mathbf{V}_{R2} = \mathbf{I}_2\mathbf{R}_2 = (6.23\angle -51.5°\ mA)(1\angle 0°\ k\Omega) = 6.23\angle -51.5°\ V$$

$$\mathbf{V}_{L2} = \mathbf{I}_2\mathbf{X}_{L2} = (6.23\angle -51.5°\ mA)(1.257\angle 90°\ k\Omega) = 7.83\angle 38.5°\ V$$

The voltage phasor diagram is shown in Figure 15–21, and the current phasor diagram is shown in Figure 15–22.

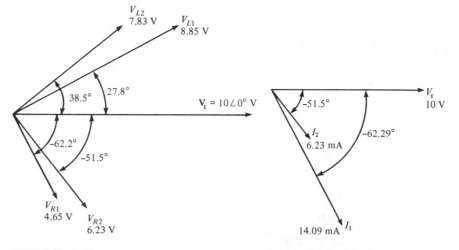

FIGURE 15–21   **FIGURE 15–22**

The two previous examples should give you a feel for how to approach the analysis of complex *RL* networks.

Review for 15–6

1. What is the total impedance of the circuit in Figure 15–20?
2. Determine the total current for the circuit in Figure 15–20.

As you know, in a purely resistive ac circuit, all of the energy that is delivered by the source is dissipated in the form of heat by the resistance. In a purely inductive ac circuit, all of the energy delivered by the source is stored by the inductor in its magnetic field during a portion of the voltage cycle and then returned to the source during another portion of the cycle so that there is no net energy loss.

When there is both resistance and inductance in a circuit, some of the energy is alternately stored and returned by the inductance, and some is dissipated by the resistance. The amount of energy loss is determined by the relative values of resistance and inductive reactance.

When the resistance is greater than the inductive reactance, more of the total energy delivered by the source is dissipated by the resistance than is stored by the inductor, and when the reactance is greater than the resistance, more of the total energy is stored and returned than is lost.

As you know, the power loss in a resistor is called the *average power.* The power in an inductor is reactive power and is expressed as

$$P_r = I^2 X_L \qquad \qquad \textbf{(15–20)}$$

The Power Triangle

The generalized power triangle for the *RL* circuit is shown in Figure 15–23. The apparent power, P_a, is the resultant of the average power, P_{avg}, and the reactive power, P_r.

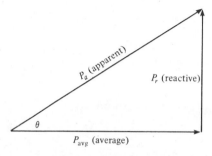

FIGURE 15–23 *Power triangle for an RL circuit.*

Recall that the power factor equals the cosine of θ ($PF = \cos \theta$). As the phase angle between the applied voltage and the total current increases, the power factor decreases, indicating an increasingly reactive circuit. The smaller the power factor, the smaller the average power is compared to the reactive power.

Example 15–10

Determine the power factor, the average power, the reactive power, and the apparent power in Figure 15–24.

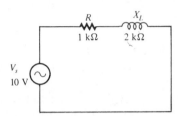

FIGURE 15–24

Solution:

The total impedance of the circuit is

$$\mathbf{Z} = R + jX_L$$

$$= \sqrt{(1\ \text{k}\Omega)^2 + (2\ \text{k}\Omega)^2}\ \angle \arctan\left(\frac{2000}{1000}\right)$$

$$= 2236 \angle 63.44°\ \Omega$$

The current magnitude is

$$I = \frac{V_s}{Z} = \frac{10\ \text{V}}{2236\ \Omega} = 4.47\ \text{mA}$$

The phase angle is indicated in the expression for $\mathbf{Z}$.

$$\theta = 63.44°$$

The power factor is therefore

$$PF = \cos \theta = \cos(63.44°) = 0.447$$

The average power is

$$P_{\text{avg}} = V_s I \cos \theta = (10\ \text{V})\,(4.47\ \text{mA})\,(0.447) = 19.99\ \text{mW}$$

The reactive power is

$$P_r = I^2 X_L = (4.47\ \text{mA})^2\,(2\ \text{k}\Omega) = 39.96\ \text{mVAR}$$

The apparent power is

$$P_a = I^2 Z = (4.47 \text{ mA})^2 (2236 \text{ }\Omega) = 44.68 \text{ mVA}$$

Review for 15–7

1. To which component in an *RL* circuit is the energy loss due?
2. Calculate the power factor when $\theta = 50$ degrees.
3. A certain *RL* circuit consists of a 470-Ω resistor and an inductive reactance of 620 Ω at the operating frequency. Determine P_{avg}, P_r, and P_a when $I = 100$ mA.

BASIC APPLICATIONS

15–8

Two basic applications of *RL* circuits are introduced in this section: *phase shift networks (lead and lag)* and *frequency selective networks (filters)*.

The *RL* Lead Network

The first type of phase shift network is the *RL* lead network, in which the output voltage leads the input voltage by a specified amount. Figure 15–25A shows a series *RL* circuit with the *output voltage taken across the inductor* (note that in the *RC* lead network, the output was taken across the resistor). The source voltage is the input, V_{in}. As you know, θ is the angle between the current and the input voltage; it is also the angle between the resistor voltage and the input voltage, because V_R and I are in phase.

Since V_L *leads* V_R by 90°, the phase angle between the inductor voltage and the input voltage is the difference between 90° and θ, as shown in Figure 15–25B. The inductor voltage is the *output;* it leads the input, thus creating a basic *lead* network.

When the input and output wave forms of the lead network are displayed on an oscilloscope, a relationship similar to that in Figure 15–26 is observed. The amount of phase difference between the input and the output is dependent on the relative values of the inductive reactance and the resistance, as is the magnitude of the output voltage.

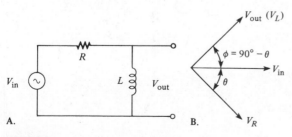

A.

B.

FIGURE 15–25 *RL lead network.*

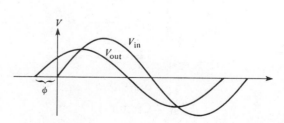

FIGURE 15–26

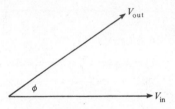

FIGURE 15–27

Phase Difference between Input and Output: The angle between V_{out} and V_{in} is designated ϕ (phi) and is developed as follows:

The polar expressions for the input voltage and the current are $V_{in} \angle 0°$ and $I \angle -\theta$, respectively. The output voltage is

$$\mathbf{V}_{out} = (I \angle -\theta)(X_L \angle 90°) = IX_L \angle (90° - \theta)$$

This shows that the output voltage is at an angle of $90° - \theta$ with respect to the input voltage. Since $\theta = \arctan(X_L/R)$, the angle between the input and output is

$$\phi = 90° - \arctan\left(\frac{X_L}{R}\right) \qquad\qquad (15\text{–}21)$$

This angle can equivalently be expressed as

$$\phi = \arctan\left(\frac{R}{X_L}\right)$$

This angle is always positive, indicating that the output voltage *leads* the input voltage, as indicated in Figure 15–27.

Example 15–11

Determine the amount of phase lead from input to output in each lead network in Figure 15–28.

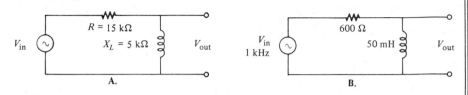

FIGURE 15–28

Solution:

Circuit A:

$$\phi = 90° - \arctan\left(\frac{X_L}{R}\right) = 90° - \arctan\left(\frac{5000}{15,000}\right)$$

$$= 90° - 18.44° = 71.56°$$

The output leads the input by 71.56°.

Circuit B:

$$X_L = 2\pi fL = 2\pi(1 \text{ kHz})(50 \text{ mH}) = 314.2 \text{ }\Omega$$

$$\phi = 90° - \arctan\left(\frac{X_L}{R}\right) = 90° - \arctan\left(\frac{314.2}{600}\right) = 62.36°$$

The output leads the input by 62.36°.

Magnitude of the Output Voltage: To evaluate the output voltage in terms of its magnitude, visualize the RL lead network as a voltage divider. A portion of the total input voltage is dropped across R and a portion across L. Since the output voltage is V_L, it can be calculated as

$$V_{out} = \left(\frac{X_L}{\sqrt{R^2 + X_L^2}}\right)V_{in} \tag{15-22}$$

Or it can be found using Ohm's law as follows:

$$V_{out} = IX_L \tag{15-23}$$

The total phasor expression for the output voltage of an RL lead network is

$$\mathbf{V}_{out} = V_{out}\angle\phi \tag{15-24}$$

Example 15-12

For the lead network in Figure 15-28B, determine the output voltage in phasor form when the input voltage has an rms value of 5 V. Sketch the input and output voltage wave forms showing the proper relationships.

Solution:

ϕ was found to be 62.36° in Example 15-11.

$$\mathbf{V}_{out} = \left(\frac{X_L}{\sqrt{R^2 + X_L^2}}\right)V_{in}\angle\phi$$

$$= \left[\frac{314.2 \text{ }\Omega}{\sqrt{(600 \text{ }\Omega)^2 + (314.2 \text{ }\Omega)^2}}\right]5\angle 62.36° \text{ V}$$

$$= 2.32\angle 62.36° \text{ V}$$

The wave forms are shown in Figure 15-29. Notice that the output voltage leads the input voltage by 62.36°.

Example 15–12 (continued)

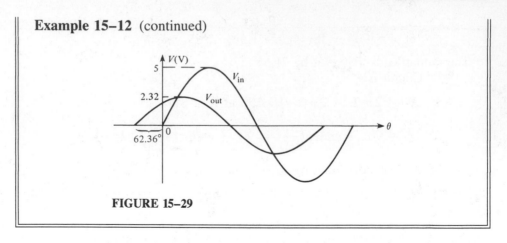

FIGURE 15–29

The *RL* Lag Network

The second basic type of phase shift network is the *RL* lag network. When the output of a series *RL* circuit is taken across the resistor rather than the inductor, as shown in Figure 15–30A, it becomes a *lag* network. Lag networks cause the phase of the output voltage to *lag* the input by a specified amount.

Phase Difference between Input and Output: In a series *RL* circuit, the current lags the input voltage. Since the output voltage is taken across the resistor, the output *lags* the input, as indicated by the phasor diagram in Figure 15–30B. The wave forms are shown in Part C.

As in the lead network, the amount of phase difference between the input and output and the magnitude of the output voltage is dependent on the relative values of the resistance and the inductive reactance. When the input voltage is assigned a reference angle of 0 degrees, the angle of the output voltage (ϕ) with respect to the input voltage equals θ, because the resistor voltage (output) and the current are in phase with each other. The expression for the angle between the input voltage and the output voltage is

$$\phi = -\arctan\left(\frac{X_L}{R}\right) \qquad\qquad \textbf{(15–25)}$$

This angle is negative because the output lags the input.

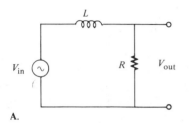

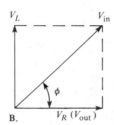

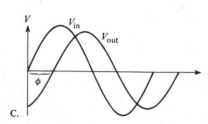

A.   B.   C.

FIGURE 15–30 *RL lag network.*

Example 15–13

Calculate the output phase angle for each circuit in Figure 15–31.

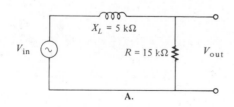

A.

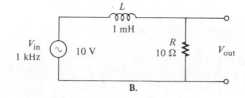

B.

FIGURE 15–31

Solution:

Circuit *A*:

$$\phi = -\arctan\left(\frac{X_L}{R}\right) = -\arctan\left(\frac{5\ k\Omega}{15\ k\Omega}\right) = -18.43°$$

The output lags the input by 18.43°.

Circuit *B*:

$$X_L = 2\pi fL = 2\pi(1\ kHz)(1\ mH) = 6.28\ \Omega$$

$$\phi = -\arctan\left(\frac{X_L}{R}\right) = -\arctan\left(\frac{6.28\ \Omega}{10\ \Omega}\right) = -32.13°$$

The output lags the input by 32.13°.

Magnitude of the Output Voltage: Since the output voltage of an *RL* lag network is taken across the resistor, the magnitude can be calculated using either the voltage divider approach or Ohm's law.

$$V_{out} = \left(\frac{R}{\sqrt{R^2 + X_L^2}}\right)V_{in} \qquad (15\text{–}26)$$

$$V_{out} = IR \qquad (15\text{–}27)$$

The expression for the output voltage in phasor form is

$$\mathbf{V}_{out} = V_{out}\angle -\phi \qquad\qquad (15\text{-}28)$$

Example 15–14

The input voltage in Figure 15–31B (Example 15–13) has an rms value of 10 V. Determine the phasor expression for the output voltage. Sketch the wave-form relationships for the input and output voltages.

Solution:

The phase angle was found to be $-32.13°$ in Example 15–13.

$$\mathbf{V}_{out} = \left(\frac{R}{\sqrt{R^2 + X_L^2}}\right) V_{in}\angle -\phi$$

$$= \left(\frac{10\ \Omega}{11.81\ \Omega}\right) 10\angle -32.13°\ V = 8.47\angle -32.13°\ V\ rms$$

The wave forms are shown in Figure 15–32.

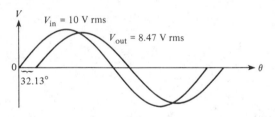

FIGURE 15–32

The *RL* Circuit as a Filter

As with *RC* circuits, series *RL* circuits also exhibit a frequency-selective characteristic and therefore act as basic filters.

Low-Pass Filter: You have seen what happens to the output magnitude and phase angle in the lag network. In terms of the filtering action, the variation of the magnitude of the output voltage as a function of frequency is of primary importance.

Figure 15–33 shows the filtering action of a series *RL* circuit using specific values for purposes of illustration. In Part A of the figure, the input is *zero* frequency (dc). Since the inductor ideally acts as a short to constant direct current, the output voltage equals the full value of the input voltage (neglecting the winding resistance). Therefore, the circuit passes *all* of the input voltage to the output (10 V in, 10 V out).

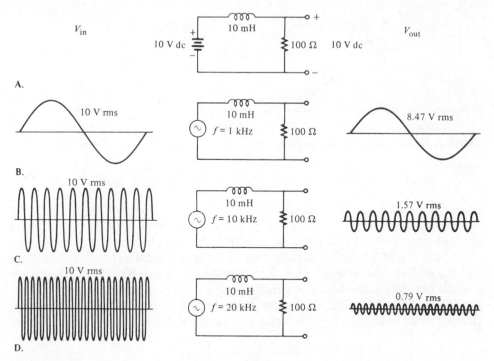

FIGURE 15–33 *Low-pass filter action of an RL circuit (phase shift from input to output is not indicated).*

In Figure 15–33B, the frequency of the input voltage has been increased to 1 kHz, causing the inductive reactance to increase to 62.83 Ω. For an input voltage of 10 V rms, the output voltage is approximately 8.47 V rms, which can be calculated using the voltage divider approach or Ohm's law.

In Figure 15–33C, the input frequency has been increased to 10 kHz, causing the inductive reactance to increase further to 628.3 Ω. For a constant input voltage of 10 V rms, the output voltage is now 1.57 V rms.

As the input frequency is increased further, the output voltage continues to decrease and approaches zero as the frequency becomes very high, as shown in Figure 15–33D. A summary of the circuit action is as follows: As the frequency of the input increases, the inductive reactance increases. Because the resistance is constant and the inductive reactance increases, the voltage across the inductor increases, and that across the resistor (output voltage) decreases. The input frequency can be increased until it reaches a value at which the reactance is so large compared to the resistance that the output voltage can be neglected, because it becomes very small compared to the input voltage.

As shown in Figure 15–33, the circuit passes dc (zero frequency) completely. As the frequency of the input increases, less of the input voltage is passed through to the output. That is, the output voltage decreases as the frequency increases. It is apparent that the lower frequencies pass through the circuit much better than the higher frequencies. This *RL* circuit is therefore a very basic form of *low-pass filter*.

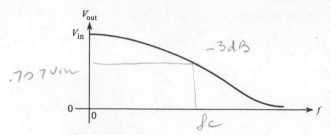

FIGURE 15–34 *Low-pass filter response curve.*

Figure 15–34 shows a response curve for the low-pass filter.

High-Pass Filter: In Figure 15–35A, the output is taken across the inductor. When the input voltage is dc (zero frequency), the output is zero volts, because the inductor ideally appears as a short across the output.

In Figure 15–35B, the frequency of the input signal has been increased to 100 Hz with an rms value of 10 V. The output voltage is 0.63 V rms. Thus, only a small percentage of the input voltage appears at the output at this frequency.

In Figure 15–35C, the input frequency is increased further to 1 kHz, causing more voltage to be developed as a result of the increase in the inductive

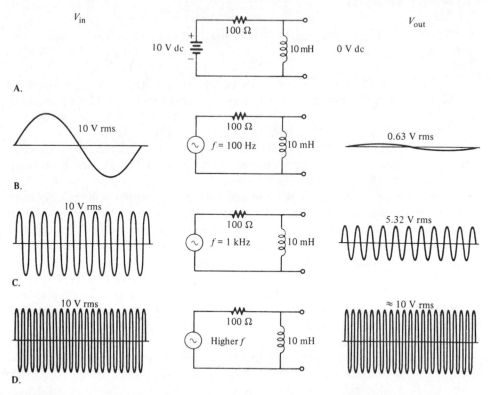

FIGURE 15–35 *High-pass filter action of an RL circuit (phase shift from input to output is not indicated).*

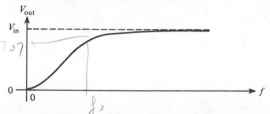

FIGURE 15–36 *High-pass filter response curve.*

reactance. The output voltage at this frequency is 5.32 V rms. As you can see, the output voltage increases as the frequency increases. A value of frequency is reached at which the reactance is very large compared to the resistance and most of the input voltage appears across the inductor, as shown in Figure 15–35D.

This circuit tends to prevent lower-frequency signals from appearing on the output but permits higher-frequency signals to pass through from input to output; thus, it is a very basic form of *high-pass filter.*

The response curve in Figure 15–36 shows that the output voltage increases and then levels off as it approaches the value of the input voltage as the frequency increases.

Review for 15–8

1. A certain *RL* lead network consists of 3.3-kΩ resistor and a 15-mH inductor. Determine the phase shift between input and output at a frequency of 5 kHz.

2. An *RL* lag network has the same component values as the lead network in Problem 1. What is the magnitude of the output voltage at 5 kHz when the input is 10 V rms?

3. When an *RL* circuit is used as a low-pass filter, across which component is the output taken?

PULSE RESPONSE OF *RL* CIRCUITS

A series *RL* circuit with an ideal pulse input is shown in Figure 15–37. We analyze the behavior of this circuit in terms of the current, resistor voltage, and inductor voltage during and after the application of the pulse.

15–9

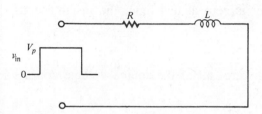

FIGURE 15–37 *Series RL circuit with an ideal pulse input.*

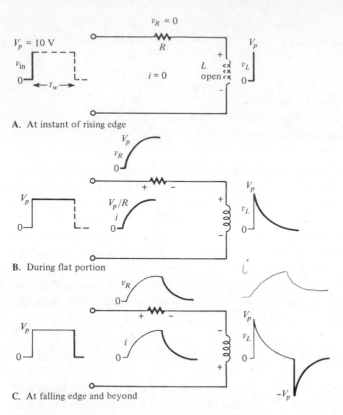

A. At instant of rising edge

B. During flat portion

C. At falling edge and beyond

FIGURE 15–38 *Pulse response for condition where the current reaches maximum value*
$(t_w \geqq 5L/R).$

As you know, both edges of an ideal pulse are considered to occur instantaneously. Two basic rules for inductor behavior will aid in analyzing *RL* circuit responses to pulse inputs: (1) the inductor appears as an open to an instantaneous change in current and as a short (ideally) to dc; and (2) the current in an inductor cannot change instantaneously — it can only change exponentially.

The ideal pulse response of an *RL* circuit is illustrated in Figure 15–38 and is described as follows. At the instant of the *rising edge* of the pulse, the inductor appears as an open; therefore, the current is zero.

$$i = 0 \text{ A}$$

Since the instantaneous current is zero at this time, the instantaneous resistor voltage is also zero.

$$v_R = iR = 0 \text{ V}$$

Because the resistor voltage is zero, all of the source voltage appears across the inductor.

$$v_L = V_p$$

where V_p is the pulse amplitude.

During the *flat portion* of the pulse, the current begins to increase exponentially toward its maximum value of V_p/R. As a result, the resistor voltage also begins to increase exponentially toward V_p. According to Kirchhoff's voltage law, the sum of the voltage drops at any instant must equal the applied voltage. Therefore, as the resistor voltage builds up, the inductor voltage must decrease correspondingly such that $v_L + v_R = V_p$.

If the pulse width of the input is sufficiently long, the current will reach its maximum value of V_p/R; if not, the current at the end of the pulse will be less than maximum. These two conditions are now discussed.

When t_w Is Greater Than 5τ

For the current to reach the maximum value, the pulse width, t_w, must be equal to or greater than five time constants ($5\tau = 5L/R$).

$$t_w \geqq 5\left(\frac{L}{R}\right)$$

As illustrated in Figure 15–38, i reaches V_p/R, v_R reaches V_p, and v_L decreases to zero at or before the end of the pulse. At the instant of the falling edge, the inductor appears as a current source equal to V_p/R, because the current in an inductor cannot change instantaneously and so remains at its previous value for an instant. Since the input pulse has fallen instantaneously to zero and the resistor voltage remains at V_p for an instant, there is an instantaneous reversal of the inductor voltage from zero to $-V_p$. Since v_R is 10 V, the inductor voltage must be -10 V in order for their sum to be zero (this satisfies Kirchhoff's voltage law). After the instant of the falling edge, the current, the resistor voltage, and the inductor voltage all go to zero exponentially in five time constants, as shown.

Example 15–15

A pulse is applied to the *RL* circuit in Figure 15–39. Determine the complete wave shapes and values for i, v_R, and v_L.

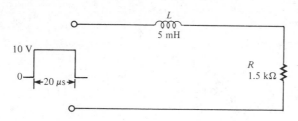

FIGURE 15–39

Solution:

The circuit time constant is

$$\tau = \frac{L}{R} = \frac{5 \text{ mH}}{1.5 \text{ k}\Omega} = 3.33 \text{ } \mu\text{s}$$

Example 15–15 (continued)

Since 5τ is less than t_w, the current will reach its maximum value and remain there until the end of the pulse.

At the rising edge of the pulse:

$$i = 0 \text{ A}$$

$$v_R = 0 \text{ V}$$

$$v_L = 10 \text{ V}$$

Since the inductor is initially an open, all of the input voltage appears across L.

During the pulse:

i increases exponentially to $V_p/R = 10 \text{ V}/1.5 \text{ k}\Omega$
$$= 6.67 \text{ mA in } 5(3.33 \text{ } \mu s)$$
$$= 16.65 \text{ } \mu s.$$
v_R increases exponentially to 10 V in 16.65 μs.
v_L decreases exponentially to zero in 16.65 μs.

At the falling edge of the pulse:

$$i = 6.67 \text{ mA}$$

$$v_R = 10 \text{ V}$$

$$v_L = -10 \text{ V}$$

After the pulse:

i decreases exponentially to zero in 16.65 μs.

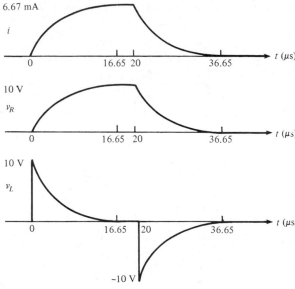

FIGURE 15–40

> v_R decreases exponentially to zero in 16.65 μs.
> v_L decreases exponentially to zero in 16.65 μs.
> The wave forms are shown in Figure 15–40.

When t_w Is Less Than 5τ

When the pulse width of the input is less than five time constants, the current does not have time to reach its maximum possible value. The amount that it increases depends on the duration of the input pulse compared to five time constants and can be determined from the following formula.

$$i = \frac{V_p}{R}(1 - e^{-t_w \, R/L}) \qquad\qquad (15\text{–}29)$$

The response of the circuit to the instantaneous rising edge is the same as described for the previous case. The difference comes during the pulse and at the falling edge, as illustrated in Figure 15–41. During the pulse, the current reaches

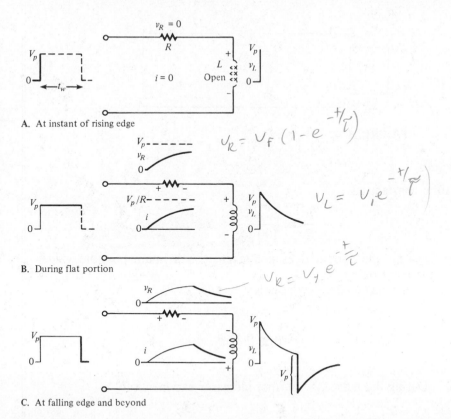

A. At instant of rising edge

$$v_R = v_F\left(1 - e^{-t/\tau}\right)$$

B. During flat portion

$$v_L = v_i e^{-t/\tau}$$

$$v_R = v_i e^{-t/\tau}$$

C. At falling edge and beyond

FIGURE 15–41 *Pulse response for the condition where the current does not reach maximum*
($t_w < 5L/R$).

a value less than V_p/R, depending on the pulse width. As a result, the resistor voltage does not reach V_p, and the inductor voltage does not reach zero.

At the instant of the falling edge, the current does not change. Thus, since the input has suddenly gone to zero, the inductor voltage instantaneously goes to a negative value equal to the resistor voltage. This, of course, is in accordance with Kirchhoff's voltage law, which states that the sum of v_L and v_R must equal the input, which is zero at the falling edge.

$$v_L + v_R = 0$$

$$v_L = -v_R$$

After the falling edge, i, v_R, and v_L all go to zero exponentially in a time equal to 5τ, as shown.

Example 15–16

Determine the pulse response in Figure 15–42. Sketch the wave forms for i, v_R, and v_L.

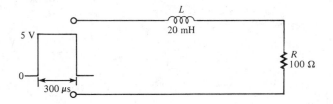

FIGURE 15–42

Solution:

The time constant is

$$\tau = \frac{L}{R} = \frac{20 \text{ mH}}{100 \text{ } \Omega} = 200 \text{ } \mu s$$

Since $5\tau > t_w$, the current does not reach its maximum possible value during the input pulse.

At the rising edge of the pulse:

$$i = 0 \text{ A}$$

$$v_R = 0 \text{ V}$$

$$v_L = 5 \text{ V}$$

During the pulse, the current increases exponentially to

$$i = \frac{V_p}{R}(1 - e^{-t_w/\tau})$$

$$= 50 \text{ mA}(1 - e^{-300 \,\mu s/200 \,\mu s}) = 38.8 \text{ mA}$$

The resistor voltage increases exponentially to

$$v_R = iR = (38.8 \text{ mA})(100 \text{ } \Omega) = 3.88 \text{ V}$$

The inductor voltage decreases exponentially to

$$v_L = V_p - v_R = 5 \text{ V} - 3.88 \text{ V} = 1.12 \text{ V}$$

At the falling edge of the pulse:

$$i = 38.8 \text{ mA}$$

$$v_R = 3.88 \text{ V}$$

$$v_L = -3.88 \text{ V}$$

The wave forms are shown in Figure 15–43.

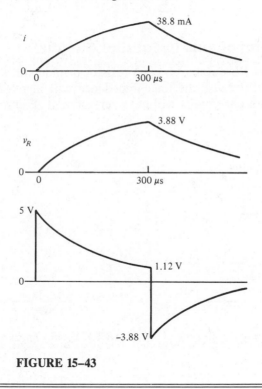

FIGURE 15–43

Review for 15–9

1. A 10-ms wide pulse is applied to a series *RL* circuit in which the time constant is 1.8 ms. Does the current reach its maximum possible value?

2. A 12-V, positive-going, 400-μs pulse is applied to a series circuit with $R = 1.5$ kΩ and $L = 50$ mH. Determine i, v_R, and v_L at the rising edge.

3. In Problem 2, what are i, v_R, and v_L at the falling edge?

TROUBLESHOOTING

15–10

Effects of an Open Inductor

The most common failure mode for inductors occurs when the winding opens as a result of excessive current or a mechanical contact failure. It is easy to see how an open coil affects the operation of a basic series *RL* circuit, as shown in Figure 15–44. Obviously, there is no current path—so the resistor voltage is zero, and total applied voltage appears across the inductor.

Effects of an Open Resistor

When the resistor is open, there is no current, and the inductor voltage is zero. The total input voltage is across the open resistor, as shown in Figure 15–45.

Open Components in Parallel Circuits

In a parallel *RL* circuit, an open resistor or inductor will cause the total current to decrease, because the total impedance will increase. Obviously, the branch with the open component will have zero current. Figure 15–46 illustrates these conditions.

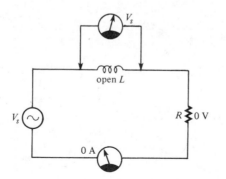

FIGURE 15–44 *Effect of an open coil.*

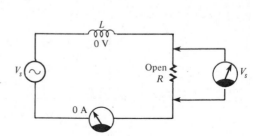

FIGURE 15–45 *Effect of an open resistor.*

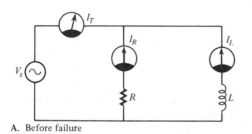

A. Before failure

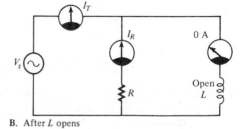

B. After *L* opens

FIGURE 15–46 *Effect of an open component in a parallel circuit.*

Effects of an Inductor with Shorted Windings

It is possible for some of the windings of coils to short together as a result of damaged insulation. This failure mode is much less likely than the open coil. Shorted windings result in a reduction in inductance, since the inductance of a coil is proportional to the square of the number of turns.

Review for 15–10

1. Describe the effect of an inductor with shorted windings on the response of a series RL circuit.

2. In the circuit of Figure 15–47, indicate whether I_T, V_{R1}, and V_{R2} increase or decrease as a result of L opening.

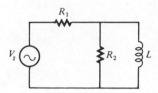

FIGURE 15–47

COMPUTER ANALYSIS

The following program provides for the computation of phase shift and output voltage as a function of frequency for an RL lead network. Inputs required are the component values, the frequency limits, and the increments of frequency.

15–11

```
10 CLS
20 PRINT "THIS PROGRAM COMPUTES THE PHASE SHIFT FROM INPUT TO"
30 PRINT "OUTPUT AND THE NORMALIZED OUTPUT VOLTAGE MAGNITUDE"
40 PRINT "AS FUNCTIONS OF FREQUENCY FOR AN RL LEAD NETWORK."
50 PRINT:PRINT:PRINT
60 INPUT "TO CONTINUE PRESS 'ENTER'";X:CLS
70 INPUT "RESISTANCE IN OHMS";R
80 INPUT "INDUCTANCE IN HENRIES";L
90 INPUT "THE LOWEST FREQUENCY IN HERTZ";FL
100 INPUT "THE HIGHEST FREQUENCY IN HERTZ";FH
110 INPUT "THE FREQUENCY INCREMENTS IN HERTZ";FI
120 CLS
130 PRINT "FREQUENCY(HZ)","PHASE SHIFT","VOUT"
140 FOR F=FL TO FH STEP FI
150 XL = 2*3.1416*F*L
160 PHI=90-ATN(XL/R)*57.3
170 VO=XL/(SQR(R[2+XL[2))
180 PRINT F,PHI,VO
190 NEXT
```

Formulas

Series RL Circuits:

$$\mathbf{X}_L = jX_L \tag{15–1}$$

$$\mathbf{Z} = R + jX_L \tag{15–2}$$

$$Z = \sqrt{R^2 + X_L^2} \tag{15–3}$$

$$\theta = \arctan\left(\frac{X_L}{R}\right) \tag{15–4}$$

$$\mathbf{Z} = \sqrt{R^2 + X_L^2} \angle \arctan\left(\frac{X_L}{R}\right) \tag{15–5}$$

$$\mathbf{V}_s = V_R + jV_L \tag{15–6}$$

$$\mathbf{V}_s = \sqrt{V_R^2 + V_L^2} \angle \arctan\left(\frac{V_L}{V_R}\right) \tag{15–7}$$

$$V_s = \sqrt{V_R^2 + V_L^2} \tag{15–8}$$

$$\theta = \arctan\left(\frac{V_L}{V_R}\right) \tag{15–9}$$

$$V_{out} = V_{in}\left(1 - e^{-t/\tau}\right)$$

Parallel RL Circuits:

$$\mathbf{Z} = \left(\frac{RX_L}{\sqrt{R^2 + X_L^2}}\right) \angle \left(90° - \arctan\left(\frac{X_L}{R}\right)\right) \tag{15–10}$$

$$Z = \frac{RX_L}{\sqrt{R^2 + X_L^2}} \tag{15–11}$$

$$\theta = 90° - \arctan\left(\frac{X_L}{R}\right) \tag{15–12}$$

$$\mathbf{B}_L = \frac{1}{X_L \angle 90°} \tag{15–13}$$

$$\mathbf{Y} = \frac{1}{Z \angle \theta} \tag{15–14}$$

$$\mathbf{Y} = G - jB_L \tag{15–15}$$

$$\mathbf{I}_T = I_R - jI_L \tag{15–16}$$

$$\mathbf{I}_T = \sqrt{I_R^2 + I_L^2} \angle -\arctan\left(\frac{I_L}{I_R}\right) \tag{15–17}$$

$$I_T = \sqrt{I_R^2 + I_L^2} \tag{15–18}$$

$$\theta = -\arctan\left(\frac{I_L}{I_R}\right) \tag{15–19}$$

Power in RL Circuits:

$$P_r = I^2 X_L \qquad\qquad (15\text{--}20)$$

Lead Network:

$$\phi = 90° - \arctan\!\left(\frac{X_L}{R}\right) \qquad\qquad (15\text{--}21)$$

$$V_{out} = \left(\frac{X_L}{\sqrt{R^2 + X_L^2}}\right) V_{in} \qquad\qquad (15\text{--}22)$$

$$V_{out} = IX_L \qquad\qquad (15\text{--}23)$$

$$\mathbf{V}_{out} = V_{out}\angle\phi \qquad\qquad (15\text{--}24)$$

Lag Network

$$\phi = -\arctan\!\left(\frac{X_L}{R}\right) \qquad\qquad (15\text{--}25)$$

$$V_{out} = \left(\frac{R}{\sqrt{R^2 + X_L^2}}\right) V_{in} \qquad\qquad (15\text{--}26)$$

$$V_{out} = IR \qquad\qquad (15\text{--}27)$$

$$\mathbf{V}_{out} = V_{out}\angle-\phi \qquad\qquad (15\text{--}28)$$

Pulse Response:

$$i = \frac{V_p}{R}\left(1 - e^{-t_w\,R/L}\right) \qquad\qquad (15\text{--}29)$$

Summary

1. A sinusoidal voltage applied to an *RL* circuit produces a sinusoidal current.
2. Current lags voltage in an *RL* circuit.
3. In a series *RL* circuit, the total impedance is the phasor sum of the resistance and the inductive reactance.
4. The resistor voltage is always in phase with the current.
5. The inductor voltage always leads the current by 90 degrees.
6. The phase angle between the applied voltage and the current is dependent on the relative values of R and X_L.
7. When $X_L = R$, the phase angle is 45 degrees.
8. The average power in an ideal inductor is zero. The power loss is due only to the winding resistance in a practical inductor.
9. An inductor stores energy on the positive half of the ac cycle and then returns that stored energy to the source on the negative half-cycle.

10. When a sinusoidal source drives an *RL* circuit, part of the total power delivered by the source is resistive power (average power), and part of it is reactive power.

11. Reactive power is the rate at which energy is stored by an inductor.

12. The total power being transferred between the source and the circuit is the combination of average power and reactive power and is called the *apparent power*. Its unit is the volt-ampere (VA).

13. The power factor is the cosine of the phase angle between the source voltage and the total current.

14. In an *RL* lead network, the output voltage is across the inductor.

15. In an *RL* lag network, the output voltage is across the resistor.

16. The pulse width must equal at least five time constants for the current in an *RL* circuit to reach its maximum possible value.

17. In an *RL* circuit, the current and voltages change exponentially in response to a pulse input.

Self-Test

1. When a sinusoidal voltage is applied to an *RL* circuit, what is the wave shape of the voltage across the inductor?

2. In a certain series *RL* circuit, R = 3.3 kΩ and L = 20 mH. Express the total impedance in rectangular form when f = 10 kHz. In polar form.

3. In Problem 2, if the frequency is halved, what is the magnitude of the total impedance? What is the phase angle?

4. In Figure 15–48, at what frequency does a 45-degree phase angle between the source voltage and the current occur?

5. In a given series *RL* circuit, the inductance is 200 μH. At 5 kHz, what value of resistance will produce a 60-degree phase angle?

6. In a certain series *RL* circuit, R = 500 Ω and X_L = 100 Ω. What is the phase angle?

7. Determine the current, resistor voltage, and inductor voltage in Figure 15–49. Express each in polar form.

8. Determine the impedance in polar form for each parallel *RL* circuit in Figure 15–50.

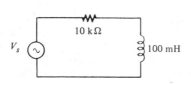

V_s 10 kΩ 100 mH

FIGURE 15–48

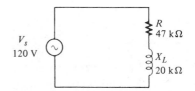

V_s
120 V

R
47 kΩ

X_L
20 kΩ

FIGURE 15–49

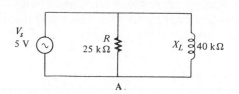

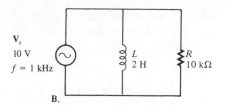

FIGURE 15–50

9. Find the total current and each branch current expressed in polar form for Figure 15–50 when $\mathbf{V}_s = 5\angle 0°$ V.

10. Determine the total impedance in Figure 15–51.

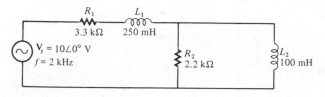

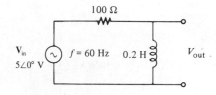

FIGURE 15–51   **FIGURE 15–52**

11. By what angle does the output lead the input in Figure 15–52?

12. What is the magnitude of the output voltage in Figure 15–52?

13. For the lag network in Figure 15–53, determine the magnitude of the output voltage and the angle by which the output lags the input.

14. For Figure 15–52, is the output voltage greater at 2 kHz or at 1 kHz?

15. Determine each of the following quantities for Figure 15–53 when the frequency is 1 kHz.
    (a) P_{avg} (b) P_r (c) P_a (d) power factor

16. A 10-V pulse is applied to a series RL circuit. The pulse width equals 1τ. To what value does the resistor voltage increase during the pulse? What are the maximum and minimum inductor voltages?

17. Determine the output voltage and sketch its shape for the circuit in Figure 15–54. A single input pulse is applied as shown.

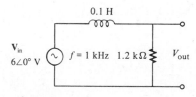

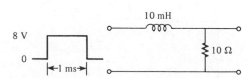

FIGURE 15–53   **FIGURE 15–54**

18. A 5-V pulse with a width of 5 μs is applied to a series *RL* circuit with $\tau = 10$ μs. Assuming that the current is initially zero, to what value will the current increase by the end of the pulse? The resistance is 5 kΩ.

Problems

Section 15–1

15–1. A 15-kHz sinusoidal voltage is applied to a series *RL* circuit. What is the frequency of *I*, V_R, and V_L?

15–2. What are the wave shapes of *I*, V_R, and V_L in Problem 15–1?

Section 15–2

15–3. Express the total impedance of each circuit in Figure 15–55 in both polar and rectangular forms.

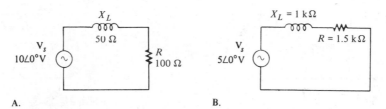

A.   B.

FIGURE 15–55

15–4. Determine the impedance magnitude and phase angle in each circuit in Figure 15–56.

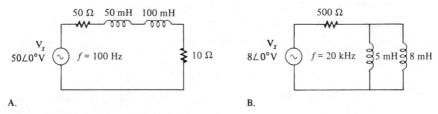

A.   B.

FIGURE 15–56

15–5. In Figure 15–57, determine the impedance at each of the following frequencies:
(a) 100 Hz   (b) 500 Hz   (c) 1 kHz   (d) 2 kHz

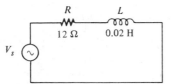

FIGURE 15–57

15-6. Determine the values of R and X_L in a series RL circuit for the following values of total impedance.

(a) $\mathbf{Z} = 20\ \Omega + j45\ \Omega$ (b) $\mathbf{Z} = 500\angle35°\ \Omega$
(c) $\mathbf{Z} = 2.5\angle72.5°\ k\Omega$ (d) $\mathbf{Z} = 998\angle45°\ \Omega$

Section 15-3

15-7. Express the current in polar form for each circuit of Figure 15-55.

15-8. Calculate the total current in each circuit of Figure 15-56, and express in polar form.

15-9. Determine θ for the circuit in Figure 15-58.

15-10. If the inductance in Figure 15-58 is doubled, does θ increase or decrease, and by how many degrees?

15-11. Sketch the wave forms for $\mathbf{V}_s$, $\mathbf{V}_R$, and $\mathbf{V}_L$ in Figure 15-58. Show the proper phase relationships.

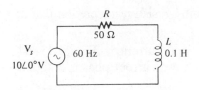

FIGURE 15-58   **FIGURE 15-59**

15-12. For the circuit in Figure 15-59, find the rms values for V_R and V_L for each of the following frequencies:

(a) 60 Hz   (b) 200 Hz   (c) 500 Hz   (d) 1 kHz

Section 15-4

15-13. What is the impedance expressed in polar form for the circuit in Figure 15-60?

15-14. Repeat Problem 15-13 for the following frequencies:

(a) 1.5 kHz   (b) 3 kHz   (c) 5 kHz   (d) 10 kHz

15-15. At what frequency does X_L equal R in Figure 15-60?   2387.32 Hz

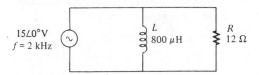
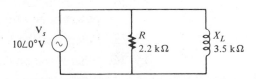

FIGURE 15-60   **FIGURE 15-61**

$\mathbf{Z} = .454 - j .285$

$\mathbf{Y} = .536 \angle -32.1$

Section 15-5

15-16. Find the total current and each branch current in Figure 15-61.

$\mathbf{Z} = 1.865 \angle 32.1$

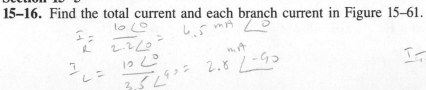

$Y = 1.78 \, \text{m}\mho - j \, 3.183 \quad I_e = 89.3 \, \text{mA} \angle 0°$
$3.647 \angle -60.8 \quad I_L = 159.1 \, \text{mA} \angle -90$
$Z = 274.2 \angle 60.8 \quad I_T = 182 \, \text{mA} \angle -60.7$

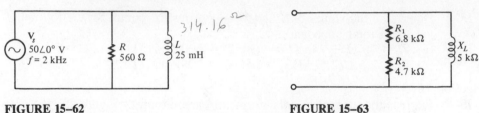

FIGURE 15–62   **FIGURE 15–63**

15–17. Determine the following quantities in Figure 15–62:
   (a) **Z**   (b) I_R (c) I_L (d) I_T (e) θ

15–18. Repeat Problem 15–17 for $R = 50 \, \Omega$ and $L = 330 \, \mu H$.

15–19. Convert the circuit in Figure 15–63 to an equivalent series form.

Section 15–6
15–20. Determine the voltages in polar form across each element in Figure 15–64. Sketch the voltage phasor diagram.

15–21. Is the circuit in Figure 15–64 predominantly resistive or predominantly inductive?

15–22. Find the current in each branch and the total current in Figure 15–64. Express the currents in polar form. Sketch the current phasor diagram.

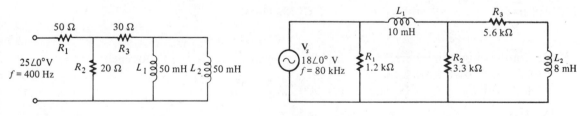

FIGURE 15–64   **FIGURE 15–65**

15–23. For the circuit in Figure 15–65, determine the following:
   (a) I_T (b) θ (c) V_{R1} (d) V_{R2}
   (e) V_{R3} (f) V_{L1} (g) V_{L2}

Section 15–7
15–24. In a certain *RL* circuit, the average power is 100 mW, and the reactive power is 340 mVAR. What is the apparent power?

15–25. Determine the average power and the reactive power in Figure 15–58.

15–26. What is the power factor in Figure 15–61?

15–27. Determine P_{avg}, P_r, P_a, and *PF* for the circuit in Figure 15–65. Sketch the power triangle.

Section 15–8
15–28. For the lag network in Figure 15–66, determine the phase lag of the output voltage with respect to the input for the following frequencies:
   (a) 1 Hz   (b) 100 Hz   (c) 1 kHz   (d) 10 kHz

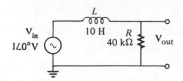

FIGURE 15-66

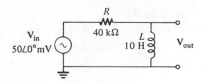

FIGURE 15-67

15-29. Draw the response curve for the circuit in Figure 15-66. Show the output voltage versus frequency in 1-kHz increments from 0 Hz to 5 kHz.

15-30. Repeat Problem 15-28 for the lead network in Figure 15-67.

15-31. Using the same procedure as in Problem 15-29, draw the response curve for Figure 15-67.

15-32. Sketch the voltage phasor diagram for each circuit in Figures 15-66 and 15-67 for a frequency of 8 kHz.

Section 15-9

15-33. Sketch the output voltage with respect to the input for the circuit in Figure 15-68.

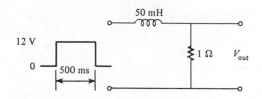

FIGURE 15-68

FIGURE 15-69

15-34. For the circuit in Figure 15-69:
  (a) What is τ?   **(b)** Sketch the output voltage.

15-35. Repeat Problem 15-34 for the circuit in Figure 15-70.

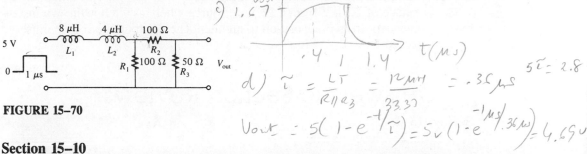

FIGURE 15-70

Section 15-10

15-36. Determine the voltage across each element in Figure 15-65 if L_1 were open.

15-37. Determine the output voltage in Figure 15-70 for each of the following failure modes:
  (a) L_1 open   **(b)** L_2 open   **(c)** R_1 open   **(d)** a short across R_2

Section 15–11

15–38. Modify the program in Section 15–11 to compute and tabulate current and average power in addition to phase shift and output voltage.

15–39. Write a program similar to the program in Section 15–11 for an *RL* lag network.

Advanced Problems

15–40. For the circuit in Figure 15–71, determine the following:
   (a) I_T **(b)** V_{L1} **(c)** V_{ab} **(d)** P_{avg}

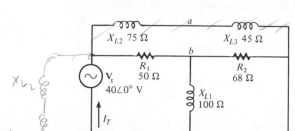

$$V_{ab} = V_a - V_b$$
$$= U \times L_3 - V_{\times L1}$$

FIGURE 15–71

15–41. Determine the phase shift and attenuation from the input to the output for the ladder network in Figure 15–72.

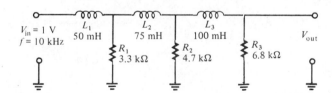

FIGURE 15–72

15–42. Design an ideal inductive switching circuit that will provide a momentary voltage of 2.5 kV from a 12-V dc source when a switch is thrown instantaneously from one position to another. The drain on the source must not exceed 1 A.

Answers to Section Reviews

Section 15–1:
1. 1 kHz. **2.** Closer to zero.

Section 15–2:
1. $R = 150\ \Omega$, $X_L = 220\ \Omega$. **2.** 33 kΩ + j50 kΩ, 59.9$\angle$56.58° kΩ.

Section 15–3:
1. 3.61 V. **2.** 56.31°. **3.** X_L increases; Z increases; θ increases.

Section 15–4:
1. 0.002 S. **2.** 0.024 S. **3.** Lags, 33.69°.

Section 15–5:
1. 32 mA. **2.** $23.32\angle-59.04°$ mA; the input voltage. **3.** $-90°$.

Section 15–6:
1. $Z = 494.6\angle59.86°\ \Omega$. **2.** $I_T = 20.22\angle-59.86°$ mA.

Section 15–7:
1. Resistor. **2.** 0.643. **3.** $P_{avg} = 4.7$ W; $P_r = 6.2$ VAR; $P_a = 7.78$ VA.

Section 15–8:
1. 81.87°. **2.** 9.9 V. **3.** Resistor.

Section 15–9:
1. Yes. **2.** $i = 0$ A, $v_R = 0$ V, $v_L = 12$ V. **3.** $i = 8$ mA, $v_R = 12$ V, $v_L = 0$ V.

Section 15–10:
1. Shorted windings reduce L and thereby reduce X_L at any given frequency. **2.** I_T decreases, V_{R1} decreases, V_{R2} increases.

In this chapter, the analysis methods learned in the last two chapters are extended to the coverage of circuits with combinations or resistive, inductive, and capacitive elements. Both series and parallel *RLC* circuits, plus combinations of both, are studied.

The concept of resonance is introduced in this chapter. Resonance is an important aspect of electronic applications. It is the basis for frequency selectivity in communications. For example, the ability of a radio or television receiver to select a certain frequency transmitted by a particular station and to eliminate frequencies from other stations is based on the principle of resonance.

Resonance occurs under certain specific conditions in circuits that have combinations of inductance and capacitance. The conditions that produce resonance and the characteristics of resonant circuits are covered in this chapter.

Specifically, you will learn:
1. How to determine the impedance of a series *RLC* circuit.
2. How to analyze a series *RLC* circuit for current and voltage relationships, as well as for power.
3. The meaning of *series resonance*.
4. The conditions that produce resonance in series *RLC* circuits.
5. How to determine the frequency, current, and impedance at resonance in a parallel *RLC* circuit.
6. How to determine the impedance of a parallel *RLC* circuit.
7. How to analyze a parallel *RLC* circuit for currents and voltage relationships.
8. The meaning of parallel resonance, and how it differs from series resonance.
9. The conditions that produce parallel resonance.
10. How to determine the frequency, current, and impedance at resonance in a parallel *RLC* circuit.
11. How to determine the bandwidth and *Q* of both series and parallel resonant circuits.

16
RLC CIRCUITS
AND RESONANCE

16–1

A series *RLC* circuit is shown in Figure 16–1. It contains resistance, inductance, and capacitance. As you know, $\mathbf{X}_L$ causes the total current to lag the applied voltage. $\mathbf{X}_C$ has the opposite effect: It causes the current to lead the voltage. Thus $\mathbf{X}_L$ and $\mathbf{X}_C$ tend to offset each other. When they are equal, they cancel, and the total reactance is zero. In any case, the magnitude of the total reactance in the series circuit is

$$X_T = X_L - X_C \tag{16–1}$$

The total impedance for the series *RLC* circuit is stated in rectangular form in Equation (16–2) and in polar form in Equation (16–3).

$$\mathbf{Z} = R + jX_L - jX_C \tag{16–2}$$

$$\mathbf{Z} = \sqrt{R^2 + (X_L - X_C)^2} \angle \arctan\left(\frac{X_T}{R}\right) \tag{16–3}$$

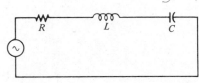

FIGURE 16–1   *Series RLC circuit.*

Example 16–1

Determine the total impedance in Figure 16–2. Express it in both rectangular and polar forms.

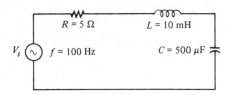

V_s ~ $f = 100$ Hz

$R = 5\ \Omega$ $L = 10$ mH

$C = 500\ \mu$F

FIGURE 16–2

Solution:

First find X_C and X_L:

$$X_C = \frac{1}{2\pi fC} = \frac{1}{2\pi(100 \text{ Hz})(500\ \mu\text{F})} = 3.18\ \Omega$$

$$X_L = 2\pi fL = 2\pi(100 \text{ Hz})(10 \text{ mH}) = 6.28\ \Omega$$

In this case, X_L is greater than X_C, and thus the circuit is more inductive than capacitive. The magnitude of the total reactance is

$$X_T = X_L - X_{C_{,}} = 6.28 \ \Omega - 3.18 \ \Omega = 3.1 \ \Omega \quad \text{inductive}$$

The impedance in rectangular form is

$$\mathbf{Z} = R + jX_L - jX_C = 5 \ \Omega + j6.28 \ \Omega - j3.18 \ \Omega$$
$$= 5 \ \Omega + j3.1 \ \Omega$$

The impedance in polar form is

$$\mathbf{Z} = \sqrt{R^2 + X_T^2} \ \angle \arctan\!\left(\frac{X_T}{R}\right)$$

$$= \sqrt{(5 \ \Omega)^2 + (3.1 \ \Omega)^2} \ \angle \arctan\!\left(\frac{3.1}{5}\right)$$

$$= 5.88\angle31.8° \ \Omega$$

(handwritten margin notes:)
$X_L \ X_C \ Z \ circuit$
ϕ
$b \ \infty \ R - jX_C \ cap$
$\uparrow \ \downarrow R + j(X_L - X_C)$
$X_L = X_C$
$\delta > \delta r \quad \infty$
$\infty \ \phi \ \infty \ inductance$

As you have seen, when the inductive reactance is greater than the capacitive reactance, the circuit appears inductive; so the current lags the applied voltage. When the capacitive reactance is greater, the circuit appears capacitive, and the current leads the applied voltage.

Review for 16–1

1. In a given series *RLC* circuit, X_C is 150 Ω and X_L is 80 Ω. What is the total reactance in ohms? Is it inductive or capacitive?

2. Determine the impedance in polar form for the circuit in Problem 1 when $R = 45 \ \Omega$. What is the magnitude of the impedance? What is the phase angle? Is the current leading or lagging the applied voltage?

ANALYSIS OF SERIES *RLC* CIRCUITS

16–2

Recall that capacitive reactance varies inversely with frequency and inductive reactance varies directly. With this in mind, you can see that for a typical series *RLC* circuit the total reactance behaves as follows: Starting at a very low frequency, X_C is high, and X_L is low, and the circuit is predominantly capacitive. As the frequency is increased, X_C decreases and X_L increases until a value is reached where $X_C = X_L$ and the two reactances cancel, making the circuit purely resistive. This condition is *series resonance* and will be studied in a later section. As the frequency is increased further, X_L becomes greater than X_C, and the circuit is predominantly inductive. The following example illustrates how the impedance and phase angle change as the input frequency is varied.

Example 16–2

For each of the following input frequencies, find the impedance in polar form for the circuit in Figure 16–3. Note the change in magnitude and phase angle.
(a) $f = 1$ kHz
(b) $f = 2$ kHz
(c) $f = 3.5$ kHz
(d) $f = 5$ kHz

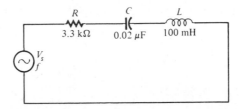

FIGURE 16–3

Solution:

(a) At $f = 1$ kHz:

$$X_C = \frac{1}{2\pi(1\ \text{kHz})(0.02\ \mu\text{F})} = 7.96\ \text{k}\Omega$$

$$X_L = 2\pi(1\ \text{kHz})(100\ \text{mH}) = 628.3\ \Omega$$

The circuit is clearly capacitive, and the impedance is

$$\mathbf{Z} = \sqrt{(3.3\ \text{k}\Omega)^2 + (7.96\ \text{k}\Omega - 628.3\ \Omega)^2}\ \angle -\arctan\left(\frac{7.33\ \text{k}\Omega}{3.3\ \text{k}\Omega}\right)$$

$$= 8.04\angle -65.76°\ \text{k}\Omega$$

(b) At $f = 2$ kHz:

$$X_C = \frac{1}{2\pi(2\ \text{kHz})(0.02\ \mu\text{F})} = 3.98\ \text{k}\Omega$$

$$X_L = 2\pi(2\ \text{kHz})(100\ \text{mH}) = 1.26\ \text{k}\Omega$$

The circuit is still capacitive, and the impedance is

$$\mathbf{Z} = \sqrt{(3.3\ \text{k}\Omega)^2 + (3.98\ \text{k}\Omega - 1.26\ \text{k}\Omega)^2}\ \angle -\arctan\left(\frac{2.72\ \text{k}\Omega}{3.3\ \text{k}\Omega}\right)$$

$$= 4.28\angle -39.5°\ \text{k}\Omega$$

(c) At $f = 3.5$ kHz:

$$X_C = \frac{1}{2\pi(3.5\ \text{kHz})(0.02\ \mu\text{F})} = 2.27\ \text{k}\Omega$$

$$X_L = 2\pi(3.5\ \text{kHz})(100\ \text{mH}) = 2.2\ \text{k}\Omega$$

The circuit is very close to being purely resistive but is still slightly capacitive.

$$\mathbf{Z} = \sqrt{(3.3 \text{ k}\Omega)^2 + (2.27 \text{ k}\Omega - 2.2 \text{ k}\Omega)^2} \angle -\arctan\left(\frac{0.07 \text{ k}\Omega}{3.3 \text{ k}\Omega}\right)$$
$$= 3.3\angle -1.2° \text{ k}\Omega$$

(d) At $f = 5$ kHz:

$$X_C = \frac{1}{2\pi(5 \text{ kHz})(0.02 \text{ }\mu\text{F})} = 1.59 \text{ k}\Omega$$

$$X_L = 2\pi(5 \text{ kHz})(100 \text{ mH}) = 3.14 \text{ k}\Omega$$

The circuit is now predominantly inductive.

$$\mathbf{Z} = \sqrt{(3.3 \text{ k}\Omega)^2 + (3.14 \text{ k}\Omega - 1.59 \text{ k}\Omega)^2} \angle \arctan\left(\frac{1.55 \text{ k}\Omega}{3.3 \text{ k}\Omega}\right)$$
$$= 3.64\angle 25.16° \text{ k}\Omega$$

Notice how the circuit changed from capacitive to inductive as the frequency increased. The phase condition changed from the current leading to the current lagging as indicated by the sign of the angle. It is interesting to note that the impedance magnitude decreased to a minimum approximately equal to the resistance and then began increasing again.

In the next example, Ohm's law is used to find the current and voltages in the series *RLC* circuit.

Example 16–3

Find the current and the voltages across each element in Figure 16–4. Express each quantity in polar form, and draw a complete voltage phasor diagram.

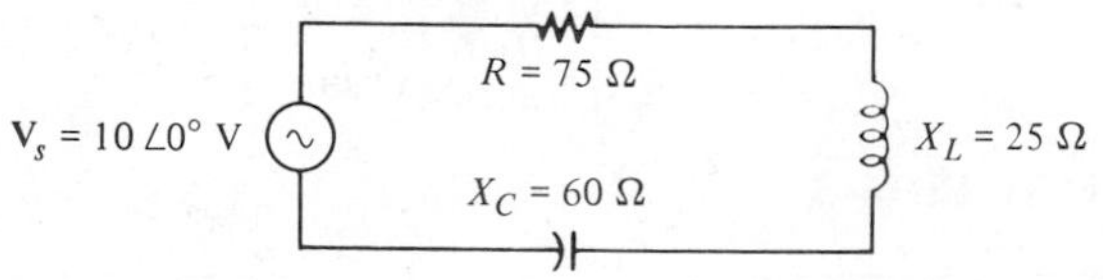

FIGURE 16–4

Solution:

First find the total impedance:

$$\mathbf{Z} = R + jX_L - jX_C = 75 \text{ }\Omega + j25 \text{ }\Omega - j60 \text{ }\Omega$$
$$= 75 \text{ }\Omega - j35 \text{ }\Omega$$

Example 16–3 (continued)

Convert to polar form for convenience in applying Ohm's law:

$$\mathbf{Z} = \sqrt{(75 \ \Omega)^2 + (35 \ \Omega)^2} \ \angle -\arctan\left(\frac{35}{75}\right)$$

$$= 82.76\angle -25° \ \Omega$$

Apply Ohm's law to find the current:

$$\mathbf{I} = \frac{\mathbf{V}_s}{\mathbf{Z}} = \frac{10\angle 0° \ \text{V}}{82.76\angle -25° \ \Omega} = 0.121\angle 25° \ \text{A}$$

Now apply Ohm's law to find the voltages across R, L, and C:

$$\mathbf{V}_R = \mathbf{IR} = (0.121\angle 25° \ \text{A})(75\angle 0° \ \Omega)$$
$$= 9.075\angle 25° \ \text{V}$$

$$\mathbf{V}_L = \mathbf{IX}_L = (0.121\angle 25° \ \text{A})(25\angle 90° \ \Omega)$$
$$= 3.025\angle 115° \ \text{V}$$

$$\mathbf{V}_C = \mathbf{IX}_C = (0.121\angle 25° \ \text{A})(60\angle -90° \ \Omega)$$
$$= 7.26\angle -65° \ \text{V}$$

The phasor diagram is shown in Figure 16–5. Notice that $\mathbf{V}_L$ is leading $\mathbf{V}_R$ by 90 degrees, and $\mathbf{V}_C$ is lagging $\mathbf{V}_R$ by 90 degrees. Also, there is a 180-degree phase difference between $\mathbf{V}_L$ and $\mathbf{V}_C$. If the current phasor were shown, it would be at the same angle as $\mathbf{V}_R$. The current is leading $\mathbf{V}_s$, the source voltage, by 25 degrees, indicating a capacitive circuit ($X_C > X_L$).

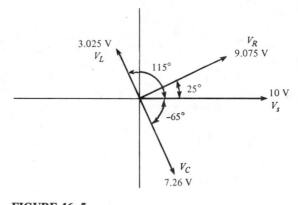

FIGURE 16–5

Review for 16–2

1. The following voltages occur in a certain series *RLC* circuit. Determine the source voltage: $\mathbf{V}_R = 24\angle 30° \ \text{V}$, $\mathbf{V}_L = 15\angle 120° \ \text{V}$, and $\mathbf{V}_C = 45\angle -60° \ \text{V}$.

2. When $R = 10\ \Omega$, $X_C = 18\ \Omega$, and $X_L = 12\ \Omega$, does the current lead or lag the applied voltage?

3. Determine the total reactance in Problem 2.

SERIES RESONANCE

16–3

In a series *RLC* circuit, *series resonance* occurs when $X_L = X_C$. The frequency at which resonance occurs is called the *resonant frequency*, f_r. Figure 16–6 illustrates the series resonant condition.

In a series resonant circuit, the total impedance is

$$\mathbf{Z} = R + jX_L - jX_C$$

Since $X_L = X_C$, the j terms cancel, and the *impedance is purely resistive*. These resonant conditions are stated in the following equations.

$$X_L = X_C \qquad\qquad \textbf{(16–4)}$$

$$Z = R \qquad\qquad \textbf{(16–5)}$$

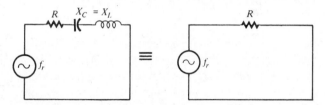

FIGURE 16–6 *Series resonance.*

Example 16–4

For the series *RLC* circuit in Figure 16–7, determine X_C and $\mathbf{Z}$ at resonance.

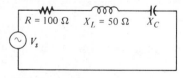

FIGURE 16–7

Solution:

$$X_L = X_C \quad \text{at the resonant frequency}$$

Thus,

$$X_C = 50\ \Omega$$

Example 16–4 (continued)

$$\mathbf{Z} = R + jX_L - jX_C = 100 \ \Omega + j50 \ \Omega - j50 \ \Omega$$
$$= 100\angle0° \ \Omega$$

The impedance is equal to the resistance because the reactances are equal in magnitude and therefore cancel.

Series Resonant Frequency

For a given series *RLC* circuit, resonance happens at only one specific frequency. A formula for this resonant frequency is developed as follows:

$$X_L = X_C.$$

Substituting the reactance formulas, we have

$$2\pi f_r L = \frac{1}{2\pi f_r C}$$

Solving for f_r:

$$(2\pi f_r L)(2\pi f_r C) = 1$$
$$4\pi^2 f_r^2 LC = 1$$
$$f_r^2 = \frac{1}{4\pi^2 LC}$$

Taking the square root of both sides:

$$f_r = \frac{1}{2\pi\sqrt{LC}} \qquad\qquad\qquad \textbf{(16–6)}$$

Example 16–5

Find the series resonant frequency for the circuit in Figure 16–8.

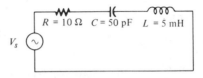

FIGURE 16–8

Solution:

$$f_r = \frac{1}{2\pi\sqrt{LC}} = \frac{1}{2\pi\sqrt{(5 \text{ mH})(50 \text{ pF})}}$$

$$= 318 \text{ kHz}$$

Series *RLC* Impedance

At frequencies below f_r, $X_C > X_L$; thus, the circuit is capacitive. At the resonant frequency, $X_C = X_L$, so the circuit is purely resistive. At frequencies above f_r, $X_L > X_C$; thus, the circuit is inductive.

The impedance magnitude is minimum at resonance $(Z = R)$ and in- creases in value above and below the resonant point. The graph in Figure 16–9 illustrates how impedance changes with frequency.

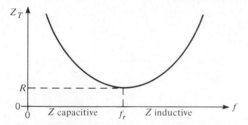

FIGURE 16–9 *Series RLC impedance as a function of frequency.*

Example 16–6

For the circuit in Figure 16–10, determine the impedance magnitude at resonance, at 1000 Hz below resonance, and at 1000 Hz above resonance.

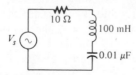

FIGURE 16–10

Solution:

$$f_r = \frac{1}{2\pi\sqrt{LC}} = \frac{1}{2\pi\sqrt{(100 \text{ mH})(0.01 \ \mu\text{F})}}$$

$$= 5.03 \text{ kHz}$$

Example 16–6 (continued)

At 1000 Hz below f_r:

$$f_r - 1 \text{ kHz} = 4.03 \text{ kHz}$$

At 1000 Hz above f_r:

$$f_r + 1 \text{ kHz} = 6.03 \text{ kHz}$$

The impedance at resonance is equal to R:

$$Z = 10 \ \Omega$$

The impedance at $f_r - 1$ kHz:

$$\begin{aligned}
Z &= \sqrt{R^2 + (X_L - X_C)^2} \\
&= \sqrt{(10 \ \Omega)^2 + (2.53 \text{ k}\Omega - 3.95 \text{ k}\Omega)^2} \\
&= 1.42 \text{ k}\Omega
\end{aligned}$$

Notice that X_C is greater than X_L; so Z is capacitive.
The impedance at $f_r + 1$ kHz:

$$\begin{aligned}
Z &= \sqrt{R^2 + (X_L - X_C)^2} \\
&= \sqrt{(10 \ \Omega)^2 + (3.79 \text{ k}\Omega - 2.64 \text{ k}\Omega)^2} \\
&= 1.15 \text{ k}\Omega
\end{aligned}$$

X_L is greater than X_C; so Z is inductive.

Current and Voltages

At the series resonant frequency, *the current is maximum* ($I_{\max} = V_s/R$). Above and below resonance, the current decreases because the impedance increases. A *response curve* showing the plot of current versus frequency is shown in Figure 16–11A.

The resistor voltage, V_R, follows the current and is maximum (equal to V_s) at resonance and zero at $f = 0$ and at $f = \infty$, as shown in Figure 16–11B. The general shapes of the V_L and V_C curves are indicated in Figure 16–11C and D. Notice that $V_C = V_s$ when $f = 0$, because the capacitor appears open.

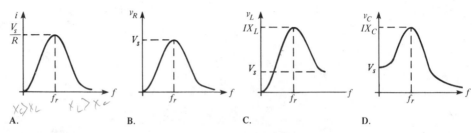

A.   B.   C.   D.

FIGURE 16–11 *Current and voltage magnitudes as a function of frequency in a series RLC circuit. (Note that the peak magnitudes are not shown to scale with respect to each other.)*

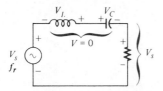

FIGURE 16–12   *Series RLC circuit at resonance.*

Also notice that V_L approaches V_s as f approaches infinity, because the inductor appears open.

The voltages are maximum at resonance but drop off above and below f_r. The voltages across L and C at resonance are exactly *equal in magnitude but 180 degrees out of phase; so they cancel.* Thus, the total voltage across both L and C is zero, and $V_R = V_s$ at resonance, as indicated in Figure 16–12. Individually, V_L and V_C can be much greater than the source voltage, as you will see later. Keep in mind that V_L and V_C are always opposite in polarity regardless of the frequency, but only at resonance are their magnitudes equal.

Example 16–7

Find I, V_R, V_L, and V_C at resonance in Figure 16–13.

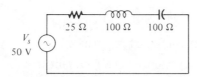

FIGURE 16–13

Solution:

At resonance, I is maximum and equal to V_s/R:

$$I = \frac{V_s}{R} = \frac{50 \text{ V}}{25 \ \Omega} = 2 \text{ A}$$

Applying Ohm's law, the following voltage magnitudes are obtained:

$$V_R = IR = (2 \text{ A})(25 \ \Omega) = 50 \text{ V}$$

$$V_L = IX_L = (2 \text{ A})(100 \ \Omega) = 200 \text{ V}$$

$$V_C = IX_C = (2 \text{ A})(100 \ \Omega) = 200 \text{ V}$$

Notice that all of the source voltage is dropped across the resistor. Also, of course, V_L and V_C are equal in magnitude but opposite in phase. This causes these voltages to cancel, making the *total* reactive voltage drop to zero.

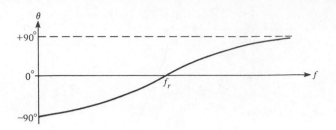

FIGURE 16–14 *Phase angle between current and voltage versus frequency in a series RLC circuit.*

Phase Characteristic

At resonance, *the current and the source voltage are in phase,* because the impedance is purely resistive. For frequencies below resonance, the circuit is capacitive ($X_C > X_L$), and the current *leads* the applied voltage. For frequencies above resonance, the circuit is inductive ($X_C < X_L$), and the current *lags* the applied voltage.

A phase plot is shown in Figure 16–14. Notice that at f_r, the phase angle, θ, is 0 degrees. As the frequency approaches zero, θ approaches -90 degrees. As the frequency becomes very high, approaching infinity, θ approaches $+90$ degrees.

Review for 16–3

1. What is the condition for series resonance?
2. Why is the current maximum at the resonant frequency?
3. Calculate the resonant frequency for $C = 1000$ pF and $L = 1000\ \mu$H.
4. In Problem 3, is the circuit inductive or capacitive at 50 kHz?

IMPEDANCE OF PARALLEL *RLC* CIRCUITS

16–4   A parallel *RLC* circuit is shown in Figure 16–15. The total impedance can be calculated using the sum-of-reciprocals method, just as was done for circuits with resistors in parallel.

$$\frac{1}{\mathbf{Z}} = \frac{1}{R\angle 0°} + \frac{1}{X_L\angle 90°} + \frac{1}{X_C\angle -90°}$$

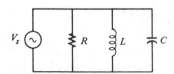

FIGURE 16–15 *Parallel RLC circuit.*

or

$$Z = \cfrac{1}{\cfrac{1}{R\angle 0°} + \cfrac{1}{X_L\angle 90°} + \cfrac{1}{X_C\angle -90°}}$$ (16–7)

Example 16–8

Find **Z** in polar form in Figure 16–16.

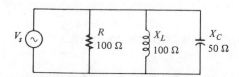

FIGURE 16–16

Solution:

$$\frac{1}{Z} = \frac{1}{R\angle 0°} + \frac{1}{X_L\angle 90°} + \frac{1}{X_C\angle -90°}$$

$$= \frac{1}{100\angle 0° \ \Omega} + \frac{1}{100\angle 90° \ \Omega} + \frac{1}{50\angle -90° \ \Omega}$$

Applying the rule for division of polar numbers:

$$\frac{1}{Z} = 0.01\angle 0° \ \text{S} + 0.01\angle -90° \ \text{S} + 0.02\angle 90° \ \text{S}$$

Recall that the sign of the denominator angle changes when dividing. Now, converting each term to its rectangular equivalent gives

$$\frac{1}{Z} = 0.01 \ \text{S} - j0.01 \ \text{S} + j0.02 \ \text{S}$$

$$= 0.01 \ \text{S} + j0.01 \ \text{S}$$

Taking the reciprocal to get **Z**, we get

$$Z = \frac{1}{0.01 \ \text{S} + j0.01 \ \text{S}}$$

$$= \frac{1}{\sqrt{(0.01 \ \text{S})^2 + (0.01 \ \text{S})^2} \ \angle \arctan\left(\dfrac{0.01}{0.01}\right)}$$

$$= \frac{1}{0.01414\angle 45° \ \text{S}}$$

$$= 70.7\angle -45° \ \Omega$$

Example 16–8 (continued)

The negative angle shows that the circuit is capacitive. This may surprise you, since $X_L > X_C$. However, in a parallel circuit, the smaller quantity has the greater effect on the total current. Just as in the case of all resistances in parallel, the smaller draws more current and has the greater effect on the total R.

　　　　In this circuit, the total current leads the total voltage by a phase angle of 45 degrees.

Conductance, Susceptance, and Admittance

As you already know, the reciprocal of resistance is conductance, G. In complex form, it is expressed as

$$\mathbf{G} = \frac{1}{R\angle 0°} = G\angle 0° \tag{16–8}$$

The reciprocal of reactance is called *susceptance*. The complex forms of capacitive susceptance and inductive susceptance are stated in the following equations.

$$\mathbf{B}_C = \frac{1}{X_C\angle -90°} = B_C\angle 90° = jB_C \tag{16–9}$$

$$\mathbf{B}_L = \frac{1}{X_L\angle 90°} = B_L\angle -90° = -jB_L \tag{16–10}$$

The reciprocal of impedance is admittance. The complex forms are

$$\mathbf{Y} = \frac{1}{Z\angle \pm\theta} = Y\angle \mp\theta = G + jB_C - jB_L \tag{16–11}$$

As you know, the unit of each of these quantities is the siemen (S).

Example 16–9

Determine the conductance, capacitive susceptance, inductive susceptance, and total admittance in Figure 16–17.

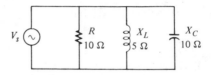

FIGURE 16–17

Solution:

$$G = \frac{1}{R\angle 0^\circ} = \frac{1}{10\angle 0^\circ\ \Omega} = 0.1\angle 0^\circ\ S$$

$$B_C = \frac{1}{X_C\angle -90^\circ} = \frac{1}{10\angle -90^\circ\ \Omega} = 0.1\angle 90^\circ\ S$$

$$B_L = \frac{1}{X_L\angle 90^\circ} = \frac{1}{5\angle 90^\circ\ \Omega} = 0.2\angle -90^\circ\ S$$

$$Y = G + jB_C - jB_L = 0.1\ S + j0.1\ S - j0.2\ S$$
$$= 0.1\ S - j0.1\ S = 0.1414\angle -45^\circ\ S$$

From **Y**, we can get **Z**:

$$Z = \frac{1}{Y} = \frac{1}{0.1414\angle -45^\circ\ S} = 7.07\angle 45^\circ\ \Omega$$

Review for 16–4

1. In a certain parallel *RLC* circuit, the capacitive reactance is 60 Ω, and the inductive reactance is 100 Ω. Is the circuit predominantly capacitive or inductive?

2. Determine the admittance of a parallel circuit in which $R = 1\ k\Omega$, $X_C = 500\ \Omega$, and $X_L = 1.2\ k\Omega$.

3. In Problem 2, what is the impedance?

ANALYSIS OF PARALLEL AND SERIES-PARALLEL *RLC* CIRCUITS

16–5

As you have seen, the smaller reactance in parallel dominates, because it results in the larger branch current. At low frequencies, the inductive reactance is less than the capacitive reactance; therefore, the circuit is inductive. As the frequency is increased, X_L increases and X_C decreases until a value is reached where $X_L = X_C$. This is the point of parallel resonance. As the frequency is increased further, X_C becomes smaller than X_L, and the circuit becomes capacitive. During this variation of frequency, the current goes from lagging to leading in its relation to the applied voltage. The following example illustrates an analysis approach to parallel *RLC* circuits.

Example 16–10

Find each branch current and the total current in Figure 16–18.

606

Example 16–10 (continued)

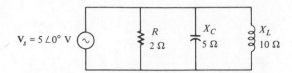

FIGURE 16–18

Solution:

Each branch current can be found using Ohm's law:

$$\mathbf{I}_R = \frac{\mathbf{V}_s}{\mathbf{R}} = \frac{5\angle 0°\text{ V}}{2\angle 0°\ \Omega}$$

$$= 2.5\angle 0°\text{ A}$$

$$\mathbf{I}_C = \frac{\mathbf{V}_s}{\mathbf{X}_C} = \frac{5\angle 0°\text{ V}}{5\angle -90°\ \Omega}$$

$$= 1\angle 90°\text{ A}$$

$$\mathbf{I}_L = \frac{\mathbf{V}_s}{\mathbf{X}_L} = \frac{5\angle 0°\text{ V}}{10\angle 90°\ \Omega}$$

$$= 0.5\angle -90°\text{ A}$$

The total current is the phasor sum of the branch currents. By Kirchhoff's law:

$$\mathbf{I}_T = \mathbf{I}_R + \mathbf{I}_C + \mathbf{I}_L$$
$$= 2.5\angle 0°\text{ A} + 1\angle 90°\text{ A} + 0.5\angle -90°\text{ A}$$
$$= 2.5\text{ A} + j1\text{ A} - j0.5\text{ A} = 2.5\text{ A} + j0.5\text{ A}$$
$$= \sqrt{(2.5\text{ A})^2 + (0.5\text{ A})^2}\ \angle\arctan\!\left(\frac{0.5}{2.5}\right)$$
$$= 2.55\angle 11.31°\text{ A}$$

The total current is 2.55 A leading V_s by 11.31°. Figure 16–19 is the current phasor diagram for the circuit.

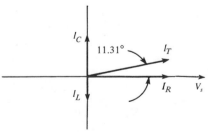

FIGURE 16–19

Series-Parallel Analysis

The analysis of series-parallel combinations basically involves the application of the methods previously covered. The following two examples illustrate typical approaches to these more complex circuits.

Example 16–11

In Figure 16–20, find the voltage across the capacitor in polar form. Is this circuit predominantly inductive or capacitive?

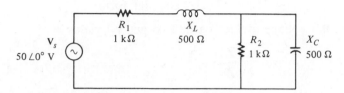

FIGURE 16–20

Solution:

We use the voltage-divider formula in this analysis. The impedance of the series combination of R_1 and X_L is called $\mathbf{Z}_1$.

$$\mathbf{Z}_1 = R_1 + jX_L$$
$$= 1000 \ \Omega + j500 \ \Omega \quad \text{in rectangular form}$$

$$\mathbf{Z}_1 = \sqrt{(1000 \ \Omega)^2 + (500 \ \Omega)^2} \angle \arctan\!\left(\frac{500}{1000}\right)$$

$$= 1118\angle 26.57° \ \Omega \quad \text{in polar form}$$

The impedance of the parallel combination of R_2 and X_C is called $\mathbf{Z}_2$.

$$\mathbf{Z}_2 = \left(\frac{R_2 X_C}{\sqrt{R_2^2 + X_C^2}}\right) \angle\left(-90° + \arctan\!\left(\frac{X_C}{R_2}\right)\right)$$

$$= \left[\frac{(1000 \ \Omega)\,(500 \ \Omega)}{\sqrt{(1000 \ \Omega)^2 + (500 \ \Omega)^2}}\right] \angle\left(-90° + \arctan\!\left(\frac{500}{1000}\right)\right)$$

$$= 447.2\angle -63.4° \ \Omega \quad \text{in polar form}$$

or

$$\mathbf{Z}_2 = 447.2 \cos(-63.4°) + j447.2 \sin(-63.4°)$$
$$= 200.24 \ \Omega - j399.87 \ \Omega \quad \text{in rectangular form}$$

Example 16–11 (continued)

The total impedance $\mathbf{Z}_T$ is

$$\mathbf{Z}_T = \mathbf{Z}_1 + \mathbf{Z}_2$$
$$= (1000 \ \Omega + j500 \ \Omega) + (200.24 \ \Omega - j399.87 \ \Omega)$$
$$= 1200.24 \ \Omega + j100.13 \ \Omega \quad \text{in rectangular form}$$

or

$$\mathbf{Z}_T = \sqrt{(1200.24 \ \Omega)^2 + (100.13 \ \Omega)^2} \ \angle \arctan\left(\frac{100.13}{1200.24}\right)$$

$$= 1204.41 \angle 4.77° \ \Omega \quad \text{in polar form}$$

Applying the voltage divider formula to get $\mathbf{V}_C$ yields

$$\mathbf{V}_C = \left(\frac{\mathbf{Z}_2}{\mathbf{Z}_T}\right)\mathbf{V}_s$$

$$= \left(\frac{447.2\angle -63.4° \ \Omega}{1204.41\angle 4.77° \ \Omega}\right)50\angle 0° \ \text{V}$$

$$= 18.57\angle -68.17° \ \text{V}$$

Therefore, V_C is 18.57 V and lags V_s by 68.17°.

The $+j$ term in $\mathbf{Z}_T$, or the positive angle in its polar form, indicates that the circuit is more inductive than capacitive. However, it is just slightly more inductive, because the angle is small. This result is surprising, because $X_C = X_L = 500 \ \Omega$. However, the capacitor is in *parallel* with a resistor; so the capacitor actually has less effect on the total impedance than does the inductor. Figure 16–21 shows the phasor relationship of $\mathbf{V}_C$ and $\mathbf{V}_s$. Although $X_C = X_L$, this circuit is not at resonance, since the j term of the total impedance is not zero due to the parallel combination of R_2 and X_C. You can see this by noting that the phase angle associated with $\mathbf{Z}_T$ is not zero.

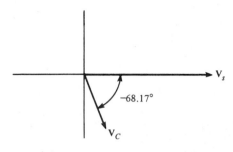

FIGURE 16–21

Example 16–12

For the reactive circuit in Figure 16–22, find the voltage at point B with respect to ground.

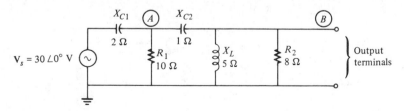

FIGURE 16–22

Solution:

The voltage at point B is the voltage across the open output terminals. Let's use the voltage divider approach. To do so, we must know the voltage at point A first; so we need to find the impedance from point A to ground as a starting point.

The parallel combination of X_L and R_2 is in series with X_{C2}. This combination is in parallel with R_1. We call this impedance from point A to ground, $\mathbf{Z}_A$. To find $\mathbf{Z}_A$, the following steps are taken:

R_2 and X_L in parallel ($\mathbf{Z}_1$):

$$\mathbf{Z}_1 = \left(\frac{R_2 X_L}{\sqrt{R_2^2 + X_L^2}}\right) \angle \left(90° - \arctan\left(\frac{X_L}{R}\right)\right)$$

$$= \left(\frac{(8\ \Omega)(5\ \Omega)}{\sqrt{(8\ \Omega)^2 + (5\ \Omega)^2}}\right) \angle \left(90° - \arctan\left(\frac{5}{8}\right)\right)$$

$$= \left(\frac{40}{9.43}\right) \angle (90° - 32°) = 4.24 \angle 58°\ \Omega$$

$\mathbf{Z}_1$ in series with $\mathbf{X}_{C2}$ ($\mathbf{Z}_2$):

$$\mathbf{Z}_2 = \mathbf{X}_{C2} + \mathbf{Z}_1$$
$$= 1 \angle -90°\ \Omega + 4.24 \angle 58°\ \Omega = -j1\ \Omega + 2.25\ \Omega + j3.6\ \Omega$$
$$= 2.25\ \Omega + j2.6\ \Omega$$

$$= \sqrt{(2.25\ \Omega)^2 + (2.6\ \Omega)^2} \angle \arctan\left(\frac{2.6}{2.25}\right)$$

$$= 3.44 \angle 49.13°\ \Omega$$

Example 16–12 (continued)

R_1 in parallel with $\mathbf{Z}_2$ ($\mathbf{Z}_A$):

$$\mathbf{Z}_A = \frac{(10\angle 0°)(3.44\angle 49.13°)}{10 + 2.25 + j2.6}$$

$$= \frac{34.4\angle 49.13°}{12.25 + j2.6}$$

$$= \frac{34.4\angle 49.13°}{12.52\angle 11.98°}$$

$$= 2.75\angle 37.15° \ \Omega$$

The simplified circuit is shown in Figure 16–23.

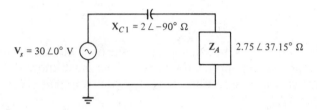

$X_{C1} = 2\angle -90° \ \Omega$

$V_s = 30\angle 0° \text{ V}$

$\mathbf{Z}_A$ $2.75\angle 37.15° \ \Omega$

FIGURE 16–23

Now, the voltage divider principle can be applied to find $\mathbf{V}_A$. The total impedance is

$$\mathbf{Z}_T = \mathbf{X}_{C1} + \mathbf{Z}_A$$
$$= 2\angle -90° \ \Omega + 2.75\angle 37.15° \ \Omega = -j2 \ \Omega + 2.19 \ \Omega + j1.66 \ \Omega$$
$$= 2.19 \ \Omega - j0.34 \ \Omega$$

$$= \sqrt{(2.19 \ \Omega)^2 + (0.34 \ \Omega)^2} \angle -\arctan\left(\frac{0.34}{2.19}\right)$$

$$= 2.22\angle -8.82° \ \Omega$$

$$\mathbf{V}_A = \left(\frac{\mathbf{Z}_A}{\mathbf{Z}_T}\right)\mathbf{V}_s$$

$$= \left(\frac{2.75\angle 37.15° \ \Omega}{2.22\angle -8.82° \ \Omega}\right)30\angle 0° \text{ V}$$

$$= 37.16\angle 45.97° \text{ V}$$

Next, $\mathbf{V}_B$ is found by dividing $\mathbf{V}_A$ down, as indicated in Figure 16–24. $\mathbf{V}_B$ is the open terminal output voltage.

$$\mathbf{V}_B = \left(\frac{\mathbf{Z}_1}{\mathbf{Z}_2}\right)\mathbf{V}_A$$

$$= \left(\frac{4.24\angle 58° \ \Omega}{3.44\angle 49.13° \ \Omega}\right) 37.16\angle 45.97° \ \text{V}$$

$$= 45.8\angle 54.84° \ \text{V}$$

Surprisingly, V_A is greater than V_s, and V_B is greater than V_A! This result is possible because of the out-of-phase relationship of the reactive voltages. Remember that X_C and X_L tend to cancel each other.

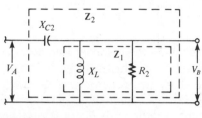

FIGURE 16–24

Review for 16–5

1. In a three-branch parallel circuit, $R = 150 \ \Omega$, $X_C = 100 \ \Omega$, and $X_L = 50 \ \Omega$. Determine the current in each branch when $V_s = 12 \ \text{V}$.

2. The impedance of a parallel RLC circuit is $2.8\angle -38.9° \ \text{k}\Omega$. Is the circuit capacitive or inductive?

PARALLEL RESONANCE

A basic parallel resonant circuit is shown in Figure 16–25. The resistance shown represents the coil's winding resistance. The parallel resonant circuit is frequently referred to as a *tank circuit*, because energy is stored by the inductance and capacitance.

16–6

Condition for Parallel Resonance

Parallel resonance occurs when $X_C = X_L$. Notice that this condition is the same as that for series resonance. It is based on the assumption that the winding

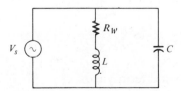

FIGURE 16–25 *Parallel resonant (tank) circuit.*

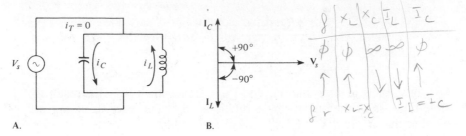

FIGURE 16–26   *Ideal parallel resonant circuit.*

resistance is much less than X_L at the resonant frequency so that there is very little energy loss. This assumption is a practical one to make for many applications.

Ideally, the current in the inductor and the current in the capacitor are equal in magnitude at the resonant frequency, but they are in opposite directions. Let's assume an ideal tank circuit with $R_W = 0$, as shown in Figure 16–26A. Since $X_C = X_L$, and since the source voltage is across both branches, $I_C = I_L$. I_C leads V_s by 90 degrees, and I_L lags V_s by 90 degrees. Thus, the currents are equal in magnitude but opposite in phase, as indicated in Figure 16–26B. Therefore, the currents cancel and the total current from the source is ideally zero, as indicated in Part A of the figure. Thus, in the ideal tank circuit, there is a current that circulates back and forth between L and C; but no current is drawn from the source.

Actually, because of the small winding resistance, some energy is lost; so the reactive currents are not *exactly* equal. Thus, there is a small current from the source. There must be a small amount of current from the source to replace the energy loss in the resistance.

Impedance

At the parallel resonant frequency, *the total impedance is maximum and appears to be purely resistive.* The total current is therefore minimum (ideally zero). A formula for the impedance of a parallel resonant circuit is developed as follows, where R_W is the winding resistance.

Using the reciprocal formula for parallel circuits, we have

$$\frac{1}{\mathbf{Z}} = \frac{1}{-jX_C} + \frac{1}{R_W + jX_L}$$

$$= j\left(\frac{1}{X_C}\right) + \frac{R_W - jX_L}{(R_W + jX_L)(R_W - jX_L)}$$

$$= j\left(\frac{1}{X_C}\right) + \frac{R_W - jX_L}{R_W^2 + X_L^2}$$

The first term plus splitting the numerator of the second term yields

$$\frac{1}{\mathbf{Z}} = j\left(\frac{1}{X_C}\right) - j\left(\frac{X_L}{R_W^2 + X_L^2}\right) + \frac{R_W}{R_W^2 + X_L^2}$$

$$\frac{1}{X_C} = \frac{X_L}{R_w^2 + X_C^2} \Rightarrow X_L X_C = R_w^2 + X_C^2$$

At resonance, Z is purely resistive; so it has no j part (the j terms in the last expression cancel). Thus, only the real part is left, as stated in the following equation for Z at resonance:

$$Z_r = \frac{R_w^2 + X_L^2}{R_w} \qquad (16\text{-}12)$$

Since the j terms in the formula preceding Equation (16–12) are equal, we can substitute $X_L X_C = R_w^2 + X_L^2$ into Equation (16–12) and get an expression for the magnitude of the total impedance in terms of R, L, and C:

$$Z_r = \frac{R_w^2 + X_L^2}{R_w} = \frac{X_L X_C}{R_w}$$

$$= \frac{(2\pi f_r L)(1/2\pi f_r C)}{R_w}$$

Canceling $2\pi f_r$ yields

$$Z_r = \frac{L}{R_w C} \qquad (16\text{-}13)$$

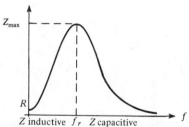

FIGURE 16–27 *Impedance of parallel resonant circuit versus frequency.*

Figure 16–27 shows that the impedance of a parallel resonant circuit is maximum at f_r and decreases above or below resonance. At frequencies below resonance, $X_L < X_C$; thus, the parallel circuit is inductive (the smaller value in parallel is predominant). At frequencies above resonance, $X_C < X_L$; thus, the circuit is capacitive. Notice that this characteristic is opposite to that of series resonant impedance.

Example 16–13

What is Z_r at resonance for the circuit in Figure 16–28?

Solution:

$$Z_r = \frac{L}{R_w C} = \frac{50 \text{ mH}}{(5 \text{ }\Omega)(33 \text{ pF})} = 303 \text{ M}\Omega$$

In an *ideal* tank circuit, R_w is zero, and Z_r is therefore infinite.

Example 16–13 (continued)

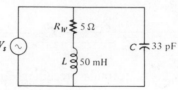

FIGURE 16–28

Resonant Frequency

The formula for the resonant frequency of the parallel *LC* circuit is developed as follows, starting with the condition $R_W^2 + X_L^2 = X_C X_L$, which was previously discussed.

$$R_W^2 + (2\pi f_r L)^2 = \frac{2\pi f_r L}{2\pi f_r C}$$

$$R_W^2 + 4\pi^2 f_r^2 L^2 = \frac{L}{C}$$

$$4\pi^2 f_r^2 L^2 = \frac{L}{C} - R_W^2$$

Solving for f_r^2,

$$f_r^2 = \frac{\left(\dfrac{L}{C}\right) - R_W^2}{4\pi^2 L^2}$$

Multiply both numerator and denominator by *C*:

$$f_r^2 = \frac{L - R_W^2 C}{4\pi^2 L^2 C}$$

$$= \frac{L - R_W^2 C}{L(4\pi^2 LC)}$$

Factoring an *L* out of the numerator and canceling gives

$$f_r^2 = \frac{1 - (R_W^2 C / L)}{4\pi^2 LC}$$

Taking the square root of both sides yields f_r:

$$f_r = \frac{\sqrt{1 - (R_W^2 C / L)}}{2\pi\sqrt{LC}} \qquad\qquad (16\text{--}14)$$

If the winding resistance is neglected and assumed to be zero, the equation reduces to the same as that for series resonance: $f_r = 1/(2\pi\sqrt{LC})$.

Example 16–14

For the tank circuit in Figure 16–28 (Example 16–13), determine the total impedance at 1000 Hz below the resonant frequency.

Solution:

$$f_r = \frac{\sqrt{1 - (R_W^2 C/L)}}{2\pi\sqrt{LC}} = 123.89 \text{ kHz}$$

At $f = f_r - 1$ kHz $= 122.89$ kHz:

$$X_L = 2\pi f_r L = 2\pi(122.89 \text{ kHz})(50 \text{ mH}) = 38.6 \text{ k}\Omega$$

$$X_C = \frac{1}{2\pi f_r C} = \frac{1}{2\pi(122.89 \text{ kHz})(33 \text{ pF})} = 39.2 \text{ k}\Omega$$

Z is determined as follows:

$$\mathbf{Z} = \frac{(jX_L)(-jX_C)}{jX_L - jX_C}$$

$$= \frac{(j38.6 \text{ k}\Omega)(-j39.2 \text{ k}\Omega)}{j38.6 \text{ k}\Omega - j39.2 \text{ k}\Omega}$$

$$= j2.52 \times 10^6 \ \Omega$$

$$= 2.52\angle 90° \text{ M}\Omega$$

This result shows that the magnitude of the impedance is considerably less than its resonant value of 303 MΩ, and the circuit is inductive.

Review for 16–6

1. At parallel resonance, $X_L = 1500$ Ω. What is X_C?

2. A parallel tank circuit has the following values: $R = 4$ Ω, $L = 50$ mH, and $C = 10$ pF. Calculate f_r and Z at resonance.

3. In a parallel resonant circuit, impedance is maximum, and the total current from the source is minimum at the resonant frequency (T or F).

4. At parallel resonance, the inductive current and the capacitive current are in phase (T or F).

BANDWIDTH OF RESONANT CIRCUITS

16–7

Series Resonant Circuit

As you learned, the current in a series *RLC* circuit is maximum at the resonant frequency and drops off on either side of this frequency. *The bandwidth is defined to be the range of frequencies for which the current is equal to or greater than 70.7% of its resonant value.* Bandwidth, sometimes abbreviated BW, is an important characteristic of a resonant circuit.

Figure 16–29 illustrates bandwidth on the response curve of a series *RLC* circuit. Notice that the frequency f_1 below f_r is the point at which the current is $0.707I_{max}$ and is commonly called the *lower cutoff frequency*. The frequency f_2 above f_r, where the current is again $0.707I_{max}$, is the *upper cutoff frequency*. Other names for f_1 and f_2 are -3 dB frequencies, *band frequencies,* and *half-power frequencies*. The significance of the latter term is discussed later in the chapter.

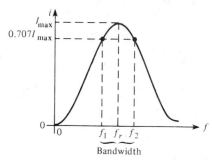

FIGURE 16–29   *Bandwidth on series resonant response curve for I.* $= f_2 - f_1$

Example 16–15

A certain series resonant circuit has a maximum current of 100 mA at the resonant frequency. What is the value of the current at the cutoff frequencies?

Solution:

Current at the cutoff frequencies is 70.7% of maximum:

$$I_{f1} = I_{f2} = 0.707I_{max}$$
$$= 0.707(100 \text{ mA}) = 70.7 \text{ mA}$$

Parallel Resonant Circuit

For a parallel resonant circuit, *the impedance is maximum at the resonant frequency;* so the total current is minimum. The bandwidth can be defined in

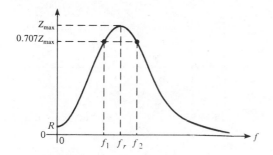

FIGURE 16–30 *Bandwidth on parallel resonant response curve for Z_T.*

rclation to the impedance curve in the same manner that the current curve was used in the series circuit. Of course, f_r is the frequency at which Z is maximum; f_1 is the lower cutoff frequency at which $Z = 0.707Z_{max}$; and f_2 is the upper cutoff frequency at which again $Z = 0.707Z_{max}$. The bandwidth is the range of frequencies between f_1 and f_2, as shown in Figure 16–30.

Formula for Bandwidth

The bandwidth for either series or parallel resonant circuits is the range of frequencies between the cutoff frequencies for which the response curve (I or Z) is 0.707 of the maximum value. Thus, the bandwidth is actually the *difference* between f_2 and f_1:

$$BW = f_2 - f_1 \qquad (16\text{--}15)$$

Ideally, f_r is the *center frequency* and can be calculated as follows:

$$f_r = \frac{f_1 + f_2}{2} \qquad (16\text{--}16)$$

Example 16–16

A resonant circuit has a lower cutoff frequency of 8 kHz and an upper cutoff frequency of 12 kHz. Determine the bandwidth and center (resonant) frequency.

Solution:

$$BW = f_2 - f_1 = 12 \text{ kHz} - 8 \text{ kHz} = 4 \text{ kHz}$$

$$f_r = \frac{f_1 + f_2}{2} = \frac{12 \text{ kHz} + 8 \text{ kHz}}{2} = 10 \text{ kHz}$$

Half-Power Frequencies

As previously mentioned, the upper and lower cutoff frequencies are sometimes called the *half-power frequencies*. This term is derived from the fact

RLC CIRCUITS AND RESONANCE

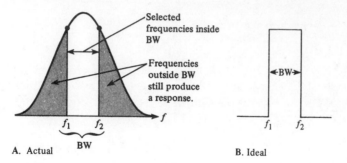

A. Actual   BW   B. Ideal

FIGURE 16–31 *Selectivity curve.*

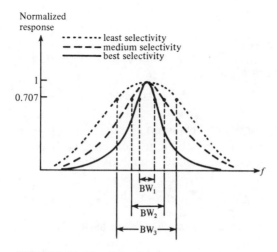

FIGURE 16–32 *Selectivity comparisons.*

that the power from the source at these frequencies is one-half the power delivered at the resonant frequency. The following steps show that this is true for a series circuit. The same result applies to a parallel circuit.

At resonance,

$$P_{max} = I_{max}^2 R$$

The power at f_1 or f_2 is

$$P_{f1} = I_{f1}^2 R = (0.707 I_{max})^2 R = (0.707)^2 I_{max}^2 R$$
$$= 0.5 I_{max}^2 R = 0.5 P_{max}$$

Selectivity

The response curves in Figures 16–29 and 16–30 are also called *selectivity curves*. Selectivity defines how well a resonant circuit responds to a certain frequency and discriminates against all others. *The smaller the bandwidth, the greater the selectivity.*

We normally assume that a resonant circuit accepts frequencies within its bandwidth and completely eliminates frequencies outside the bandwidth. Such is not actually the case, however, because signals with frequencies outside the bandwidth are not completely eliminated. Their magnitudes, however, are greatly reduced. The further the frequencies are from the cutoff frequencies, the greater is the reduction, as illustrated in Figure 16–31A. An ideal selectivity curve is shown in Figure 16–31B.

As you can see in Figure 16–31, another factor that influences selectivity is the *sharpness* of the slopes of the curve. The faster the curve drops off at the cutoff frequencies, the more selective the circuit is, because it responds only to the frequencies within the bandwidth. Figure 16–32 shows a general comparison of three response curves with varying degrees of selectivity.

Review for 16–7

1. What is the bandwidth when $f_1 = 2.2$ MHz and $f_2 = 1.8$ MHz?

2. For a resonant circuit with the cutoff frequencies in Problem 1, what is the center frequency?

3. The power at resonance is 1.8 W. What is the power at the upper cutoff frequency?

QUALITY FACTOR (Q) OF RESONANT CIRCUITS

16–8

The quality factor (Q) is the ratio of the reactive power in the inductor to the resistive power in the winding resistance of the coil or resistance in series with the coil. It is a ratio of the power in L to the power in R. The quality factor is an important factor in resonant circuits. A formula for Q is developed as follows:

$$Q = \frac{\text{energy stored}}{\text{energy lost}}$$

$$= \frac{\text{reactive power}}{\text{resistive power}}$$

$$= \frac{I^2 X_L}{I^2 R}$$

Since, in a series circuit, I is the same in L and R, the I^2 terms cancel, leaving

$$Q = \frac{X_L}{R} \qquad\qquad (16\text{–}17)$$

When the resistance is just the winding resistance of the coil, the circuit Q and the coil Q are the same. Since Q varies with frequency because X_L varies, we are interested mainly in Q at resonance. Note that Q is a ratio of like units and, therefore, has no unit itself.

$Q = \dfrac{V/s}{X_v} \Big/ \dfrac{V/s}{R} = \dfrac{R}{X_L}$

Example 16–17

Determine Q at resonance for each circuit in Figure 16–33.

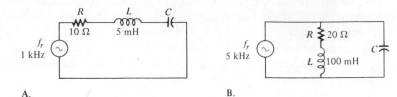

A.

B.

FIGURE 16–33

Solution:

Circuit *A*:

$$X_L = 2\pi f_r L$$
$$= 31.4 \ \Omega$$

$$Q = \frac{X_L}{R} = \frac{31.4 \ \Omega}{10 \ \Omega} = 3.14$$

Circuit *B*:

$$X_L = 3.14 \ \text{k}\Omega$$

$$Q = \frac{X_L}{R} = \frac{3.14 \ \text{k}\Omega}{20 \ \Omega} = 157$$

Q "Magnifies" Applied Voltage in Series *RLC* Circuits

In a series *RLC* circuit, the voltage across L and across C at resonance depends on the resonant Q. At resonance, the total source voltage is dropped across R, and the combined voltage across X_L and X_C is zero, because $\mathbf{V}_L$ and $\mathbf{V}_C$ are equal in magnitude but opposite in phase. By Ohm's law, $V_L = IX_L$ and $V_C = IX_C$. Figure 16–34 illustrates these voltages. The following steps result in expressions for V_L and V_C at resonance in terms of Q:

$$Q = \frac{X_L}{R}$$

or

$$X_L = QR$$

Substituting QR for X_L in the expression $V_L = IX_L$, $= Vc$

$$V_L = IQR = IRQ$$

Since $IR = V_s$ at resonance

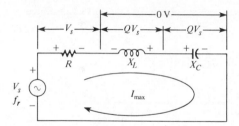

FIGURE 16–34 *Voltages in a series resonant circuit. Instantaneous polarities are shown.*

$$V_L = QV_s \qquad (16\text{–}18)$$

The same holds true for V_C:

$$V_C = QV_s \qquad (16\text{–}19)$$

These equations tell us that the circuit Q times the source voltage is the voltage across L or C at resonance. If Q is high, then V_L and V_C can be much larger than V_s. Keep in mind, however, that V_L and V_C are out of phase and cancel, leaving a net reactive voltage of zero.

Example 16–18

Find V_L and V_C in the circuit of Figure 16–35 at the resonant frequency.

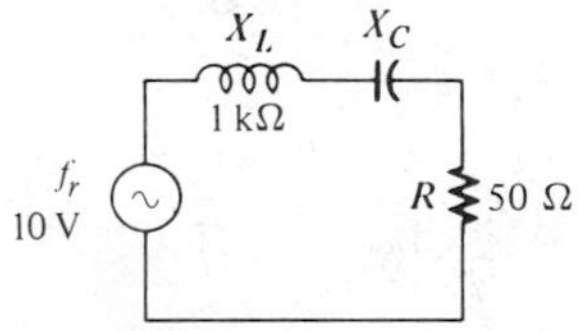

FIGURE 16–35

Solution:

$$Q = \frac{X_L}{R} = \frac{1000\ \Omega}{50\ \Omega} = 20$$

$$V_L = V_C = QV_s = 20(10\ \text{V}) = 200\ \text{V}$$

Q Affects Bandwidth

A higher value of circuit Q results in a smaller bandwidth. A lower value of Q causes a larger bandwidth. A formula for the bandwidth of a resonant circuit in terms of Q is stated in the following equation:

$$\text{BW} = \frac{f_r}{Q} \qquad (16\text{–}20)$$

Example 16–19

What is the bandwidth of each circuit in Figure 16–36?

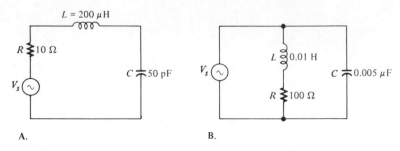

A.   B.

FIGURE 16–36

Solution:

Circuit *A*:

$$f_r = \frac{\sqrt{1 - (R_W^2 C / L)}}{2\pi\sqrt{LC}} \cong \frac{1}{2\pi\sqrt{LC}}$$

$$= 1.59 \text{ MHz}$$

$$Q = \frac{X_L}{R} = \frac{2 \text{ k}\Omega}{10 \text{ }\Omega}$$

$$= 200$$

$$\text{BW} = \frac{f_r}{Q} = \frac{1.59 \text{ MHz}}{200}$$

$$= 7.95 \text{ kHz}$$

Circuit *B*:

$$f_r = \frac{\sqrt{1 - (R_W^2 C / L)}}{2\pi\sqrt{LC}} \cong \frac{1}{2\pi\sqrt{LC}}$$

$$= 22.5 \text{ kHz}$$

$$Q = \frac{X_L}{R} = \frac{1.41 \text{ k}\Omega}{100 \text{ }\Omega}$$

$$= 14.1$$

$$\text{BW} = \frac{f_r}{Q} = \frac{22.5 \text{ kHz}}{14.1}$$

$$= 1.6 \text{ kHz}$$

Review for 16–8

1. What is Q when $X_L = 200$ and $R = 10 \ \Omega$?

2. A 5-V source is applied to a series *RLC* circuit with a Q of 10 at the resonant frequency. What are V_L and V_C?

3. Does a larger Q mean a smaller or a larger bandwidth?

COMPUTER ANALYSIS

The following program computes the impedance and phase angle over a specified frequency range for a series *RLC* circuit with specified parameters. The resonant frequency and bandwidth are also computed.

16–9

```
10 CLS
20 PRINT "THE FOLLOWING PARAMETERS ARE COMPUTED FOR A SERIES RLC CIRCUIT:"
30 PRINT :PRINT"IMPEDANCE"
40 PRINT "PHASE ANGLE"
50 PRINT "RESONANT FREQUENCY"
60 PRINT "BANDWIDTH"
70 PRINT:PRINT:PRINT
80 INPUT "TO CONTINUE PRESS 'ENTER'";X:CLS
90 INPUT "THE VALUE OF R IN OHMS";R
100 INPUT "THE VALUE OF C IN FARADS";C
110 INPUT "THE VALUE OF L IN HENRIES";L
120 FR=1/(2*3.1416*SQR(L*C))
130 Q=2*3.1416*FR*L/R
140 BW=FR/Q
150 PRINT "RESONANT FREQUENCY = ";FR;"HZ"
160 PRINT "BANDWIDTH = ";BW;"HZ"
170 PRINT:PRINT:PRINT
180 INPUT "FOR IMPEDANCE, PHASE ANGLE, AND RESPONSE CONDITION, PRESS 'ENTER";X
190 INPUT "THE MINIMUM FREQUENCY IN HERTZ";FL
200 INPUT "THE MAXIMUM FREQUENCY IN HERTZ";FH
210 INPUT "THE INCREMENTS OF FREQUENCY IN HERTZ";FI
220 PRINT "FREQUENCY(HZ)", "IMPEDANCE", "PHASE ANGLE", "CONDITION"
230 FOR F = FL TO FH STEP FI
240 XC = 1/(2*3.1416*F*C)
250 XL = 2*3.1416*F*L
260 Z = SQR(R[2 + (XL-XC)[2)
270 THETA = ATN((XL-XC)/R)
280 IF F<FR THEN C$ = "BELOW RESONANCE"
290 IF F>FR THEN C$ = "ABOVE RESONANCE"
300 PRINT F, Z, THETA, C$
310 NEXT
320 PRINT "RESONANT FREQUENCY = ";FR;"HZ
330 PRINT "BANDWIDTH = ";BW;"HZ
```

Formulas

Series RLC Circuits:

$$X_T = X_L - X_C \qquad\qquad (16\text{–}1)$$

$$\mathbf{Z} = R + jX_L - jX_C \qquad\qquad (16\text{–}2)$$

$$\mathbf{Z} = \sqrt{R^2 + (X_L - X_C)^2} \angle \arctan\!\left(\frac{X_T}{R}\right) \qquad (16\text{--}3)$$

Series Resonance:

$$X_L = X_C \qquad (16\text{--}4)$$

$$Z = R \qquad (16\text{--}5)$$

$$f_r = \frac{1}{2\pi\sqrt{LC}} \qquad (16\text{--}6)$$

Parallel RLC Circuits:

$$\mathbf{Z} = \frac{1}{\dfrac{1}{R\angle 0^\circ} + \dfrac{1}{X_L\angle 90^\circ} + \dfrac{1}{X_C\angle -90^\circ}} \qquad (16\text{--}7)$$

$$\mathbf{G} = \frac{1}{R\angle 0^\circ} = G\angle 0^\circ \qquad (16\text{--}8)$$

$$\mathbf{B}_C = \frac{1}{X_C\angle -90^\circ} = B_C\angle 90^\circ = jB_C \qquad (16\text{--}9)$$

$$\mathbf{B}_L = \frac{1}{X_L\angle 90^\circ} = B_L\angle -90^\circ = -jB_L \qquad (16\text{--}10)$$

$$\mathbf{Y} = \frac{1}{Z\angle \pm\theta} = Y\angle \mp\theta = G + jB_C - jB_L \qquad (16\text{--}11)$$

Parallel Resonance:

$$Z = \frac{R_W^2 + X_L^2}{R_W} \qquad (16\text{--}12)$$

$$Z_r = \frac{L}{R_W C} \qquad (16\text{--}13)$$

$$f_r = \frac{\sqrt{1 - (R_W^2 C/L)}}{2\pi\sqrt{LC}} \qquad (16\text{--}14)$$

$$\mathrm{BW} = f_2 - f_1 \qquad (16\text{--}15)$$

$$f_r = \frac{f_1 + f_2}{2} \qquad (16\text{--}16)$$

Quality Factor, Q:

$$Q = \frac{X_L}{R} \qquad (16\text{--}17)$$

$$V_L = QV_s \qquad\qquad (16\text{–}18)$$

$$V_C = QV_s \qquad\qquad (16\text{–}19)$$

$$\text{BW} = \frac{f_r}{Q} \qquad\qquad (16\text{–}20)$$

Summary

1. X_L and X_C have opposing effects in an RLC circuit.
2. In a series RLC circuit, the larger reactance determines the net reactance of the circuit.
3. At series resonance, the inductive and capacitive reactances are equal.
4. The impedance of a series RLC circuit is purely resistive at resonance.
5. In a series RLC circuit, the current is maximum at resonance.
6. The reactive voltages V_L and V_C cancel at resonance in a series RLC circuit, because they are equal in magnitude and 180 degrees out of phase.
7. In a parallel RLC circuit, the smaller reactance determines the net reactance of the circuit.
8. In a parallel resonant circuit, the impedance is maximum at the resonant frequency.
9. A parallel resonant circuit is commonly called a *tank circuit*.
10. The impedance of a parallel resonant circuit is purely resistive at resonance.
11. The bandwidth of a series resonant circuit is the range of frequencies for which the current is $0.707I_{max}$ or greater.
12. The bandwidth of a parallel resonant circuit is the range of frequencies for which the impedance is $0.707Z_{max}$ or greater.
13. The cutoff frequencies are the frequencies above and below resonance where the circuit response is 70.7% of the maximum response.
14. The quality factor, Q, of a circuit is a ratio of stored energy to lost energy.
15. At series resonance, the inductive voltage and the capacitive voltage each equal Q times the applied voltage.

Self-Test

1. Does the current in the RLC circuit in Figure 16–37 (page 626) lead or lag the source voltage? Why?
2. Determine the impedance in polar form in Figure 16–37.
3. Find each of the following in Figure 16–37.
   (a) $\mathbf{I}$ (b) $\mathbf{V}_R$ (c) $\mathbf{V}_L$ (d) $\mathbf{V}_C$ (e) P_{avg} (f) $P_r(\text{net})$
4. At series resonance, if $X_L = 500\ \Omega$, what is X_C?

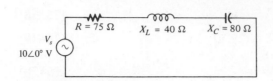

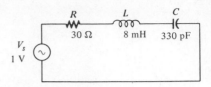

FIGURE 16–37   FIGURE 16–38

5. A certain series *RLC* circuit has $R = 15\ \Omega$, $L = 20$ mH, and $C = 12$ pF. Determine the resonant frequency. What is the total impedance at resonance?

6. In Problem 5, what are the impedance and phase angle 10 kHz above resonance?

7. Find the following in Figure 16–38 at the resonant frequency:
   (a) I (b) f_r (c) V_R (d) V_L (e) V_C (f) θ

8. Determine the impedance in polar form for the circuit in Figure 16–39.

9. Find the current in each branch of Figure 16–39.

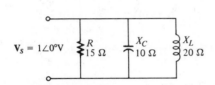

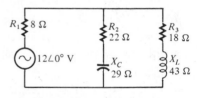

FIGURE 16–39   FIGURE 16–40

10. Determine the following in Figure 16–40.
    (a) $\mathbf{Z}_T$ (b) $\mathbf{I}_T$ (c) $\mathbf{I}_L$ (d) $\mathbf{I}_C$

11. Determine f_r and Z at resonance for the tank circuit in Figure 16–41.

12. In Figure 16–41, how much current is drawn from the source at resonance?

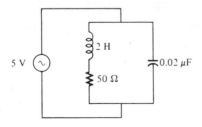

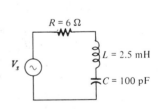

FIGURE 16–41   FIGURE 16–42

13. Determine Q and the bandwidth of the circuit in Figure 16–42.

14. A tank circuit has a lower cutoff frequency of 8 kHz and a bandwidth of 2 kHz. What is the resonant frequency?

Problems

Section 16–1

16–1. A certain series *RLC* circuit has the following values: $R = 10\ \Omega$, $C = 0.05\ \mu F$, and $L = 5$ mH. Determine the impedance in polar form. What is the net reactance? The source frequency is 5 kHz.

16–2. Find the impedance in Figure 16–43, and express it in polar form.

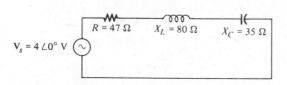

$$47 + j80 - j35$$
$$47 + j45 = 65.07 \angle 43.75$$

FIGURE 16–43

16–3. If the frequency of the source voltage in Figure 16–43 is doubled from the value that produces the indicated reactances, how does the impedance change?

Section 16–2

16–4. For the circuit in Figure 16–43, find I_T, V_R, V_L, and V_C in polar form.

16–5. Sketch the voltage phasor diagram for the circuit in Figure 16–43.

Section 16–3

16–6. Find X_L, X_C, Z, and I at the resonant frequency in Figure 16–44.

16–7. A certain series resonant circuit has a maximum current of 50 mA and a V_L of 100 V. The applied voltage is 10 V. What is Z? What are X_L and X_C?

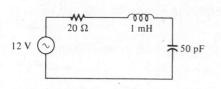

FIGURE 16–44

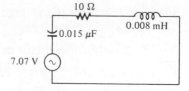

FIGURE 16–45

16–8. For the *RLC* circuit in Figure 16–45, determine the resonant frequency and the cutoff frequencies.

16–9. What is the value of the current at the half-power points in Figure 16–45?

16–10. Determine the phase angle between the applied voltage and the current at the cutoff frequencies in Figure 16–45. What is the phase angle at resonance?

Section 16–4

16–11. Express the impedance of the circuit in Figure 16–46 in polar form.

16–12. Is the circuit in Figure 16–46 capacitive or inductive? Explain.

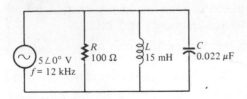

FIGURE 16–46

Section 16–5

16–13. For the circuit in Figure 16–46, find all the currents and voltages in polar form.

16–14. Find the total impedance for each circuit in Figure 16–47.

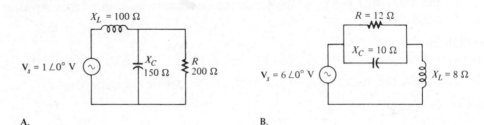

A.

B.

FIGURE 16–47

16–15. For each circuit in Figure 16–47, determine the phase angle between the source voltage and the total current.

16–16. Determine the voltage across each element in Figure 16–48, and express each in polar form.

16–17. Convert the circuit in Figure 16–48 to an equivalent series form.

16–18. What is the current through R_2 in Figure 16–49?

16–19. In Figure 16–49, what is the phase angle between I_2 and the source voltage?

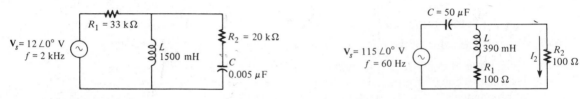

FIGURE 16–48   **FIGURE 16–49**

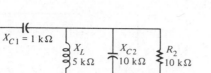

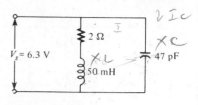

FIGURE 16–50   **FIGURE 16–51**

16–20. Determine the total resistance and the total reactance in Figure 16–50.

16–21. Find the current through each component in Figure 16–50. Find the voltage across each component.

Section 16–6

16–22. What is the impedance of an ideal parallel resonant circuit (no resistance in either branch)?

16–23. Find Z at resonance and f_r for the tank circuit in Figure 16–51.

16–24. How much current is drawn from the source in Figure 16–51 at resonance? What are the inductive current and the capacitive current at the resonant frequency?

Section 16–7

16–25. At resonance, $X_L = 2 \text{ k}\Omega$ and $R_W = 25 \ \Omega$ in a parallel RLC circuit. The resonant frequency is 5 kHz. Determine the bandwidth.

16–26. If the lower cutoff frequency is 2400 Hz and the upper cutoff frequency is 2800 Hz, what is the bandwidth? What is the resonant frequency?

16–27. In a certain RLC circuit, the power at resonance is 2.75 W. What is the power at the lower cutoff frequency?

Section 16–8

16–28. The Q of a series resonant circuit is 25, and the source voltage is 15 V. Determine V_L and V_C at resonance. What is V_R?

16–29. What values of L and C should be used in a tank circuit to obtain a resonant frequency of 8 kHz? The bandwidth must be 800 Hz. The winding resistance of the coil is 10 Ω.

16–30. A parallel resonant circuit has a Q of 50 and a BW of 400 Hz. If Q is doubled, what is the bandwidth for the same f_r?

Section 16–9

16–31. Write a computer program to compute f_r, Q, and BW of a specified tank circuit with L, C, and R_W as the input variables.

16–32. Write a program to compute the impedance and phase shift of a specified parallel tank circuit over a specified frequency range.

Advanced Problems

16–33. Determine if there is a value of C that will make $V_{ab} = 0$ V in Figure 16–52 (page 630). If not, explain.

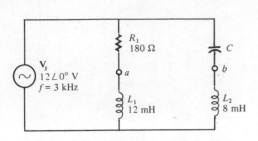

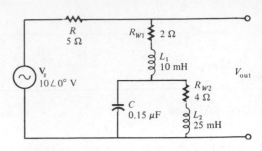

FIGURE 16–52   **FIGURE 16–53**

16–34. If the value of C is 0.2 μF, how much current flows through a 100-Ω resistor connected from a to b in Figure 16–52?

16–35. Determine the resonant frequencies and the output voltage at each frequency in Figure 16–53.

Answers to Section Reviews

Section 16–1:
1. 70 Ω, capacitive. **2.** 83.22$\angle$−57.26° Ω, 83.22 Ω, −57.26°, leading.

Section 16–2:
1. 38.42$\angle$−21.34° V. **2.** Leads. **3.** 6 Ω.

Section 16–3:
1. $X_L = X_C$. **2.** The impedance is minimum. **3.** 159 kHz. **4.** Capacitive.

Section 16–4:
1. Capacitive. **2.** 0.00154$\angle$49.4° S. **3.** 649.4$\angle$−49.4° Ω.

Section 16–5:
1. $I_R = 80$ mA, $I_C = 120$ mA, $I_L = 240$ mA. **2.** Capacitive.

Section 16–6:
1. 1500 Ω. **2.** $f_r = 225$ kHz, $Z = 1250$ MΩ. **3.** T. **4.** F.

Section 16–7:
1. 400 kHz. **2.** 1.99 MHz. **3.** 0.9 W.

Section 16–8:
1. 20. **2.** 50 V. **3.** Smaller BW.

Several important circuit theorems were introduced in Chapter 8 with emphasis on their applications in the analysis of dc circuits. In this chapter, these theorems are covered in relation to the analysis of ac circuits with reactive elements.

In this chapter, you will learn:

1. How to use the superposition theorem to analyze multiple-source ac circuits.
2. How to apply Thevenin's theorem to circuits with ac sources and reactive components.
3. How to apply Norton's theorem to circuits with ac sources and reactive components.
4. How to use Millman's theorem to reduce parallel voltage sources and impedances to a single equivalent voltage source and impedance.
5. How the maximum power transfer theorem applies to reactive circuits.

17

CIRCUIT THEOREMS IN ac ANALYSIS

17-1

The *superposition theorem* was introduced in Chapter 8 for dc circuit analysis. In this section, the superposition theorem is applied to circuits with ac sources and reactive elements. The theorem can be stated as follows:

The current in any given branch of a multiple-source circuit can be found by determining the currents in that particular branch produced by each source acting alone, with all other sources replaced by their internal impedances. The total current in the branch is the phasor sum of the individual source currents in that branch.

The procedure for the application of the superposition theorem is:

1. Leave one of the sources in the circuit, and reduce all others to zero. Reduce voltage sources to zero by placing a theoretical short between the terminals; any internal series impedance remains.

2. Find the current in the branch of interest produced by the one remaining source.

3. Repeat steps 1 and 2 for each source in turn. When complete, you will have a number of current values equal to the number of sources in the circuit.

4. Add the individual current values as phasor quantities.

The following examples illustrate this procedure.

Example 17-1

Find the current in R of Figure 17-1 using the superposition theorem.

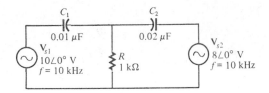

FIGURE 17-1

Solution:

First, by zeroing V_{s2}, find the current in R due to V_{s1}, as indicated in Figure 17-2.

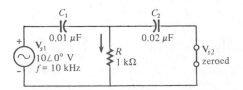

FIGURE 17-2

634

Looking from V_{s1}, we know that the impedance is

$$Z = X_{C1} + \frac{RX_{C2}}{R + X_{C2}}$$

$$X_{C1} = \frac{1}{2\pi(10 \text{ kHz})(0.01 \ \mu\text{F})} = 1.59 \text{ k}\Omega$$

$$X_{C2} = \frac{1}{2\pi(10 \text{ kHz})(0.02 \ \mu\text{F})} = 796 \ \Omega$$

$$Z = 1.59\angle-90° \text{ k}\Omega + \frac{(1\angle0° \text{ k}\Omega)(795\angle-90° \ \Omega)}{1 \text{ k}\Omega - j795 \ \Omega}$$

$$= 1.59\angle-90° \text{ k}\Omega + 622.3\angle-51.51° \ \Omega$$
$$= -j1.59 \text{ k}\Omega + 387.3 \ \Omega - j487.1 \ \Omega$$
$$= 387.3 \ \Omega - j2077 \ \Omega = 2112.8\angle-79.44° \ \Omega$$

The total current from source 1 is:

$$I_{s1} = \frac{V_{s1}}{Z} = \frac{10\angle0° \text{ V}}{2112.8\angle-79.44° \ \Omega} = 4.73\angle79.44 \text{ mA}$$

Using the current divider formula, the current through R due to V_{s1} is

$$I_{R1} = \left(\frac{X_{C2}\angle-90°}{R - jX_{C2}}\right)I_{s1}$$

$$= \left(\frac{796\angle-90° \ \Omega}{1 \text{ k}\Omega - j796 \ \Omega}\right)(4.73\angle79.44 \text{ mA})$$

$$= (0.663\angle-51.48° \ \Omega)(4.73\angle79.44° \text{ mA})$$
$$= 3.14\angle27.96° \text{ mA}$$

Next, find the current in R due to source V_{s2} by zeroing V_{s1}, as shown in Figure 17–3.

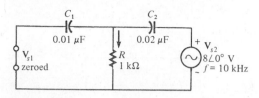

FIGURE 17–3

Looking from V_{s2}, we see that the impedance is

$$Z = X_{C2} + \frac{RX_{C1}}{R + X_{C1}}$$

$$X_{C1} = 1.59 \text{ k}\Omega$$

$$X_{C2} = 796 \ \Omega$$

Example 17–1 (continued)

$$Z = 796\angle -90° \ \Omega + \frac{(1\angle 0° \ \text{k}\Omega)(1.59\angle -90° \ \text{k}\Omega)}{1 \ \text{k}\Omega - j1.59 \ \text{k}\Omega}$$

$$= 796\angle -90° \ \Omega + 846.5\angle -32.17° \ \Omega$$
$$= -j796 \ \Omega + 716.54 \ \Omega - j450.71 \ \Omega$$
$$= 716.54 \ \Omega - j1246.71 \ \Omega = 1437.95\angle -60.11° \ \Omega$$

The total current from source 2 is

$$\mathbf{I}_{s2} = \frac{\mathbf{V}_{s2}}{\mathbf{Z}} = \frac{8\angle 0° \ \text{V}}{1437.95\angle -60.11° \ \Omega}$$

$$= 5.56\angle 60.11° \ \text{mA}$$

Using the current divider formula, the current through R due to V_{s2} is

$$\mathbf{I}_{R2} = \left(\frac{X_{C1}\angle -90°}{R - jX_{C1}}\right)\mathbf{I}_{s2}$$

$$= \frac{1.59\angle -90° \ \text{k}\Omega}{1.88\angle -57.83° \ \text{k}\Omega}(5.56\angle 60.11° \ \text{mA})$$

$$= 4.70\angle 27.94° \ \text{mA}$$

The two individual resistor currents are now added in phasor form:

$$\mathbf{I}_{R1} = 3.14\angle 27.96° \ \text{mA} = 2.77 \ \text{mA} + j1.47 \ \text{mA}$$

$$\mathbf{I}_{R2} = 4.70\angle 27.94° \ \text{mA} = 4.15 \ \text{mA} + j2.20 \ \text{mA}$$

$$\mathbf{I}_{R} = \mathbf{I}_{R1} + \mathbf{I}_{R2} = 6.92 \ \text{mA} + j3.67 \ \text{mA} = 7.83\angle 27.94° \ \text{mA}$$

Example 17–2

Find the coil current in Figure 17–4.

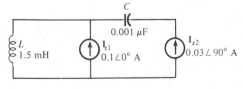

FIGURE 17–4

Solution:

First find the current through the inductor due to current source I_{s1} by replacing source I_{s2} with an open, as shown in Figure 17–5. As you can see, the entire 0.1 A from the current source I_{s1} is through the coil.

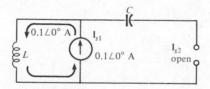

FIGURE 17–5

Next, find the current through the inductor due to current source I_{s2} by replacing source I_{s1} with an open, as indicated in Figure 17–6. Notice that all of the 0.03 A from source I_{s2} is through the coil.

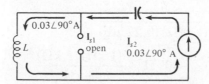

FIGURE 17–6

To get the total inductor current, the two individual currents are superimposed and added as phasor quantities.

$$\mathbf{I}_L = \mathbf{I}_{L1} + \mathbf{I}_{L2}$$
$$= 0.1\angle 0° \text{ A} + 0.03\angle 90° \text{ A} = 0.1 \text{ A} + j0.03 \text{ A}$$
$$= 0.104\angle 16.7° \text{ A}$$

Example 17–3

Find the total current in the resistor, R_L, in Figure 17–7.

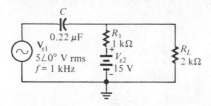

FIGURE 17–7

Solution:

First, find the current through R_L due to source V_{s1} by zeroing source V_{S2}, as shown in Figure 17–8.

Looking from V_{s1}, we see that the impedance is

$$\mathbf{Z} = \mathbf{X}_C + \frac{\mathbf{R}_1 \mathbf{R}_L}{\mathbf{R}_1 + \mathbf{R}_L}$$

Example 17–3 (continued)

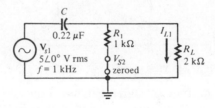

FIGURE 17–8

$$X_C = \frac{1}{2\pi(1 \text{ kHz})(0.22 \ \mu\text{F})} = 723 \ \Omega$$

$$\mathbf{Z} = 723\angle-90° \ \Omega + \frac{(1 \text{ k}\Omega)(2 \text{ k}\Omega)}{3 \text{ k}\Omega}$$

$$= -j723 \ \Omega + 667 \ \Omega = 983.68\angle-47.31° \ \Omega$$

The total current from source 1 is

$$\mathbf{I}_{s1} = \frac{\mathbf{V}_{s1}}{\mathbf{Z}} = \frac{5\angle0° \text{ V}}{983.68\angle-47.31° \ \Omega} = 5.08\angle47.31° \text{ mA}$$

Using the current divider approach, the current in R_L due to V_{s1} is

$$\mathbf{I}_{L1} = \left(\frac{R_1}{R_1 + R_L}\right)\mathbf{I}_{s1}$$

$$= \left(\frac{1 \text{ k}\Omega}{3 \text{ k}\Omega}\right)(5.08\angle47.31° \text{ mA})$$

$$= 1.69\angle47.31° \text{ mA}$$

Next, find the current in R_L due to the dc source V_{S2} by zeroing V_{s1}, as shown in Figure 17–9:

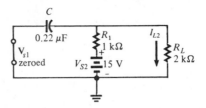

FIGURE 17–9

The impedance as seen by V_{S2} is

$$Z = R_1 + R_L = 3 \text{ k}\Omega$$

The current produced by V_{S2} is

$$I_{L2} = \frac{V_{S2}}{Z} = \frac{15 \text{ V}}{3 \text{ k}\Omega}$$

$$= 5 \text{ mA dc}$$

By superposition, the total current in R_L is $1.69\angle47.31°$ mA riding on a dc level of 5 mA, as indicated in Figure 17–10.

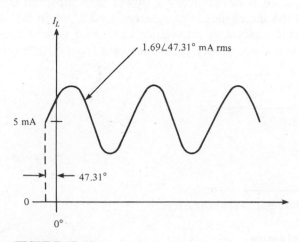

FIGURE 17–10

Review for 17–1

1. If two equal currents are in opposing directions at any instant of time in a given branch of a circuit, what is the net current at that instant?

2. Why is the superposition theorem useful in the analysis of multiple-source circuits?

3. Using the superposition theorem, find the magnitude of the current through R in Figure 17–11.

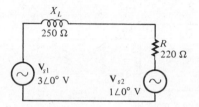

FIGURE 17–11

THEVENIN'S THEOREM

17–2

Thevenin's theorem, as applied to ac circuits, provides a method for reducing any circuit to an equivalent form that consists of an *equivalent ac voltage source* in series with an *equivalent impedance*.

The form of Thevenin's equivalent circuit is shown in Figure 17–12. Regardless of how complex the original circuit is, it can always be reduced to this equivalent form. The equivalent voltage source is designated $\mathbf{V}_{th}$; the equivalent impedance is designated $\mathbf{Z}_{th}$ (lower-case subscript denotes ac quantity). Notice that the impedance is represented by a block in the circuit diagram. This is because the equivalent impedance can be of several forms: purely resistive, purely capacitive, purely inductive, or a combination of resistance and a reactance.

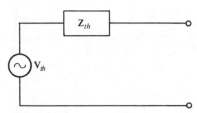

FIGURE 17–12   *Thevenin's equivalent circuit.*

Equivalency

Figure 17–13A shows a block diagram that represents an ac circuit of any given complexity. This circuit has two output terminals, *A* and *B*. A load impedance, $\mathbf{Z}_L$, is connected to the terminals. The circuit produces a certain voltage, $\mathbf{V}_L$, and a certain current, $\mathbf{I}_L$, as illustrated.

By Thevenin's theorem, the circuit in the block can be reduced to an *equivalent* form, as indicated by the dashed lines of Figure 17–13B. The term

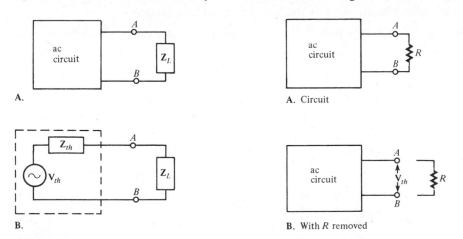

A.

B.

FIGURE 17–13 *An ac circuit of any complexity can be reduced to a Thevenin equivalent for analysis purposes.*

A. Circuit

B. With *R* removed

FIGURE 17–14   *How V_{th} is determined.*

equivalent means that when the same value of load is connected to both the original circuit and Thevenin's equivalent circuit, the load voltages and currents are equal for both. Therefore, as far as the load is concerned, there is no difference between the original circuit and Thevenin's equivalent circuit. The load "sees" the same current and voltage regardless of whether it is connected to the original circuit or to the Thevenin equivalent.

Thevenin's Equivalent Voltage (V_{th})

As you have seen, the equivalent voltage, V_{th}, is one part of the complete Thevenin equivalent circuit. V_{th} is defined as *the open circuit voltage between two specified points in a circuit.*

To illustrate, assume that an ac circuit of some type has a resistor connected between two defined points, A and B, as shown in Figure 17–14A. We wish to find the Thevenin equivalent circuit for the circuit as "seen" by R. V_{th} is the voltage across the points A and B, *with R removed,* as shown in Part B of the figure. The circuit is viewed from the open terminals AB, and R is considered external to the circuit for which the Thevenin equivalent is to be found. The following three examples show how to find V_{th}.

Example 17–4

Determine V_{th} for the circuit external to R_L in Figure 17–15.

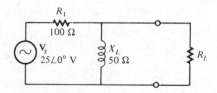

FIGURE 17–15

Solution:

Remove R_L and determine the voltage from A to B (V_{th}). In this case, the voltage from A to B is the same as the voltage across X_L. This is determined using the voltage divider method:

$$V_L = \left(\frac{X_L \angle 90°}{R_1 + jX_L}\right) V_s$$

$$= \left(\frac{50\angle 90°\ \Omega}{111.8\angle 26.57°\ \Omega}\right) 25\angle 0°\ V$$

$$= 11.18\angle 63.43°\ V$$

$$V_{th} = V_{AB} = V_L = 11.18\angle 63.43°\ V$$

Example 17–5

For the circuit in Figure 17–16, determine the Thevenin voltage as seen by R_L.

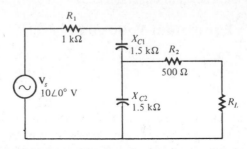

FIGURE 17–16

Solution:

Thevenin's voltage for the circuit between terminals A and B is the voltage that appears across A and B with R_L removed from the circuit.

There is no voltage drop across R_2 because the open terminals AB prevent current through it. Thus, $\mathbf{V}_{AB}$ is the same as $\mathbf{V}_{C2}$ and can be found by the voltage divider formula:

$$\mathbf{V}_{AB} = \mathbf{V}_{C2} = \left(\frac{X_{C2}\angle -90°}{R_1 - jX_{C1} - jX_{C2}}\right)\mathbf{V}_s$$

$$= \left(\frac{1.5\angle -90° \ k\Omega}{1 \ k\Omega - j3 \ k\Omega}\right)10\angle 0° \ V$$

$$= \left(\frac{1.5\angle -90° \ k\Omega}{3.16\angle -71.57° \ k\Omega}\right)10\angle 0° \ V$$

$$= 4.75\angle -18.43° \ V$$

$$\mathbf{V}_{th} = \mathbf{V}_{AB} = 4.75\angle -18.43° \ V$$

Example 17–6

For Figure 17–17, find $\mathbf{V}_{th}$ for the circuit external to R_L.

Solution:

First remove R_L and determine the voltage across the resulting open terminals, which is $\mathbf{V}_{th}$. We find $\mathbf{V}_{th}$ by applying the voltage divider formula to X_C and R:

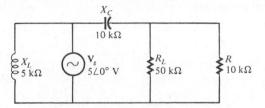

FIGURE 17–17

$$\mathbf{V}_{th} = \mathbf{V}_R = \left(\frac{R \angle 0°}{R - jX_C}\right)\mathbf{V}_s$$

$$= \left(\frac{10\angle 0° \text{ k}\Omega}{10 \text{ k}\Omega - j 10 \text{ k}\Omega}\right) 5\angle 0° \text{ V}$$

$$= \left(\frac{10\angle 0° \text{ k}\Omega}{14.14\angle -45° \text{ k}\Omega}\right) 5\angle 0° \text{ V}$$

$$= 3.54\angle 45° \text{ V}$$

Notice that L has no effect on the result, since the 5-V source appears across C and R in combination.

Thevenin's Equivalent Impedance ($\mathbf{Z}_{th}$)

The previous examples illustrated how to find only one part of a Thevenin equivalent circuit. Now, we turn our attention to determining the Thevenin equivalent impedance, $\mathbf{Z}_{th}$. As defined by Thevenin's theorem, $\mathbf{Z}_{th}$ *is the total impedance appearing between two specified terminals in a given circuit with all sources replaced by their internal impedances (ideally zeroed).* Thus, when we wish to find $\mathbf{Z}_{th}$ between any two terminals in a circuit, first all the voltage sources are shorted (ideally, neglecting any internal impedance), and all the current sources are opened (ideally, to represent infinite impedance). Then the total impedance between the two terminals is determined. The following three examples illustrate how to find $\mathbf{Z}_{th}$.

Example 17–7

Find $\mathbf{Z}_{th}$ for the part of the circuit in Figure 17–18 that is external to R_L. This is the same circuit used in Example 17–4.

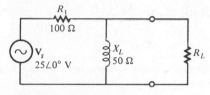

FIGURE 17–18

Example 17–7 (continued)

Solution:

First, reduce V_s to zero by shorting it, as shown in Figure 17–19.

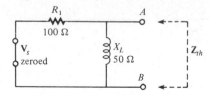

FIGURE 17–19

Looking in at terminals A and B, R and X_L are in parallel. Thus,

$$\mathbf{Z}_{th} = \frac{(R \angle 0°)(X_L \angle 90°)}{R + jX_L}$$

$$= \frac{(100 \angle 0° \ \Omega)(50 \angle 90° \ \Omega)}{100 \ \Omega + j50 \ \Omega}$$

$$= \frac{(100 \angle 0° \ \Omega)(50 \angle 90° \ \Omega)}{111.8 \angle 26.57° \ \Omega}$$

$$= 44.72 \angle 63.43° \ \Omega$$

Example 17–8

For the circuit in Figure 17–20, determine $\mathbf{Z}_{th}$ as seen by R_L. This is the same circuit used in Example 17–5.

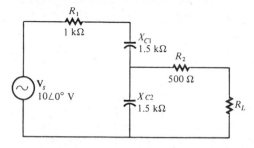

FIGURE 17–20

Solution:

First, zero the voltage source, as shown in Figure 17–21.

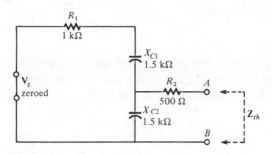

FIGURE 17–21

Looking from terminals A and B, C_2 appears in parallel with the series combination of R_1 and C_1. This entire combination is in series with R_2. The calculation for $\mathbf{Z}_{th}$ is as follows:

$$\mathbf{Z}_{th} = R_2\angle 0° + \frac{(X_{C2}\angle -90°)\,(R_1 - jX_{C1})}{R_1 - jX_{C1} - jX_{C2}}$$

$$= 500\angle 0°\ \Omega + \frac{(1.5\angle -90°\ \text{k}\Omega)\,(1\ \text{k}\Omega - j1.5\ \text{k}\Omega)}{1\ \text{k}\Omega - j3\ \text{k}\Omega}$$

$$= 500\angle 0°\ \Omega + \frac{(1.5\angle -90°\ \text{k}\Omega)\,(1.8\angle -56.31°\ \text{k}\Omega)}{3.16\angle -71.57°\ \text{k}\Omega}$$

$$= 500\angle 0°\ \Omega + 854\angle -74.74°\ \Omega$$

$$= 500\ \Omega + 225\ \Omega - j824\ \Omega$$

$$= 725\ \Omega - j824\ \Omega$$

$$= 1098\angle -48.66°\ \Omega$$

Example 17–9

For the circuit in Figure 17–22, determine $\mathbf{Z}_{th}$ for the portion of the circuit external to R_L. This is the same circuit as in Example 17–6.

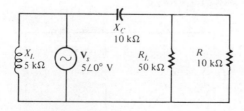

FIGURE 17–22

Example 17–9 (continued)

Solution:

With the voltage source zeroed, X_L is effectively out of the circuit. R and C appear in parallel when viewed from the open terminals, as indicated in Figure 17–23. $\mathbf{Z}_{th}$ is calculated as follows:

$$\mathbf{Z}_{th} = \frac{(R\angle 0°)(X_C\angle -90°)}{R - jX_C}$$

$$= \frac{(10\angle 0° \text{ k}\Omega)(10\angle -90° \text{ k}\Omega)}{14.14\angle -45° \text{ k}\Omega}$$

$$= 7.07\angle -45° \text{ k}\Omega$$

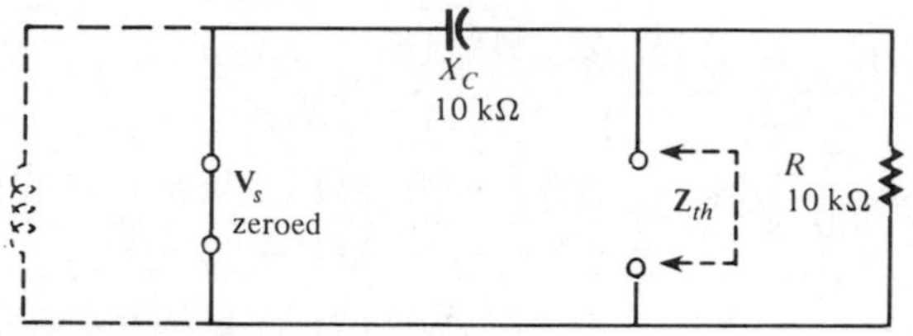

FIGURE 17–23

The Thevenin Equivalent Circuit

The previous examples have shown how to find the two equivalent components of a Thevenin circuit, $\mathbf{V}_{th}$ and $\mathbf{Z}_{th}$. Keep in mind that $\mathbf{V}_{th}$ and $\mathbf{Z}_{th}$ can be found for any circuit. Once these equivalent values are determined, they must be connected in *series* to form the Thevenin equivalent circuit. The following examples use the previous examples to illustrate this final step.

Example 17–10

Draw the Thevenin equivalent circuit for the original circuit in Figure 17–24 that is external to R_L.

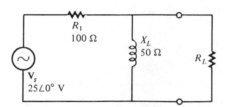

FIGURE 17–24

Solution:

We found in Examples 17–4 and 17–7 that $\mathbf{V}_{th} = 11.18\angle63.43°$ V and $\mathbf{Z}_{th} = 44.72\angle63.43°$ Ω.
    In rectangular form, the impedance is

$$\mathbf{Z}_{th} = 20\ \Omega + j40\ \Omega$$

This form indicates that the impedance is a 20-Ω resistor in series with a 40-Ω inductor. The Thevenin equivalent circuit is shown in Figure 17–25.

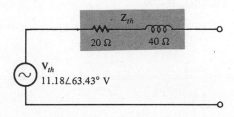

FIGURE 17–25

Example 17–11

For the original circuit in Figure 17–26, sketch the Thevenin equivalent circuit. This is the circuit used in Examples 17–5 and 17–8.

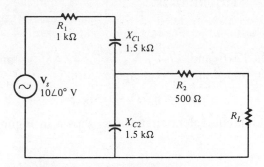

FIGURE 17–26

Solution:

In Examples 17–5 and 17–8, $\mathbf{V}_{th}$ was found to be $4.75\angle-18.43°$ V, and $\mathbf{Z}_{th}$ was found to be $1098\angle-48.66°$ Ω. In rectangular form, $\mathbf{Z}_{th} = 725\ \Omega - j824\ \Omega$. The Thevenin equivalent circuit is shown in Figure 17–27.

Example 17–11 (continued)

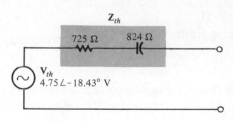

FIGURE 17–27

Example 17–12

For the original circuit in Figure 17–28, determine the Thevenin equivalent circuit as seen by R_L.

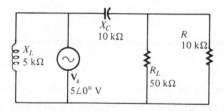

FIGURE 17–28

Solution:

From Examples 17–6 and 17–9, $\mathbf{V}_{th} = 3.54 \angle 45°$ V, and $\mathbf{Z}_{th} = 7.07 \angle -45°$ kΩ. The impedance in rectangular form is

$$\mathbf{Z}_{th} = 5 \text{ k}\Omega - j5 \text{ k}\Omega$$

Thus, the Thevenin equivalent circuit is as shown in Figure 17–29.

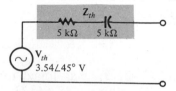

FIGURE 17–29

Summary of Thevenin's Theorem

Remember that the Thevenin equivalent circuit is always of the series form regardless of the original circuit that it replaces. The significance of Thevenin's theorem is that the equivalent circuit can replace the original circuit as far as any external load is concerned. Any load connected between the terminals of a Thevenin equivalent circuit experiences the same current and voltage as if it were connected to the terminals of the original circuit.

A summary of steps for applying Thevenin's theorem follows:

1. Open the two terminals between which you want to find the Thevenin circuit. This is done by removing the component from which the circuit is to be viewed.

2. Determine the voltage across the two open terminals.

3. Determine the impedance viewed from the two open terminals with all sources zeroed (voltage sources replaced with shorts and current sources replaced with opens).

4. Connect V_{th} and Z_{th} in series to produce the complete Thevenin equivalent circuit.

Review for 17–2

1. What are the two basic components of a Thevenin equivalent ac circuit?

2. For a certain circuit, $Z_{th} = 25 \ \Omega - j50 \ \Omega$, and $V_{th} = 5\angle0°$ V. Sketch the Thevenin equivalent circuit.

3. For the circuit in Figure 17–30, find the Thevenin equivalent looking from terminals AB.

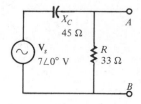

FIGURE 17–30

Like Thevenin's theorem, Norton's theorem provides a method of reducing a more complex circuit to a simpler, more manageable form. The basic difference is that Norton's theorem gives an equivalent current source (rather than a voltage source) in parallel (rather than series) with an equivalent impedance. The form of Norton's equivalent circuit is shown in Figure 17–31. Regardless of how complex

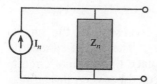

FIGURE 17–31 *Norton equivalent circuit.*

the original circuit is, it can be reduced to this equivalent form. The equivalent
current source is designated I_n, and the equivalent impedance is Z_n (lower-case
subscript denotes ac quantity).

Norton's theorem shows you how to find I_n and Z_n. Once they are
known, simply connect them in parallel to get the complete Norton equivalent
circuit.

Norton's Equivalent Current Source (I_n)

I_n is one part of the Norton equivalent circuit; Z_n is the other part. I_n is
defined as *the short circuit current between two specified points in a given
circuit.* Any load connected between these two points effectively "sees" a current
source I_n in parallel with Z_n.

To illustrate, suppose that the circuit shown in Figure 17–32 has a load
resistor connected to points A and B, as indicated in Part A. We wish to find the
Norton equivalent for the circuit external to R_L. To find I_n, calculate the current
between points A and B with those terminals shorted, as shown in Part B. The
following example shows how to find I_n.

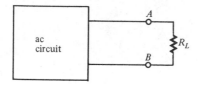

A. Circuit with load resistor

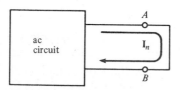

B. Short circuit current is I_n.

FIGURE 17–32 *How I_n is determined.*

Example 17–13

In Figure 17–33, determine $\mathbf{I}_n$ for the circuit as "seen" by the load resistor. The dashed lines identify the portion of the circuit to be Nortonized.

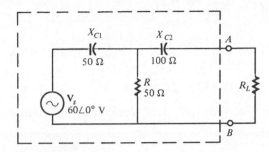

FIGURE 17–33

Solution:

First, short the terminals A and B, as shown in Figure 17–34. $\mathbf{I}_n$ is the current through the short and is calculated as follows:

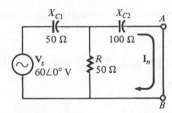

FIGURE 17–34

The total impedance viewed from the source is

$$\mathbf{Z} = \mathbf{X}_{C1} + \frac{\mathbf{R}\mathbf{X}_{C2}}{\mathbf{R} + \mathbf{X}_{C2}}$$

$$= 50\angle-90°\ \Omega + \frac{(50\angle0°\ \Omega)\,(100\angle-90°\ \Omega)}{50\ \Omega - j100\ \Omega}$$

$$= 50\angle-90°\ \Omega + 44.72\angle-26.56°\ \Omega$$

$$= -j50\ \Omega + 40\ \Omega - j20\ \Omega$$

$$= 40\ \Omega - j70\ \Omega$$

$$= 80.62\angle-60.26°\ \Omega$$

The total current from the source is

$$\mathbf{I}_s = \frac{\mathbf{V}_s}{\mathbf{Z}} = \frac{60\angle0°\ \text{V}}{80.62\angle-60.26°\ \Omega} = 0.74\angle60.26°\ \text{A}$$

Example 17–13 (continued)

Applying the current divider formula to get $\mathbf{I}_n$ (the current through the short):

$$\mathbf{I}_n = \left(\frac{\mathbf{R}}{\mathbf{R} + \mathbf{X}_{C2}}\right)\mathbf{I}_s$$

$$= \left(\frac{50\angle 0°\ \Omega}{50\ \Omega - j100\ \Omega}\right)(0.744\angle 60.26°\ \text{A})$$

$$= 0.333\angle 123.69°\ \text{A}$$

This is the value for the equivalent Norton current source.

Norton's Equivalent Impedance ($\mathbf{Z}_n$)

$\mathbf{Z}_n$ is defined the same as $\mathbf{Z}_{th}$ was: It is the total impedance appearing between two specified terminals of a given circuit viewed from the open terminals with all sources zeroed.

Example 17–14

Find $\mathbf{Z}_n$ for the circuit in Figure 17–33 (Example 17–13) viewed from the open terminals AB.

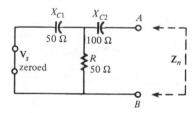

FIGURE 17–35

Solution:

First reduce $\mathbf{V}_s$ to zero, as indicated in Figure 17–35. Looking in at terminals AB, C_2 is in series with the parallel combination of R and C_1. Thus,

$$\mathbf{Z}_n = \mathbf{X}_{C2} + \frac{\mathbf{R}\mathbf{X}_{C1}}{\mathbf{R} + \mathbf{X}_{C1}}$$

$$= 100\angle -90°\ \Omega + \frac{(50\angle 0°\ \Omega)(50\angle -90°\ \Omega)}{50\ \Omega - j50\ \Omega}$$

$$= 100\angle -90°\ \Omega + 35.36\angle -45°\ \Omega$$

$$= -j100 \ \Omega + 25 \ \Omega - j25 \ \Omega$$
$$= 25 \ \Omega - j125 \ \Omega$$

The Norton equivalent impedance is a 25-Ω resistor in series with a 125-Ω capacitive reactance.

The previous two examples have shown how to find the two equivalent components of a Norton equivalent circuit. Keep in mind that these values can be found for any given ac circuit. Once these values are known, they are connected in parallel to form the Norton equivalent circuit, as the following example illustrates.

Example 17–15

Sketch the complete Norton equivalent circuit for the original circuit in Figure 17–33 (Example 17–13).

Solution:

From Examples 17–13 and 17–14:

$$\mathbf{I}_n = 0.333 \angle 123.69° \text{ A}$$

$$\mathbf{Z}_n = 25 \ \Omega - j125 \ \Omega$$

The Norton equivalent circuit is shown in Figure 17–36.

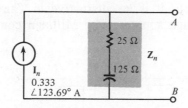

FIGURE 17–36

Summary of Norton's Theorem

Any load connected between the terminals of a Norton equivalent circuit will have the same current through it and the same voltage across it as it would when connected to the terminals of the original circuit. A summary of steps for theoretically applying Norton's theorem is as follows:

1. Short the two terminals between which the Norton circuit is to be determined.

2. Determine the current through the short. This is I_n.

3. Determine the impedance between the two open terminals with all sources zeroed. This is Z_n.

4. Connect I_n and Z_n in parallel.

Review for 17–3

1. For a given circuit, $I_n = 5\angle 0°$ mA, and $Z_n = 150\ \Omega + j100\ \Omega$. Draw the Norton equivalent circuit.

2. Find the Norton circuit as seen by R_L in Figure 17–37.

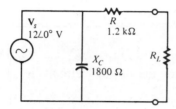

FIGURE 17–37

MILLMAN'S THEOREM

17–4   Millman's theorem permits any number of parallel branches consisting of voltage sources and impedances to be reduced to a single equivalent voltage source and equivalent impedance. It is an alternative to Thevenin's theorem for the case of all parallel voltage sources. The Millman conversion is illustrated in Figure 17–38.

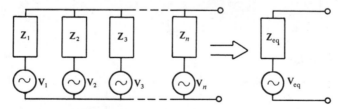

FIGURE 17–38   *The Millman conversion.*

Millman's Equivalent Voltage (V_{eq}) and Equivalent Impedance (Z_{eq})

Millman's theorem provides a formula for calculating the equivalent voltage, V_{eq}. To find V_{eq}, convert each of the parallel voltage sources into current sources, as shown in Figure 17–39.

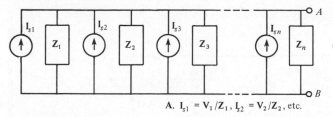

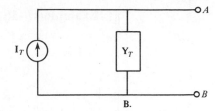

A. $I_{s1} = V_1/Z_1$, $I_{s2} = V_2/Z_2$, etc.

B.

FIGURE 17–39

In Figure 17–39B, the total current from the parallel current sources is

$$I_T = I_{s1} + I_{s2} + I_{s3} + \cdots + I_{sn}$$

The total admittance between terminals AB is

$$Y_T = Y_1 + Y_2 + Y_3 + \cdots + Y_n$$

where $Y_T = 1/Z_T$, $Y_1 = 1/Z_1$, and so on. (Note that n in this equation represents the number of elements and does not represent a Norton quantity.) Remember that current sources are effectively open (ideally). Therefore, by Millman's theorem, the *equivalent impedance* is the total impedance, Z_T.

$$Z_{eq} = \frac{1}{Y_T} = \frac{1}{\dfrac{1}{Z_1} + \dfrac{1}{Z_2} + \cdots + \dfrac{1}{Z_n}} \qquad (17\text{–}1)$$

By Millman's theorem, the *equivalent voltage* is $I_T Z_{eq}$, where I_T is expressed as follows:

$$I_T = \frac{V_1}{Z_1} + \frac{V_2}{Z_2} + \cdots + \frac{V_n}{Z_n}$$

The following is the formula for the equivalent voltage:

$$V_{eq} = \frac{V_1/Z_1 + V_2/Z_2 + \cdots + V_n/Z_n}{1/Z_1 + 1/Z_2 + \cdots + 1/Z_n}$$

$$V_{eq} = \frac{V_1/Z_1 + V_2/Z_2 + \cdots + V_n/Z_n}{Y_{eq}} \qquad (17\text{–}2)$$

Equations (17–1) and (17–2) are the two Millman formulas.

Example 17–16

Use Millman's theorem to find the voltage across R_L and the current through R_L in Figure 17–40.

656

Example 17–16 (continued)

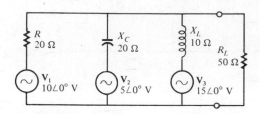

FIGURE 17–40

Solution:

Apply Millman's theorem as follows:

$$\mathbf{Z}_{eq} = \cfrac{1}{\cfrac{1}{R\angle 0°} + \cfrac{1}{X_C\angle -90°} + \cfrac{1}{X_L\angle 90°}}$$

$$= \cfrac{1}{\cfrac{1}{20\angle 0°\ \Omega} + \cfrac{1}{20\angle -90°\ \Omega} + \cfrac{1}{10\angle 90°\ \Omega}}$$

$$= \cfrac{1}{0.05\angle 0° + 0.05\angle 90° + 0.1\angle -90°}$$

$$= \cfrac{1}{0.05 + j0.05 - j0.1}$$

$$= \cfrac{1}{0.05 - j0.05}$$

$$= \cfrac{1}{0.0707\angle -45°}$$

$$= 14.14\angle 45°\ \Omega = 10\ \Omega + j10\ \Omega$$

$$\mathbf{Y}_{eq} = 0.707\angle -45°\ \text{S}$$

$$\mathbf{V}_{eq} = \cfrac{\cfrac{V_1\angle 0°}{R\angle 0°} + \cfrac{V_2\angle 0°}{X_C\angle -90°} + \cfrac{V_3\angle 0°}{X_L\angle 90°}}{\mathbf{Y}_{eq}}$$

$$= \cfrac{\cfrac{10\angle 0°\ \text{V}}{20\angle 0°\ \Omega} + \cfrac{5\angle 0°\ \text{V}}{20\angle -90°\ \Omega} + \cfrac{15\angle 0°\ \text{V}}{10\angle 90°\ \Omega}}{0.0707\angle -45°\ \text{S}}$$

$$= \cfrac{0.5\angle 0° + 0.25\angle 90° + 1.5\angle -90°}{0.0707\angle -45°\ \text{S}}$$

$$= \frac{1.35\angle-68.2°}{0.0707\angle-45° \text{ S}}$$

$$= 19.1\angle-23.2° \text{ V}$$

The single equivalent voltage source is shown in Figure 17–41.

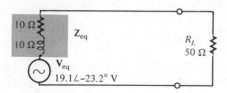

FIGURE 17–41

$\mathbf{I}_L$ and $\mathbf{V}_L$ are calculated as follows:

$$\mathbf{I}_L = \frac{\mathbf{V}_{eq}}{\mathbf{Z}_{eq} + \mathbf{R}_L}$$

$$= \frac{19.1\angle-23.2° \text{ V}}{10 \text{ } \Omega + j10 \text{ } \Omega + 50 \text{ } \Omega}$$

$$= \frac{19.1\angle-23.2° \text{ V}}{60.83\angle9.46° \text{ } \Omega}$$

$$= 0.314\angle-32.66° \text{ A}$$

$$\mathbf{V}_L = \mathbf{I}_L \mathbf{R}_L = (0.314\angle-32.66° \text{ A})(50\angle0° \text{ } \Omega)$$

$$= 15.7\angle-32.66° \text{ V}$$

Review for 17–4

1. To what type of circuit does Millman's theorem apply?

2. Find the load current in Figure 17–42 using Millman's theorem.

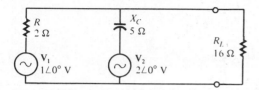

FIGURE 17–42

MAXIMUM POWER TRANSFER THEOREM

17–5

The maximum power transfer theorem as applied to ac circuits states that when a load is connected to a circuit, *maximum power is transferred to the load when the*

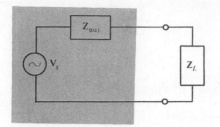

FIGURE 17–43 *Equivalent circuit with load.*

load impedance is the complex conjugate of the circuit's output impedance. The complex conjugate of $R - jX_C$ is $R + jX_L$, where the resistances and the reactances are equal in magnitude. The output impedance is effectively Thevenin's equivalent impedance viewed from the output terminals. When $\mathbf{Z}_L$ is the complex conjugate of $\mathbf{Z}_{out}$, maximum power is transferred from the circuit to the load with a power factor of 1. An equivalent circuit with its output impedance and load is shown in Figure 17–43.

Example 17–17 shows that maximum power occurs when the impedances are *conjugately* matched.

Example 17–17

The circuit to the left of terminals A and B in Figure 17–44 provides power to the load, $\mathbf{Z}_L$. It is the Thevenin equivalent of a more complex circuit. Calculate the average power delivered to the load for each of the following frequencies: 10 kHz, 30 kHz, 50 kHz, 80 kHz, and 100 kHz.

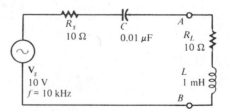

FIGURE 17–44

Solution:

For $f = 10$ kHz:

$$X_C = \frac{1}{2\pi(10 \text{ kHz})(0.01 \ \mu\text{F})} = 1.59 \text{ k}\Omega$$

$$X_L = 2\pi(10 \text{ kHz})(1 \text{ mH}) = 62.8 \ \Omega$$

$$Z_T = \sqrt{(R_s + R_L)^2 + (X_L - X_C)^2}$$

$$= \sqrt{(20 \ \Omega)^2 + (1.53 \text{ k}\Omega)^2}$$

$$= 1.53 \text{ k}\Omega$$

$$I = \frac{V_s}{Z_T} = \frac{10\ \text{V}}{1.53\ \text{k}\Omega} = 6.54\ \text{mA}$$

$$P_L = I^2 R_L = (6.54\ \text{mA})^2 (10\ \Omega)$$
$$= 0.428\ \text{mW}$$

For $f = 30$ kHz:

$$X_C = \frac{1}{2\pi(30\ \text{kHz})\,(0.01\ \mu\text{F})} = 530.5\ \Omega$$

$$X_L = 2\pi(30\ \text{kHz})\,(1\ \text{mH}) = 188.5\ \Omega$$

$$Z_T = \sqrt{(20\ \Omega)^2 + (342\ \Omega)^2}$$
$$= 342.6\ \Omega$$

$$I = \frac{V_s}{Z_T} = \frac{10\ \text{V}}{342.6\ \Omega} = 29.2\ \text{mA}$$

$$P_L = I^2 R_L = (29.2\ \text{mA})^2 (10\ \Omega)$$
$$= 8.53\ \text{mW}$$

For $f = 50$ kHz:

$$X_C = \frac{1}{2\pi(50\ \text{kHz})\,(0.01\ \mu\text{F})} = 318.3\ \Omega$$

$$X_L = 2\pi(50\ \text{kHz})\,(1\ \text{mH}) = 314.2\ \Omega$$

Note that X_C and X_L are very close to being equal and make the impedances complex conjugates. The actual frequency at which $X_L = X_C$ is 50.33 kHz.

$$Z_T = \sqrt{(20\ \Omega)^2 + (4.1\ \Omega)^2}$$
$$= 20.42\ \Omega$$

$$I = \frac{V_s}{Z_T} = \frac{10\ \text{V}}{20.42\ \Omega} = 489.7\ \text{mA}$$

$$P_L = I^2 R_L = (489.7\ \text{mA})^2 (10\ \Omega)$$
$$= 2.5\ \text{W}$$

For $f = 80$ kHz:

$$X_C = \frac{1}{2\pi(80\ \text{kHz})\,(0.01\ \mu\text{F})} = 198.9\ \Omega$$

$$X_L = 2\pi(80\ \text{kHz})\,(1\ \text{mH}) = 502.7\ \Omega$$

$$Z_T = \sqrt{(20\ \Omega)^2 + (303.8\ \Omega)^2}$$
$$= 304.5\ \Omega$$

$$I = \frac{V_s}{Z_T} = \frac{10\ \text{V}}{304.5\ \Omega} = 32.8\ \text{mA}$$

Example 17–17   (continued)

$$P_L = I^2 R_L = (32.8 \text{ mA})^2 (10 \ \Omega)$$
$$= 10.76 \text{ mW}$$

For $f = 100$ kHz:

$$X_C = \frac{1}{2\pi(100 \text{ kHz})(0.01 \ \mu\text{F})} = 159.2 \ \Omega$$

$$X_L = 2\pi(100 \text{ kHz})(1 \text{ mH}) = 628.3 \ \Omega$$

$$Z_T = \sqrt{(20 \ \Omega)^2 + (469.1 \ \Omega)^2}$$
$$= 469.5 \ \Omega$$

$$I = \frac{V_s}{Z_T} = \frac{10 \text{ V}}{469.5 \ \Omega} = 21.3 \text{ mA}$$

$$P_L = I^2 R_L = (21.3 \text{ mA})^2 (10 \ \Omega)$$
$$= 4.54 \text{ mW}$$

As you can see from the results, the average power to the load peaks at the frequency at which the load impedance is the complex conjugate of the output impedance (when the reactances are equal in magnitude). A graph of the average load power versus frequency is shown in Figure 17–45.

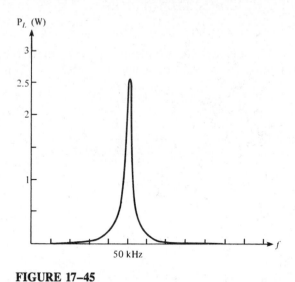

FIGURE 17–45

Review for 17–5

1. If the output impedance of a certain driving circuit is $50 \ \Omega - j10 \ \Omega$, what value of load impedance will result in the maximum average power to the load?

2. For the circuit in Problem 1, how much average power is delivered to the load when the load impedance is the complex conjugate of the output impedance and when the load current is 2 A?

COMPUTER ANALYSIS

The following program permits the computation and tabulation of average power transferred to a load for specified output and load impedances over a defined range of frequencies. The maximum possible average power and the frequency at which it is realized is computed and displayed also.

17–6

```
10 CLS
20 PRINT "POWER TRANSFER":PRINT
30 PRINT "YOU SPECIFY THE OUTPUT IMPEDANCE OF A CIRCUIT AND THE"
40 PRINT "LOAD IMPEDANCE BOTH IN TERMS OF RESISTANCE AND THE"
50 PRINT "REACTIVE COMPONENT VALUES AND A RANGE OF FREQUENCIES."
60 PRINT "THE COMPUTER WILL DETERMINE THE AVERAGE POWER TO THE LOAD"
70 PRINT "FOR EACH FREQUENCY INCREMENT IN THE RANGE.  ALSO, THE"
80 PRINT "MAXIMUM AVERAGE LOAD POWER AND THE FREQUENCY AT WHICH IT"
90 PRINT "OCCURS IS DETERMINED."
100 PRINT:PRINT:PRINT
110 INPUT "TO CONTINUE PRESS 'ENTER'";X:CLS
120 INPUT "OUTPUT RESISTANCE IN OHMS";RO
130 INPUT "OUTPUT CAPACITANCE IN FARADS";CO
140 INPUT "OUTPUT INDUCTANCE IN HENRIES";LO
150 INPUT "LOAD RESISTANCE IN OHMS";RL
160 INPUT "LOAD CAPACITANCE IN FARADS";CL
170 INPUT "LOAD INDUCTANCE IN HENRIES";LL
180 INPUT "SOURCE VOLTAGE IN VOLTS";VS
190 PRINT
200 INPUT "MINIMUM FREQUENCY IN HERTZ";FL
210 INPUT "MAXIMUM FREQUENCY IN HERTZ";FH
220 INPUT "FREQUENCY INCREMENTS IN HERTZ";FI
230 CLS
240 PRINT "FREQUENCY(HZ)", "AVERAGE LOAD POWER(W)"
250 FOR F=FL TO FH STEP FI
260 IF CO=0 THEN GOTO 290
270 XO=2*3.1416*F*LO-(1/(2*3.1416*F*CO))
280 GOTO 300
290 XO=2*3.1416*F*LO
300 IF CL=0 THEN GOTO 330
310 XL=2*3.1416*F*LL-(1/(2*3.1416*F*CL))
320 GOTO 340
330 XL=2*3.1416*F*LL
340 XT=XO+XL
350 R=RO+RL
360 I=VS/(SQR(R[2+XT[2))
370 PLAVG=I[2*RL
380 PRINT F,PLAVG
390 NEXT
400 IMAX=VS/R
410 PLMAX=IMAX[2*RL
420 IF LO+LL=0 OR CO+CL=0 GOTO 470
430 IF CO>0 AND CL>0 GOTO 460
440 FM=1/(2*3.1416*(SQR((LO+LL)*(CO+CL))))
450 GOTO470
460 FM=1/(2*3.1416*(SQR((LO+LL)*(CO*CL/(CO+CL)))))
470 PRINT "THE MAXIMUM LOAD POWER IS ";PLMAX;"W
480 IF (LO>0 OR LL>0)AND(CO>0 OR CL>0) THEN 520
490 IF (LO=0 AND LL=0)AND(CO=0 AND CL=0) THEN 530
500 IF LO=0 AND LL=0 THEN 540
510 IF CO=0 AND CL=0 THEN 550
```

```
520 PRINT "THE FREQUENCY AT WHICH MAXIMUM POWER OCCURS IS";FM;"HZ":END
530 PRINT "POWER IS NOT A FUNCTION OF FREQUENCY.":END
540 PRINT "THE FREQUENCY AT WHICH MAXIMUM POWER OCCURS IS INFINITY.":END
550 PRINT "THE FREQUENCY AT WHICH MAXIMUM POWER OCCURS IS ZERO.":END
```

Formulas

Millman's Theorem:

$$\mathbf{Z}_{eq} = \frac{1}{\mathbf{Y}_T} = \frac{1}{\dfrac{1}{\mathbf{Z}_1} + \dfrac{1}{\mathbf{Z}_2} + \cdots + \dfrac{1}{\mathbf{Z}_n}} \qquad (17\text{–}1)$$

$$\mathbf{V}_{eq} = \frac{\mathbf{V}_1/\mathbf{Z}_1 + \mathbf{V}_2/\mathbf{Z}_2 + \cdots + \mathbf{V}_n/\mathbf{Z}_n}{\mathbf{Y}_{eq}} \qquad (17\text{–}2)$$

Summary

1. The superposition theorem is useful for the analysis of multiple-source circuits.

2. Thevenin's theorem provides a method for the reduction of any ac circuit to an equivalent form consisting of an equivalent voltage source in series with an equivalent impedance.

3. The term *equivalency,* as used in Thevenin's and Norton's theorems, means that when a given load impedance is connected to the equivalent circuit, it will have the same voltage across it and the same current through it as when it is connected to the original circuit.

4. Norton's theorem provides a method for the reduction of any ac circuit to an equivalent form consisting of an equivalent current source in parallel with an equivalent impedance.

5. Millman's theorem provides a method for the reduction of parallel voltage sources and impedances to a single equivalent voltage source and an equivalent impedance.

6. Maximum average power is transferred to a load when the load impedance is the complex conjugate of the output impedance of the driving circuit.

Self-Test

1. In Figure 17–46, use the superposition theorem to find the total current in R.

2. In Figure 17–46, what is the total current in the C branch?

3. Reduce the circuit external to R_L in Figure 17–47 to its Thevenin equivalent.

4. Using Thevenin's theorem, find the current in R_L for the circuit in Figure 17–48.

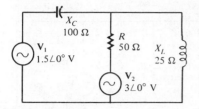

FIGURE 17–46

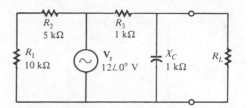

FIGURE 17–47

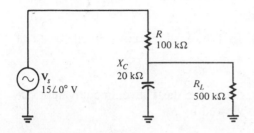

FIGURE 17–48

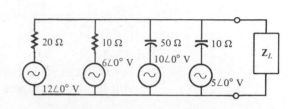

FIGURE 17–49

5. Reduce the circuit in Figure 17–48 to its Norton equivalent as seen by R_L.

6. Use Millman's theorem to simplify the circuit (external to Z_L) in Figure 17–49 to a single voltage source and impedance.

7. What value of Z_L in Figure 17–49 is required for maximum average power transfer to the load?

Problems

Section 17–1

17–1. Using the superposition method, calculate the current in the right-most branch of Figure 17–50.

17–2. Use the superposition theorem to find the current in and the voltage across the R_2 branch of Figure 17–50.

17–3. Using the superposition theorem, solve for the current through R_1 in Figure 17–51.

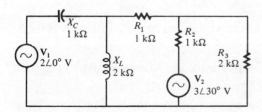

FIGURE 17–50

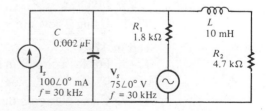

FIGURE 17–51

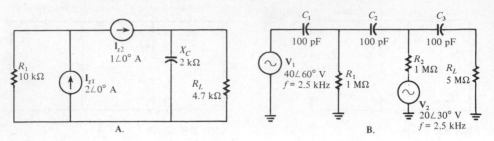

FIGURE 17–52

17–4. Using the superposition theorem, find the load current in each circuit of Figure 17–52.

Section 17–2

17–5. For each circuit in Figure 17–53, determine the Thevenin equivalent circuit for the portion of the circuit viewed by R_L.

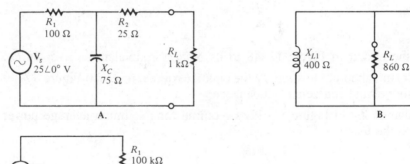

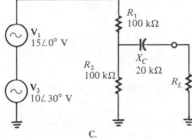

FIGURE 17–53

17–6. Using Thevenin's theorem, determine the current through the load, R_L, in Figure 17–54.

17–7. Using Thevenin's theorem, find the voltage across R_4 in Figure 17–55.

Section 17–3

17–8. For each circuit in Figure 17–53, determine the Norton equivalent as seen by R_L.

17–9. Using Norton's theorem, find the current through the load resistor, R_L, in Figure 17–54.

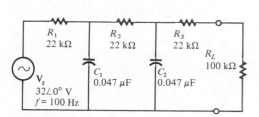

FIGURE 17–54

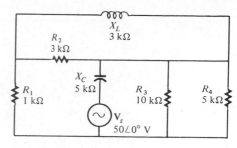

FIGURE 17–55

17–10. Using Norton's theorem, find the voltage across R_4 in Figure 17–55.

Section 17–4

17–11. Apply Millman's theorem to the circuit of Figure 17–56.

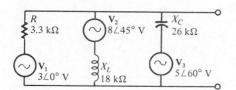

FIGURE 17–56

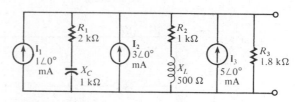

FIGURE 17–57

17–12. Use Millman's theorem and source conversions to reduce the circuit in Figure 17–57 to a single voltage source and impedance.

Section 17–5

17–13. For each circuit in Figure 17–58, maximum average power is to be transferred to the load R_L. Determine the appropriate value for the load impedance in each case.

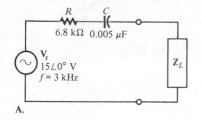

A.

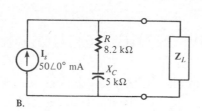

B.

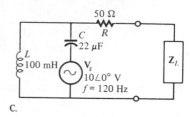

C.

FIGURE 17–58

17–14. Determine $\mathbf{Z}_L$ for maximum power in Figure 17–59 (page 666).

Section 17–6

17–15. Write a computer program to convert a specified Thevenin equivalent circuit to a Norton equivalent.

17–16. Write a computer program implementing Millman's theorem.

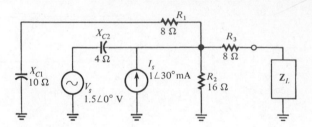

FIGURE 17–59

Advanced Problems

17–17. Use the superposition theorem to find the capacitor current in Figure 17–60.

17–18. Simplify the circuit external to R_3 in Figure 17–61 to its Thevenin equivalent.

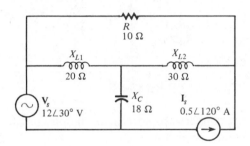

FIGURE 17–60

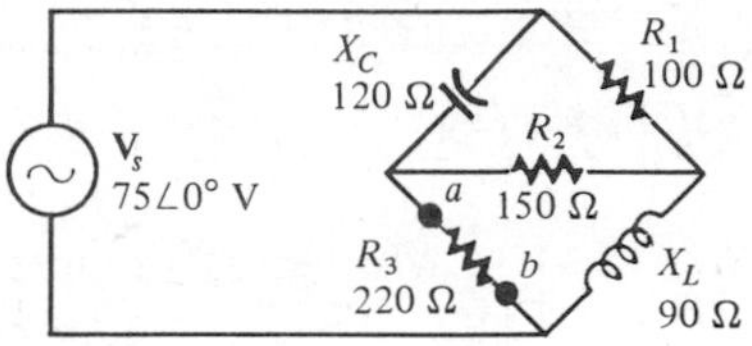

FIGURE 17–61

17–19. A load is to be connected in the place of R_2 in Figure 17–61 to achieve maximum power transfer. Determine the type of load, and express it in rectangular form.

Answers to Section Reviews

Section 17–1:
1. 0. **2.** The circuit can be analyzed one source at a time. **3.** 12 mA.

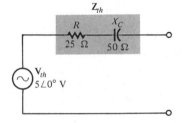

FIGURE 17–62

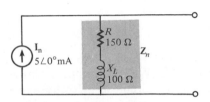

FIGURE 17–63

Section 17–2:
1. Equivalent voltage source and equivalent series impedance. **2.** See Figure 17–62. **3.** $Z_{th} = 21.45 - j15.73$ Ω, $V_{th} = 4.14\angle53.75°$ V.

Section 17–3:
1. See Figure 17–63. **2.** $Z_n = R\angle0° = 1.2\angle0°$ kΩ, $I_n = 10\angle0°$ mA.

Section 17–4:
1. Parallel voltage sources with series impedances. **2.** $I_L = 66.5\angle19.1°$ mA.

Section 17–5:
1. $Z_L = 50$ $\Omega + j10$ Ω. **2.** 200 W.

Filters are widely used in electronics for selecting certain frequencies and rejecting others. In this chapter, *passive filters* are discussed. Passive filters use various combinations of resistors, capacitors, and inductors. You have already seen in previous chapters how basic *RC* and *RL* circuits can be used as filters. In this chapter, you will see that passive filters can be placed in four general categories according to their function: *low-pass, high-pass, band-pass,* and *band-stop.* Within each functional category, there are several common types to be examined.

Specifically, you will learn:

1. The definition of *pass band* in relation to low-pass filters.
2. How a capacitor to ground provides a common type of low-pass filter often used in power supplies.
3. The concept of a *bypass capacitor* in electronic applications.
4. The meaning of the term *choke,* often used for an inductor, and how a series choke acts as a low-pass filter.
5. How various combinations of *L* and *C* form low-pass filters.
6. How a choke to ground provides high-pass filter action.
7. How a series capacitor acts as a coupling capacitor in high-pass applications.
8. How various combinations of *L* and *C* form high-pass filters.
9. How basic band-pass filters are formed using high-pass, low-pass, and resonant circuits.
10. How basic band-stop filters are formed using high-pass, low-pass, and resonant circuits.
11. How to use the decibel (dB) measurement.
12. How to generate a Bode plot of a filter's response.

18
FILTERS

18-1

As you already know, a low-pass filter allows signals with lower frequencies to pass from input to output while rejecting higher frequencies. A block diagram and a general response curve for a low-pass filter appear in Figure 18–1.

Pass Band

The range of low frequencies passed by a low-pass filter within a specified limit is called the *pass band* of the filter. The point considered to be the *upper end* of the pass band is at the *cutoff frequency, f_c*, as illustrated in Figure 18–1B. The cutoff frequency is the frequency at which the filter's output voltage is 70.7% of the maximum. The filter cutoff frequency is sometimes called the *break frequency*. The output is often said to be *down 3 dB* at this frequency. This term is a commonly used one that you should understand; so we discuss decibel (dB) measurement in Section 18–5.

Capacitor to Ground

A capacitor connected from some point in a circuit to ground is a simple but commonly used low-pass filter. The basic concept is illustrated in Figure 18–2.

In many circuits you will find that a *time-varying* voltage is superimposed on a *dc* voltage to form a pulsating or varying dc voltage. This case is graphically illustrated at point *A* in Figure 18–2.

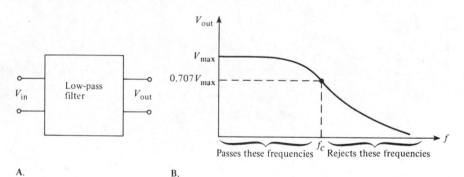

A.   B.

FIGURE 18–1   *Low-pass filter block diagram and response curve.*

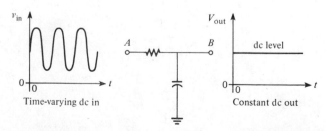

FIGURE 18–2   *Capacitor to ground as an ideal low-pass filter.*

In many circuits the *variation* of the dc voltage is undesirable and must be eliminated. The capacitor presents a *low-impedance* path (low X_C) to ground to the time-varying voltage because of its frequency. However, the capacitor looks like an *open* to the constant dc voltage. As a result, the time-varying voltage is effectively shorted to ground (assuming $X_C \cong 0$), and the constant dc remains on the output. This type of arrangement is often used in electronic power supplies to eliminate the variations on the dc voltage, which are a natural result of *rectification,* the process used for converting ac to dc.

Example 18–1

In Figure 18–3, the output voltage of the power supply rectifier is a pulsating dc with a frequency of 120 Hz. Choose a capacitor value that presents a maximum of 10 Ω to the 120-Hz voltage, and show the capacitor connected in the proper place.

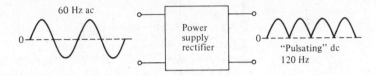

60 Hz ac

Power supply rectifier

"Pulsating" dc
120 Hz

FIGURE 18–3

Solution:

$$X_C \leqq 10 \ \Omega \text{ by requirement}$$

$$X_C = \frac{1}{2\pi f C}$$

$$C = \frac{1}{2\pi f X_C} \geqq \frac{1}{2\pi (120 \text{ Hz}) (10 \ \Omega)}$$

$$\geqq 133 \ \mu\text{F}$$

In practice, you should use the next larger standard value and connect it as shown in Figure 18–4.

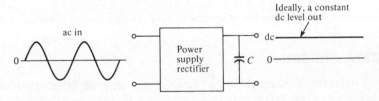

Ideally, a constant
dc level out

ac in

Power supply rectifier

dc

C

FIGURE 18–4

Example 18–2

In a transistor amplifier, it is required that the signal (ac) voltage at a certain point (the emitter) be eliminated and the constant dc voltage be retained for proper operation of the circuit. Determine the value of a filter capacitor to be connected from point A to ground in Figure 18–5 so that its reactance is no greater than $0.1R_E$ at a frequency of 10 kHz. In this application the capacitor is called a *bypass capacitor*.

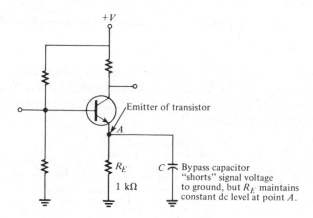

FIGURE 18–5

Solution:

$$X_C = 0.1R_E = 0.1(1000 \ \Omega)$$
$$= 100 \ \Omega$$

$$C = \frac{1}{2\pi f X_C} = \frac{1}{2\pi(10 \text{ kHz})(100 \ \Omega)}$$
$$= 0.159 \ \mu\text{F}$$

X_C provides a parallel path to ground for the signal voltage, but it presents an open to dc. Therefore, only the dc voltage remains at point A because most of the signal is shorted to ground through the capacitor. That is, very little signal voltage is developed across the capacitor because of its low X_C.

Choke in Series

An inductor is often called a *choke,* especially in filter applications. A choke in series between two points in a circuit can be used as a low-pass filter, as illustrated in Figure 18–6. At point A in the figure, there is a varying dc voltage. When X_L is high compared to R at the frequency of the time-varying signal, only

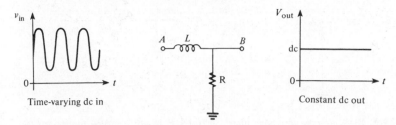

FIGURE 18–6 *Series choke as an ideal low-pass filter.*

the constant dc will get through to the output at point B. That is, practically all of the varying voltage is dropped across the X_L and none across the resistor. Practically all of the constant dc is dropped across R and none across X_L.

The *reduction* of an unwanted frequency or frequencies by a filter is called *attenuation,* a commonly used term that applies to all types of filters.

Example 18–3

Determine the output voltage magnitude for the low-pass filter in Figure 18–7. The input is a 2-V peak-to-peak sine wave riding on a 5-V constant dc level. $X_L = 1000\ \Omega$ at the frequency of the sine wave.

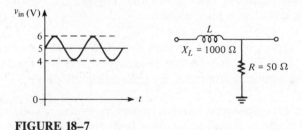

FIGURE 18–7

Solution:

The sine wave will be attenuated by the filter, but the constant dc level of 5 V will pass unattenuated.

We can find the magnitude of the sinusoidal output voltage by applying the voltage divider formula:

$$V_{out} = \left(\frac{R}{\sqrt{R^2 + X_L^2}}\right)V_{in} = \left(\frac{50\ \Omega}{\sqrt{(50\ \Omega)^2 + (1000\ \Omega)^2}}\right)2\ V$$

$$= 0.1\ V\ pp$$

The total output voltage is a 0.1-V peak-to-peak sine wave riding on the 5-V dc level. As you can see, the varying part of the input has been almost eliminated (filtered out).

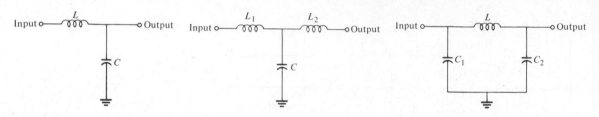

FIGURE 18–8 *Inverted-L low-pass filter.*   **FIGURE 18–9** *T-type low-pass filter.*   **FIGURE 18–10** *π-type low-pass filter.*

Other Types of Low-Pass Filters

Several configurations of low-pass filters other than the capacitor to ground and the series choke are common and are discussed in this section.

Inverted-L Type: Figure 18–8 shows a capacitor and an inductor connected in what is often called an *inverted-L* configuration. The name, of course, comes from the shape of the inverted letter *L* formed by the two components when they are drawn in a schematic.

The inductive reactance of the choke acts to block higher frequencies. The capacitive reactance acts to short higher frequencies to ground. Thus, only frequencies below the cutoff frequency are passed without significant attenuation. This type of filter is more effective than the *RC* or *RL* types because the high-frequency response decreases at a faster rate beyond f_c. We discuss frequency response in more detail in a later section.

T-Type: The T-type low-pass filter uses two chokes and a capacitor, as shown in Figure 18–9. This filter has certain advantages over the inverted-L type because the filtering action is improved by the additional choke.

π-Type: The π-type low-pass filter is shown in Figure 18–10. The additional capacitor again improves filtering over that in the inverted-L type. Notice that in all of these configurations, the choke is in series from input to output and the capacitors are connected to ground.

Review for 18–1

1. In a certain low-pass filter, f_c = 2.5 kHz. What is its pass band?

2. In a certain low-pass filter, R = 100 Ω and X_C = 2 Ω at a frequency, f_1. Determine $\mathbf{V}_{\text{out}}$ at f_1 when $\mathbf{V}_{\text{in}}$ = 5∠0° V rms.

HIGH-PASS FILTERS

18–2   A high-pass filter allows signals with higher frequencies to pass from input to output while rejecting lower frequencies. A block diagram and a general response curve for a high-pass filter are shown in Figure 18–11.

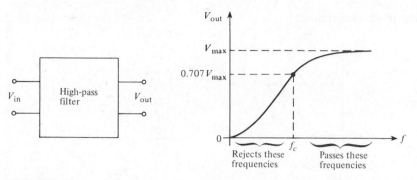

FIGURE 18–11 *High-pass filter block diagram and response curve.*

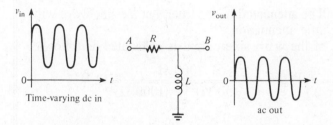

FIGURE 18–12 *Choke to ground as an ideal high-pass filter.*

The frequency considered to be the *lower* end of the band is called the *cutoff frequency*. Just as in the low-pass filter, it is the frequency at which the output is 70.7% of the maximum, as indicated in the figure.

Choke to Ground

A choke connected from some point in a circuit to ground acts as a simple high-pass filter. The basic concept is illustrated in Figure 18–12.

We use the same example as we used for the low-pass filter. A time-varying voltage in Figure 18–12 is superimposed on a constant dc voltage at point A. The choke presents a high impedance to the sine wave between point B and ground. However, the choke looks like a short to the constant dc voltage; so it is shorted to ground. Thus, only the sine wave is left at point B, the output. Notice that the output voltage now varies about a zero volt level at B. This circuit has filtered out the zero frequency (dc) and has allowed the higher-frequency signal to pass from input to output.

Example 18–4

Determine the output voltage magnitude for the high-pass filter in Figure 18–13. The input is a 2-V peak-to-peak sine wave riding on a 4-V constant dc level. $X_L = 1000$ Ω at the frequency of the sine wave. Neglect the coil's winding resistance.

Example 18–4 (continued)

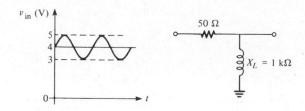

FIGURE 18–13

Solution:

The constant dc level will be attenuated by the filter, but the sine wave will pass through with very little attenuation.

The magnitude of the wave sine output is calculated as follows:

$$V_{out} = \left(\frac{X_L}{\sqrt{R^2 + X_L^2}}\right)V_{in} = \left(\frac{1000 \ \Omega}{\sqrt{(50 \ \Omega)^2 + (1000 \ \Omega)^2}}\right)2 \ V$$

$$= 1.998 \ V$$

The output is a 1.998-V peak-to-peak sine wave (almost the same as the input), varying about zero volts. The constant 4-V dc has been filtered out.

Series Capacitor

A capacitor connected between two points in a circuit acts as a high-pass filter, as illustrated in Figure 18–14. To illustrate, we again use a sine wave superimposed on a constant dc level. When X_C is low compared to the resistance, the sine wave passes through to point *B*, the output. However, the capacitor blocks the constant dc. Thus, we have a high-pass filter.

A very common application of the series capacitor high-pass filter is as a *coupling capacitor* between audio amplifier stages. It is used to pass the amplified audio signal and block the constant dc level (bias voltage) from one stage to the next, as illustrated in Figure 18–15.

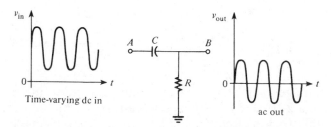

FIGURE 18–14 *Series capacitor as a high-pass filter.*

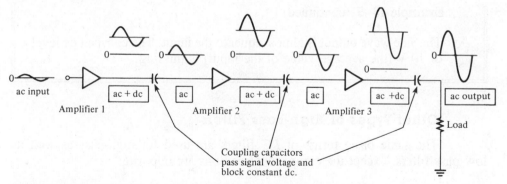

FIGURE 18–15 *Example of a coupling capacitor application.*

Example 18–5

(a) In Figure 18–16, find the value of C so that X_C is ten times less than R at an input frequency of 10 kHz.

(b) If a 5-V sine wave with a dc level of 10 V is applied, what is the output voltage magnitude?

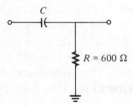

FIGURE 18–16

Solution:

(a) The value of C is determined as follows:

$$X_C = 0.1R$$
$$= 60 \ \Omega$$

$$C = \frac{1}{2\pi f X_C} = \frac{1}{2\pi(10 \text{ kHz})(60 \ \Omega)}$$

$$= 0.265 \ \mu\text{F}$$

(b) The magnitude of the sine wave output is determined as follows:

$$V_{\text{out}} = \left(\frac{R}{\sqrt{R^2 + X_C^2}}\right) V_{\text{in}} = \left(\frac{600 \ \Omega}{\sqrt{(600 \ \Omega)^2 + (60 \ \Omega)^2}}\right) 5 \text{ V}$$

$$= 4.98 \text{ V}$$

$$dB = 20 \log \frac{V_{out}}{V_{in}} = 20 \log \left(\frac{.707 \, V_{in}}{V_{in}} \right) = -3 \, dB$$

$$= 10 \log \frac{P_{out}}{P_{in}}$$

Example 18–5 (continued)

The sine wave output is almost equal to the input. The constant dc level of 10 V (the average value of the input) is missing.

Other Types of High-Pass Filters

The same basic forms of *LC* filters are used for high-pass as well as low-pass filters, except the component positions are opposite.

Inverted-L Type: Figure 18–17 shows an inverted-L high-pass filter. At lower frequencies the capacitive reactance is large and the inductive reactance is small. Thus, most of the input voltage is dropped across the capacitor and very little across the choke at lower frequencies. Therefore, the output voltage is much less than the input. As the frequency is increased, X_C becomes less and X_L becomes greater, causing the output voltage to increase. Thus, high frequencies are passed while low frequencies are attenuated.

T-Type: A T-type high-pass filter uses two capacitors and a choke, as shown in Figure 18–18.

π-Type: The π-type high-pass filter is shown in Figure 18–19. Notice that in all of the high-pass configurations, the capacitors are in series between input and output, and the chokes go to ground. The capacitors can be viewed as shorts to high frequencies, thus passing them from input to output, but as open to low frequencies, thus blocking them. The chokes can be thought of as open to high frequencies and as shorts to ground for low frequencies.

Review for 18–2

1. The maximum output voltage of a high-pass filter is 1 V. What is V_{out} at the cutoff frequency?

2. In a certain high-pass filter, $\mathbf{V}_{in} = 10\angle 0°$ V, $R = 1$ kΩ, and $X_L = 15$ kΩ. Determine $\mathbf{V}_{out}$.

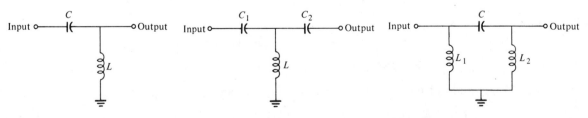

FIGURE 18–17 *Inverted-L high-pass filter.*

FIGURE 18–18 *T-type high-pass filter.*

FIGURE 18–19 *π-type high-pass filter.*

A band-pass filter allows a certain *band* of frequencies to pass and attenuates or rejects all frequencies below and above the pass band. A typical band-pass response curve is shown in Figure 18–20.

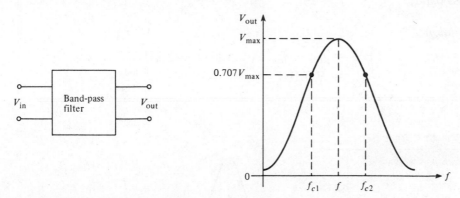

FIGURE 18–20   *Typical band-pass response curve.*

Low-Pass/High-Pass Filter

A combination of a low-pass and a high-pass filter can be used to form a band-pass filter, as illustrated in Figure 18–21.

If the cutoff frequency of the low-pass ($f_{c(l)}$) is higher than the cutoff of the high-pass ($f_{c(h)}$), the responses overlap. Thus, all frequencies except those between $f_{c(h)}$ and $f_{c(l)}$ are eliminated, as shown in Figure 18–22.

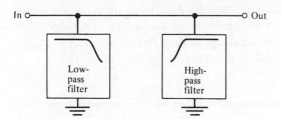

FIGURE 18–21   *Low-pass and high-pass filters used to form a band-pass filter.*

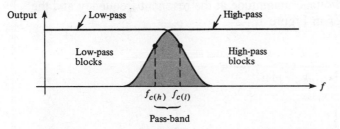

FIGURE 18–22   *Overlapping response curves of high-pass/low-pass filter.*

Example 18–6

A high-pass filter with f_c = 2 kHz and a low-pass filter with f_c = 2.5 kHz are used to construct a band-pass filter. What is the bandwidth of the pass band?

Solution:

$$\text{BW} = f_{c(l)} - f_{c(h)} = 2.5 \text{ kHz} - 2 \text{ kHz}$$
$$= 500 \text{ Hz}$$

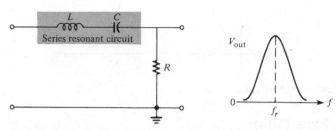

FIGURE 18–23 *Series resonant band-pass filter.*

Series Resonant Filter

A type of series resonant band-pass filter is shown in Figure 18–23. As you learned in the last chapter, a series resonant circuit has *minimum impedance* and *maximum current* at the resonant frequency, f_r. Thus, most of the input voltage is dropped across the resistor at the resonant frequency. Therefore, the output across R has a band-pass characteristic with a maximum output at the frequency of resonance. The bandwidth is determined by the circuit Q.

Example 18–7

Determine the output voltage magnitude at the resonant frequency and the bandwidth for the filter in Figure 18–24.

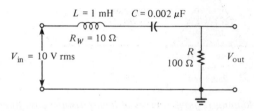

FIGURE 18–24

Solution:

At f_r, the impedance of the resonant circuit is equal to the winding resistance, R_W. By the voltage divider formula,

$$V_{out} = \left(\frac{R}{R + R_W}\right)V_{in} = \left(\frac{100\ \Omega}{110\ \Omega}\right)10\ V$$

$$= 9.09\ V$$

The resonant frequency is

$$f_r = \frac{1}{2\pi\sqrt{LC}}$$

$$= 112.5\ kHz$$

$f_1 = f_r - \frac{1}{2} BW$

$= 103.7$

The circuit Q is

$$Q = \frac{X_L}{R_T} = \frac{707\ \Omega}{110\ \Omega}$$

$f_2 = f_r + \frac{1}{2} BW$

$= 121.3$

$$= 6.4$$

The bandwidth is

$$BW = \frac{f_r}{Q} = \frac{112.5\ kHz}{6.4}$$

$$= 17.6\ kHz$$

Parallel Resonant Filter

A type of band-pass filter using a parallel resonant circuit is shown in Figure 18–25. Recall that a parallel resonant circuit has *maximum impedance* at resonance. The circuit in Figure 18–25 acts as a voltage divider. At resonance, the impedance of the tank is much greater than R. Thus, most of the input voltage is across the tank, producing a maximum output voltage at the resonant frequency.

For frequencies above or below resonance, the tank impedance drops off, and more of the input voltage is across R. As a result, the output voltage across the tank drops off, creating a band-pass characteristic.

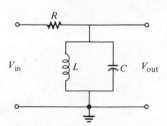

FIGURE 18–25 *Parallel resonant band-pass filter.*

Example 18–8

What is the center frequency of the filter in Figure 18–26?

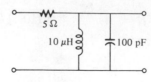

FIGURE 18–26

Solution:

The center frequency of the filter is its resonant frequency:

$$f_r \cong \frac{1}{2\pi\sqrt{LC}} = \frac{1}{2\pi\sqrt{(10 \ \mu H)(100 \ pF)}}$$

$$= 5.03 \ \text{MHz}$$

Review for 18–3

1. For a band-pass filter, $f_{c(h)} = 30.2$ kHz and $f_{c(l)} = 29.8$ kHz. What is the bandwidth?

2. A parallel resonant band-pass filter has the following values: $R = 15 \ \Omega$, $L = 50 \ \mu H$, and $C = 470$ pF. Determine the approximate center frequency.

BAND-STOP FILTERS

18–4 A band-stop filter is essentially the opposite of a band-pass in terms of the responses. It allows *all* frequencies to pass except those lying within a certain *stop band*. A general band-stop response curve is shown in Figure 18–27.

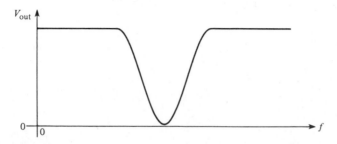

FIGURE 18–27 *General band-stop response curve.*

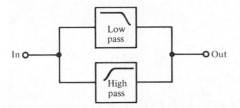

FIGURE 18-28 *Low-pass and high-pass filters used to form a band-stop filter.*

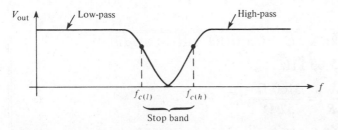

FIGURE 18-29 *Band-stop response curve.*

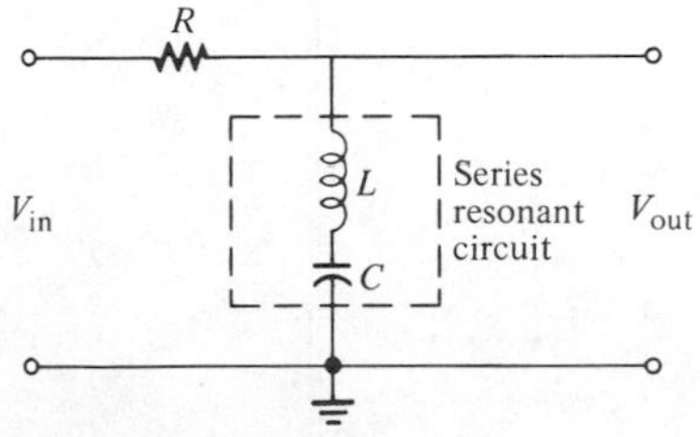

FIGURE 18-30 *Series resonant band-stop filter.*

Low-Pass/High-Pass Filter

A band-stop filter can be formed from a low-pass and a high-pass filter, as shown in Figure 18-28.

If the low-pass cutoff frequency $f_{c(l)}$ is set lower than the high-pass cutoff frequency $f_{c(h)}$, a band-stop characteristic is formed as illustrated in Figure 18-29.

Series Resonant Filter

A series resonant circuit used in a band-stop configuration is shown in Figure 18-30. Basically, it works as follows: At the resonant frequency, the impedance is minimum, and therefore the output voltage is minimum. Most of the input voltage is dropped across R. At frequencies above and below resonance, the impedance increases, causing more voltage across the output.

Example 18-9

Find the output voltage magnitude at f_r and the bandwidth in Figure 18-31.

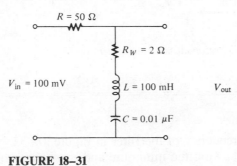

FIGURE 18-31

Example 18–9 (continued)

Solution:

Since $X_L = X_C$ at resonance, then

$$V_{out} = \left(\frac{R_W}{R + R_W}\right)V_{in} = \left(\frac{2\ \Omega}{52\ \Omega}\right)100\ \text{mV}$$

$$= 3.85\ \text{mV}$$

$$f_r = \frac{1}{2\pi\sqrt{LC}} = \frac{1}{2\pi\sqrt{(100\ \text{mH})(0.01\ \mu\text{F})}}$$

$$= 5.03\ \text{kHz}$$

$$Q = \frac{X_L}{R} = \frac{3160\ \Omega}{52\ \Omega}$$

$$= 60.8$$

$$\text{BW} = \frac{f_r}{Q} = \frac{5.03\ \text{kHz}}{60.8}$$

$$= 82.7\ \text{Hz}$$

Parallel Resonant Filter

A parallel resonant circuit used in a band-stop configuration is shown in Figure 18–32. At the resonant frequency, the tank impedance is maximum, and so most of the input voltage appears across it. Very little voltage is across R at resonance. As the tank impedance decreases above and below resonance, the output voltage increases.

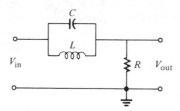

FIGURE 18–32 *Parallel resonant band-stop filter.*

Example 18–10

Find the center frequency of the filter in Figure 18–33. Sketch the output response curve showing the minimum and maximum voltages.

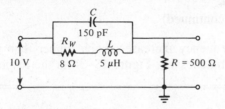

FIGURE 18–33

Solution:

$$f_r = \frac{\sqrt{1 - \dfrac{R_w^2 C}{L}}}{2\pi\sqrt{LC}} = \frac{\sqrt{1 - \dfrac{(64)(150 \text{ pF})}{5 \text{ }\mu\text{H}}}}{2\pi\sqrt{(5 \text{ }\mu\text{H})(150 \text{ pF})}}$$

$$= 5.79 \text{ MHz}$$

From Equation (16–13), at the resonant frequency,

$$Z_{\text{tank}} = \frac{L}{R_w C}$$

$$= \frac{5 \text{ }\mu\text{H}}{(8 \text{ }\Omega)(150 \text{ pF})}$$

$$= 4.167 \text{ k}\Omega \quad \text{purely resistive}$$

We use the voltage divider formula to find the *minimum* output voltage magnitude:

$$V_{\text{out(min)}} = \left(\frac{R}{R + Z_{\text{tank}}}\right) V_{\text{in}} = \left(\frac{500 \text{ }\Omega}{4.667 \text{ k}\Omega}\right) 10 \text{ V}$$

$$= 1.07 \text{ V}$$

At zero frequency, the impedance of the tank is R_w because $X_C = \infty$ and $X_L = 0 \text{ }\Omega$. Therefore, the maximum output voltage is

$$V_{\text{out(max)}} = \left(\frac{R}{R + R_w}\right) V_{\text{in}} = \left(\frac{500 \text{ }\Omega}{508 \text{ }\Omega}\right) 10 \text{ V}$$

$$= 9.84 \text{ V}$$

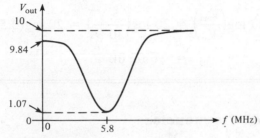

FIGURE 18–34

Example 18–10 (continued)

As the frequency increases much higher than f_r, X_C approaches 0 Ω, and V_{in} approaches V_{in}. Figure 18–34 is the response curve.

Review for 18–4

1. How does a band-stop filter differ from a band-pass filter?

2. Name three basic ways to construct a band-stop filter.

FILTER RESPONSE CHARACTERISTICS

18–5

In this section we examine the filter response curve in more detail and identify some important characteristics associated with basic *RC* and *RL* filters.

Decibel (dB) Measurement

The decibel (dB) is a logarithmic measurement of voltage or power ratios which can be used to express the input-to-output relationship of a filter. The following equation expresses a *voltage* ratio in decibels:

$$dB = 20 \log\left(\frac{V_{out}}{V_{in}}\right) \qquad (18-1)$$

The following equation is the decibel formula for a *power* ratio:

$$dB = 10 \log\left(\frac{P_{out}}{P_{in}}\right) \qquad (18-2)$$

Example 18–11

The output voltage of a filter is 5 V and the input is 10 V. Express the voltage ratio in decibels.

Solution:

$$20 \log\left(\frac{V_{out}}{V_{in}}\right) = 20 \log\left(\frac{5 \text{ V}}{10 \text{ V}}\right) = 20 \log(0.5)$$

$$= -6.02 \text{ dB}$$

The −3 dB Frequencies

The output of a filter is said to be *down 3 dB* at the cutoff frequencies. Actually, this frequency is the point at which the output voltage is 70.7% of the

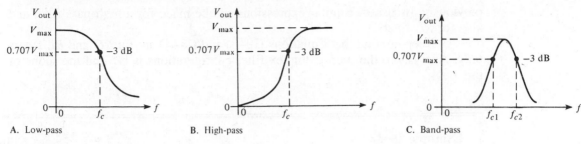

A. Low-pass   B. High-pass   C. Band-pass

FIGURE 18–35 *Response curves showing −3 dB points.*

maximum voltage, as shown in Figure 18–35A for a low-pass filter. The same applies to high-pass and band-pass filters, as shown in Parts B and C of the figure.

We can show that the 70.7% point is the same as 3 dB below maximum (or −3 dB) as follows. The maximum voltage is the zero dB reference:

$$20 \log\left(\frac{0.707 V_{max}}{V_{max}}\right) = 20 \log(0.707)$$

"Bode plot"

$$= -3 \text{ dB}$$

Formula for Cutoff Frequencies

For a basic RC filter, $X_C = R$ at the cutoff frequency f_c. This condition can be written as $1/(2\pi f_c C) = R$. Solving for f_c, we get

$$f_c = \frac{1}{2\pi RC} \tag{18-3}$$

Also, for a basic RL filter, $X_L = R$ at the cutoff frequency. This condition can be stated as $2\pi f_c L = R$. Solving for f_c, we get

$$f_c = \frac{1}{2\pi\left(\dfrac{L}{R}\right)} \tag{18-4}$$

$-1 = 2 \log \frac{V_{ut}}{V_{in}}$

$-\frac{1}{2} = \log \frac{V_{i}}{V_{in}}$

The condition that the reactance equals the resistance at the cutoff frequency can be shown as follows: When $X_C = R$, the output voltage magnitude can be expressed as $X_C V_{in}/\sqrt{R^2 + X_C^2}$ by the voltage divider formula. This formula is for a low-pass RC filter.

If $X_C = R$, then

$$\frac{X_C V_{in}}{\sqrt{R^2 + X_C^2}} = \frac{R V_{in}}{\sqrt{R^2 + R^2}} = \frac{R V_{in}}{\sqrt{2R^2}}$$

$$= \frac{R V_{in}}{R\sqrt{2}} = \frac{V_{in}}{\sqrt{2}} = 0.707 V_{in}$$

These calculations show that the output is 70.7% of the input when $X_C = R$. The frequency at which this occurs is, by definition, the cutoff frequency, as we have previously discussed. Similar expressions can be made for a high-pass filter and also for RL filters.

Keep in mind that Equations (18–3) and (18–4) are for RC and RL filters only. Analysis of the more complex filter configurations is beyond the scope of this book.

Example 18–12

Determine the cutoff frequency for the low-pass RC filter in Figure 18–36.

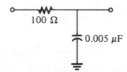

FIGURE 18–36

Solution:

$$f_c = \frac{1}{2\pi RC} = \frac{1}{2\pi (100 \ \Omega)(0.005 \ \mu F)}$$

$$= 318.3 \text{ kHz}$$

This result means that the output voltage is 3 dB below V_{in} at this frequency (V_{out} has a maximum value of V_{in}). Also, the bandwidth of this filter goes from zero frequency to 318.3 kHz, or BW = 318.3 kHz.

"Roll-Off" of the Response Curve

The dashed lines in Figure 18–37 show an actual response curve for a low-pass filter. The maximum output is defined to be 0 dB as a reference. Zero decibels corresponds to $V_{out} = V_{in}$, because 20 log(V_{out}/V_{in}) = 20 log 1 = 0 dB. The output drops from 0 dB to −3 dB at the cutoff frequency and then continues to decrease at a *fixed* rate. This pattern of decrease is called the *roll-off* of the frequency response. The solid line shows an ideal output response that is considered to be "flat" out to f_c. The output then decreases at the fixed rate.

As you have seen, the output voltage of a low-pass filter decreases by 3 dB when the frequency is increased to the critical value f_c. As the frequency continues to increase, the output voltage continues to decrease. In fact, for each tenfold increase in frequency above f_c, there is a 20-dB reduction in the output, as shown in the following steps.

Let's take a frequency that is ten times the critical frequency ($f = 10f_c$). Since $R = X_C$ at f_c, then $R = 10X_C$ at $10f_c$ because of the inverse relationship of X_C and f.

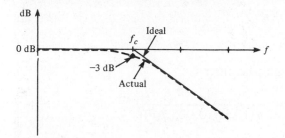

FIGURE 18–37 *Actual and ideal response curves for a low-pass filter.*

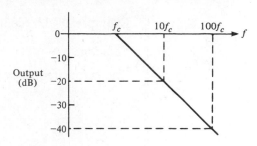

FIGURE 18–38 *Frequency roll-off for a low-pass RC filter (Bode plot).*

The attenuation of the RC circuit is developed as follows:

$$\frac{V_{out}}{V_{in}} = \frac{X_C}{\sqrt{R^2 + X_C^2}} = \frac{X_C}{\sqrt{(10X_C)^2 + X_C^2}}$$

$$= \frac{X_C}{\sqrt{100X_C^2 + X_C^2}} = \frac{X_C}{\sqrt{X_C^2(100 + 1)}}$$

$$= \frac{X_C}{X_C\sqrt{101}} = \frac{1}{\sqrt{101}} \cong \frac{1}{10} = 0.1$$

The dB attenuation is

$$20 \log\left(\frac{V_{out}}{V_{in}}\right) = 20 \log(0.1) = -20 \text{ dB}$$

A tenfold change in frequency is called a *decade*. So, for the RC network, the output voltage is reduced by 20 dB for each decade increase in frequency. A similar result can be derived for the high-pass network. The roll-off is a constant 20 dB/decade for a basic RC or RL filter. Figure 18–38 shows a frequency response plot on a semilog scale, where each interval on the horizontal axis represents a tenfold increase in frequency. This response curve is called a *Bode plot*. It can be done also for a high-pass filter, as shown in Example 18–13.

Example 18–13

Make a Bode plot for the filter in Figure 18–39 for three decades of frequency. Use semilog graph paper.

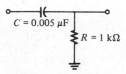

FIGURE 18–39

Example 18–13 (continued)

Solution:

The cutoff frequency for this high-pass filter is

$$f_c = \frac{1}{2\pi RC} = \frac{1}{2\pi(1 \text{ k}\Omega)(0.005 \text{ }\mu\text{F})}$$

$$= 31.8 \text{ kHz}$$

The idealized Bode plot is shown with the solid line on the semilog graph in Figure 18–40. The approximate actual response curve is shown with the dashed lines. Notice first that the horizontal scale is logarithmic and the vertical scale is linear. The frequency is on the logarithmic scale, and the filter output in decibels is on the vertical.

The output is flat beyond f_c (31.8 kHz). As the frequency is reduced below f_c, the output drops at a 20 dB/decade rate. Thus, for the ideal curve, every time the frequency is reduced by ten, the output is reduced by 20 dB. A slight variation from this occurs in actual practice. The output is actually at −3 dB rather than 0 dB at the cutoff frequency.

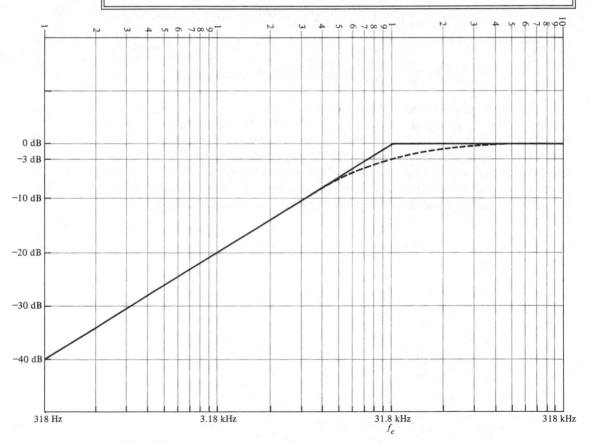

FIGURE 18–40 *Bode plot for Example 18–13.*

Review for 18–5

1. The cutoff frequency of a certain low-pass filter is 1.8 kHz. If $R = 1$ kΩ, what is the value of C?

2. $V_{out} = 400$ mV, and $V_{in} = 1.2$ V. Express the ratio V_{out}/V_{in} in dB.

COMPUTER ANALYSIS

18–6

With this program, the response of low-pass RC and high-pass RC filters can be computed. The critical or center frequency is calculated, and the output voltages are displayed for a specified range of input frequencies. Other inputs required are the component values and the input voltage.

```
10 CLS
20 PRINT "PLEASE SELECT AN OPTION BY NUMBER."
30 PRINT
40 PRINT "(1) RC LOW-PASS ANALYSIS"
50 PRINT "(2) RC HIGH-PASS ANALYSIS"
60 INPUT X
70 ON X GOTO 80   ,290
80 CLS
90 PRINT "RC LOW-PASS ANALYSIS"
100 PRINT
110 INPUT "RESISTANCE IN OHMS";R
120 INPUT "CAPACITANCE IN FARADS";C
130 INPUT "INPUT VOLTAGE IN VOLTS";VIN
140 PRINT
150 INPUT "LOWEST FREQUENCY IN HERTZ";FL
160 INPUT "HIGHEST FREQUENCY IN HERTZ";FH
170 INPUT "FREQUENCY INCREMENTS IN HERTZ";FI
180 CLS
190 PRINT "FREQUENCY(HZ)","VOUT"
200 FOR F=FL TO FH STEP FI
210 XC=1/(2*3.1416*F*C)
220 V=(XC/(SQR(R*R+XC*XC)))*VIN
230 PRINT F, V
240 NEXT
250 FC=1/(2*3.1416*R*C)
260 PRINT "THE CRITICAL FREQUENCY IS";FC;"HZ"
270 PRINT:PRINT:INPUT "TO RETURN TO MENU, ENTER 1";X
280 ON X GOTO10
290 CLS
300 PRINT "RC HIGH-PASS ANALYSIS"
310 PRINT
320 INPUT "RESISTANCE IN OHMS";R
330 INPUT "CAPACITANCE IN FARADS";C
340 INPUT "INPUT VOLTAGE IN VOLTS";VIN
350 PRINT
360 INPUT "LOWEST FREQUENCY IN HERTZ";FL
370 INPUT "HIGHEST FREQUENCY IN HERTZ";FH
380 INPUT "FREQUENCY INCREMENTS IN HERTZ";FI
390 CLS
400 PRINT "FREQUENCY(HZ)","VOUT(V)"
410 FOR F=FL TO FH STEP FI
420 XC=1/(2*3.1416^F^C)
430 V=(R/(SQR(R*R+XC*XC)))*VIN
440 PRINT F,V
450 NEXT
460 FC=1/(2*3.1416*R*C)
470 PRINT "THE CRITICAL FREQUENCY IS";FC;"HZ"
480 PRINT:PRINT:INPUT "TO RETURN TO MENU, ENTER 1";X
490 ON X GOTO 10
```

Formulas

$$dB = 20 \log\left(\frac{V_{out}}{V_{in}}\right) \qquad \qquad \textbf{(18–1)}$$

$$dB = 10 \log\left(\frac{P_{out}}{P_{in}}\right) \qquad \qquad \textbf{(18–2)}$$

$$f_c = \frac{1}{2\pi RC} \qquad \qquad \textbf{(18–3)}$$

$$f_c = \frac{1}{2\pi\left(\dfrac{L}{R}\right)} \qquad \qquad \textbf{(18–4)}$$

[handwritten left margin:] series RL
$X_L = R$
$2\pi f_c L = R$
$f_c = \dfrac{R}{2\pi L}$

[handwritten:] RC

[handwritten:] series RL

Summary

1. A capacitor to ground forms a basic type of low-pass filter.
2. A choke between input and output forms a basic type of low-pass filter.
3. A choke to ground forms a basic type of high-pass filter.
4. A capacitor between input and output forms a basic type of high-pass filter.
5. A band-pass filter passes frequencies between the lower and upper cutoff frequencies and rejects all others.
6. A band-stop filter rejects frequencies between its lower and upper cutoff frequencies and passes all others.
7. The bandwidth of a resonant filter is determined by the quality factor (Q) of the circuit and the resonant frequency.
8. Cutoff frequencies are also called -3 dB frequencies.
9. The output voltage is 70.7% of its maximum at the cutoff frequencies.
10. The roll-off rate of a basic RC or RL filter is 20 dB per decade.

Self-Test

1. The maximum output voltage of a given low-pass filter is 10 V. What is the output voltage at the cutoff frequency?
2. A sine wave with a peak-to-peak value of 15 V is applied to an RC low-pass filter. The average value of the sine wave is 10 V. If the reactance at the sine wave frequency is assumed to be zero, what is the output voltage?
3. The same signal in Problem 2 is applied to an RC high-pass filter. If the reactance is zero at the signal frequency, what is the output voltage of the filter?
4. Calculate the value of C used as a *bypass* if its reactance must be at least ten times smaller than the resistance of 2.5 kΩ at 1 kHz.

5. A low-pass RL filter has $X_L = 100 \ \Omega$ and $R = 5 \ \Omega$ at the input frequency. What is the exact output voltage if the input is a 5-V peak-to-peak sine wave riding on an 8-V constant dc level? Assume a winding resistance of $0 \ \Omega$.

6. Sketch an inverted-L type of low-pass filter and high-pass filter.

7. Sketch a π-type low-pass filter and high-pass filter.

8. Sketch a series resonant band-pass filter and a series resonant band-stop filter.

9. Sketch a parallel resonant band-pass filter and a parallel resonant band-stop filter.

10. The resonant frequency of a band-pass filter is 5 kHz. The circuit Q is 20. What is the bandwidth?

11. The output voltage of a filter is 4 V, and the input is 12 V. Express this relationship in dB.

12. The output power of a circuit is 5 W, and the input power is 10 W. What is the dB power ratio?

13. What is the cutoff frequency of an RC filter with $R = 1 \ k\Omega$ and $C = 0.02 \ \mu F$?

14. What is the cutoff frequency of an RL filter with $R = 5 \ k\Omega$ and $L = 0.01 \ mH$?

Problems

Section 18–1

18–1. In a certain low-pass filter, $X_C = 500 \ \Omega$ and $R = 2 \ k\Omega$. What is the output voltage when the input is 10 V rms?

18–2. A certain low-pass filter has a cutoff frequency of 3 kHz. Determine which of the following frequencies are passed and which are rejected:
 (a) 100 Hz   **(b)** 1 kHz   **(c)** 2 kHz   **(d)** 3 kHz   **(e)** 5 kHz

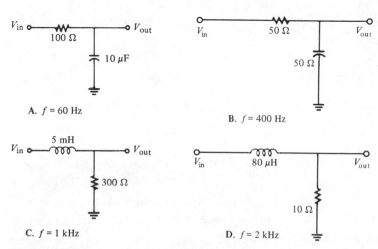

A. $f = 60$ Hz

B. $f = 400$ Hz

C. $f = 1$ kHz

D. $f = 2$ kHz

FIGURE 18–41

18–3. Determine the output voltage of each filter in Figure 18–41 (page 693) at the specified frequency when $V_{in} = 10$ V.

18–4. What is f_c for each filter in Figure 18–41? Determine the output voltage at f_c in each case when $V_{in} = 5$ V.

18–5. For the filter in Figure 18–42, calculate the value of C required for each of the following cutoff frequencies:
(a) 60 Hz   (b) 500 Hz   (c) 1 kHz   (d) 5 kHz

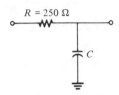

$R = 250 \, \Omega$

C

FIGURE 18–42

Section 18–2

18–6. In a high-pass filter, $X_C = 500 \, \Omega$ and $R = 2$ kΩ. What is the output voltage when $V_{in} = 10$ V rms?

18–7. A high-pass filter has a cutoff frequency of 50 Hz. Determine which of the following frequencies are passed and which are rejected:
(a) 1 Hz   (b) 20 Hz   (c) 50 Hz   (d) 60 Hz   (e) 30 kHz

18–8. Determine the output voltage of each filter in Figure 18–43 at the specified frequency when $V_{in} = 10$ V.

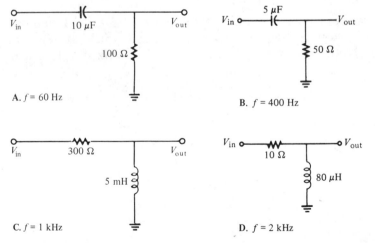

V_{in} 10 μF V_{out}

100 Ω

A. $f = 60$ Hz

5 μF

V_{in} V_{out}

50 Ω

B. $f = 400$ Hz

V_{in} 300 Ω V_{out}

5 mH

C. $f = 1$ kHz

V_{in} 10 Ω V_{out}

80 μH

D. $f = 2$ kHz

FIGURE 18–43

18–9. What is f_c for each filter in Figure 18–43? Determine the output voltage at f_c in each case ($V_{in} = 10$ V).

Section 18–3

18–10. Determine the center frequency for each filter in Figure 18–44.

A.

B.

FIGURE 18–44

18–11. Assuming that the coils in Figure 18–44 have a winding resistance of 10 Ω, find the bandwidth for each filter.

18–12. What are the upper and lower cutoff frequencies for each filter in Figure 18–44?

18–13. For each filter in Figure 18–45, find the center frequency of the pass band.

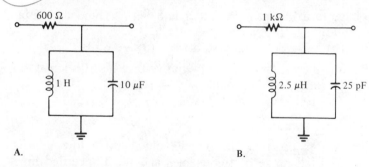

A.

B.

FIGURE 18–45

18–14. If the coils in Figure 18–45 have a winding resistance of 4 Ω, what is the output voltage at resonance when $V_{in} = 120$ V?

Section 18–4

18–15. Determine the center frequency for each filter in Figure 18–46.

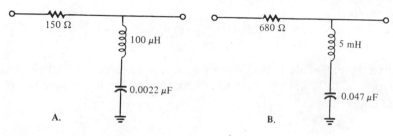

A.

B.

FIGURE 18–46

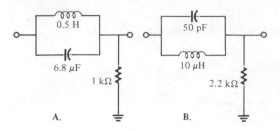

FIGURE 18–47

18–16. For each filter in Figure 18–47, find the center frequency of the stop band.

18–17. If the coils in Figure 18–47 have a winding resistance of 8 Ω, what is the output voltage at resonance when V_{in} = 50 V?

Section 18–5

18–18. For each following case, express the voltage ratio in dB:
(a) V_{in} = 1 V, V_{out} = 1 V   **(b)** V_{in} = 5 V, V_{out} = 3 V
(c) V_{in} = 10 V, V_{out} = 7.07 V   **(d)** V_{in} = 25 V, V_{out} = 5 V

18–19. The input voltage to a low-pass RC filter is 8 V rms. Find the output voltage at the following dB levels:
(a) -1 dB   **(b)** -3 dB   **(c)** -6 dB   **(d)** -20 dB

18–20. For a basic RC low-pass filter, find the output voltage in dB relative to a 0-dB input for the following frequencies (f_c = 1 kHz):
(a) 10 kHz   **(b)** 100 kHz   **(c)** 1 MHz

Section 18–6

18–21. Develop a program to compute the output voltage for a low-pass RL circuit over a specified range of frequencies and for a specified input voltage. The winding resistance is to be treated as an input variable along with values for R and L.

18–22. Modify the program in Problem 18–21 to compute the attenuation of the filter, and express it in dB for a specified range of frequencies.

Advanced Problems

18–23. Design a band-pass filter using a parallel resonant circuit to meet the following specifications:
(a) BW = 500 Hz
(b) Q = 40
(c) $I_{C(max)}$ = 20 mA, $V_{C(max)}$ = 2.5 V

18–24. Determine the values of L_1 and L_2 in Figure 18–48 to pass a signal with a frequency of 1200 kHz and stop (reject) a signal with a frequency of 456 kHz.

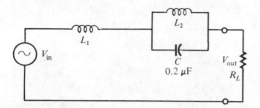

FIGURE 18–48

Answers to Section Reviews

Section 18–1:
1. 0 Hz to 2.5 kHz. **2.** $0.1\angle -88.85°$ V rms.

Section 18–2:
1. 0.707 V. **2.** $9.97\angle 3.81°$ V.

Section 18–3:
1. 400 Hz. **2.** 1.04 MHz.

Section 18–4:
1. It *rejects* a certain band of frequencies. **2.** High-pass/low-pass combination, series resonant circuit, and parallel resonant circuit.

Section 18–5:
1. 0.088 μF. **2.** -9.54 dB.

In Chapter 13, we studied self-inductance. In this chapter, we study the application of another inductive effect called *mutual inductance,* which occurs when two or more coils are in close proximity. Mutual inductance is the basis for transformer action. Because there is no electrical contact between the two magnetically coupled coils that form the basic transformer, the transfer of energy from one coil to the other can be achieved in a situation of complete electrical isolation. There are many uses for the transformer, and a basic understanding of its operation is essential.

Specifically, in this chapter, you will learn:

1. The meaning of *mutual inductance.*
2. How a basic transformer operates.
3. What the turns ratio is and how it affects the transformer's operation.
4. How to determine the polarity of the voltages in a transformer.
5. How a transformer steps up the input voltage.
6. How a transformer steps down the input voltage.
7. The meaning of *primary winding* and *secondary winding.*
8. The effects of connecting a load to the secondary winding.
9. The meaning of *reflected impedance* and how to determine the impedance seen by the primary voltage source.
10. How a transformer can be used for impedance matching.
11. How a transformer can be used as an electrical isolation device.
12. The various types of transformer configurations.
13. How a transformer is constructed.
14. How to interpret transformer ratings.

19
TRANSFORMERS

19–1

When two coils are placed close to each other, as depicted in Figure 19–1, a changing magnetic field produced by current in one coil will cause an induced voltage in the second coil. This induced voltage occurs because the magnetic lines of force (flux) of coil 1 "cut" the winding of coil 2. The two coils are thereby *magnetically linked* or *coupled*. It is important to notice that there is no electrical connection between the coils; therefore they are *electrically isolated*.

The mutual inductance L_M is a measure of how much voltage is induced in coil 2 as a result of a change in current in coil 1. Mutual inductance is measured in henries (H), just as self-inductance is. A greater L_M means that there is a greater induced voltage in coil 2 for a given change in current in coil 1.

There are several factors that determine L_M: the coefficient of coupling (k), the inductance of coil 1 (L_1), and the inductance of coil 2 (L_2).

Coefficient of Coupling

The coefficient of coupling k between two coils is the ratio of the lines of force (flux) produced by coil 1 linking coil 2 (ϕ_{12}) to the total flux produced by coil 1 (ϕ_1):

$$k = \frac{\phi_{12}}{\phi_1} \qquad\qquad (19\text{–}1)$$

For example, if half of the total flux produced by coil 1 links coil 2, then $k = 0.5$. A greater value of k means that more voltage is induced in coil 2 for a certain rate of change of current in coil 1. Note that *k has no units. Recall that the unit of magnetic lines of force (flux) is the weber, abbreviated Wb.*

The coefficient k depends on the physical closeness of the coils and the type of core material on which they are wound. Also, the construction and shape of the cores are factors.

Formula for Mutual Inductance

The three factors influencing L_M (k, L_1, and L_2) are shown in Figure 19–2. The formula for L_M is

$$L_M = k\sqrt{L_1 L_2} \qquad\qquad (19\text{–}2)$$

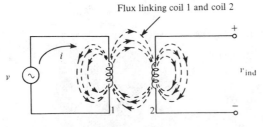

FIGURE 19–1 *Two magnetically coupled coils.*

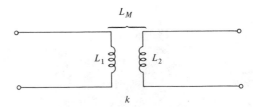

FIGURE 19–2 *The mutual inductance of two coils.*

Example 19–1

One coil produces a total magnetic flux of 50 μWb, and 20 μWb link coil 2. What is k?

Solution:

$$k = \frac{20 \ \mu\text{Wb}}{50 \ \mu\text{Wb}} = 0.4$$

Example 19–2

Two coils are wound on a single core, and the coefficient of coupling is 0.3. The inductance of coil 1 is 10 μH, and the inductance of coil 2 is 15 μH. What is L_M?

Solution:

$$L_M = k\sqrt{L_1 L_2} = 0.3\sqrt{(10 \ \mu\text{H})(15 \ \mu\text{H})}$$
$$= 3.67 \ \mu\text{H}$$

Review for 19–1

1. Define *mutual inductance*.

2. Two 50-mH coils have $k = 0.9$. What is L_M?

3. If k is increased, what happens to the voltage induced in one coil as a result of a current change in the other coil?

THE BASIC TRANSFORMER

A *basic transformer* is an electrical device constructed of two coils with mutual inductance. A schematic diagram of a transformer is shown in Figure 19–3A. As

19–2

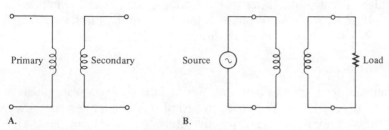

A.   B.

FIGURE 19–3   *The basic transformer.*

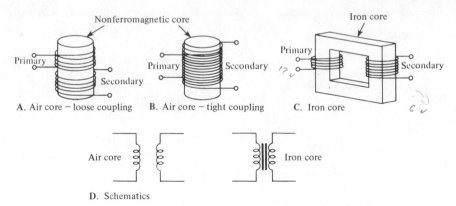

FIGURE 19–4 *Basic types of transformers and schematic symbols.*

shown, one coil is called the *primary winding,* and the other is called the *secondary winding.* The source voltage is applied to the primary, and the load is connected to the secondary, as shown in Figure 19–3B.

Typical transformers are wound on a common core in several ways. The core can be either an air core or an iron core. Figure 19–4A and B shows the primary and secondary coils on a nonferromagnetic cylindrical form as examples of air core transformers. In Part A, the windings are separated; in Part B, they overlap for *tighter coupling* (higher k). Figure 19–4C illustrates an iron core transformer. The iron core increases the coefficient of coupling. Standard schematic symbols are shown for both types in Part D.

Turns Ratio

An important characteristic of a transformer is its *turns ratio. The turns ratio is the number of turns in the secondary winding, N_s, divided by the number of turns in the primary winding, N_p:*

$$\text{Turns ratio} = \frac{N_s}{N_p} \tag{19–3}$$

Direction of Windings

The directions in which the primary and secondary windings are wound on the core determine the polarity of the induced secondary voltage with respect to the primary voltage. The ac voltage across the secondary can be either in phase with the primary voltage or 180° out of phase with it, depending on the winding directions. Figure 19–5 shows how winding direction affects polarity.

Dot Convention

In the schematic for a transformer, dots are often used to indicate polarities, as illustrated in Figure 19–6. The rules for using this convention are simple: A positive at the primary dot causes a positive at the secondary dot (Part A), and a negative at the primary dot causes a negative at the secondary dot (Part B).

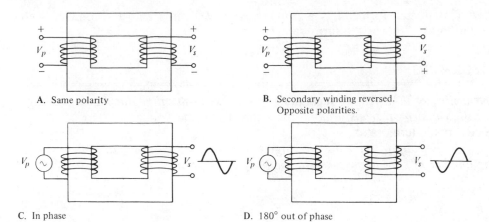

A. Same polarity

B. Secondary winding reversed.
Opposite polarities.

C. In phase

D. 180° out of phase

FIGURE 19–5 *How winding directions determine polarity.*

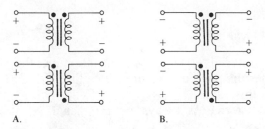

A.

B.

FIGURE 19–6 *Illustration of dot convention for polarity determination.*

Example 19–3

A transformer primary has 100 turns, and the secondary has 400 turns. What is the turns ratio?

Solution:

$$N_s = 400 \quad \text{and} \quad N_p = 100$$

$$\text{Turns ratio} = \frac{N_s}{N_p} = \frac{400}{100}$$

$$= 4$$

Review for 19–2

1. Define *turns ratio*.

2. How does the type of core affect a transformer?

3. Why are the directions of the winding important?

4. A certain transformer has a primary with 500 turns and a secondary with 250 turns. What is the turns ratio?

STEP-UP TRANSFORMERS

19–3   *A transformer in which the secondary voltage is greater than the primary voltage is called a step-up transformer.* The amount that the voltage is stepped up depends on the turns ratio.

   The ratio of secondary voltage to primary voltage is equal to the ratio of the number of secondary turns to the number of primary turns:

$$\frac{V_s}{V_p} = \frac{N_s}{N_p} \qquad\qquad (19\text{--}4)$$

where V_s is the secondary voltage and V_p is the primary voltage. From Equation (19–4), we get

$$V_s = \left(\frac{N_s}{N_p}\right)V_p \qquad\qquad (19\text{--}5)$$

This equation shows that the secondary voltage is equal to the turns ratio times the primary voltage. This condition assumes that the coefficient of coupling is 1. A good iron core transformer approaches this value.

   The turns ratio for a step-up transformer is always *greater* than 1.

Example 19–4

The transformer in Figure 19–7 has a 200-turn primary winding and a 600-turn secondary winding. What is the voltage across the secondary?

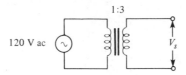

FIGURE 19–7

Solution:

$$N_p = 200 \quad \text{and} \quad N_s = 600$$

$$\text{Turns ratio} = \frac{N_s}{N_p} = \frac{600}{200} = 3$$

$$V_s = 3V_p = 3(120 \text{ V})$$
$$= 360 \text{ V}$$

Note that the turns ratio of 3 is indicated on the schematic as $1:3$, meaning that there are 3 secondary turns for each primary turn.

Review for 19–3

1. What does a step-up transformer do?
2. If the turns ratio is 5, how much greater is the secondary voltage than the primary voltage?
3. When 240 V ac are applied to a transformer with a turns ratio of 10, what is the secondary voltage?

STEP-DOWN TRANSFORMERS

19–4

A transformer in which the secondary voltage is less than the primary voltage is called a step-down transformer. The amount by which the voltage is stepped down depends on the turns ratio. Equation (19–5) applies also to a step-down transformer; the turns ratio, however, is always *less* than 1.

Example 19–5

The transformer in Figure 19–8 has 50 turns in the primary and 10 turns in the secondary. What is the secondary voltage?

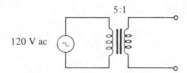

FIGURE 19–8

Solution:

$$N_p = 50 \quad \text{and} \quad N_s = 10$$

$$\text{Turns ratio} = \frac{N_s}{N_p} = \frac{10}{50}$$

$$= 0.2$$

$$V_s = 0.2V_p = 0.2(120 \text{ V})$$

$$= 24 \text{ V}$$

$P = R_p I_p^2 = R_s I_s^2$

$P_p \to V$

Review for 19–4

1. What does a step-down transformer do?

2. One hundred twenty volts ac are applied to the primary of a transformer with a turns ratio of 0.5. What is the secondary voltage?

3. A primary voltage of 120 V ac is reduced to 12 V ac. What is the turns ratio?

LOADING THE SECONDARY

19–5
When a load resistor is connected to the secondary, as shown in Figure 19–9, secondary current will flow because of the voltage induced in the secondary coil. It can be shown that the ratio of the primary current I_p to the secondary current I_s is equal to the turns ratio, as expressed in the following equation:

$$\frac{I_p}{I_s} = \frac{N_s}{N_p} \qquad (19–6)$$

A manipulation of this equation gives Equation (19–7), which shows that I_s is equal to I_p times the *reciprocal* of the turns ratio:

$$I_s = \left(\frac{N_p}{N_s}\right) I_p \qquad (19–7)$$

Thus, for a step-up transformer, in which N_s/N_p is greater than 1, the secondary current is less than the primary current. For a step-down transformer, N_s/N_p is less than 1, and I_s is greater than I_p.

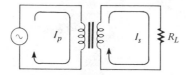

FIGURE 19–9

Example 19–6

The transformers in Figure 19–10A and B have loaded secondaries. If the primary current is 100 mA in each case, how much current flows through the load?

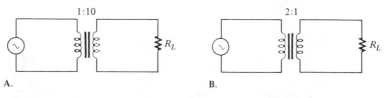

A.   B.

FIGURE 19–10

Solution:
Part A:

$$I_s = \left(\frac{N_p}{N_s}\right)I_p = 0.1(100 \text{ mA})$$

$$= 10 \text{ mA}$$

Part B:

$$I_s = \left(\frac{N_p}{N_s}\right)I_p = 2(100 \text{ mA})$$

$$= 200 \text{ mA}$$

Primary Power Equals Secondary Power

In a transformer the secondary power is the same as the primary power regardless of the turns ratio, as shown in the following steps:

$$P_p = V_p I_p \quad \text{and} \quad P_s = V_s I_s$$

$$I_s = \left(\frac{N_p}{N_s}\right)I_p \quad \text{and} \quad V_s = \left(\frac{N_s}{N_p}\right)V_p$$

By substitution, we obtain

$$P_s = \left(\frac{N_p}{N_s}\right)\left(\frac{N_s}{N_p}\right)V_p I_p$$

Canceling yields

$$P_s = V_p I_p = P_p$$

$$P = \frac{P}{I} \qquad R^2 I = R^2 I$$

Review for 19–5

1. If the turns ratio of a transformer is 2, is the secondary current greater than or less than the primary current? By how much?

2. A transformer has 100 primary turns and 25 secondary turns, and I_p is 0.5 A. What is the value of I_s?

3. In Problem 2, how much primary current is necessary to produce a secondary load current of 10 A?

REFLECTED IMPEDANCE

19–6

When an impedance is connected across the secondary of a transformer, to a source connected to the primary, it "appears" to have a value that is dependent on the turns ratio. That is, the source "sees" a certain impedance *reflected* from

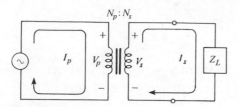

$N_p : N_s$

FIGURE 19–11 *Circuit for derivation of reflected impedance.*

the secondary into the primary circuit. This reflected impedance is determined as follows:

The impedance magnitude in the primary circuit of Figure 19–11 is $Z_p = V_p/I_p$. The impedance magnitude in the secondary circuit is $Z_L = V_s/I_s$. From the previous sections, we know that $V_s/V_p = N_s/N_p$ and $I_p/I_s = N_s/N_p$. Using these relationships, we find a formula for Z_p in terms of Z_L as follows:

$$\frac{Z_P}{Z_L} = \frac{V_p/I_p}{V_s/I_s} = \left(\frac{V_p}{V_s}\right)\left(\frac{I_s}{I_p}\right) = \left(\frac{N_p}{N_s}\right)\left(\frac{N_p}{N_s}\right)$$

$$= \frac{N_p^2}{N_s^2} = \left(\frac{N_p}{N_s}\right)^2$$

Solving for Z_p, we get

$$Z_p = \left(\frac{N_p}{N_s}\right)^2 Z_L \tag{19–8}$$

Equation (19–8) tells us that *the reflected impedance in the square of the reciprocal of the turns ratio times the load impedance.*

Example 19–7

Figure 19–12 shows a source that is transformer-coupled to a load resistor of 100 Ω. The transformer has a turns ratio of 4. What is the reflected resistance seen by the source?

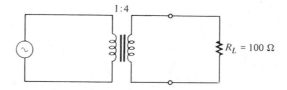

1:4

$R_L = 100\ \Omega$

FIGURE 19–12

Solution:

In this case, the load impedance is a resistor. Therefore, the reflected resistance is determined by Equation (19–8):

$$R_p = \left(\frac{1}{4}\right)^2 R_L = \left(\frac{1}{16}\right)(100\ \Omega)$$

$$= 6.25\ \Omega$$

The source sees a resistance of 6.25 Ω just as if it were connected directly, as shown in the equivalent circuit of Figure 19–13.

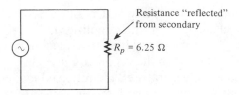

Resistance "reflected" from secondary

$R_p = 6.25\ \Omega$

FIGURE 19–13

Example 19–8

If a transformer is used in Figure 19–12 having 40 primary turns and 10 secondary turns, what is the reflected resistance?

Solution:

Turns ratio = 0.25

$$R_p = \left(\frac{1}{0.25}\right)^2 (100\ \Omega) = (4)^2 (100\ \Omega)$$

$$= 1600\ \Omega$$

This result illustrates the difference that the turns ratio makes.

Review for 19–6

1. Define *reflected impedance*.
2. What transformer characteristic determines the reflected impedance?
3. A given transformer has a turns ratio of 10, and the load is 50 Ω. How much impedance is reflected into the primary?
4. What is the turns ratio required to reflect a 4-Ω load resistance into the primary as 400 Ω?

IMPEDANCE MATCHING

19–7

Transformers are often used to match a load to a source impedance for *maximum power transfer.* An example is illustrated in Figure 19–14 where an audio ampli-

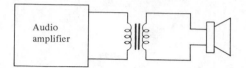

FIGURE 19–14 *Impedance matching an amplifier to a speaker.*

fier is *transformer-coupled* to a speaker. It is often desirable to have the maximum possible power available from the amplifier delivered to the speaker. The amplifier is the source in this case, and it is connected to the primary. The speaker is the load, and it is connected to the secondary.

To get maximum power transfer, the reflected load impedance "seen" by the amplifier must equal the output or source resistance of the amplifier. In other words, the impedance of the load must be *matched* to that of the source. We achieve this matching by choosing the proper turns ratio for the transformer.

Example 19–9

An amplifier has an 800-Ω output resistance. In order to provide maximum power to an 8 Ω speaker, what turns ratio must we use in the coupling transformer?

Solution:

$$Z_p = \left(\frac{1}{\text{turns ratio}}\right)^2 Z_L$$

Solve for the turns ratio as follows:

$$\left(\frac{1}{\text{turns ratio}}\right)^2 = \frac{Z_p}{Z_L}$$

$$\frac{1}{\text{turns ratio}} = \sqrt{\frac{Z_p}{Z_L}}$$

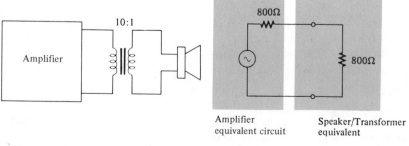

FIGURE 19–15

$$\text{Turns ratio} = \sqrt{\frac{Z_L}{Z_p}} = \sqrt{\frac{8\ \Omega}{800\ \Omega}}$$

$$= \frac{\sqrt{1}}{\sqrt{100}} = \frac{1}{10} = 0.1$$

The diagram and its equivalent reflected circuit are shown in Figure 19–15.

Review for 19–7

1. What does *impedance matching* mean?
2. What is the advantage of matching the load impedance to the impedance of a source?
3. A transformer has 100 primary turns and 50 secondary turns. What is the reflected impedance with 100 Ω across the secondary?

THE TRANSFORMER AS AN ISOLATION DEVICE

dc Isolation

19–8

As illustrated in Figure 19–16, if a direct current is made to flow in a transformer primary, nothing happens in the secondary, because a *changing* current in the primary is necessary to induce a voltage in the secondary. Therefore, the transformer serves to *isolate* the secondary from any dc in the primary.

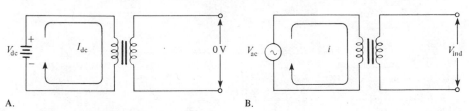

A.   B.

FIGURE 19–16 *dc isolation and ac coupling.*

In a typical application, a transformer can be used to keep the dc voltage on the output of an amplifier stage from affecting the dc bias of the next amplifier. Only the ac signal is coupled through the transformer from one stage to the next, as Figure 19–17 (page 712) illustrates.

Power Line Isolation

Transformers are often used to isolate the 60-Hz, 120-V ac power line from a piece of electronic equipment, such as a TV set or any test instrument that operates from the 60-Hz ac power.

The reason for using a transformer to couple the 60-Hz ac to the equipment is to prevent a possible shock hazard if the "hot" side (120 V ac) of the

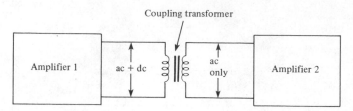

FIGURE 19–17 *Amplifier stages with transformer coupling for dc isolation.*

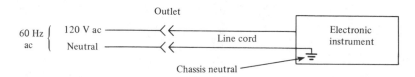

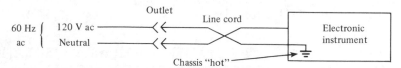

FIGURE 19–18 *Instrument powered without transformer isolation.*

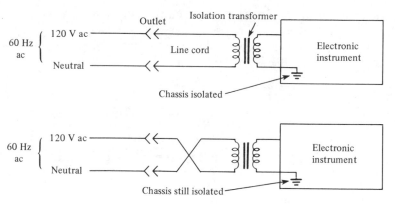

FIGURE 19–19 *Power line isolation.*

power line is connected to the equipment chassis. This condition is possible if the line cord socket can be plugged into the outlet either way. Figure 19–18 illustrates this situation.

A transformer can prevent this hazardous condition as illustrated in Figure 19–19. With such isolation, there is no way of directly connecting the 120-V ac line to the instrument ground, no matter how the power cord is plugged into the outlet.

Many TV sets, for example, do not have isolation transformers for reasons of economy. When working on the chassis, you should exercise care by using an external isolation transformer or by plugging into the outlet so that chassis ground is not connected to the 120-V side of the outlet.

Review for 19-8

1. Name two applications of a transformer as an isolation device.

2. Will a transformer operate with a dc input?

TAPPED TRANSFORMERS

19-9

A schematic diagram of a transformer with a *center-tapped* secondary winding is shown in Figure 19–20A. The center tap (CT) is equivalent to two secondary windings with half the total voltage across each.

The voltages between either end of the secondary and the center tap are, at any instant, equal in magnitude but opposite in polarity, as illustrated in Figure 19–20B. Here, for example, at some instant on the sine wave voltage, the polarity across the entire secondary is as shown (top end +, bottom −). At the center tap, the voltage is less positive than the top end but more positive than the bottom end of the secondary. Therefore, measured *with respect to the center tap*, the top end of the secondary is positive, and the bottom end is negative. This center-tapped feature is used in power supply rectifiers in which the ac voltage is converted to dc.

Some transformers have taps on the secondary other than at the center. Also, when a transformer is to be used on a power line, several taps are sometimes brought out near one end of the primary. When connection is made to one of these taps instead of to the end of the winding, the turns ratio is adjusted to overcome line voltages that are slightly too high or too low. Example diagrams are shown in Figure 19–21.

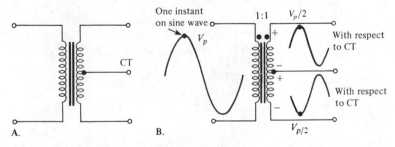

FIGURE 19–20 *A transformer with a center-tapped secondary.*

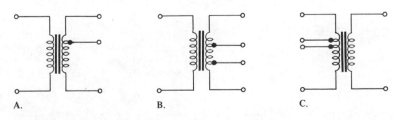

FIGURE 19–21 *Tapped transformers.*

Review for 19-9

1. What is a center tap?

2. How are the voltage phases at each end of the secondary related to the CT?

MULTIPLE-WINDING TRANSFORMERS

19—10 The basic transformer with one primary and one secondary is the simplest type of multiple-winding transformer; it has two windings. Other transformers may have more than one primary or more than one secondary.

Multiple Primaries

Some transformers are designed to operate from either 120-V ac or 240-V ac lines. These transformers usually have two primary windings, each of which is designed for 120 V ac. When the two are connected in series, the transformer can be used for 240-V ac operation, as illustrated in Figure 19–22.

Multiple Secondaries

More than one secondary can be wound on a common core. Transformers with several secondaries are often used to achieve several voltages by either stepping up or stepping down the primary voltage. These types are commonly used in power supply applications in which several voltage levels are required for the operation of an electronic instrument.

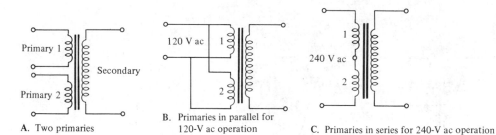

A. Two primaries

B. Primaries in parallel for 120-V ac operation

C. Primaries in series for 240-V ac operation

FIGURE 19–22 *Multiple-primary transformer.*

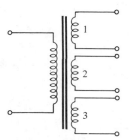

FIGURE 19–23 *Multiple-secondary transformer.*

A typical schematic of a multiple-secondary transformer is shown in Figure 19–23; this transformer has three secondaries. Sometimes you will find combinations of multiple-primary, multiple-secondary, and tapped transformers all in one unit.

Example 19–10

The transformer shown in Figure 19–24 has the numbers of turns (denoted as T) indicated. One of the secondaries is also center-tapped. If 120 V ac are connected to the primary, determine each secondary voltage and the voltages with respect to CT on secondary 2.

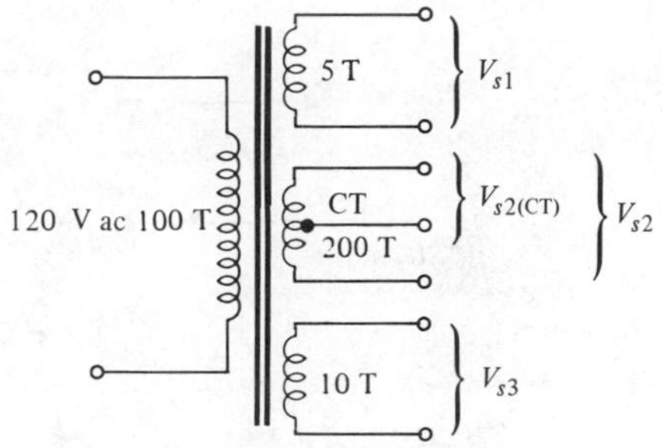

FIGURE 19–24

Solution:

$$V_{s1} = \left(\frac{5}{100}\right) 120 \text{ V} = 6 \text{ V}$$

$$V_{s2} = \left(\frac{200}{100}\right) 120 \text{ V} = 240 \text{ V}$$

$$V_{s2(CT)} = \frac{240 \text{ V}}{2} = 120 \text{ V}$$

$$V_{s3} = \left(\frac{10}{100}\right) 120 \text{ V} = 12 \text{ V}$$

Review for 19–10

1. A certain transformer has two secondaries. The turns ratio from primary to secondary 1 is 10. The turns ratio from the primary to the other secondary is 0.2. If 240 V ac are applied to the primary, what are the secondary voltages?

AUTOTRANSFORMERS

19–11
In an autotransformer, *one winding* serves as both the primary and the secondary. The winding is tapped at the proper points to achieve the desired turns ratio for stepping up or stepping down the voltage.

    The autotransformer is normally smaller and lighter than a typical multiple-winding type. However, since the primary and the secondary are the same winding, there is no electrical isolation between the windings. Autotransformers are used quite often in TV receivers. Figure 19–25A shows a diagram for a step-up autotransformer, and Part B shows a step-down autotransformer diagram.

    A *variac* is an adjustable autotransformer in which the output (secondary) voltage can be varied by means of an adjustable contact tap.

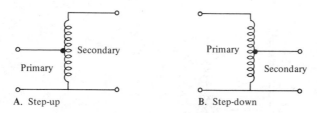

A. Step-up   B. Step-down

FIGURE 19–25 *Autotransformers.*

Review for 19–11

1. What is the difference between an autotransformer and a multiple-winding transformer?

2. What is a variac?

TRANSFORMER CONSTRUCTION

19–12
Two methods of construction for iron core transformers are shown in Figure 19–26. The core is normally made from *laminated* sheets of ferromagnetic material separated by thin insulating sheets to reduce losses in the core.

    Construction of a slug core transformer is shown in Figure 19–27. An air core transformer often has this type of construction.

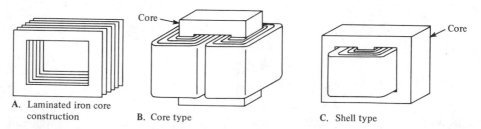

A. Laminated iron core construction   B. Core type   C. Shell type

FIGURE 19–26 *Construction of iron core transformer.*

FIGURE 19–27 *Slug core transformer.*

Transformer Ratings

Typically, a transformer rating is specified as, for example, 2 kVA, 500/50, 60 Hz. The 500 and the 50 can be either secondary or primary voltages. If 500 is the primary voltage, then 50 is the secondary voltage, and vice versa. The 2-kVA value is the *apparent power rating*.

Let us assume, for example, that in this transformer, 50 V is the secondary voltage. In this case the load current is $I_L = P_a/V_s = 2$ kVA/50 V $= 40$ A. If, on the other hand, 500 V is the secondary voltage, then $I_L = P_a/V_s = 2$ kVA/500 V $= 4$ A. These are the maximum currents that the secondary can handle in either case.

The reason that the power rating is in volt-amperes (VA) rather than average power (watts) is as follows: If the transformer load is purely capacitive, for example, the average power delivered to the load is zero. However, the current for $V_s = 500$ V and $X_C = 100 \ \Omega$ at 60 Hz is $I_L = 500$ V/100 $\Omega = 5$ A. This current exceeds the maximum that the secondary can handle, and it can cause damage to the transformer. Thus, you can see that it is meaningless to specify average power.

Figure 19–28 shows some typical types of transformers.

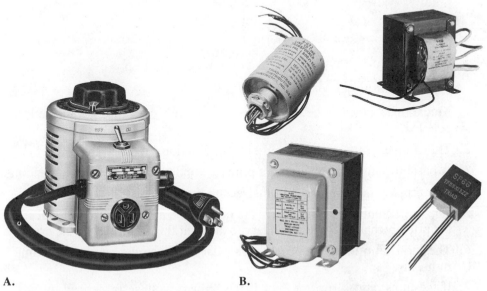

A. B.

FIGURE 19–28 *Typical transformers.* **A.** *Variable autotransformer. (Courtesy of The Superior Electric Company)* **B.** *Fixed. (Courtesy of Litton Triad-Utrad)*

Review for 19–12

1. What is laminated core construction?
2. What is the difference between construction of a shell type of transformer and that of a core type?

Formulas

$$k = \frac{\phi_{12}}{\phi_1} \qquad (19\text{--}1)$$

$$L_M = k\sqrt{L_1 L_2} \qquad (19\text{--}2)$$

$$\text{Turns ratio} = \frac{N_s}{N_p} \qquad (19\text{--}3)$$

$$\frac{V_s}{V_p} = \frac{N_s}{N_p} \qquad (19\text{--}4)$$

$$V_s = \left(\frac{N_s}{N_p}\right) V_p \qquad (19\text{--}5)$$

$$\frac{I_p}{I_s} = \frac{N_s}{N_p} \qquad (19\text{--}6)$$

$$I_s = \left(\frac{N_p}{N_s}\right) I_p \qquad (19\text{--}7)$$

$$Z_p = \left(\frac{N_p}{N_s}\right)^2 Z_L \qquad (19\text{--}8)$$

Summary

1. Mutual inductance is the inductance between two coils that are magnetically coupled.
2. Transformer action is based on mutual induction.
3. A basic transformer consists of a primary winding and a secondary winding.
4. The primary is the input, and the secondary is the output.
5. The coefficient of coupling (k) is the ratio of the amount of flux produced by the primary that links the secondary to the total flux produced by the primary.
6. The turns ratio is the number of secondary turns to the number of primary turns.
7. If the turns ratio is greater than 1, the primary voltage is stepped up.
8. If the turns ratio is less than 1, the primary voltage is stepped down.

9. With a loaded secondary, the current in the secondary is inversely related to the turns ratio.

10. The primary "sees" a certain impedance reflected from the secondary. The value of this impedance depends on the square of the reciprocal of the turns ratio.

11. For maximum power transfer to a load, the reflected load impedance must equal the source impedance, a condition called *impedance matching*.

12. A transformer can be used as an isolation device to keep dc from getting from the primary to the secondary or to isolate power line ground from equipment ground.

13. An autotransformer has one tapped winding used for both primary and secondary.

Self-Test

1. In a given transformer, 90% of the flux produced by the primary coil links the secondary. What is the value of the coefficient of coupling, k?

2. For the transformer in Problem 1, the inductance of the primary is 10 μH, and the inductance of the secondary is 5 μH. Determine L_M.

3. What is the turns ratio of a transformer having 12 primary turns and 36 secondary turns? Is it a step-up or a step-down transformer?

4. Determine the polarity of the secondary voltage for each transformer in Figure 19–29.

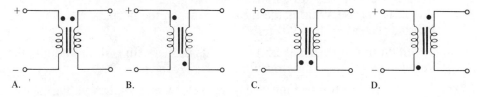

A.   B.   C.   D.

FIGURE 19–29

5. If 120 V ac are connected to the primary of a transformer with 10 primary turns and 15 secondary turns, what is the secondary voltage?

6. With 50 V ac across the primary, what is the secondary voltage if the turns ratio is 0.5?

7. If 240 V ac are applied to a transformer primary, and the secondary voltage is 60 V, what is the turns ratio?

8. If a load is connected to the transformer in Problem 7, and if I_p is 0.25 A, what is I_s?

9. What is the reflected impedance in Figure 19–30?

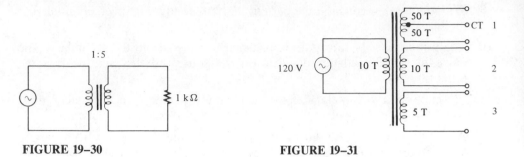

FIGURE 19–30   FIGURE 19–31

10. If the source resistance in Figure 19–30 is 10 Ω, what must R_L be for maximum power?

11. Determine the voltage for each secondary in Figure 19–31.

12. In Figure 19–31, what is the voltage from each end of the upper secondary to the center tap? Are the polarities the same?

Problems

Section 19–1
19–1. What is the mutual inductance when $k = 0.75$, $L_p = 1 \ \mu H$, and $L_s = 4 \ \mu H$?

19–2. Determine the coefficient of coupling when $L_M = 1 \ \mu H$, $L_p = 8 \ \mu H$, and $L_s = 2 \ \mu H$.

Section 19–2
19–3. What is the turns ratio of a transformer having 250 primary turns and 1000 secondary turns? What is the turns ratio when the primary has 400 turns and the secondary has 100 turns?

19–4. A certain transformer has 25 turns in its primary. In order to double the voltage, how many turns must be in the secondary?

19–5. For each transformer in Figure 19–32, sketch the secondary voltage showing its relationship to the primary voltage. Also indicate the amplitude.

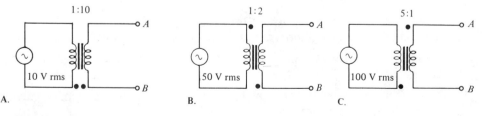

FIGURE 19–32

Section 19–3
19–6. To step 240 V ac up to 720 V, what must the turns ratio be?

19–7. The primary of a transformer has 120 V ac across it. What is the secondary voltage if the turns ratio is 5?

19–8. How many primary volts must be applied to a transformer with a turns ratio of 10 to obtain a secondary voltage of 60 V ac?

Section 19–4
19–9. To step 120 V down to 30 V, what must the turns ratio be?

19–10. The primary of a transformer has 1200 V across it. What is the secondary voltage if the turns ratio is 0.2?

19–11. How many primary volts must be applied to a transformer with a turns ratio of 0.1 to obtain a secondary voltage of 6 V ac?

Section 19–5
19–12. Determine I_s in Figure 19–33. What is the value of R_L?

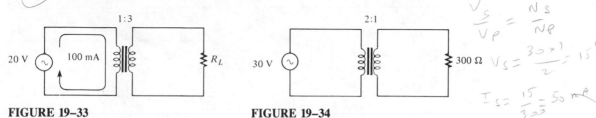

FIGURE 19–33   **FIGURE 19–34**

$$\frac{V_s}{V_p} = \frac{N_s}{N_p}$$

$$V_s = \frac{30 \times 1}{2} = 15$$

$$I_s = \frac{15}{3 \infty} = 50\ mA$$

$$I_p = \frac{50 \times 1}{2} = 25\ mA$$

$$P = 30 \times 25\ mA$$
$$= 750\ mW$$

19–13. Determine the following quantities in Figure 19–34:
    (a) Primary current   **(b)** Secondary current
    (c) Secondary voltage   **(d)** Power in the load

Section 19–6
19–14. What is the load resistance as seen by the source in Figure 19–35?

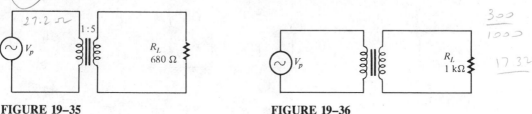

$$\frac{300}{1000}$$
$$17\ 32$$

FIGURE 19–35   **FIGURE 19–36**

19–15. What must the turns ratio be in Figure 19–36, in order to reflect 300 Ω into the primary?

Section 19–7
19–16. For the circuit in Figure 19–37 (page 722), find the turns ratio required to deliver maximum power to the 4-Ω speaker.

19–17. In Figure 19–37, what is the maximum power delivered to the speaker?

Section 19–8
19–18. What is the voltage across the load in each circuit of Figure 19–38?

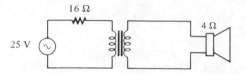

FIGURE 19–37

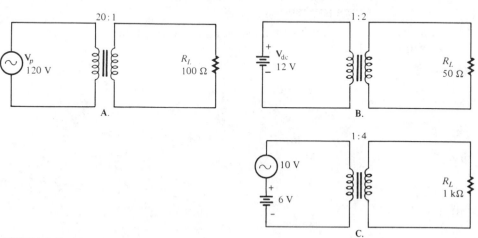

FIGURE 19–38

19–19. Determine the unknown meter readings in Figure 19–39.

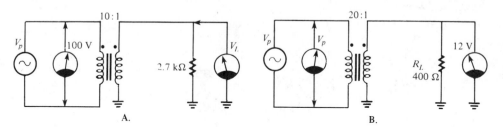

FIGURE 19–39

Section 19–9
19–20. Determine each unknown voltage indicated in Figure 19–40.

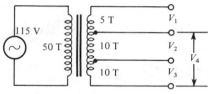

FIGURE 19–40

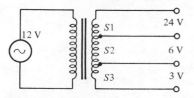

FIGURE 19–41

19–21. Using the indicated secondary voltages in Figure 19–41, determine the turns ratio of the primary to each tapped section.

Section 19–10

19–22. In Figure 19–42, each primary can accommodate 120 V ac. How should the primaries be connected for 240-V ac operation? Determine each secondary voltage.

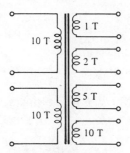

FIGURE 19–42

Section 19–11

19–23. Find the secondary voltage for each autotransformer in Figure 19–43.

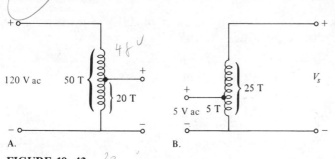

A.   B.

FIGURE 19–43

Section 19–12

19–24. A certain transformer is rated at 1 kVA. It operates on 60 Hz, 120 V ac. The secondary voltage is 600 V.
  (a) What is the maximum load current?
  (b) What is the smallest R_L that you can drive?
  (c) What is the largest capacitor that can be connected as a load?

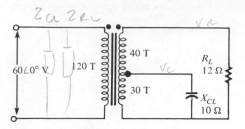

FIGURE 19-44

19-25. What is the kVA rating of a transformer that can handle a maximum load current of 10 A with a secondary voltage of 2.5 kV?

Advanced Problems

19-26. For the loaded, tapped-secondary transformer in Figure 19-44, determine the following:
   (a) All load voltages and currents
   (b) The impedance looking into the primary

19-27. A certain transformer is rated at 5 kVA, 2400/120 V, at 60 Hz.
   (a) What is the turns ratio if the 120 V is the secondary voltage?
   (b) What is the current rating of the secondary if 2400 V is the primary voltage?
   (c) What is the current rating of the primary if 2400 V is the primary voltage?

Answers to Section Reviews

Section 19-1:
1. The inductance between two coils. **2.** 45 mH. **3.** It increases.

Section 19-2:
1. Number of turns in secondary divided by number of turns in primary. **2.** It influences k. **3.** They determine polarities. **4.** 0.5.

Section 19-3:
1. Increases primary voltage. **2.** Five times greater. **3.** 2400 V.

Section 19-4:
1. Decreases primary voltage. **2.** 60 V. **3.** 0.1.

Section 19-5:
1. Less; half. **2.** 2 A. **3.** 2.5 A.

Section 19-6:
1. The impedance in the secondary reflected into the primary. **2.** The turns ratio. **3.** 0.5 Ω. **4.** 0.1.

Section 19-7:
1. Making the load impedance equal the source impedance. **2.** Maximum power is delivered to the load. **3.** 400 Ω.

Section 19–8:
1. dc isolation and power line isolation. 2. No.

Section 19–9:
1. A contact at the center of a winding. 2. Each is opposite to the other in polarity.

Section 19–10:
1. 2400 V, 48 V.

Section 19–11:
1. An autotransformer has only one winding. 2. A variable autotransformer.

Section 19–12:
1. Thin sheets of a ferromagnetic material separated by sheets of insulating material. 2. The shell type has the windings on one leg of the core.

$19-26-$

$$\frac{1}{Z_P} = \frac{1}{Z_{CL}} + \frac{1}{a_{RL}}$$

$$\frac{1}{Z_P} = \frac{1}{(\frac{N_P}{N_S})^2 X_{CL}} + \frac{1}{(\frac{N_P}{N_S})^2 R_L} = \frac{1}{(\frac{120}{30})10\Omega} + \frac{1}{(\frac{120}{70})^2 12\Omega} =$$

$$Z_P = 28.9\Omega$$

$$V_{X_C} = \left(\frac{30}{120}\right)(6\angle 0) = 15\angle 0$$

$$V_{R_L} = \left(\frac{70}{120}\right) 6\angle 0 =$$

$$I_{CL} = \frac{V_{X_C}}{X_{CL}} =$$

$$I_{R_L} = \frac{V_{R_L}}{R_L} =$$

In our coverage of ac analysis up to this point, only *single-phase* sinusoidal sources have been considered. In Chapter 11, it was mentioned that a sinusoidal voltage can be generated by rotating a conductor at a constant velocity in a magnetic field; in Chapter 13, the basic principles of electromagnetic induction were introduced.

In this chapter, the basic generation of single-phase sinusoidal wave forms is examined and extended to the generation of polyphase voltages. The advantages of polyphase systems in power applications are covered, and various types of three-phase connections and power measurement are introduced.

In this chapter, you will learn:

1. The basic principle of ac generators.
2. The factors that determine the frequency and amplitude of a generated voltage.
3. The differences of single-phase, two-phase, and three-phase generators.
4. The advantages of polyphase voltages over single phase.
5. The various types of three-phase generator configurations.
6. The relationships of currents and voltages in a three-phase system.
7. How to analyze three-phase systems.
8. The difference between a balanced and an unbalanced load.
9. Two methods for measuring power in three-phase systems.
10. How a basic power distribution system works.

20

POLYPHASE SYSTEMS IN POWER APPLICATIONS

20–1   A sinusoidal voltage can be generated by rotating a conductor formed into a loop at constant velocity through a uniform magnetic field, as illustrated in Figure 20–1A. When the conductor moves at a right angle (90° or 270°) to the magnetic lines of force, maximum voltage is induced. This condition occurs at two points in the rotation, as shown in Part B. The polarities of the induced voltages at these two points are opposite. When the conductor moves parallel (0° or 180°) with the magnetic field, zero voltage is induced in the conductor. This condition also occurs at two points in the rotation, as indicated. As the conductor rotates from the points of maximum induced voltage to the points of zero induced voltage, the angle with respect to the magnetic field is changing at a constant rate, causing the voltage to change according to the following formula:

$$v = V_{max}\sin\theta \qquad\qquad (20\text{–}1)$$

As the conductor rotates through a full revolution (360°), one full cycle of a sinusoidal voltage is induced, as indicated in Part C of the figure. As you can see, the induced voltage across the conductor is, at any instant during a revolution, dependent on the maximum voltage during that revolution and on the angle at which the conductor is moving through the magnetic field.

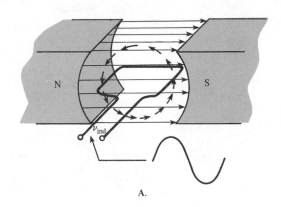

A.

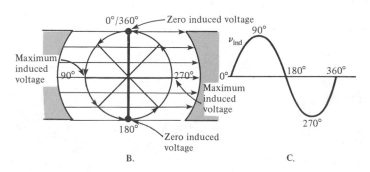

B.   C.

FIGURE 20–1   *Induced voltage produced by a conductive loop rotating in a magnetic field.*

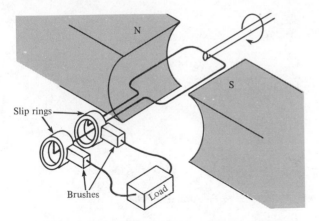

FIGURE 20–2   *A simple generator.*

A Simple ac Generator

To make practical the concept just discussed, there must be a means of mechanically rotating the conductive loop through the magnetic field and of electrically connecting the ends of the loop across a load in order to utilize the induced voltage. Figure 20–2 shows a shaft attached to the conductor such that when the shaft is rotated by mechanical power, the conductor is forced to rotate. The two open ends of the conductor are connected to conductive rings, called *slip rings,* that are free to rotate with the conductor. These slip rings make continuous connection with stationary contacts called *brushes*. The load is then connected across the brushes as shown so that the induced voltage causes current, thus delivering power to the load.

Factors That Affect Frequency

You have seen that one revolution of the conductor through the magnetic field in the basic ac generator (commonly called an *alternator*) produces one cycle of induced sinusoidal voltage. It is obvious that the rate at which the conductor is rotated determines the time for completion of one cycle. For example, if the conductor completes 60 revolutions in one second, the period of the resulting sine wave is 1/60 second. This corresponds to a frequency of 60 Hz. So the faster the conductor rotates, the higher the resulting frequency of the induced voltage, as illustrated in Figure 20–3 on page 730.

Another way of achieving a higher frequency is to increase the number of magnetic poles. In the previous discussion, two magnetic poles were used to illustrate the ac generator principle. During one revolution, the conductor passes under a north pole and a south pole, thus producing one cycle of a sine wave. When four magnetic poles are used instead of two, as shown in Figure 20–4, one cycle is generated during *one-half* a revolution. This doubles the frequency for the same rate of rotation.

An expression for frequency in terms of the number of poles and the number of revolutions per second (rps) is as follows:

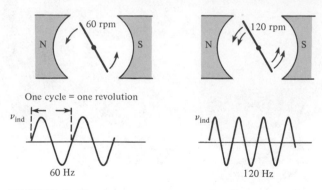

One cycle = one revolution

FIGURE 20–3 *Frequency is directly proportional to the rate of rotation.*

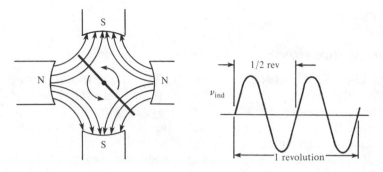

FIGURE 20–4 *Four poles achieve higher frequency than two for the same rps.*

$$f = (\text{number of pole pairs}) (\text{rps}) \qquad\qquad (20\text{–}2)$$

Example 20–1

A four-pole generator has a rotation speed of 100 rps. Determine the frequency of the output voltage.

Solution:

$$f = (\text{number of pole pairs}) (\text{rps}) = 2(100) = 200 \text{ Hz}$$

Factors That Affect Voltage Amplitude

Recall from Chapter 13 that the voltage induced in a conductor depends on the number of turns (N) and the rate of change with respect to the magnetic field ($d\phi/dt$). Therefore, when the speed of rotation of the conductor is increased, not only the frequency of the induced voltage increases — so does the magnitude. Since the frequency value is normally fixed, the most practical method of increasing the amount of induced voltage is to increase the number

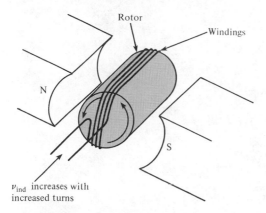

v_{ind} increases with
increased turns

FIGURE 20–5 *Multiple turns on a rotor.*

of turns of the conductor, as illustrated in Figure 20–5. The turns are wound on
a rotating drum called a *rotor*.

A Basic Two-Phase Generator

Figure 20–6A shows a second separate conductor loop added to the basic
generator in Figure 20–1 with the two loops separated by 90 degrees. Both loops
are mounted on the same rotor and therefore rotate at the same speed. Loop *A*
is 90 degrees ahead of loop *B* in the direction of rotation. As they rotate, two
induced sinusoidal voltages are produced that are 90 degrees apart in phase, as
shown in Part B of the figure.

A Basic Three-Phase Generator

Figure 20–7A shows a generator with three separate conductor loops
placed at 120-degree intervals around the rotor. This configuration generates three

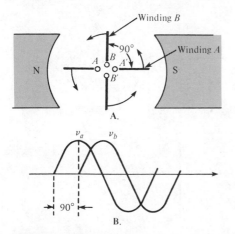

FIGURE 20–6 *Basic two-phase generator.*

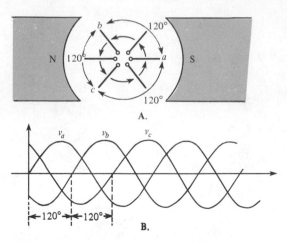

A.

B.

FIGURE 20–7 *Basic three-phase generator.*

sinusoidal voltages that are separated from each other by phase angles of 120 degrees, as shown in Part B.

A Practical ac Generator

The purpose of using the simple ac generators (alternators) presented up to this point was to illustrate the basic concept of the generation of sinusoidal voltages from the rotation of a conductor in a magnetic field. Most practical ac generators produce three-phase ac using a configuration that is somewhat different from that previously discussed. The basic principle, however, is the same. The reasons for the use of three-phase ac are examined in the next section.

A basic two-pole, three-phase alternator is shown in Figure 20–8. Most practical alternators are of this basic form. Rather than using a permanent magnet

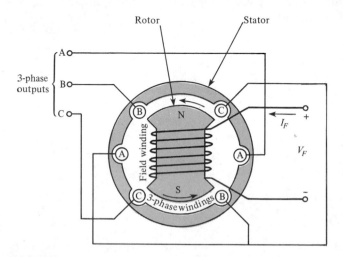

FIGURE 20–8 *Basic two-pole, three-phase alternator.*

in a fixed position, a rotating electromagnet is used. The electromagnet is created by passing direct current (I_F) through a winding around the rotor, as shown. These windings are called *field windings*. The direct current is applied through a brush and slip ring assembly. The stationary outer portion of the alternator is called the *stator*. Three separate windings are placed 120 degrees apart around the stator; three-phase voltages are induced in these windings as the magnetic field rotates, as indicated in Part B of Figure 20–7.

Review for 20–1

1. Describe the basic principle used in ac generators.

2. A two-pole, single-phase generator rotates at 400 rps. What frequency is produced?

3. How many separate armature windings are required in a three-phase alternator?

ADVANTAGES OF POLYPHASE IN POWER APPLICATIONS

There are several advantages of using polyphase alternators to deliver power to a load over using a single-phase system to deliver power to a given load. These considerations are now discussed.

20–2

The Copper Advantage

A Single-Phase System: Figure 20–9 is a simplified representation of a single-phase alternator connected to a resistive load. The coil symbol represents the alternator winding.

A single-phase sinusoidal voltage is induced in the winding and applied to the 60-Ω load, as indicated in Figure 20–10. The resulting load current is

$$\mathbf{I}_L = \frac{120\angle 0° \text{ V}}{60\angle 0° \text{ }\Omega} = 2\angle 0° \text{ A}$$

The total current that must be delivered by the alternator to the load is $2\angle 0°$ A. This means that the two conductors carrying current to and from the load must *each* be capable of handling 2 A; thus, the *total copper cross section* must handle 4 A. (The copper cross section is a measure of the *total* amount of wire required based on its physical size as related to its diameter.) The total load power is $I_L^2 R_L = 240$ W.

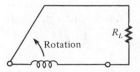

FIGURE 20–9 *Simplified representation of a single-phase alternator connected to a resistive load.*

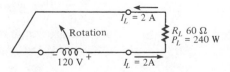

FIGURE 20–10 *Single-phase example.*

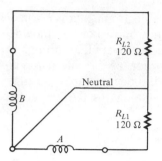

FIGURE 20–11 *Simplified representation of a two-phase alternator with each phase connected to a 120-Ω load.*

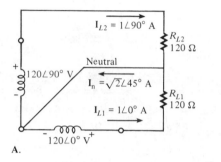   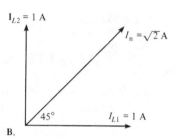

A.   **B.**

FIGURE 20–12 *Two-phase example.*

A Two-Phase System: Figure 20–11 shows a simplified representation of a two-phase alternator connected to two 120-Ω load resistors. The coils drawn at a 90-degree angle represent the armature windings spaced 90 degrees apart. An equivalent single-phase system would be required to feed two 120-Ω resistors in parallel, thus creating a 60-Ω load.

The voltage across load R_{L1} is $120\angle 0°$ V, and the voltage across load R_{L2} is $120\angle 90°$ V, as indicated in Figure 20–12A. The current in R_{L1} is

$$\mathbf{I}_{L1} = \frac{120\angle 0° \text{ V}}{120\angle 0° \text{ }\Omega} = 1\angle 0° \text{ A}$$

and the current in R_{L2} is

$$\mathbf{I}_{L2} = \frac{120\angle 90° \text{ V}}{120\angle 0° \text{ }\Omega} = 1\angle 90° \text{ A}$$

Notice that the two load resistors are connected to a common or neutral conductor, which provides a return path for the currents. Since the two load currents are 90 degrees out of phase with each other, the current in the neutral conductor is the *phasor sum* of the load currents.

$$I_n = \sqrt{I_{L1}^2 + I_{L2}^2} = \sqrt{2} \text{ A} = 1.414 \text{ A}$$

Figure 20–12B shows the phasor diagram for the currents in the two-phase system.

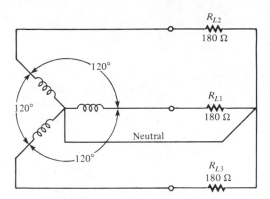

FIGURE 20–13 *A simplified representation of a three-phase alternator with each phase connected to a 180-Ω load.*

Three conductors are required in this system to carry current to and from the loads. Two of the conductors must be capable of handling 1 A each, and the neutral conductor must handle 1.414 A. The total copper cross section must handle 1 A + 1 A + 1.414 A = 3.414 A. This is *less* copper cross section than was required in the single-phase system to deliver the same amount of total load power ($P_{L1} + P_{L2} = I_{L1}^2 R_{L1} + I_{L2}^2 R_{L2}$ = 120 W + 120 W = 240 W).

A Three-Phase System: Figure 20–13 shows a simplified representation of a three-phase alternator connected to three 180-Ω resistive loads. An equivalent single-phase system would be required to feed three 180-Ω resistors in parallel, thus creating an effective load resistance of 60 Ω. The coils represent the alternator windings separated by 120 degrees.

The voltage across R_{L1} is $120\angle 0°$ V, the voltage across R_{L2} is $120\angle +120°$ V, and the voltage across R_{L3} is $120\angle -120°$ V, as indicated in Figure 20–14A. The current from each winding to its respective load is as follows:

$$\mathbf{I}_{L1} = \frac{120\angle 0° \text{ V}}{180\angle 0° \text{ }\Omega} = 0.667\angle 0° \text{ A}$$

$$\mathbf{I}_{L2} = \frac{120\angle 120° \text{ V}}{180\angle 0° \text{ }\Omega} = 0.667\angle +120° \text{ A}$$

$$\mathbf{I}_{L3} = \frac{120\angle -120° \text{ V}}{180\angle 0° \text{ }\Omega} = 0.667\angle -120° \text{ A}$$

The total load power is

$$P_{tot} = I_{L1}^2 R_{L1} + I_{L2}^2 R_{L2} + I_{L3}^2 R_{L3} = 240 \text{ W}$$

This is the same total load power as delivered by both the single-phase and the two-phase systems previously discussed.

Notice that four conductors, including the neutral, are required to carry the currents to and from the loads. The current in each of the three conductors is 0.667 A, as indicated in the figure. The current in the neutral conductor is the phasor sum of the three load currents and is equal to *zero,* as shown in the

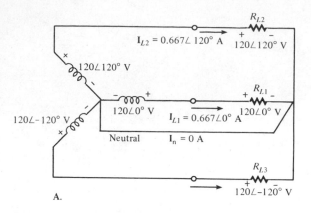

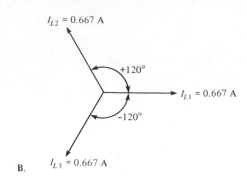

FIGURE 20–14   *Three-phase example.*

following equation, with reference to the phasor diagram in Figure 20–14B. This condition, where all loads are equal and the neutral current is zero, is called a *balanced load* condition.

$$\mathbf{I}_{L1} + \mathbf{I}_{L2} + \mathbf{I}_{L3} = 0.667\angle 0° \text{ A} + 0.667\angle +120° \text{ A} + 0.667\angle -120° \text{ A}$$
$$= 0.667 \text{ A} - 0.3335 \text{ A} + j0.578 \text{ A} - 0.3335 \text{ A} - j0.578 \text{ A}$$
$$= 0.667 \text{ A} - 0.667 \text{ A}$$
$$= 0 \text{ A}$$

The total copper cross section must handle $0.667 \text{ A} + 0.667 \text{ A} + 0.667 \text{ A} + 0 \text{ A} = 2 \text{ A}$. This result shows that considerably less copper is required to deliver the same load power with a three-phase system than is required for either the single-phase or the two-phase systems. The amount of copper is an important consideration in power distribution systems.

Example 20–2

Compare the total copper cross sections in terms of current-carrying capacity for single-phase and three-phase 120-V systems with effective load resistances of 12 Ω.

Solution:

Single-phase system: The load current is

$$I_L = \frac{120 \text{ V}}{12 \text{ }\Omega} = 10 \text{ A}$$

The conductor to the load must carry 10 A, and the conductor from the load must also carry 10 A. The total copper cross section, therefore, must be sufficient to handle 20 A.

Three-phase system: For an effective load resistance of 12 Ω, the three-phase alternator feeds three load resistors of 36 Ω each. The current in each load resistor is

$$I_L = \frac{120 \text{ V}}{36 \ \Omega} = 3.33 \text{ A}$$

Each of the three conductors feeding the balanced load must carry 3.33 A, and the neutral current is zero. Therefore, the total copper cross section must be sufficient to handle 9.99 A. This is significantly less than for the single-phase system with an equivalent load.

The Advantage of Constant Power

A second advantage of polyphase systems over single-phase is that polyphase systems produce a constant amount of power in the load.

A Single-Phase System: As shown in Figure 20–15, the load power fluctuates as the square of the sinusoidal voltage divided by the resistance. It changes from a maximum of $V_{R(\max)}^2/R_L$ to a minimum of zero at a frequency equal to twice that of the voltage.

A Polyphase System: The power wave form across one of the load resistors in a two-phase system is 90 degrees out-of-phase with the power wave form across the other, as shown in Figure 20–16. When the instantaneous power to one load is maximum, the other is minimum. In between the maximum and minimum points, one power is increasing while the other is decreasing. Careful examination of the power wave forms shows that when two instantaneous values are added, the sum is always constant and equal to $V_{R(\max)}^2/R_L$.

In a three-phase system, the total power delivered to the load resistors is also constant for the same basic reason as for the two-phase system. The sum of the instantaneous voltages is always the same; therefore, the power has a constant value. A constant load power means a uniform conversion of mechanical to electrical energy, which is an important consideration in many power applications.

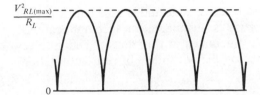

FIGURE 20–15 *Single-phase load power ($\sin^2$ curve).*

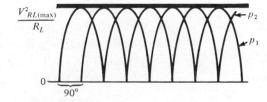

FIGURE 20–16 *Two-phase power.*

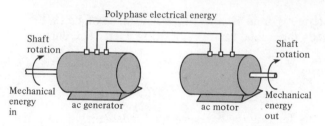

FIGURE 20–17 *Simple example of mechanical-to-electrical-to-mechanical energy conversion.*

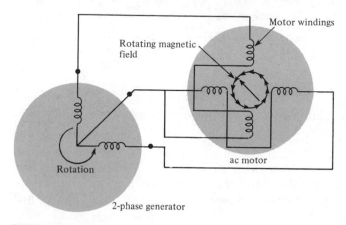

FIGURE 20–18 *A two-phase generator producing a constant rotating magnetic field in an ac motor.*

The Advantage of a Constant, Rotating Magnetic Field

In many applications, ac generators are used to drive ac motors for conversion of electrical energy to mechanical energy in the form of shaft rotation in the motor. The original energy for operation of the generator can come from any of several sources such as hydroelectric, steam, etc. Figure 20–17 illustrates the basic concept.

When a polyphase generator is connected to the motor windings as depicted in Figure 20–18, where two-phase is used for purposes of illustration, a magnetic field is created within the motor that has a constant flux density and that rotates at the frequency of the two-phase sine wave, The motor's rotor is pulled around at a constant rotational velocity by the rotating magnetic field, producing a constant shaft rotation.

Three-phase, of course, has the same advantage and is used universally in the field of power distribution. Single-phase is unsuitable for such applications, because it produces a magnetic field that fluctuates in flux density and reverses direction during each cycle without providing the advantage of constant rotation.

Review for 20–2

1. List three advantages of polyphase systems over single-phase systems.

2. Which advantage(s) are most important in mechanical-to-electrical energy conversions?

3. Which advantage(s) are most important in electrical-to-mechanical conversions?

THREE-PHASE GENERATOR CONFIGURATIONS

The Y-Connected Generator

20–3

In introduction of three-phase systems in previous sections, the Y-connection was used for illustration. We now examine this configuration further. A second configuration, called the *delta* (Δ), is introduced later.

A Y-connected system can be either a *three-wire* or, when the neutral is used, a *four-wire* system, as shown in Figure 20–19, connected to a generalized load represented by the block. Recall that when the loads are perfectly balanced, the neutral current is zero; therefore, the neutral conductor is unnecessary. However, in cases where the loads are not equal (unbalanced), a neutral wire is essential to provide a return current path, because the neutral current has a nonzero value.

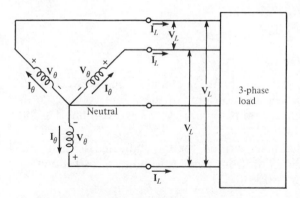

FIGURE 20–19 *Y-connected generator.*

The voltages across the generator windings are called *phase voltages* (V_θ), and the currents through the windings are called *phase currents* (I_θ). Also, the currents in the lines connecting the generator windings to the load are called *line currents* (I_L), and the voltages across the lines are called the *line voltages* (V_L). Note that the magnitude of each line current is equal to the corresponding phase current in the Y-connected circuit.

$$I_L = I_\theta \qquad\qquad (20\text{–}3)$$

In Figure 20–20, the line terminations of the windings are designated *a*, *b*, and *c*, and the neutral point is designated "n." These letters are added as subscripts to the phase and line currents to indicate the phase with which each is associated. The phase voltages are also designated in the same manner. Notice that the phase voltages are always positive at the terminal end of the winding and are negative at the neutral point. The line voltages are from one winding terminal

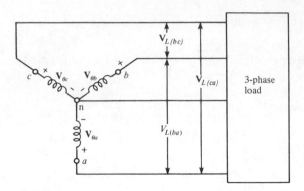

FIGURE 20–20 *Phase voltages and line voltages in a Y-connected system.*

to another, as indicated by the double-letter subscripts. For example, $\mathbf{V}_{L(ab)}$ is the line voltage from a to b, and $\mathbf{V}_{L(ba)}$ is the line voltage from b to a.

Figure 20–21A shows a phasor diagram for the phase voltages. By rotation of the phasors, as shown in Part B, $V_{\theta a}$ is given a reference angle of zero for analysis purposes, and the polar expressions for the phasor voltages are as follows.

$$\mathbf{V}_{\theta a} = V_{\theta a} \angle 0°$$

$$\mathbf{V}_{\theta b} = V_{\theta b} \angle +120°$$

$$\mathbf{V}_{\theta c} = V_{\theta c} \angle -120°$$

There are three line voltages: one between a and b, one between a and c, and another between b and c. The relationships of these line voltages to the phase voltages are developed with the aid of Figure 20–22 and Kirchhoff's voltage law.

Kirchhoff's voltage equation is written for the indicated loop in Figure 20–22A using a clockwise direction from point b to point a, as follows:

$$\mathbf{V}_{L(ba)} - \mathbf{V}_{\theta a} + \mathbf{V}_{\theta b} = 0$$

$$\mathbf{V}_{L(ba)} = \mathbf{V}_{\theta a} - \mathbf{V}_{\theta b}$$

$$= V_{\theta a} \angle 0° - V_{\theta b} \angle 120°$$

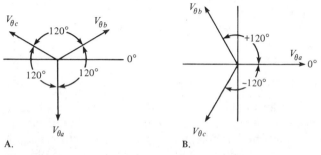

A.

B.

FIGURE 20–21 *Phase voltage diagram.*

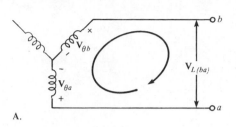

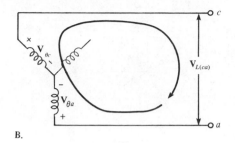

A.

B.

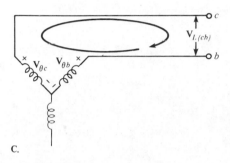

C.

FIGURE 20–22 *Development of the relationship between V_L and V_θ.*

All the phase voltages are equal, so

$$V_{\theta a} = V_{\theta b} = V_{\theta c} = V_\theta$$

Therefore,

$$\begin{aligned}
\mathbf{V}_{L(ba)} &= V_\theta - (-0.5V_\theta + j0.866V_\theta) \\
&= 1.5V_\theta - j0.866V_\theta \\
&= \sqrt{(1.5V_\theta)^2 + (0.866V_\theta)^2} \angle -\arctan\left(\frac{0.866}{1.5}\right) \\
&= \sqrt{3}\ V_\theta \angle -30°
\end{aligned}$$

Next, Kirchhoff's voltage equation is written for the indicated loop in Figure 20–22B.

$$\begin{aligned}
\mathbf{V}_{L(ca)} &= \mathbf{V}_{\theta c} - \mathbf{V}_{\theta a} \\
&= V_{\theta c}\angle -120° - V_{\theta a}\angle 0° \\
&= (-0.5V_\theta - j0.866V_\theta) - V_\theta \\
&= -1.5V_\theta - j0.866V_\theta \\
&= \sqrt{(1.5V_\theta)^2 + (0.866V_\theta)^2} \angle -\left[180 - \arctan\left(\frac{0.866}{1.5}\right)\right] \\
&= \sqrt{3}\ V_\theta \angle -150°
\end{aligned}$$

Finally, Kirchhoff's voltage equation is written for the loop, as shown in Figure 20–22C.

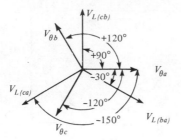

FIGURE 20–23 *Phasor diagram for the phase voltages and line voltages in a three-phase system.*

$$\mathbf{V}_{L(cb)} = \mathbf{V}_{\theta c} - \mathbf{V}_{\theta b}$$
$$= V_{\theta c}\angle -120° - V_{\theta b}\angle +120°$$
$$= (-0.5V_\theta - j0.866V_\theta) - (-0.5V_\theta + j0.866V_\theta)$$
$$= j0.866V_\theta + j0.866V_\theta$$
$$= \sqrt{3}\ V_\theta\angle +90°$$

You can see from the above results that the magnitude of each line voltage is equal to $\sqrt{3}$ times the phase voltage.

$$V_L = \sqrt{3}\ V_\theta \qquad\qquad (20\text{--}4)$$

The line voltage phasor diagram is shown in Figure 20–23 superimposed on the phasor diagram for the phase voltages. Notice that there is a phase angle of 30 degrees between each line voltage and the nearest phase voltage and that the line voltages are 120 degrees apart.

Example 20–3

The instantaneous position of a certain ac generator is shown in Figure 20–24. If each phase voltage has a magnitude of 120 V rms, determine the magnitude of each line voltage, and sketch the phasor diagram.

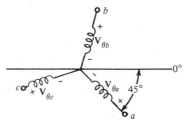

FIGURE 20–24

Solution:

The magnitude of each line voltage is

$$V_L = \sqrt{3}\, V_\theta = \sqrt{3}\,(120\ \text{V}) = 207.85\ \text{V}$$

The phasor diagram for the given instantaneous generator position is shown in Figure 20–25.

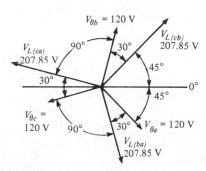

FIGURE 20–25

The Δ-Connected Generator

In the Y-connected generator, two voltage magnitudes are available at the terminals in the four-wire system: the phase voltage and the line voltage. Also, in the Y-connected generator, the line current is equal to the phase current. Keep these characteristics in mind as you examine the Δ-connected generator.

The windings of a three-phase generator can be rearranged to form a Δ-connected generator, as shown in Figure 20–26. By examination of this diagram, it is apparent that the magnitudes of the line voltages and phase voltages are equal, but the line currents do not equal the phase currents.

Since this is a three-wire system, only a single voltage magnitude is available, expressed as:

$$V_L = V_\theta \qquad\qquad (20\text{–}5)$$

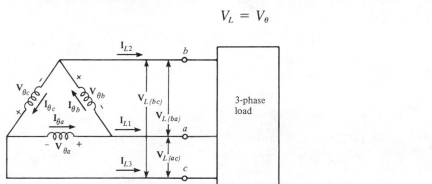

FIGURE 20–26 *Δ-connected generator.*

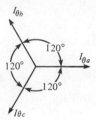

FIGURE 20–27   *Phase current diagram.*

All of the phase voltages are equal; thus, the line voltages are expressed in polar form as follows:

$$\mathbf{V}_{L(ac)} = V_\theta \angle 0°$$

$$\mathbf{V}_{L(ba)} = V_\theta \angle +120°$$

$$\mathbf{V}_{L(bc)} = V_\theta \angle -120°$$

The phasor diagram for the phase currents is shown in Figure 20–27, and the polar expressions for each current are as follows:

$$\mathbf{I}_{\theta a} = I_{\theta a} \angle 0°$$

$$\mathbf{I}_{\theta b} = I_{\theta b} \angle +120°$$

$$\mathbf{I}_{\theta c} = I_{\theta c} \angle -120°$$

The relationship between line current and phase currents for a balanced load (all phase currents equal) is developed as follows by applying Kirchhoff's current law at each node ($I_{\theta a} = I_{\theta b} = I_{\theta c}$).

$$\mathbf{I}_{L1} + \mathbf{I}_{\theta b} - \mathbf{I}_{\theta a} = 0$$

$$\begin{aligned}
\mathbf{I}_{L1} &= \mathbf{I}_{\theta a} - \mathbf{I}_{\theta b} \\
&= I_{\theta a} \angle 0° - I_{\theta b} \angle +120° \\
&= I_\theta - (-0.5 I_\theta + j0.866 I_\theta) \\
&= 1.5 I_\theta - j0.866 I_\theta \\
&= \sqrt{(1.5 I_\theta)^2 + (0.866 I_\theta)^2} \ \angle -\arctan\!\left(\frac{0.866}{1.5}\right) \\
&= \sqrt{3}\, I_\theta \angle -30°
\end{aligned}$$

Next, $\mathbf{I}_{L2}$ is

$$\begin{aligned}
\mathbf{I}_{L2} &= I_{\theta b} \angle +120° - I_{\theta c} \angle -120° \\
&= (-0.5 I_\theta + j0.866 I_\theta) - (-0.5 I_\theta - j0.866 I_\theta) \\
&= j0.866 I_\theta + j0.866 I_\theta \\
&= \sqrt{3}\, I_\theta \angle +90°
\end{aligned}$$

Finally, $\mathbf{I}_{L3}$ is

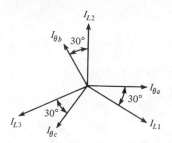

FIGURE 20–28 *Phasor diagram of phase currents and line currents.*

$$\begin{aligned}
\mathbf{I}_{L3} &= I_{\theta c}\angle-120° - I_{\theta a}\angle 0° \\
&= (-0.5I_\theta - j0.866I_\theta) - I_\theta \\
&= -1.5I_\theta - j0.866I_\theta \\
&= \sqrt{(1.5I_\theta)^2 + (0.866I_\theta)^2} \angle-\left(180 - \arctan\left(\frac{0.866}{1.5}\right)\right) \\
&= \sqrt{3}\,I_\theta\angle-150°
\end{aligned}$$

These results show that the relationship of the line current magnitude to the phase current magnitude is

$$I_L = \sqrt{3}\,I_\theta \qquad\qquad (20\text{--}6)$$

Also notice that there is a 30-degree angle between each line current and the nearest phase current, as shown in the phasor diagram of Figure 20–28.

Example 20–4

The three-phase Δ-connected generator represented in Figure 20–29 is driving a balanced load such that each phase current is 10 A in magnitude. When $\mathbf{I}_{\theta a} = 10\angle+30°$ A, determine the following:

    (a) The polar expressions for the other phase currents.
    (b) The polar expressions for each of the line currents.
    (c) Sketch the complete current phasor diagram.

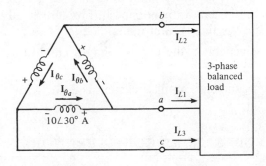

FIGURE 20–29

Example 20–4 (continued)

Solution:

(a) The phase currents are separated by 120 degrees; therefore

$$\mathbf{I}_{\theta b} = 10\angle(30° + 120°) = 10\angle +150° \text{ A}$$
$$\mathbf{I}_{\theta c} = 10\angle(30° - 120°) = 10\angle -90° \text{ A}$$

(b) The line currents are separated from the nearest phase current by 30 degrees; therefore

$$\mathbf{I}_{L1} = \sqrt{3}\, I_{\theta a}\angle(30° - 30°) = 17.32\angle 0° \text{ A}$$
$$\mathbf{I}_{L2} = \sqrt{3}\, I_{\theta b}\angle(150° - 30°) = 17.32\angle +120° \text{ A}$$
$$\mathbf{I}_{L3} = \sqrt{3}\, I_{\theta c}\angle(-90° - 30°) = 17.32\angle -120° \text{ A}$$

(c) The phasor diagram is shown in Figure 20–30.

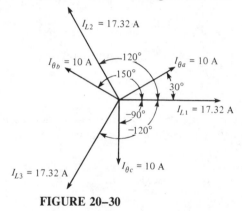

FIGURE 20–30

Review for 20–3

1. In a certain three-wire, Y-connected generator, the phase voltages are 1 kV. Determine the magnitude of the line voltages.

2. In the Y-connected generator mentioned in Problem 1, all the phase currents are 5 A. What are the line current magnitudes?

3. In a Δ-connected generator, the phase voltages are 240 V. What are the line voltages?

4. In a Δ-connected generator, a phase current is 2 A. Determine the magnitude of the line current.

ANALYSIS OF THREE-PHASE SOURCE/LOAD CONFIGURATIONS

20–4   As with the generator connections, the load can be either in a Y or a Δ configuration. A Y-connected load is shown in Figure 20–31A, and a Δ-connected load

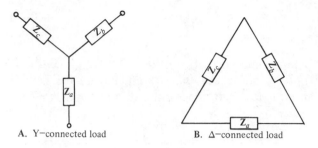

FIGURE 20–31 *Three-phase loads.*

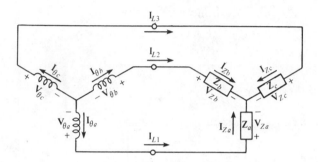

FIGURE 20–32 *A Y-connected source feeding a Y-connected load.*

is shown in Part B. The blocks represent the load impedances, which can be resistive, reactive, or both.

In this section, four source/load configurations are examined: a Y-connected source driving a Y-connected load (Y-Y system), a Y-connected source driving a Δ-connected load (Y-Δ system), a Δ-connected source driving a Δ-connected load (Δ-Δ system), and a Δ-connected source driving a Y-connected load (Δ-Y system).

The Y-Y System

Figure 20–32 shows a Y-connected source driving a Y-connected load. The load can be a balanced load, such as a three-phase motor where $Z_a = Z_b = Z_c$, or it can be three independent single-phase loads where, for example, Z_a is a lighting circuit, Z_b is a heater, and Z_c is an air-conditioning compressor.

An important feature of a Y-connected source is that two different values of three-phase voltage are available: the phase voltage and the line voltage. For example, in the standard power distribution system, a three-phase transformer can be considered a source of three-phase voltage supplying 120 V and 208 V. In order to utilize a phase voltage of 120 V, the loads are connected in the Y configuration. Later, you will see that a Δ-connected load is used for the 208-V line voltages.

Notice in the Y-Y system in Figure 20–32 that the phase current, the line current, and the load current are all equal in each phase. Also, each load voltage equals the corresponding phase voltage. These relationships are expressed as follows and are true for either a balanced or an unbalanced load.

$$I_\theta = I_L = I_Z \qquad\qquad (20\text{--}7)$$

$$V_\theta = V_Z \qquad\qquad (20\text{--}8)$$

where V_Z and I_Z are the load voltage and current.

For a balanced load, all the phase currents are equal, and the neutral current is zero. For an unbalanced load, each phase current is different, and the neutral current is, therefore, nonzero.

Example 20–5

In the Y-Y system of Figure 20–33, determine the following:
(a) Each load current
(b) Each line current
(c) Each phase current
(d) Neutral current
(e) Each load voltage

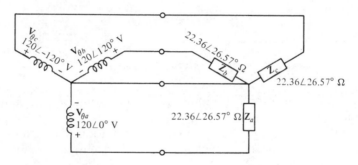

FIGURE 20–33

Solution:

This system has a balanced load.

(a) $\mathbf{Z}_a = \mathbf{Z}_b = \mathbf{Z}_c = 20\ \Omega + j10\ \Omega = 22.36\angle 26.57°\ \Omega$

$$\mathbf{I}_{Za} = \frac{\mathbf{V}_{\theta a}}{\mathbf{Z}_a} = \frac{120\angle 0°\ \text{V}}{22.36\angle 26.57°\ \Omega} = 5.37\angle -26.57°\ \text{A}$$

$$\mathbf{I}_{Zb} = \frac{\mathbf{V}_{\theta b}}{\mathbf{Z}_b} = \frac{120\angle -120°\ \text{V}}{22.36\angle 26.57°\ \Omega} = 5.37\angle -146.57°\ \text{A}$$

$$\mathbf{I}_{Zc} = \frac{\mathbf{V}_{\theta c}}{\mathbf{Z}_c} = \frac{120\angle +120°\ \text{V}}{22.36\angle 26.57°\ \Omega} = 5.37\angle +93.43°\ \text{A}$$

(b) $\mathbf{I}_{L1} = 5.37\angle -26.57°\ \text{A}$

$\mathbf{I}_{L2} = 5.37\angle -146.57°\ \text{A}$

$\mathbf{I}_{L3} = 5.37\angle +93.43°\ \text{A}$

(c) $\mathbf{I}_{\theta a} = 5.37\angle -26.57°$ A

$\mathbf{I}_{\theta b} = 5.37\angle -146.57°$ A

$\mathbf{I}_{\theta c} = 5.37\angle +93.43°$ A

(d) $\mathbf{I}_n = \mathbf{I}_{Za} + \mathbf{I}_{Zb} + \mathbf{I}_{Zc}$

$= 5.37\angle -26.57°$ A $+ 5.37\angle -146.57°$ A $+ 5.37\angle +93.43°$ A

$= (4.8$ A $- j2.4$ A$) + (-4.48$ A $- j2.96$ A$)$

$+ (-0.32$ A $+ j5.36$ A$)$

$= 0$ A

If the load impedances were not equal (balanced load), the neutral current would have a nonzero value.

(e) The load voltages are equal to the corresponding source phase voltages:

$$\mathbf{V}_{Za} = 120\angle 0°\ \mathrm{V}$$

$$\mathbf{V}_{Zb} = 120\angle -120°\ \mathrm{V}$$

$$\mathbf{V}_{Zc} = 120\angle +120°\ \mathrm{V}$$

The Y-Δ System

Figure 20–34 shows a Y-connected source feeding a Δ-connected load. An important feature of this configuration is that each phase of the load has the full *line voltage* across it.

$$V_Z = V_L \tag{20–9}$$

The line currents equal the corresponding phase currents, and each line current divides into two load currents, as indicated. For a *balanced* load ($Z_a = Z_b = Z_c$), the expression for the current in each load is developed as follows:

$$\mathbf{I}_{L1} = \mathbf{I}_{Za} - \mathbf{I}_{Zb}$$
$$= I_{Za}\angle 0° - I_{Zb}\angle -120°$$

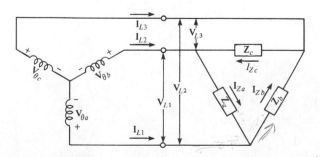

FIGURE 20–34 *A Y-connected source feeding a Δ-connected load.*

$$= I_{Za} - [I_{Zb} \cos(-120°) + jI_{Zb} \sin(-120°)]$$
$$= I_{Za} + 0.5I_{Zb} + j0.866I_{Zb}$$

Since $I_{Za} = I_{Zb} = I_{Zc} = I_Z$ and $I_{L1} = I_{L2} = I_{L3} = I_L$ for a balanced load,

$$\mathbf{I}_L = 1.5I_Z + j0.866I_Z$$

Therefore,

$$I_L = \sqrt{(1.5I_Z)^2 + (0.866I_Z)^2}$$
$$= \sqrt{(1.5)^2 + (0.866)^2}\sqrt{I_Z^2}$$
$$I_L = \sqrt{3}\,I_Z \qquad\qquad (20\text{--}10)$$

Example 20–6

Determine the load voltages and load currents in Figure 20–35, and show their relationship in a phasor diagram.

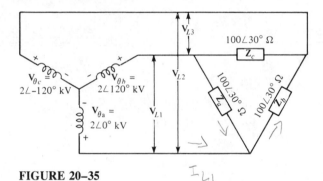

FIGURE 20–35

Solution:

$$\mathbf{V}_{Za} = \mathbf{V}_{L1} = \mathbf{V}_{\theta a} + \mathbf{V}_{\theta b}$$
$$= 2\angle 0°\text{ kV} + 2\angle +120°\text{ kV}$$
$$= 2\text{ kV} - 1\text{ kV} + j1.732\text{ kV}$$
$$= 1\text{ kV} + j1.732\text{ kV}$$
$$= \sqrt{(1\text{ kV})^2 + (1.732\text{ kV})^2}\angle\arctan\left(\frac{1.732}{1}\right)$$
$$= 2\angle 60°\text{ kV}$$

$$\mathbf{V}_{Zb} = \mathbf{V}_{L2} = \mathbf{V}_{\theta a} - \mathbf{V}_{\theta c}$$
$$= 2\angle 0°\text{ kV} + 2\angle -120°\text{ kV}$$
$$= 2\text{ kV} - 1\text{ kV} - j1.732\text{ kV}$$
$$= 1\text{ kV} - j1.732\text{ kV}$$
$$= 2\angle -60°\text{ kV}$$

$$\begin{aligned}
\mathbf{V}_{Zc} = \mathbf{V}_{L3} &= \mathbf{V}_{\theta b} + \mathbf{V}_{\theta c} \\
&= 2\angle +120° \text{ kV} + 2\angle -120° \text{ kV} \\
&= -1 \text{ kV} + j1.732 \text{ kV} - 1 \text{ kV} - j1.732 \text{ kV} \\
&= 2\angle 180° \text{ kV}
\end{aligned}$$

The load currents are:

$$\begin{aligned}
\mathbf{I}_{Za} &= \frac{\mathbf{V}_{Za}}{\mathbf{Z}_a} \\
&= \frac{2\angle 60° \text{ kV}}{100\angle 30° \text{ }\Omega} \\
&= 20\angle 30° \text{ A}
\end{aligned}$$

$$\begin{aligned}
\mathbf{I}_{Zb} &= \frac{\mathbf{V}_{Zb}}{\mathbf{Z}_b} \\
&= \frac{2\angle -60° \text{ kV}}{100\angle 30° \text{ }\Omega} \\
&= 20\angle -90° \text{ A}
\end{aligned}$$

$$\begin{aligned}
\mathbf{I}_{Zc} &= \frac{\mathbf{V}_{Zc}}{\mathbf{Z}_c} \\
&= \frac{2\angle 180° \text{ kV}}{100\angle 30° \text{ }\Omega} \\
&= 20\angle 150° \text{ A}
\end{aligned}$$

The phasor diagram is shown in Figure 20–36.

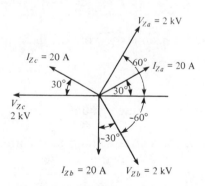

FIGURE 20–36

The Δ-Y System

Figure 20–37 shows a Δ-connected source feeding a Y-connected balanced load. By examination of the figure, you can see that the line voltages are

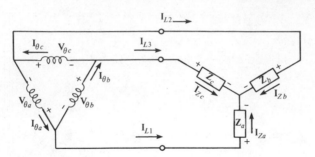

FIGURE 20–37 *A Δ-connected source feeding a Y-connected load.*

equal to the corresponding phase voltages of the source. Also, each phase voltage equals the difference of the corresponding load voltages, as you can see by the polarities.

Each load current equals the corresponding line current. The sum of the load currents is zero, because the load is balanced; thus, there is no need for a neutral return.

The relationship between the load voltages and the corresponding phase voltages (and line voltages) is developed as follows.

$$\mathbf{V}_{\theta a} = \mathbf{V}_{L1} = \mathbf{V}_{Za} - \mathbf{V}_{Zb}$$

Since $V_{Za} = V_{Zb} = V_{Zc} = V_Z$ in a balanced load, the expression can be written as

$$
\begin{aligned}
\mathbf{V}_{\theta a} = \mathbf{V}_{L1} &= V_Z \angle 0° - V_Z \angle 120° \\
&= V_Z - [V_Z \cos(120°) + jV_Z \sin(120°)] \\
&= V_Z + 0.5V_Z - j0.866V_Z \\
&= 1.5V_Z - j0.866V_Z
\end{aligned}
$$

The magnitude is

$$V_{\theta a} = \sqrt{3}\, V_Z \quad \angle\theta - 3^{\circ}$$

This relationship is true for each phase voltage; therefore

$$V_Z = \frac{V_\theta}{\sqrt{3}} \tag{20–11}$$

The magnitude of the load voltage equals $1/\sqrt{3}$ times the corresponding phase voltage.

The line currents and corresponding load currents are equal, and for a balanced load, the sum of the load currents is zero.

$$\mathbf{I}_L = \mathbf{I}_Z \tag{20–12}$$

As you can see in Figure 20–37, each line current is the difference of two phase currents.

$$\mathbf{I}_{L1} = \mathbf{I}_{\theta a} - \mathbf{I}_{\theta b}$$

$$\mathbf{I}_{L2} = \mathbf{I}_{\theta c} - \mathbf{I}_{\theta a}$$

$$\mathbf{I}_{L3} = \mathbf{I}_{\theta b} - \mathbf{I}_{\theta c}$$

Example 20–7

Determine the currents and voltages in the balanced load and the magnitude of the line voltages in Figure 20–38.

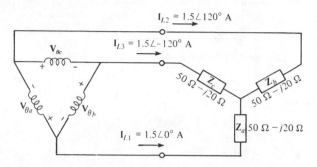

FIGURE 20–38

Solution:

The load currents equal the specified line currents.

$$\mathbf{I}_{Za} = \mathbf{I}_{L1} = 1.5\angle 0° \text{ A}$$
$$\mathbf{I}_{Zb} = \mathbf{I}_{L2} = 1.5\angle +120° \text{ A}$$
$$\mathbf{I}_{Zc} = \mathbf{I}_{L3} = 1.5\angle -120° \text{ A}$$

The load voltages are

$$\begin{aligned}\mathbf{V}_{Za} &= \mathbf{I}_{Za}\,\mathbf{Z}_a \\ &= (1.5\angle 0° \text{ A})\,(50\ \Omega - j20\ \Omega) \\ &= (1.5\angle 0° \text{ A})\,(53.85\angle -21.8°\ \Omega) \\ &= 80.78\angle -21.8° \text{ V}\end{aligned}$$

$$\begin{aligned}\mathbf{V}_{Zb} &= \mathbf{I}_{Zb}\,\mathbf{Z}_b \\ &= (1.5\angle +120° \text{ A})\,(53.85\angle -21.8°\ \Omega) \\ &= 80.78\angle +98.2° \text{ V}\end{aligned}$$

$$\begin{aligned}\mathbf{V}_{Zc} &= \mathbf{I}_{Zc}\,\mathbf{Z}_c \\ &= (1.5\angle -120° \text{ A})\,(53.85\angle -21.8°\ \Omega) \\ &= 80.78\angle -141.8° \text{ V}\end{aligned}$$

The magnitude of the line voltages is

$$V_L = \sqrt{3}\,V_Z = \sqrt{3}\,(80.78 \text{ V}) = 139.92 \text{ V}$$

The Δ-Δ System

Figure 20–39 shows a Δ-connected source driving a Δ-connected load. Notice that the load voltage, line voltage, and source phase voltage are all equal

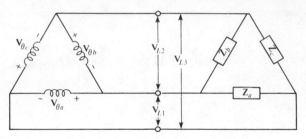

FIGURE 20–39   *A Δ-connected source feeding a Δ-connected load.*

for a given phase.

$$V_{\theta a} = V_{L1} = V_{Za}$$
$$V_{\theta b} = V_{L2} = V_{Zb}$$
$$V_{\theta c} = V_{L3} = V_{Zc}$$

Of course, when the load is balanced, all the voltages are equal, and a general expression can be written.

$$V_\theta = V_L = V_Z \qquad \qquad \textbf{(20–13)}$$

For a balanced load and equal source phase voltages, it can be shown in a manner similar to the one used previously that

$$I_L = \sqrt{3}\, I_Z \qquad \qquad \textbf{(20–14)}$$

Example 20–8

Determine the magnitude of the load currents and the line currents in Figure 20–40.

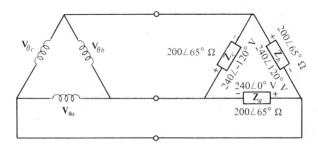

FIGURE 20–40

Solution:

$$V_{Za} = V_{Zb} = V_{Zc} = 240 \text{ V}$$

The magnitude of the load currents is

$$I_{Za} = I_{Zb} = I_{Zc} = \frac{V_{Za}}{Z_a}$$

$$= \frac{240 \text{ V}}{200 \text{ } \Omega} = 1.2 \text{ A}$$

The magnitude of the line currents is

$$I_L = \sqrt{3} \, I_Z = \sqrt{3} \, (1.2 \text{ A}) = 2.08 \text{ A}$$

Review for 20–4

1. List the four types of three-phase source/load configurations.
2. In a certain Y-Y system, the source phase currents each have a magnitude of 3.5 A. What is the magnitude of each load current for a balanced load condition?
3. In a given Y-Δ system, $V_L = 220$ V. Determine V_Z.
4. Determine the line voltages in a balanced Δ-Y system when the magnitude of the source phase voltages is 60 V.
5. Determine the magnitude of the load currents in a balanced Δ-Δ system having a line current magnitude of 3.2 A.

POWER IN THREE-PHASE SYSTEMS

20–5

Each phase of a balanced three-phase load has an equal amount of power. Therefore, the total average load power is three times the power in each phase of the load.

$$P_T = 3V_Z I_Z \cos \theta \qquad (20\text{–}15)$$

where V_Z and I_Z are the voltage and current associated with each phase of the load, and $\cos \theta$ is the power factor.

Recall that in a balanced Y-connected system,

$$V_L = \sqrt{3} \, V_Z \qquad \text{and} \qquad I_L = I_Z$$

and in a balanced Δ-connected system,

$$V_L = V_Z \qquad \text{and} \qquad I_L = \sqrt{3} \, I_Z$$

When either of these relationships is substituted into Equation (20–15), the total average power for both Y- and Δ-connected systems is

$$P_T = \sqrt{3} \, V_L I_L \cos \theta \qquad (20\text{–}16)$$

$$= 3(V_L)\left(\frac{I_L}{\sqrt{3}}\right) \cos \theta$$

Example 20–9

In a certain Δ-connected balanced load, the line voltages are 250 V and the impedances are $50\angle 30°$ Ω. Determine the total load power.

Solution:

In a Δ-connected system, $V_Z = V_L$ and $I_L = \sqrt{3}\, I_Z$. The load current magnitudes are

$$I_Z = \frac{V_Z}{Z} = \frac{250 \text{ V}}{50 \text{ }\Omega} = 5 \text{ A}$$

and

$$I_L = \sqrt{3}\, I_Z = \sqrt{3}\,(5) \text{ A} = 8.66 \text{ A}$$

The power factor is

$$\cos\theta = \cos 30° = 0.866$$

The total power is

$$\begin{aligned} P_T &= \sqrt{3}\, V_L I_L \cos\theta \\ &= \sqrt{3}\,(250 \text{ V})\,(8.66 \text{ A})\,(0.866) = 3247 \text{ W} \end{aligned}$$

Power Measurement

Power is measured in three-phase systems using wattmeters. The wattmeter uses a basic electrodynamometer-type movement (see Chapter 10) consisting of two coils. One coil is used to measure the current, and the other is used to measure the voltage. The needle of the meter is deflected proportionally to the current through a load and the voltage across the load, thus indicating power. Figure 20–41 shows a basic wattmeter symbol and the connections for measuring power in a load. The resistor in series with the voltage coil limits the current through the coil to a small amount proportional to the voltage across the coil.

In many three-phase loads, it is difficult to connect a wattmeter such that the voltage coil is across the load or such that the current coil is in series with the load because of inaccessibility of points within the load.

Power can be measured easily in a three-phase load of either the Y or the Δ type by using three wattmeters connected as shown in Figure 20–42. This is sometimes known as the *three-wattmeter method*.

Notice that the voltage coils of the wattmeters are effectively connected in a Y-configuration and that the current coils are in series in each line. The voltage across each wattmeter voltage coil is $V_L/\sqrt{3}$, and the current through each wattmeter current coil is I_L. The power indicated by each wattmeter is

$$P = \left(\frac{V_L}{\sqrt{3}}\right) I_L \cos\theta$$

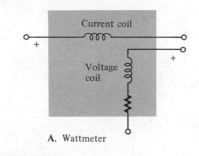

A. Wattmeter

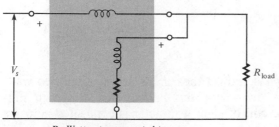

V_s R_{load}

B. Wattmeter connected to measure
load power

FIGURE 20–41

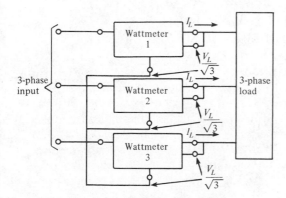

I_L Wattmeter 1 $\frac{V_L}{\sqrt{3}}$

3-phase input I_L Wattmeter 2   3-phase load $\frac{V_L}{\sqrt{3}}$

I_L Wattmeter 3 $\frac{V_L}{\sqrt{3}}$

FIGURE 20–42   *Three-wattmeter method.*

Since, in a balanced load, each wattmeter produces an identical reading, the total power measured by the three wattmeters is found by multiplying the individual readings by three.

$$P_T = 3P = 3\left(\frac{V_L}{\sqrt{3}}\right)I_L \cos \theta = \sqrt{3}\, V_L I_L \cos \theta$$

Notice that this expression is the same as for the total power in either a Y- or Δ-connected load (Equation 20–16). So, the three-wattmeter configuration gives the total load power for a Y or Δ load. Although we have used a balanced load to illustrate this concept, the power measurement is valid for unbalanced loads as well.

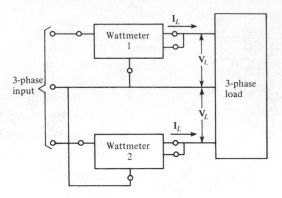

FIGURE 20–43 *Two-wattmeter method.*

Another method of three-phase power measurement uses only two watt-meters. The connections for this *two-wattmeter method* are shown in Figure 20–43. Notice that the voltage coil of each wattmeter is connected across a line voltage and that the current coil has a line current flowing through it. It can be shown that the algebraic sum of the two wattmeter readings equals the total power in the Y- or Δ-connected load as follows.

For the Y-connected balanced load in Figure 20–44,

$$\mathbf{V}_{L1} = V_{Za}\angle 0° - V_{Zc}\angle -120°$$
$$= V_Z + 0.5V_Z + j0.866V_Z$$
$$= 1.5V_Z + j0.866V_Z$$
$$= \sqrt{3}\ V_Z\angle 30°$$

and

$$\mathbf{V}_{L2} = V_{Za}\angle 0° - V_{Zb}\angle 120°$$
$$= 1.5V_Z - j0.866V_Z$$
$$= \sqrt{3}\ V_Z\angle -30°$$

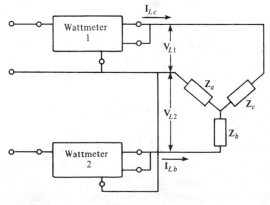

FIGURE 20–44 *Measurement of total power in a Y-connected balanced load.*

This shows that the line voltage is displaced from the load voltage by 30°. Also, there is a phase angle θ between the line current (same as load current in Y) and the load voltage. The line current through the current coil of wattmeter 1 is therefore displaced from the line voltage across the voltage coil of wattmeter 1 by an angle of $(\theta + 30°)$. Also, the line current through the current coil of wattmeter 2 is displaced from the line voltage across the voltage coil of wattmeter 2 by an angle of $(\theta - 30°)$.

The power indicated by wattmeter 1 is therefore

$$P_1 = V_{L1} I_{L1} \cos(\theta + 30°)$$

Since all the line voltages are equal and all the line currents are equal, the general expression is

$$P_1 = V_L I_L \cos(\theta + 30°) \tag{20-17}$$

and the power indicated by wattmeter 2 is

$$P_2 = V_{L2} I_{L2} \cos(\theta - 30°)$$

The general expression is

$$P_2 = V_L I_L \cos(\theta - 30°) \tag{20-18}$$

Adding P_1 and P_2 gives the total power:

$$P_T = P_1 + P_2 \tag{20-19}$$

$$P_T = P_1 + P_2 = V_L I_L [\cos(\theta + 30°) + \cos(\theta - 30°)]$$

Using a trigonometric identity, the cosine terms are expanded as follows:

$$P_T = V_L I_L (\cos \theta \cos 30° - \sin \theta \sin 30° + \cos \theta \cos 30° + \sin \theta \sin 30°)$$
$$= V_L I_L (2 \cos \theta \cos 30°)$$

Since $2 \cos 30° = 1.73 = \sqrt{3}$,

$$P_T = \sqrt{3} \, V_L I_L \cos \theta$$

This is the total average power, as was shown in Equation 20–16. Thus, the two-wattmeter method measures the total power as accurately as the three-wattmeter method.

It can be shown in a similar way that the two-wattmeter method works on a balanced Δ-connected load as well as on the balanced Y-connected load. Also, it can be shown that either method of power measurement can be applied to unbalanced loads as well as balanced loads. These proofs are left as exercises at the end of the chapter.

Example 20–10

Show both a three-wattmeter and a two-wattmeter connection for measuring the total load power in Figure 20–45. Determine the power measured by each individual meter in both cases.

Example 20–10 (continued)

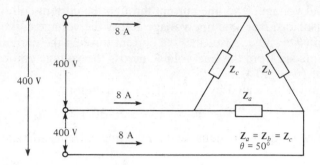

FIGURE 20–45

Solution:

The two- and three-wattmeter configurations are shown in Figure 20–46.

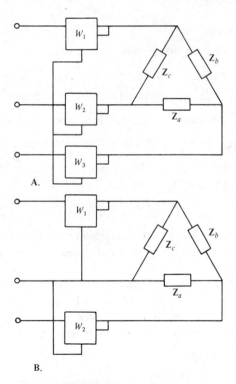

A.

B.

FIGURE 20–46

In the three-wattmeter circuit, each meter reads as follows:

$$P = V_L \left(\frac{I_L}{\sqrt{3}} \right) \cos \theta$$

$$= (400 \text{ V}) \left(\frac{8 \text{ A}}{\sqrt{3}}\right) \cos 50°$$

$$= (400 \text{ V}) (4.62 \text{ A}) (0.643)$$

$$= 1188.3 \text{ W}$$

The total power is

$$P_T = 3(1188.3 \text{ W}) = 3564.9 \text{ W}$$

In the two-wattmeter circuit, each meter reads as follows:

$$P_1 = V_L I_L \cos(\theta + 30°)$$
$$= (400 \text{ V}) (8 \text{ A}) \cos 80°$$
$$= (400 \text{ V}) (8 \text{ A}) (0.174)$$
$$= 556.8 \text{ W}$$

$$P_2 = V_L I_L \cos(\theta - 30°)$$
$$= (400 \text{ V}) (8 \text{ A}) \cos 20°$$
$$= (400 \text{ V}) (8 \text{ A}) (0.94)$$
$$= 3008 \text{ W}$$

The total power is

$$P_T = P_1 + P_2 = 556.8 \text{ W} + 3008 \text{ W} = 3564.8 \text{ W}$$

Notice that the total power measured by both methods is the same (the small difference is due to rounding off).

Review for 20–5

1. $V_L = 30$ V, $I_L = 1.2$ A, and the power factor is 0.257. What is the total power in a balanced Y-connected load? In a balanced Δ-connected load?

2. Three wattmeters connected to measure the power in a certain balanced load indicate a total of 2678 W. How much power does each meter measure?

POWER TRANSMISSION AND DISTRIBUTION SYSTEMS

Three-phase systems are used for the transmission and distribution of power to industrial and residential users. In this section, a typical power distribution system and its components are examined.

20–6

Three-Phase Transformers

Three-phase transformers are used to step up voltages for energy transmission to a customer and to step down voltages to a required level at the point of energy use.

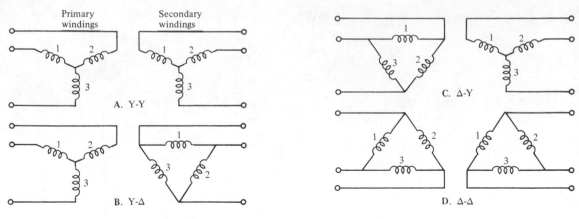

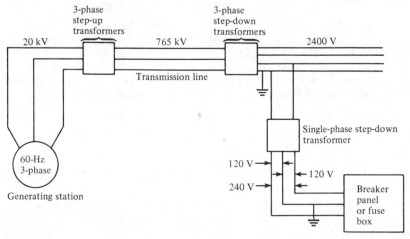

FIGURE 20–47   *Three-phase transformers.*

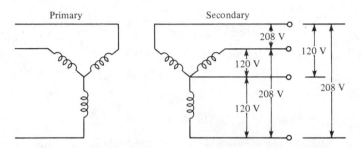

FIGURE 20–48   *Example of three-phase transformer voltages.*

FIGURE 20–49   *Simplified power generation and distribution system. Voltages shown are approximate magnitudes for a typical system.*

The three-phase windings can be connected in any one of the four configurations discussed in the last section: Y-Y, Y-Δ, Δ-Y, and Δ-Δ. In Figure 20–47, the left group of windings represents the primary, and the right group of windings represents the secondary.

In a typical power distribution system, a Y-connected generator is connected to a step-up transformer with a Δ-connected primary and a Y-connected secondary. The Y-connected secondary acts as a source to feed a high-voltage transmission line. Often, several step-up transformers are used to raise the voltage to the transmission line level. At the end of the transmission line, step-down transformers are used to reduce the high voltage for local distribution. For example, a Y-connected secondary can supply a user with 120-V single-phase and 208-V three-phase as illustrated in Figure 20–48.

A typical complete power generation and distribution system is shown in Figure 20–49.

Review for 20–6

1. List the major components of a power distribution system.

2. Does the secondary of a power transformer feeding a transmission line act as a source or a load to the transmission line?

Formulas

$$v = V_{max}\sin \theta \qquad \text{(20–1)}$$

$$f = (\text{number of pole pairs})(\text{rps}) \qquad \text{(20–2)}$$

Y generator:

$$I_L = I_\theta \qquad \text{(20–3)}$$

$$V_L = \sqrt{3}\,V_\theta \qquad \text{(20–4)}$$

Δ generator:

$$V_L = V_\theta \qquad \text{(20–5)}$$

$$I_L = \sqrt{3}\,I_\theta \qquad \text{(20–6)}$$

Y-Y system:

$$I_\theta = I_L = I_Z \qquad \text{(20–7)}$$

$$V_\theta = V_Z \qquad \text{(20–8)}$$

Y-Δ system:

$$V_Z = V_L \qquad \text{(20–9)}$$

$$I_L = \sqrt{3}\,I_Z \qquad \text{(20–10)}$$

Δ-Y system:

$$V_Z = \frac{V_\theta}{\sqrt{3}} \qquad (20\text{--}11)$$

$$I_L = I_Z \qquad (20\text{--}12)$$

Δ-Δ system:

$$V_\theta = V_L = V_Z \qquad (20\text{--}13)$$

$$I_L = \sqrt{3}\,I_Z \qquad (20\text{--}14)$$

3-phase power:

$$P_T = 3V_Z I_Z \cos\theta \qquad (20\text{--}15)$$

$$P_T = \sqrt{3}\,V_L I_L \cos\theta \qquad (20\text{--}16)$$

2-wattmeter method:

$$P_1 = V_L I_L \cos(\theta + 30°) \qquad (20\text{--}17)$$

$$P_2 = V_L I_L \cos(\theta - 30°) \qquad (20\text{--}18)$$

$$P_T = P_1 + P_2 \qquad (20\text{--}19)$$

Summary

1. A sinusoidal voltage is generated when a conductive loop rotates in a constant magnetic field. This forms a simple ac generator.

2. The speed of rotation of the conductive loop determines the frequency of the induced sinusoidal voltage.

3. A practical way of increasing the magnitude of the induced voltage is to increase the number of turns in the conductive loop.

4. A simple two-phase generator consists of two conductive loops, separated by 90 degrees, rotating in a magnetic field.

5. A simple three-phase generator consists of three conductive loops separated by 120 degrees.

6. Three advantages of polyphase systems over single-phase systems are a smaller copper cross section for the same power delivered to the load, constant power delivered to the load, and a constant, rotating magnetic field.

7. In a Y-connected generator, $I_L = I_\theta$ and $V_L = \sqrt{3}\,V_\theta$.

8. In a Y-connected generator, there is a 30-degree difference between each line voltage and the nearest phase voltage.

9. In a Δ-connected generator, $V_L = V_\theta$ and $I_L = \sqrt{3}\,I_\theta$.

10. In a Δ-connected generator, there is a 30-degree difference between each line current and the nearest phase current.

11. In a Y-Y source/load system, $I_\theta = I_L = I_z$ and $V_\theta = V_z$.

12. In a Y-Δ system, $V_z = V_L$ and $I_L = \sqrt{3} \, I_z$.

13. In a Δ-Y system, $V_z = V_\theta/\sqrt{3}$ and $I_L = I_z$.

14. In a Δ-Δ system, $V_\theta = V_L = V_z$ and $I_L = \sqrt{3} \, I_z$.

15. A balanced load is one in which all the impedances are equal.

16. Power is measured in a three-phase load using either the three-wattmeter method or the two-wattmeter method.

Self-Test

1. An ac generator produces a sinusoidal voltage with a maximum value of 208 V. What is the instantaneous value at a phase point of 35 degrees?

2. The conductive loop on the rotor of a simple two-pole, single-phase generator rotates at a rate of 250 rps. What is the frequency of the induced output voltage?

3. If the rotational speed of the generator in Problem 2 is increased to 800 rps, would the output voltage amplitude increase or decrease?

4. Describe the purpose of the field windings in an ac generator.

5. A single-phase voltage of $30\angle0°$ V is applied across a load impedance of $15\angle22°$ Ω. Determine the load current in polar form.

6. The phase current produced by a certain Y-connected generator is 12 A. What is the corresponding line current?

7. If each phase voltage in a Y-connected generator is 180 V, what is the magnitude of the line voltages?

8. A certain Δ-connected generator produces phase voltages of 30 V. What are the line voltages in terms of their magnitude?

9. A loaded Δ-connected generator produces phase currents of 5 A. Determine the line currents.

10. A certain Y-Y system produces phase currents of 15 A. Determine the line current and load current magnitudes.

11. The phase voltages produced by the generator in a certain Y-Y system have a magnitude of 1.5 kV. Determine the magnitudes of the line voltages and the load voltages.

12. In a particular Y-Δ system, the load voltages are each 75 V in magnitude, and the load currents are 3.3 A. Determine the line voltages and currents.

13. The source phase voltages of a Δ-Y system are 220 V. Find the load voltage magnitude.

14. A balanced three-phase system exhibits a value of load voltage and load current of 145 V and 2.5 A, respectively. Determine the total load power when $\theta = 30°$.

Problems

Section 20–1

20–1. The output of an ac generator has a maximum value of 250 V. At what angle is the instantaneous value equal to 75 V?

20–2. A certain four-pole generator has a speed of rotation of 3600 rpm. What is the frequency of the voltage produced by this generator?

Section 20–2

20–3. A single-phase generator feeds a load consisting of a 200-Ω resistor and a capacitor with a reactance of 175 Ω. The generator produces a voltage of 100 V. Determine the magnitude of the load current and its phase relation to the generator voltage.

20–4. In a two-phase system, the two currents flowing through the lines connecting the generator to the load are 3.8 A. Determine the current in the neutral line.

20–5. A certain three-phase unbalanced load in a four-wire system has currents of $2\angle 20°$ A, $3\angle 140°$ A, and $1.5\angle -100°$ A. Determine the current in the neutral line.

Section 20–3

20–6. Determine the line voltages in Figure 20–50.

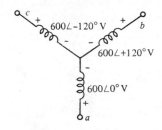

FIGURE 20–50   **FIGURE 20–51**

20–7. Determine the line currents in Figure 20–51.

20–8. Develop a complete current phasor diagram for Figure 20–51.

Section 20–4

20–9. Determine the following quantities for the Y-Y system in Figure 20–52.
(a) Line voltages   (b) Phase currents   (c) Line currents
(d) Load currents   (e) Load voltages

20–10. Repeat Problem 20–9 for the system in Figure 20–53, and also find the neutral current.

20–11. Repeat Problem 20–9 for the system in Figure 20–54.

20–12. Repeat Problem 20–9 for the system in Figure 20–55.

20–13. Determine the line voltages and load currents for the system in Figure 20–56.

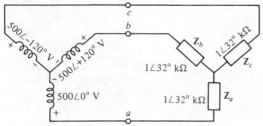

FIGURE 20–52

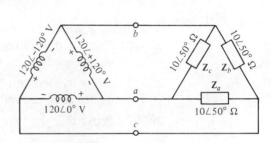

FIGURE 20–53

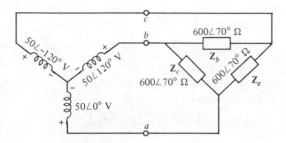

FIGURE 20–54

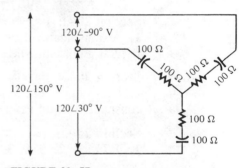

FIGURE 20–55

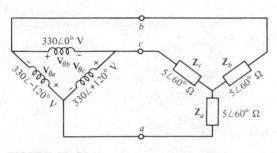

FIGURE 20–56

FIGURE 20–57

Section 20–5

20–14. The power in each phase of a balanced three-phase system is 1200 W. What is the total power?

20–15. Determine the load power in Figures 20–52 through 20–56.

20–16. Find the total load power in Figure 20–57.

20–17. Using the three-wattmeter method for the system in Figure 20–57, how much power does each wattmeter indicate?

20–18. Repeat Problem 20–17 using the two-wattmeter method.

Section 20–6

20–19. A certain Y-Y three-phase step-up transformer has a turns ratio of 10 for each phase. If the input line voltages are 100 V, what voltages are available on the outputs?

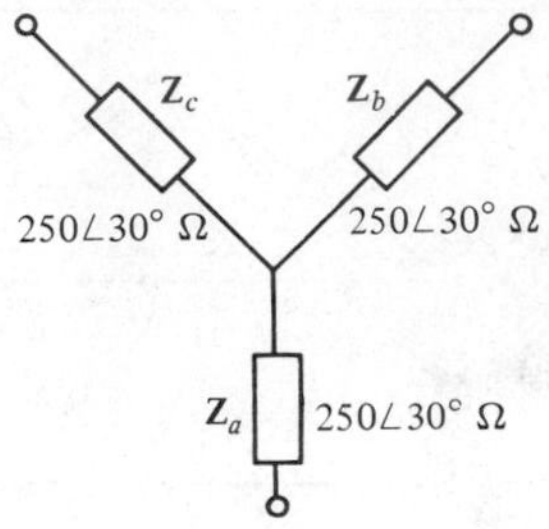

FIGURE 20-58

Advanced Problems

20-20. Convert the Y-connected load in Figure 20-58 to an equivalent Δ-connected load.

20-21. Show by derivation that the two-wattmeter method applies to a Δ-connected load as well as a Y-connected load.

20-22. Prove that the two-wattmeter method is valid for unbalanced loads.

Answers to Section Reviews

Section 20-1:
1. A sinusoidal voltage is induced when a conductive loop is rotated in a magnetic field at a constant speed. **2.** 400 Hz. **3.** 3.

Section 20-2:
1. Less copper cross section to conduct current, constant power to load, and constant, rotating magnetic field. **2.** Constant power. **3.** Constant field.

Section 20-3:
1. 1.73 kV. **2.** 5 A. **3.** 240 V. **4.** 3.46 A.

Section 20-4:
1. Y-Y, Y-Δ, Δ-Y, and Δ-Δ. **2.** 3.5 A. **3.** 220 V. **4.** 60 V. **5.** 1.85 A.

Section 20-5:
1. 27.76 W, 27.76 W. **2.** 892.67 W.

Section 20-6:
1. Generator, three-phase transformers, transmission line, load. **2.** Source.

In Chapter 10, several basic dc measurement instruments were introduced. In this chapter, several basic ac instruments are discussed.

Specifically, you will learn:

1. The basic difference between dc and ac meters.
2. Several specialized measurement instruments.
3. The instruments that are used to generate ac signals.
4. The basics of oscilloscopes and how to use them.

21
ac
MEASUREMENTS

21-1

Meters

The basic meter movements were discussed in Chapter 10 in relation to dc measurements. The d'Arsonval movement is restricted to dc only, because a unidirectional current is required to produce an upscale deflection as a result of the fixed magnetic field. For ac measurement of current or voltage, additional circuitry is required. This additional circuitry consists of a *rectifier,* which converts ac to dc. There are two types of rectifiers: half-wave and full-wave.

A half-wave rectifier is shown in block form in Figure 21-1A with its input and output. The ac input is converted to *pulsating dc* on every positive half-cycle. The full-wave rectifier converts the ac to pulsating dc on both the positive and the negative half-cycles, as indicated in Figure 21-1B.

Basically, in an ac meter the rectifier precedes the meter movement. The movement responds to the *average value* of the pulsating dc. The scale can be calibrated to show rms, average, or peak values, because these relationships are fixed mathematically, as you have learned. Figure 21-2 shows a basic meter with a full-wave rectifier for converting ac to dc.

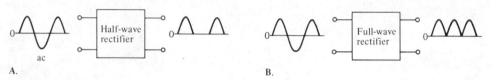

FIGURE 21-1 *Converting ac to pulsating dc.*

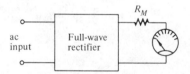

FIGURE 21-2 *Basic ac voltmeter.*

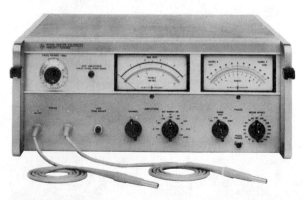

FIGURE 21-3 *A vector voltmeter. (Courtesy of Hewlett-Packard Company)*

The electrodynamometer movement can be used to measure both dc and ac quantities with no additional circuitry. A change in current direction does not alter the upscale deflection, because both the stationary and the movable coils experience a reversal of their magnetic fields.

The Vector Voltmeter

Figure 21–3 shows a vector meter that provides both the magnitude and phase of a measured voltage. With two-channel capability, a meter of this type can indicate the phase difference between two input signals. This type of meter is usually restricted to frequencies in the 1-MHz to 1000-MHz range.

The Vector Impedance Meter

An impedance meter such as the one shown in Figure 21–4 is used to make precision measurements of impedance magnitude and phase angle.

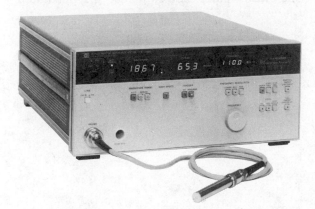

FIGURE 21–4 *A vector impedance meter. (Courtesy of Hewlett-Packard Company)*

Review for 21–1

1. Explain why a rectifier is necessary when a d'Arsonval movement is used for ac measurements.
2. Why is the electrodynamometer movement applicable to ac?

SIGNAL GENERATORS

21–2

The signal generator is an instrument that produces electrical signals for use in testing or controlling electronic circuits and systems. There are a variety of signal generators, ranging from special-purpose instruments that produce only one type of wave form in a limited frequency range, to programmable instruments that produce a wide range of frequencies and a variety of wave forms. In this section, we discuss several important types.

Audio Frequency Generators

The audible range of frequencies is typically from about 20 Hz to about 20 kHz. As a minimum requirement, an audio frequency (af) generator must provide sine wave voltages within this range. Many af generators or *audio oscillators*, however, have frequency ranges much greater than the actual audible range. A typical laboratory af generator produces frequencies from less than 1 Hz to greater than 1 MHz.

Figure 21–5 shows one type of audio generator. Notice that there is a range switch for selecting the desired frequency range. The frequency control sets the exact frequency within the selected range. The amplitude control adjusts the output voltage. Once it is set, it should remain essentially constant over the frequency range of the instrument. The maximum specified output voltage occurs when there is no load connected across the output terminals of the generator.

Many af generators have an *output impedance* of 600 Ω. Thus, if a load resistance of 600 Ω is connected to the output terminals, the voltage will be half of its open terminal value. Smaller load values further reduce the maximum achievable output amplitude.

FIGURE 21–5 *Audio oscillator. (Courtesy of Tektronix, Inc.)*

Radio Frequency Generators

Many radio frequency (rf) generators cover frequencies ranging from 30 kHz to 3000 MHz. The lower end of the range, of course, overlaps with the range of many af generators. Most rf generators produce at least two types of output signal: *continuous-wave* and *modulated*.

A continuous-wave or CW signal is a single-frequency sine wave with a steady amplitude. A modulated signal is a sine wave with an amplitude that varies sinusoidally at a much lower frequency, called *amplitude modulation* (AM). The two types of signals are illustrated in Figure 21–6.

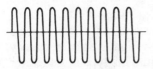

A. Continuous wave (CW)

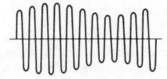

B. Amplitude modulation (AM)

FIGURE 21–6 *CW and amplitude-modulated rf signals.*

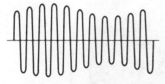

Lower percentage of modulation

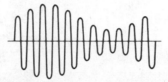

Higher percentage of modulation

FIGURE 21–7 *Variations in modulation.*

There are many types of rf generators. All have provisions for adjusting the frequency over the specified range and for setting the output voltage amplitude. The percentage of modulation can also be set, as illustrated in Figure 21–7. Some rf generators also have provisions for frequency modulation (FM). Figure 21–8 shows two typical rf generators.

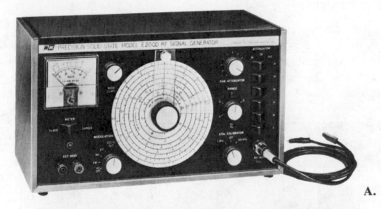

A.

B.

FIGURE 21–8 *Typical rf signal generators. (**A,** courtesy of B&K-Precision Test Instruments, Dynascan Corp. **B,** courtesy of Hewlett-Packard Company)*

Sweep Generators

Sweep generators are a specialized type of rf generator that can produce a continuously varying frequency, called a *sweep frequency*. That is, the frequency changes continuously from a lower limit to an upper limit. This type of generator is widely used in testing of the band-pass circuits in communications receivers such as TV and radar. A typical sweep generator is shown in Figure 21–9.

FIGURE 21–9 *Microwave sweep oscillator. (Courtesy of Hewlett-Packard Company)*

Pulse Generators

Pulse generators are nonsinusoidal generators used extensively in testing digital circuits and systems. A typical pulse generator produces output pulses with variable pulse width, frequency, and amplitude. Many also have provisions for adjusting the rise time and the fall time of the pulses. Figure 21–10 shows one type of pulse generator.

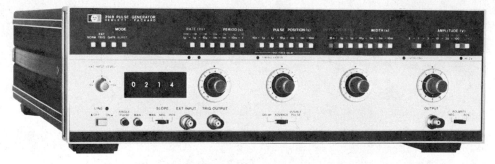

FIGURE 21–10 *Pulse generator. (Courtesy of Hewlett-Packard Company)*

Multipurpose Instruments

Figure 21–11 shows a multipurpose instrument that is actually several instruments in one. It uses a modular plug-in arrangement that allows the instruments to be interchanged for various measurement requirements. It also provides the convenience of having several instruments in one package for excellent portability. Therefore, this type of unit is ideal for field use.

FIGURE 21–11 *Multipurpose test system with plug-in instruments. (Courtesy of Tektronix, Inc.)*

Function Generators

A function generator is a versatile instrument that provides several types of output wave forms, including *square waves, triangular waves, ramps,* and *sine waves*. A function switch on the front panel selects the type of output wave form desired. Other controls select frequency, amplitude, and slope of ramps. Typical function generators are shown in Figure 21–12A and B. Part C of the figure shows the wave forms that are normally produced by many function generators.

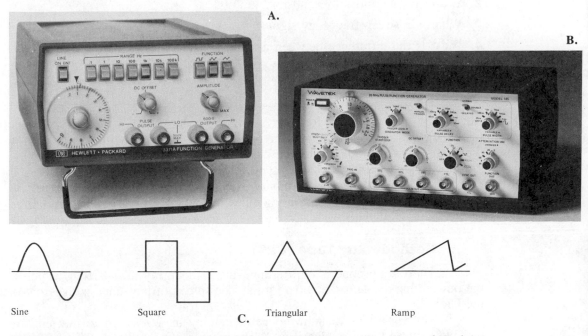

FIGURE 21–12 *Function generators and typical output wave forms. (**A,** courtesy of Hewlett-Packard Company. **B,** courtesy of Wavetek)*

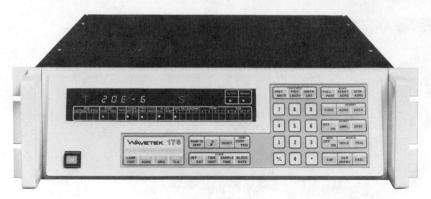

FIGURE 21-13 *Programmable wave-form generator. (Courtesy of Wavetek)*

Programmable Wave-Form Generators

With this type of generator, any wave form can be programmed in and then generated as an output. Irregular and standard wave forms can be programmed and stored. A typical programmable wave-form generator is shown in Figure 21-13.

Review for 21-2

1. Typically, what is the audible range of frequencies?
2. What do "af" and "rf" stand for?
3. What is a sweep frequency signal?
4. What is the difference between a pulse generator and a function generator?

THE OSCILLOSCOPE

21-3 The oscilloscope, or *scope* for short, is one of the most widely used and versatile test instruments. It displays on a screen the actual wave shape of a voltage from which amplitude, time, and frequency measurements can be made.

Figure 21-14 shows two typical oscilloscopes. The one in Part A is a simpler and less expensive model. The one in Part B has better performance characteristics and plug-in modules that provide a variety of specialized functions.

Cathode Ray Tube (CRT)

The oscilloscope is built around the cathode ray tube (CRT), which is the device that displays the wave forms. The screen of the scope is the front of the CRT.

The CRT is a vacuum device containing an *electron gun* that emits a narrow, focused *beam* of electrons. A phosphorescent coating on the face of the

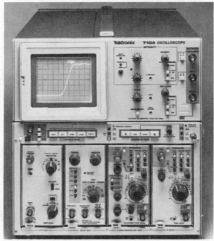

A.   B.

FIGURE 21–14 *Oscilloscopes. (Courtesy of Tektronix, Inc.)*

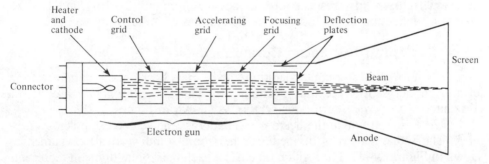

FIGURE 21–15 *Basic construction of CRT.*

tube forms the *screen*. The beam is electronically focused and accelerated so that it strikes the screen, causing light to be emitted at the point of impact.

Figure 21–15 shows the basic construction of a CRT. The *electron gun* assembly contains a *heater,* a *cathode,* a *control grid,* and *accelerating* and *focusing grids*. The heater carries current that indirectly heats the cathode. The heated cathode emits electrons. The amount of voltage on the control grid determines the flow of electrons and thus the *intensity* of the beam. The electrons are accelerated by the accelerating grid and are focused by the focusing grid into a narrow beam that converges at the screen. The beam is further accelerated to a high speed after it leaves the electron gun by a high voltage on the *anode* surfaces of the CRT.

Deflection of the Beam

The purpose of the *deflection plates* in the CRT is to produce a "bending" or deflection of the electron beam. This deflection allows the position of the point

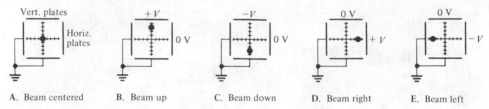

FIGURE 21–16 *Deflection of electron beam.*

of impact on the screen to be varied. There are two sets of deflection plates. One set is for *vertical deflection,* and the other set is for *horizontal deflection.*

Figure 21–16 shows a front view of the CRT's deflection plates. One plate from each set is normally grounded as shown. If there is *no voltage* on the other plates, as in Figure 21–16A, the beam is not deflected and hits the center of the screen. If a *positive* voltage is on the vertical plate, the beam is attracted upward, as indicated in Part B of the figure. Remember that opposite charges attract. If a *negative* voltage is applied, the beam is deflected downward because like charges repel, as shown in Part C.

Likewise, a positive or a negative voltage on the horizontal plate deflects the beam right or left, respectively, as shown in Figure 21–16D and E. The amount of deflection is proportional to the amount of voltage on the plates.

Sweeping the Beam Horizontally

In normal oscilloscope operation, the beam is horizontally deflected from left to right across the screen at a certain rate. This *sweeping action* produces a horizontal line or *trace* across the screen, as shown in Figure 21–17.

The *rate* at which the beam is swept across the screen establishes a *time base.* The scope screen is divided into horizontal (and vertical) divisions, as shown in Figure 21–17. For a given time base, each horizontal division represents a fixed interval of time. For example, if the beam takes 1 second for a full left-to-right sweep, then each division represents 0.1 second. All scopes have provisions for selecting various sweep rates, which will be discussed later.

The actual sweeping of the beam is accomplished by application of a *sawtooth voltage* across the horizontal plates. The basic idea is illustrated in Figure 21–18. When the sawtooth is at its maximum negative peak, the beam is deflected to its left-most screen position. This deflection is due to maximum repulsion from the right deflection plate.

As the sawtooth voltage *increases,* the beam moves toward the center of the screen. When the sawtooth voltage is *zero,* the beam is at the center of the screen, because there is no repulsion or attraction from the plate.

As the voltage increases *positively,* the plate attracts the beam, causing it to move toward the right side of the screen. At the positive peak of the sawtooth, the beam is at its right-most screen position.

The rate at which the sawtooth goes from negative to positive is determined by its frequency. This rate in turn establishes the *sweep rate* of the beam.

When the sawtooth makes the abrupt change from positive back to negative, the beam is rapidly returned to the left side of the screen, ready for another

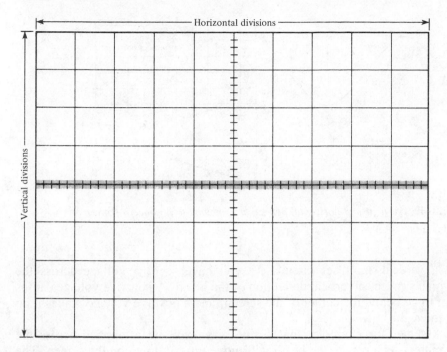

FIGURE 21–17 *Scope screen with horizontal trace (8 cm x 10 cm).*

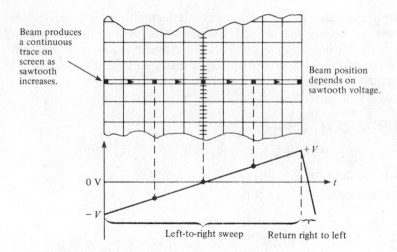

FIGURE 21–18   *Sweeping the beam across the screen.*

sweep. During this "flyback" time, the beam is *blanked* out and thus does not produce a trace on the screen.

How a Wave-Form Pattern Is Produced

The main purpose of the scope is to display the wave form of a voltage under test. To do so, we apply the voltage under test across the *vertical plates*

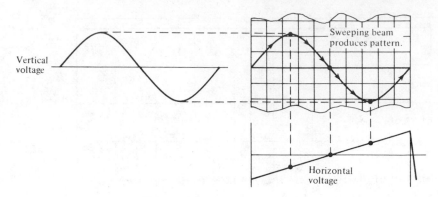

FIGURE 21–19 *Vertical and horizontal deflection combined to produce a pattern on the screen.*

through a vertical amplifier circuit. As you have seen, a voltage across the vertical plates causes a vertical deflection of the beam. A negative voltage causes the beam to go below the center of the screen, and a positive voltage makes it go above center.

Assume, for example, that a sine wave voltage is applied across the vertical plates. As a result, the beam will move up and down on the screen. The amount that the beam goes above or below center depends on the peak value of the sine wave voltage.

At the same time that the beam is being deflected vertically, it is also sweeping horizontally, causing the vertical voltage wave form to be traced out across the screen, as illustrated in Figure 21–19. All scopes provide for the calibrated adjustment of the vertical deflection, so each vertical division represents a known amount of voltage.

Review for 21–3

1. What is the basic purpose of an oscilloscope?

2. What does "CRT" stand for?

3. Name the basic components of a CRT.

OSCILLOSCOPE CONTROLS

21–4   There are a wide variety of oscilloscopes available, ranging from relatively simple instruments with limited capabilities to much more sophisticated models that provide a variety of optional functions and precision measurements. Regardless of their complexity, however, all scopes have certain operational features in common. In this section, we examine the most common front panel controls. Each control and its basic function are described. Figure 21–20 shows a representative oscilloscope front panel.

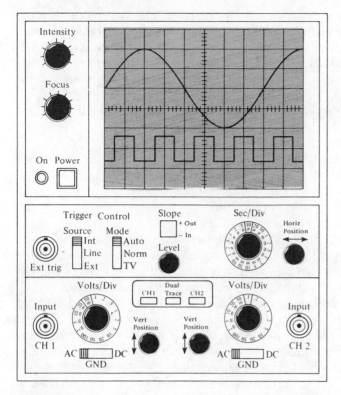

FIGURE 21–20 *Representative dual-trace oscilloscope front panel.*

Screen: In the upper portion of Figure 21–20 is the CRT screen. There are 8 vertical divisions and 10 horizontal divisions indicated with grid lines or graticules. A standard screen size is 8 cm × 10 cm. The screen is coated with phosphor that emits light when struck by the electron beam.

Power Switch and Light: This switch turns the power on and off to the scope. The light indicates when the power is on.

Intensity: This control knob varies the brightness of the trace on the screen. Caution should be used so that the intensity is not left too high for an extended period of time, especially when the beam forms a motionless dot on the screen. Damage to the screen can result from excessive intensity.

Focus: This control focuses the beam so that it converges to a tiny point at the screen. An out-of-focus condition results in a fuzzy trace.

Horizontal Position: This control knob adjusts the neutral horizontal position of the beam. It is used to reposition horizontally a wave-form display for more convenient viewing or measurement.

Seconds/Division: This selector switch sets the horizontal sweep rate. It is the *time base control*. The switch selects the time interval that is to be represented by each horizontal division in seconds, milliseconds, or microseconds. The setting in Figure 21–20 is at 10 μs. Thus, each of the ten horizontal divisions represents 10 μs; so there are 100 μs from the extreme left of the screen to the extreme right.

One cycle of the displayed sine wave covers eight horizontal divisions. Therefore, the period of the sine wave is (8 div)(10 μs/div) = 80 μs. From this, the frequency can be calculated as $f = 1/T = 1/80 \ \mu s = 12.5$ kHz. If the sec/div switch is moved to a different setting, the displayed sine wave will change correspondingly. If it is moved to a lower time setting, fewer cycles will be displayed. If it is moved to a higher time setting, more cycles will be displayed.

Trigger Control: These controls allow the beam to be triggered from various selected sources. The triggering of the beam causes it to begin its sweep across the screen. It can be triggered from an internally generated signal derived from an input signal, or from the line voltage, or from an externally applied trigger signal. The *modes* of triggering are auto, normal, and TV. In the auto mode, sweep occurs in the absence of an adequate trigger signal. In the normal mode, a trigger signal must be present for the sweep to occur. The TV mode provides triggering on the TV field or TV line signals. The *slope* switch allows the triggering to occur on either the positive-going slope or the negative-going slope of the trigger wave form. The *level* control selects the amplitude point on the trigger signal at which the triggering occurs.

Basically, the trigger controls provide for synchronization of the sweep wave form and the input signal wave form. As a result, the display of the input signal is stable on the screen, rather than appearing to drift across the screen.

Volts/Division: The example scope in Figure 21–20 is a *dual-trace* type, which allows two wave forms to be displayed simultaneously. Many scopes have only single-trace capability. Notice that there are two identical volts/div selectors. There is a set of controls for each of the two *input channels*. Only one is described here, but the same applies to the other as well.

The volts/div selector switch sets the number of volts to be represented by each division on the *vertical* scale. For example, the displayed sine wave is applied to channel 1 and covers four vertical divisions from the positive peak to the negative peak. The volts/div switch for channel 1 is set at 50 mV, which means that each vertical division represents 50 mV. Therefore, the peak-to-peak value of the sine wave is (4 div)(50 mV/div) = 200 mV. If a lower setting were selected, the displayed wave would cover more vertical divisions. If a higher setting were selected, the displayed wave would cover fewer vertical divisions.

Notice that there is a set of three switches for selecting channel 1 (CH 1), channel 2 (CH 2), or dual trace. Either input signal can be displayed separately, or both can be displayed as illustrated.

Vertical Position: The two vertical position controls move the traces up or down for easier measurement or observation.

ac-gnd-dc Switch: This switch, located below the volts/div control, allows the input signal to be ac coupled, dc coupled, or grounded. The ac coupling eliminates any dc component on the input signal. The dc coupling permits dc values to be displayed. The ground position allows a zero volt reference to be established on the screen.

Input: The signals to be displayed are connected into the channel 1 and channel 2 input connectors. This connection is normally done via a special *probe* that minimizes the loading effect of the scope's input resistance and capacitance on the circuit being measured.

Review for 21–4

1. Name the common controls found on all oscilloscopes.
2. Which control adjusts the vertical deflection?
3. Which control adjusts the sweep rate of the horizontal trace?

Summary

1. ac must first be rectified (converted to pulsating dc) when a d'Arsonval movement is used in a meter.
2. The electrodynamometer movement can be used to measure both dc and ac.
3. Signal generators can be classified broadly as audio frequency (af), radio frequency (rf), and nonsinusoidal.
4. Oscilloscopes display a graphical picture of the wave shape of a voltage.
5. The CRT (cathode ray tube) is the basic element of an oscilloscope.

Self-Test

1. Which type of meter movement, d'Arsonval or electrodynamometer, is applicable to both dc and ac measurements?
2. What is a rectifier circuit?
3. What quantities does a vector voltmeter measure?
4. The volts/div switch on an oscilloscope is set at 0.1 V. The wave form displayed extends over 5 vertical divisions from its positive to its negative peak. What is its peak-to-peak value?
5. A certain oscilloscope has 10 horizontal divisions. When the time/div switch is set at 1 μs, how much time is represented by the full 10 divisions?
6. How many cycles of a 1-kHz sine wave are displayed on a scope screen when the time/div switch is set at 0.1 ms? At 1 ms? Assume that the screen is 10 divisions wide.

Problems

21–1. Determine how many cycles of a 5-kHz sine wave are displayed on the screen for each of the following time/div settings. Assume 10 horizontal divisions.

    (a) 10 μs   **(b)** 0.1 ms   **(c)** 1 ms

21–2. Determine the peak and rms values and the frequency of the sine wave displayed in Figure 21–21 for the control settings of 0.5 volt/div and 10 microseconds/div.

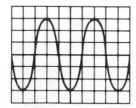

FIGURE 21–21

21–3. On a typical 8 × 10 division scope screen, you expect to observe a 1-V rms wave with a frequency of 2.5 kHz. What are the minimum settings of the volts/division and time/division controls so that at least one full cycle of the complete wave form is displayed?

Answers to Section Reviews

Section 21–1:
1. The rectifier is necessary to convert ac to pulsating dc, because the d'Arsonval movement will produce a scale deflection only for unidirectional current. **2.** Because it does not have a fixed magnetic field as does the d'Arsonval.

Section 21–2:
1. 20 Hz to 20 kHz. **2.** Audio frequency and radio frequency. **3.** The frequency continuously changes between two limits. **4.** A function generator produces several types of wave forms; a pulse generator produces only rectangular pulses.

Section 21–3:
1. To display a voltage wave form. **2.** Cathode ray tube. **3.** Electron gun, deflection plates, anode, screen, and glass envelope.

Section 21–4:
1. Power switch, intensity, focus, vertical input, position, volts/division, time/division, trigger. **2.** Volts/division. **3.** Time/division.

APPENDICES

TABLE OF STANDARD 10% RESISTOR VALUES (commercially available)

APPENDIX A

Ohms (Ω)				Kilohms (kΩ)		Megohms (MΩ)	
1.0	10	100	1000	10	100	1.0	10
1.2	12	120	1200	12	120	1.2	12
1.5	15	150	1500	15	150	1.5	15
1.8	18	180	1800	18	180	1.8	18
2.2	22	220	2200	22	220	2.2	22
2.7	27	270	2700	27	270	2.7	
3.3	33	330	3300	33	330	3.3	
3.9	39	390	3900	39	390	3.9	
4.7	47	470	4700	47	470	4.7	
5.6	56	560	5600	56	560	5.6	
6.8	68	680	6800	68	680	6.8	
8.2	82	820	8200	82	820	8.2	

Batteries are an important source of dc voltage. They are available in two basic categories: the *wet cell* and the *dry cell*. A battery generally is made up of several individual cells.

A cell consists basically of two *electrodes* immersed in an *electrolyte*. A voltage is developed between the electrodes as a result of the *chemical* action between the electrodes and the electrolyte. The electrodes typically are two dissimilar metals, and the electrolyte is a chemical solution.

Simple Wet Cell

Figure B–1 shows a simple copper-zinc (Cu-Zn) chemical cell. One electrode is made of copper, the other of zinc. These electrodes are immersed in a solution of water and hydrochloric acid (HCl), which is the electrolyte.

Positive hydrogen ions (H^+) and negative chlorine ions (Cl^-) are formed when the HCl ionizes in the water. Since zinc is more active than hydrogen, zinc atoms leave the zinc electrode and form zinc ions (Zn^{++}) in the solution. When a zinc ion is formed, two excess electrons are left on the zinc electrode, and two hydrogen ions are displaced from the solution. These two hydrogen ions will migrate to the copper electrode, take two electrons from the copper, and form a molecule of hydrogen gas (H_2). As a result of this reaction, a negative charge develops on the zinc electrode, and a positive charge develops on the copper electrode, creating a potential difference or voltage between the two electrodes.

In this copper-zinc cell, the hydrogen gas given off at the copper electrode tends to form a layer of bubbles around the electrodes, insulating the copper from the electrolyte. This effect, called *polarization,* results in a reduction in the voltage produced by the cell. Polarization can be remedied by the addition of an agent to the electrolyte to remove hydrogen gas or by the use of an electrolyte that does not form hydrogen gas.

Lead-Acid Cell: The positive electrode of a lead-acid cell is lead peroxide (PbO_2), and the negative electrode is spongy lead (Pb). The electrolyte is sulfuric acid (H_2SO_4) in water. Thus, the lead-acid cell is classified as a wet cell.

Two positive hydrogen ions ($2H^+$) and one negative sulfate ion (SO_4^{--}) are formed when the sulfuric acid ionizes in the water. Lead ions (Pb^{++}) from both electrodes displace the hydrogen ions in the electrolyte solution. When the lead ion from the spongy lead electrode enters the solution, it combines with a sulfate ion (SO_4^{--}) to form lead sulfate ($PbSO_4$), and it leaves two excess electrons on the electrode.

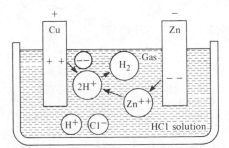

FIGURE B–1 *Simple chemical cell.*

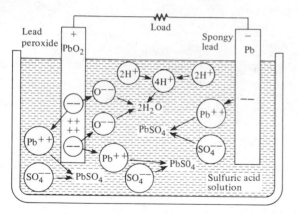

FIGURE B–2 *Chemical reaction in a discharging lead-acid cell.*

FIGURE B–3 *Recharging a lead-acid cell.*

When a lead ion from the lead peroxide electrode enters the solution, it also leaves two excess electrons on the electrode and forms lead sulfate in the solution. However, because this electrode is lead peroxide, two free oxygen atoms are created when a lead atom leaves and enters the solution as a lead ion. These two oxygen atoms take four electrons from the lead peroxide electrode and become oxygen ions (O^{--}). This process creates a deficiency of two electrons on this electrode (there were initially two excess electrons).

The two oxygen ions ($2O^{--}$) combine in the solution with four hydrogen ions ($4H^+$) to produce two molecules of water ($2H_2O$). This process dilutes the electrolyte over a period of time. Also, there is a buildup of lead sulfate on the electrodes. These two factors result in a reduction in the voltage produced by the cell and necessitate periodic recharging.

As you have seen, for each departing lead ion, there is an excess of two electrons on the spongy lead electrode, and there is a deficiency of two electrons on the lead peroxide electrode. Therefore, the lead peroxide electrode is positive, and the spongy lead electrode is negative. This chemical reaction is pictured in Figure B–2.

As mentioned, the dilution of the electrolyte by the formation of water and lead sulfate requires that the lead-acid cell be recharged to *reverse* the chemical process. A chemical cell that can be recharged is called a *secondary cell*. One that cannot be recharged is called a *primary cell*.

The cell is recharged by connection of an external voltage source to the electrodes, as shown in Figure B–3. The formula for the chemical reaction in a lead-acid cell is as follows:

$$Pb + PbO_2 + 2H_2SO_4 \longrightarrow 2PbSO_4 + 2H_2O$$

Dry Cell

In the dry cell, some of the disadvantages of a liquid electrolyte are overcome. Actually, the electrolyte in a typical dry cell is not dry but rather is in the form of a moist paste. This electrolyte is a combination of granulated carbon, powdered manganese dioxide, and ammonium chloride solution.

A typical carbon-zinc dry cell is illustrated in Figure B–4. The zinc container or can is dissolved by the electrolyte. As a result of this reaction, an excess of electrons accumulates on the container, making it the negative electrode.

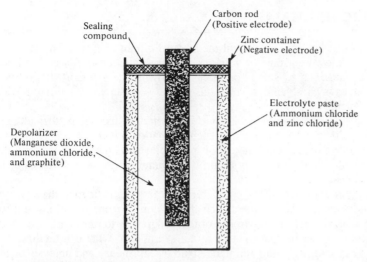

FIGURE B–4 *Simplified construction of dry cell.*

The hydrogen ions in the electrolyte take electrons from the carbon rod, making it the positive electrode. Hydrogen gas is formed near the carbon electrode, but this gas is eliminated by reaction with manganese dioxide (called a *depolarizing agent*). This depolarization prevents bursting of the container due to gas formation. Because the chemical reaction is not reversible, the carbon-zinc cell is a primary cell.

Types of Chemical Cells

Although only two common types of battery cells have been discussed, there are several types, listed in Table B–1.

TABLE B–1 *Types of battery cells.*

Type	+ electrode	− electrode	Electrolyte	Volts	Comments
Carbon-zinc	Carbon	Zinc	Ammonium and zinc chloride	1.5	Dry, primary
Lead-acid	Lead peroxide	Spongy lead	Sulfuric acid	2.0	Wet, secondary
Manganese-alkaline	Manganese dioxide	Zinc	Potassium hydroxide	1.5	Dry, primary or secondary
Mercury	Zinc	Mercuric oxide	Potassium hydroxide	1.3	Dry, primary
Nickel-cadmium	Nickel	Cadmium hydroxide	Potassium hydroxide	1.25	Dry, secondary
Nickel-iron (Edison cell)	Nickel oxide	Iron	Potassium hydroxide	1.36	Wet, secondary

A COMPUTER PROGRAM FOR CIRCUIT ANALYSIS

APPENDIX C

In Chapter 9, you learned how to use several methods of circuit analysis. The analyses were limited to two unknown currents or voltages and two equations. You also learned how to set up three equations to solve for three unknowns. It is important that you understand the principles of these analysis methods so that you can apply them to a wide variety of circuit problems.

The more unknown quantities there are in a circuit, the more difficult and tedious it becomes to calculate the solutions. In such a situation, *computer-aided analysis* is extremely helpful. If you understand the circuit principles and know how to apply the various analysis methods, then you can program a computer to do the tedious calculations and give you almost immediate answers.

Although a thorough coverage of computer programming is beyond the scope of this book, a computer program is presented for solving simultaneous equations and for doing mesh or node analysis. The main purpose of this program is to familiarize you with the use of a computer as a circuit analysis tool by an example of what can be done.

When you learn programming and understand circuit theory and analysis methods, the possibilities are almost limitless for use of the computer as an analysis and design tool. The computer does, however, have one limitation: It cannot think for you.

The Program

The program presented here can do three operations:

1. It can solve two or three simultaneous equations for unknown currents or voltages.
2. It can solve for two or three loop currents using the mesh analysis method when the resistor values and voltages are entered.
3. It can solve for two or three unknown node voltages using the node analysis method when the resistor values and voltages sources are entered.

This program is written in Level II BASIC for the TRS-80 computer. It can be run on many other BASIC computers with perhaps only minor changes. If you wish to convert to other BASIC "dialects," see *The Basic Handbook: An Encyclopedia of the Basic Computer Language,* by David A. Lien. This book is available from Compu-Soft Publishing, A Division of CompuSoft, Inc., 8643 Navajo Rd., San Diego, California 92119.

How to Enter the Program in the Computer

Refer to your computer manual for instructions on entering the program into the computer. Once the program is entered and running properly, you should store it on magnetic tape or disk to eliminate the task of entering it from the keyboard each time you want to use it. Again, refer to your manual for this procedure. Comments and instructional statements are used throughout the program, making it completely self-explanatory. You will find it very useful to use this program in conjunction with the problems in Chapter 9.

The program, of course, is limited and is not a universal circuit analysis program. Its purpose is to illustrate the potential of the computer in circuit analysis.

Program for Mesh/Node Analysis

```
10    CLEAR(1000): DIMA(20,20)
20    CLS
30    PRINT @ 384, CHR$(23) "PROGRAM FOR MESH/NODE ANALYSIS"
40    FOR X=1 TO 5000: NEXT: CLS
50    GO TO 530
60    PRINT "THIS PROGRAM WILL SOLVE FOR 2 OR 3 UNKNOWN CURRENTS
      OR VOLTAGES. FIRST YOU MUST SET UP THE EQUATIONS USING THE
      MESH ANALYSIS OR THE NODE ANALYSIS METHOD."
70    PRINT "FOR 2 UNKNOWNS, THE EQUATIONS MUST BE IN THE FOLLOWING
      FORMAT."
80    PRINT TAB(20) "AX+BY=C"
90    PRINT TAB(20) "DX+EY=F"
100   PRINT "FOR 3 UNKNOWNS, THE EQUATIONS MUST BE IN THE FOLLOWING
      FORMAT."
110   PRINT TAB(20) "AX+BY+CZ=D"
120   PRINT TAB(20) "EX+FY+GZ=H": PRINT TAB(20) "IX+JY+KZ=L"
130   INPUT "WHEN YOUR EQUATIONS ARE READY, PRESS 'ENTER'"; W: CLS
140   INPUT "HOW MANY UNKNOWNS";N
150   IF N=2 THEN A(3,3)=1: C$="AX+BY=C"
160   IF N=3 THEN C$="AX+BY+CZ=D"
170   PRINT "THE EQUATIONS MUST BE IN THE FORM"; C$
180   INPUT "ARE THE UNKNOWNS VOLTAGE OR CURRENT"; A$
190   CLS
200   INPUT "ARE THE UNITS VOLTS OR AMPS"; B$
210   FOR I=1 TO N
220   PRINT "THE COEFFICIENTS FOR EQUATION "; I
230   FOR J=1 TO N
240   PRINT "WHAT IS THE COEFFICIENT FOR "; A$; J
250   INPUT A(I,J)
260   NEXT J
270   INPUT "WHAT IS THE VALUE OF THE CONSTANT"; A(I,4)
280   CLS
290   NEXT I
300   CLS
310   PRINT "FOR VERIFICATION, THE COEFFICIENTS AND CONSTANTS OF
      THE EQUATIONS ARE"
320   FOR I=1 TO N
330   FOR J=1 TO N
340   PRINT A(I,J),
350   NEXT J
360   PRINT A(I,4)
370   NEXT I
380   PRINT "IF INCORRECT TYPE NO"
390   INPUT "IF CORRECT TYPE YES"; N$: IF N$="NO" THEN 20
400   D=A(1,1)*(A(2,2)*A(3,3)-A(3,2)*A(2,3))-A(1,2)*(A(2,1)*A(3,3)
      -A(3,1)*A(2,3))+A(1,3)*(A(2,1)*A(3,2)-A(3,1)*A(2,2))
410   N(1)=A(1,4)*(A(2,2)*A(3,3)-A(3,2)*A(2,3))-A(1,2)*(A(2,4)*
      A(3,3)-A(3,4)*A(2,3))+A(1,3)*(A(2,4)*A(3,2)-A(3,4)*A(2,2))
420   N(2)=A(1,1)*(A(2,4)*A(3,3)-A(3,4)*A(2,3))-A(1,4)*(A(2,1)*
      A(3,3)-A(3,1)*A(2,3))+A(1,3)*(A(2,1)*A(3,4)-A(3,1)*A(2,4))
430   N(3)=A(1,1)*(A(2,2)*A(3,4)-A(3,2)*A(2,4))-A(1,2)*(A(2,1)*
      A(3,4)-A(3,1)*A(2,4))+A(1,4)*(A(2,1)*A(3,2)-A(3,1)*A(2,2))
440   CLS
450   FOR I=1 TO N
460   PRINT @ 275+64*I, A$;I;"=";N(I)/D; B$
470   NEXT
```

```
480    FOR X=10 TO 100: SET(X,10): NEXT
490    FOR X=10 TO 100: SET(X,30): NEXT
500    FOR Y=10 TO 30: SET(10,Y): NEXT
510    FOR Y=10 TO 30: SET(100,Y): NEXT
520    GO TO 520
530    PRINT "****THIS PROGRAM PROVIDES 3 OPTIONS FOR CIRCUIT
       ANALYSIS ****"
540    PRINT
550    PRINT "OPTION 1: SOLUTION OF 2 OR 3 SIMULTANEOUS MESH OR NODE
       EQUATIONS. USE THIS OPTION WHEN YOU HAVE ALREADY SET UP THE
       EQUATIONS."
560    PRINT
570    PRINT "OPTION 2: MESH ANALYSIS FOR A 2 OR 3 LOOP CIRCUIT.
       THIS OPTION ALLOWS YOU TO ENTER THE CIRCUIT COMPONENT VALUES
       AFTER YOU HAVE DEFINED THE LOOPS, FOR RESISTIVE CIRCUITS ONLY.
580    PRINT
590    PRINT "OPTION 3: NODE ANALYSIS FOR CIRCUIT WITH 2 OR 3
       UNKNOWN NODES. THIS OPTION ALLOWS YOU TO ENTER THE CIRCUIT
       COMPONENT VALUES AFTER YOU HAVE DEFINED THE NODES AT WHICH
       THE VOLTAGES ARE NOT KNOWN. FOR RESISTIVE CIRCUITS ONLY."
600    PRINT
610    PRINT "IF YOU ARE READY TO SELECT YOUR OPTION PRESS 'ENTER'":
       INPUT W
620    CLS: PRINT "TO SOLVE SIMULTANEOUS EQUATIONS TYPE 1"
630    PRINT "TO USE MESH ANALYSIS TYPE 2"
640    PRINT "TO USE NODE ANALYSIS TYPE 3"
650    INPUT Q: CLS: ON Q GO TO 60, 660, 1010
660    PRINT "*************** MESH ANALYSIS INSTRUCTIONS ***************
       : A$="I"
670    PRINT
680    PRINT "1. ASSIGN LOOPS IN YOUR CIRCUIT."
690    PRINT
700    PRINT "2. ASSIGN LOOP CURRENTS CLOCKWISE."
710    PRINT
720    PRINT "3. VOLTAGE SOURCES - TO + IN THE DIRECTION OF CURRENT
       ARE POSITIVE, OTHERWISE NEGATIVE."
730    PRINT
740    PRINT "TO CONTINUE PRESS 'ENTER'": INPUT W
750    CLS
760    PRINT "HOW MANY LOOPS?"
770    INPUT N
780    IF N=2 THEN A(3,3)=1
790    FOR I= 1 TO N
800    FOR J= I TO N
810    S=1
820    IF I=J PRINT "HOW MANY RESISTORS IN LOOP "; I ELSE PRINT "HOW
       MANY RESISTORS COMMON TO LOOP ";I; "AND LOOP ";J: S=-1
830    INPUT M
840    FOR K=1 TO M
850    INPUT "RESISTOR VALUE"; R
860    A(I,J)=A(I,J)+R
870    NEXT K
880    A(I,J)=S*A(I,J)
890    A(J,I)=A(I,J)
900    CLS
910    NEXT J
920    PRINT "HOW MANY VOLTAGE SOURCES IN LOOP "; I
930    INPUT P
940    FOR L=1 TO P: IF P=0 THEN E=0
950    IF P>0 THEN INPUT "VOLTAGE SOURCE VALUE WITH PROPER SIGN"; E
960    A(I,4)=A(I,4)+E
```

```
970    NEXT L
980    CLS
990    NEXT I
1000   GO TO 300
1010   PRINT "************** NODE ANALYSIS INSTRUCTIONS **************"
1020   PRINT
1030   PRINT "1. IDENTIFY AND NUMBER THE UNKNOWN NODES."
1040   PRINT
1050   PRINT "2. VOLTAGE SOURCES WITH + TERMINAL TOWARD NODE ARE
       POSITIVE. OTHERS ARE NEGATIVE."
1060   PRINT
1070   PRINT
1080   PRINT "TO CONTINUE PRESS 'ENTER'": INPUT W
1090   A$="V"
1100   CLS
1110   PRINT "HOW MANY NODES?"
1120   INPUT N
1130   IF N=2 THEN A(3,3)=1
1140   FOR I=1 TO N
1150   FOR J=I TO N
1160   S=1
1170   IF I=J PRINT "HOW MANY RESISTORS ARE CONNECTED TO NODE "; I
       ELSE PRINT "HOW MANY RESISTORS BETWEEN NODE "; I; "AND NODE ";
       J; S=-1
1180   INPUT M
1190   FOR K=1 TO M
1200   INPUT "RESISTOR VALUE";R
1210   A(I,J)=A(I,J)+1/R
1220   NEXT K
1230   A(I,J)=S*A(I,J)
1240   A(J,I)=A(I,J)
1250   CLS
1260   NEXT J
1270   PRINT "HOW MANY VOLTAGE SOURCES CONNECTED TO NODE ";I;"THROUGH
       A SERIES RESISTOR?"
1280   INPUT P
1290   FOR L=1 TO P: IF P=0 THEN E=0 AND R=1
1300   IF P>0 THEN INPUT "VOLTAGE SOURCE VALUE WITH PROPER SIGN"; E
1310   IF P>0 THEN INPUT "SERIES RESISTOR VALUE";R
1320   A(I,4)=A(I,4)+E/R
1330   NEXT L
1340   CLS
1350   NEXT I
1360   CLS
1370   GO TO 300
```

rms (EFFECTIVE) VALUE OF A SINE WAVE

APPENDIX $\mathbf{D}$

The abbreviation "rms" stands for the *root mean square* process by which this value is derived. In the process, we first square the equation of a sine wave:

$$v^2 = V_p^2 \sin^2\theta$$

Next, we obtain the mean or average value of v^2 by dividing the area under a half-cycle of the curve by π (see Figure D–1). The area is found by integration and trigonometric identities:

$$V_{avg}^2 = \frac{\text{area}}{\pi}$$

$$= \frac{1}{\pi} \int_0^\pi V_p^2 \sin^2\theta \, d\theta$$

$$= \frac{V_p^2}{2\pi} \int_0^\pi (1 - \cos 2\theta) \, d\theta$$

$$= \frac{V_p^2}{2\pi} \int_0^\pi 1 \, d\theta - \frac{V_p^2}{2\pi} \int_0^\pi (-\cos 2\theta) \, d\theta$$

$$= \frac{V_p^2}{2\pi} (\theta - \tfrac{1}{2} \sin 2\theta)_0^\pi$$

$$= \frac{V_p^2}{2\pi} (\pi - 0)$$

$$= \frac{V_p^2}{2}$$

Finally, the square root of V_{avg}^2 is V_{rms}:

$$V_{rms} = \sqrt{V_{avg}^2}$$

$$= \sqrt{V_p^2/2}$$

$$= \frac{V_p}{\sqrt{2}}$$

$$= 0.707 V_p$$

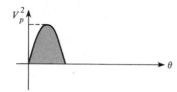

FIGURE D–1

AVERAGE VALUE OF A HALF-CYCLE SINE WAVE

The average value of a sine wave is determined for a half-cycle because the average over a full-cycle is zero.

The equation for a sine wave is

$$\nu = V_p \sin \theta$$

The average value of the half-cycle is the area under the curve divided by the distance of the curve along the horizontal axis (see Figure E–1):

$$V_{avg} = \frac{\text{area}}{\pi}$$

To find the area, we use integral calculus:

$$V_{avg} = \frac{1}{\pi} \int_0^{\pi} V_p \sin \theta \, d\theta$$

$$= \frac{V_p}{\pi} (-\cos \theta) \Big|_0^{\pi}$$

$$= \frac{V_p}{\pi} [-\cos \pi - (-\cos 0)]$$

$$= \frac{V_p}{\pi} [-(-1) - (-1)]$$

$$= \frac{V_p}{\pi} (2)$$

$$= \frac{2}{\pi} V_p$$

$$= 0.637 V_p$$

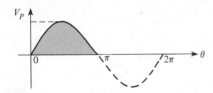

FIGURE E–1

REACTANCE DERIVATIONS

APPENDIX $\mathbf{F}$

Derivation of Capacitive Reactance

$$\theta = 2\pi ft = \omega t$$

$$i = C\frac{dV}{dt} = C\frac{d(V_p \sin \theta)}{dt}$$

$$= C\frac{d(V_p \sin \omega t)}{dt}$$

$$= \omega C(V_p \cos \omega t)$$

$$I_{rms} = \omega C V_{rms}$$

$$X_C = \frac{V_{rms}}{I_{rms}} = \frac{V_{rms}}{\omega C V_{rms}} = \frac{1}{\omega C}$$

$$= \frac{1}{2\pi fC}$$

Derivation of Inductive Reactance

$$v = L\frac{di}{dt} = L\frac{d(I_p \sin \omega t)}{dt}$$

$$= \omega L(I_p \cos \omega t)$$

$$V_{rms} = \omega L I_{rms}$$

$$X_L = \frac{V_{rms}}{I_{rms}} = \frac{\omega L I_{rms}}{I_{rms}} = \omega L$$

$$= 2\pi fL$$

POLAR/RECTANGULAR CONVERSION WITH THE CALCULATOR (TI-55-II)

Polar to Rectangular

APPENDIX G

1. Enter the magnitude.
2. Press $\boxed{\text{x} \rightleftarrows \text{y}}$.
3. Enter the angle in degrees (select DEG mode with $\boxed{\text{DRG}}$).
4. Press $\boxed{\text{2nd}}$, then $\boxed{\text{P} \leftrightarrow \text{R}}$.
5. The j value is displayed.
6. Press $\boxed{\text{x} \rightleftarrows \text{y}}$ to display the real value.

Rectangular to Polar

1. Enter the real value.
2. Press $\boxed{\text{x} \rightleftarrows \text{y}}$.
3. Enter the j value.
4. Select DEG mode with $\boxed{\text{DRG}}$.
5. Press $\boxed{\text{INV}}$, then $\boxed{\text{2nd}}$, then $\boxed{\text{P} \leftrightarrow \text{R}}$.
6. The angle is displayed.
7. Press $\boxed{\text{x} \rightleftarrows \text{y}}$ to display the magnitude.

NOTE: This procedure is specifically for the TI-55-II calculator but is easily adapted to other scientific calculators (consult your calculator guide).

GLOSSARY

Admittance   A measure of the ability of a reactive circuit to permit current. The reciprocal of impedance.

Alternating current (ac)   Current that reverses direction in response to a change in voltage polarity.

Ammeter   An electrical instrument used to measure current.

Ampere (A or amp)   The unit of electrical current.

Ampere-hour (Ah)   The measure of the capacity of a battery to supply electrical current.

Amplification   The process of increasing the power, voltage, or current of an electrical signal.

Amplifier   An electronic circuit having the capability of amplification and designed specifically for that purpose.

Amplitude   The voltage or current value of an electrical signal which, in some cases, implies the maximum value.

Analog   Characterizing a linear process in which a variable takes on a *continuous* set of values within a given range.

Apparent power   The power that *appears* to be being delivered to a reactive circuit. The product of volts and amperes with units of VA (volt-amperes).

Arc tangent   An inverse trigonometric function meaning "the angle whose tangent is." Also called *inverse tangent*.

Asynchronous   Having no fixed time relationship, such as two wave forms that are not related to each other in terms of their time variations.

Atom   The smallest particle of an element possessing the unique characteristics of that element.

Attenuation   The process of reducing the power, voltage, or current value of an electrical signal. It can be thought of as negative amplification.

Audio   Related to ability of the human ear to detect sound. Audio frequencies (af) are those that can be heard by the human ear, typically from 20 Hz to 20 kHz.

Autotransformer   A transformer having only one coil for both its primary and its secondary.

Average power   The average rate of energy consumption. In an electrical circuit, average power occurs only in the resistance and represents a net energy loss.

AWG   American Wire Gage, a standardization of wire sizes according to the diameter of the wire.

Band-pass   The characteristic of a certain type of filter whereby frequencies within a certain range are passed through the circuit and all others are blocked.

Band-stop   The characteristics of a certain type of filter whereby frequencies within a certain range are blocked from passing through the circuit.

Base   One of the semiconductor regions in a bipolar transistor.

Baseline   The normal level of a pulse wave form. The level in the absence of a pulse.

Battery   An energy source that uses a chemical reaction to convert chemical energy into electrical energy.

Bias   The application of a dc voltage to a diode, transistor, or other electronic device to produce a desired mode of operation.

Bode plot   An idealized graph of the gain, in dB, versus frequency. Used to illustrate graphically the response of an amplifier or filter circuit.

Branch   One of the current paths in a parallel circuit.

Capacitance   The ability of a capacitor to store electrical charge.

Capacitor   An electrical device possessing the property of capacitance.

Cascade   An arrangement of circuits in which the output of one circuit becomes the input to the next.

Center tap (CT)   A connection at the midpoint of the secondary of a transformer.

Charge   An electrical property of matter in which an attractive force or a repulsive force is created between two particles. Charge can be positive or negative.

Chassis   The metal framework or case upon which an electrical circuit or system is constructed.

Chip   A tiny piece of semiconductor material upon which an integrated circuit is constructed.

Choke   An inductor. The term is more commonly used in connection with inductors used to block or *choke off* high frequencies.

Circuit   An interconnection of electrical components designed to produce a desired result.

Circuit breaker   A resettable protective electrical device used for interrupting current in a circuit when the current has reached an excessive level.

Circular mil (CM)   The unit of the cross-sectional area of a wire. A wire with a diameter of 0.001 inch has a cross-sectional area of one circular mil.

Closed circuit   A circuit with a complete current path.

Coefficient of coupling   A constant associated with transformers that specifies the magnetic field in the secondary as a result of that in the primary.

Coil   A common term for an inductor or for the primary or secondary winding of a transformer.

Common   A term sometimes used for the ground or reference point in a circuit.

Computer   An electronic system that can process data and perform calculations at a very fast rate. It operates under the direction of a *stored program*.

Conductance   The ability of a circuit to allow current. It is the reciprocal of resistance. The units are siemens (S).

Conductor   A material that allows electrical current to flow with relative ease. An example is copper.

Core   The material within the windings of an inductor that influences the electromagnetic characteristics of the inductor.

Coulomb (C)   The unit of electrical charge.

CRT   Cathode ray tube.

Current   The rate of flow of electrons in a circuit.

Cycle   The repetition of a pattern.

Decibel (dB)   The unit of the logarithmic expression of a ratio, such as power or voltage gain.

Degree   The unit of angular measure corresponding to $1/360$ of a complete revolution.

Derivative   The rate of change of a function, determined mathematically.

Dielectric   The insulating material used between the plates of a capacitor.

Differentiator   An *RC* or *RL* circuit that produces an output which approaches the mathematical derivative of the input.

Digital   Characterizing a nonlinear process in which a variable takes on discrete values within a given range.

Diode   An electronic device that permits current flow in only one direction. It can be a semiconductor or a tube.

Direct current (dc)   Current that flows in only one direction.

DMM   Digital multimeter.

Duty cycle   A characteristic of a pulse wave form that indicates the percentage of time that a pulse is present during a cycle.

DVM   Digital voltmeter.

Effective value   A measure of the heating effect of a sine wave. Also known as the rms (root mean square) value.

Electrical   Related to the use of electrical voltage and current to achieve desired results.

Electromagnetic   Related to the production of a magnetic field by an electrical current in a conductor.

Electron   The basic particle of electrical charge in matter. The electron possesses negative charge.

Electronic   Related to the movement and control of free electrons in semiconductors or vacuum devices.

Element   One of the unique substances that make up the known universe. Each element is characterized by a unique atomic structure.

Emitter   One of the three regions in a bipolar transistor.

Energy   The ability to do work.

Exponent   The number of times a given number is multiplied by itself. The exponent is called the *power,* and the given number is called the *base*.

Falling edge   The negative-going transition of a pulse.

Fall time   The time interval required for a pulse to change from 90% of its amplitude to 10% of its amplitude.

Farad (F)   The unit of capacitance.

Field   The invisible forces that exist between oppositely charged particles (electric field) or between the north and south poles of a magnet or electromagnet (magnetic field).

Filter   A type of electrical circuit that passes certain frequencies and rejects all others.

Flux   The lines of force in a magnetic field.

Flux density   The amount of flux per unit area in a magnetic field.

Free electron   A valence electron that has broken away from its parent atom and is free to move from atom to atom within the atomic structure of a material.

Frequency   A measure of the rate of change of a periodic function. The electrical unit of frequency is hertz (Hz).

Function generator   An electronic test instrument that is capable of producing several types of electrical wave forms, such as sine, triangular, and square waves.

Fuse   A protective electrical device that burns open when excessive current flows in a circuit.

Gain   The amount by which an electrical signal is increased or amplified.

Generator   A general term for any of several types of devices or instruments that are sources for electrical signals.

Germanium   A semiconductor material.

Giga   A prefix used to designate 10^9 (one thousand million).

Ground   In electrical circuits, the common or reference point. It can be chassis ground or earth ground.

Harmonics   The frequencies contained in a composite wave form that are integer multiples of the repetition frequency or fundamental frequency of the wave form.

Henry (H)   The unit of inductance.

Hertz (Hz)   The unit of frequency. One hertz is one cycle per second.

High-pass   The characteristics of a certain type of filter whereby higher frequencies are passed and lower frequencies are rejected.

Hypotenuse   The longest side of a right triangle.

Impedance   The total opposition to current in a reactive circuit. The unit is ohms (Ω).

Induced voltage   Voltage produced as a result of a changing magnetic field.

Inductance   The property of an inductor whereby a change in current causes the inductor to produce an opposing voltage.

Inductor   An electrical device having the property of inductance. Also known as a *coil* or a *choke*.

Infinite   Having no bounds or limits.

Input   The voltage, current, or power applied to an electrical circuit to produce a desired result.

Instantaneous value   The value of a variable at a given instant in time.

Insulator   A material that does not allow current under normal conditions.

Integrated circuit (IC)   A type of circuit in which all of the components are constructed on a single, tiny piece of semiconductor material.

Integrator   A type of *RC* or *RL* circuit that produces an output which approaches the mathematical integral of the input.

Joule (J)   The unit of energy.

Kilo   A prefix used to designate 10^3 (one thousand).

Kilowatt-hour (kWh)   A common unit of energy used mainly by utility companies.

Kirchhoff's laws   A set of circuit laws that describe certain voltage and current relationships in a circuit.

Lag   A condition of the phase or time relationship of wave forms in which one wave form is behind the other in phase or time.

Lead   A wire or cable connection to an electrical or electronic device or instrument. Also, a condition of the phase or time relationship of wave forms in which one wave form is ahead of the other in phase or time.

Leading edge   The first step or transition of a pulse.

Linear   Characterized by a straight-line relationship.

Load   The device upon which work is performed.

Logarithm   The exponent to which a base number must be raised to produce a given number. For example, the logarithm of 100 is 2 ($10^2 = 100$).

Loop A closed path in a circuit.

Low-pass The characteristics of a certain type of filter whereby lower frequencies are passed and higher frequencies are rejected.

Magnetic Related to or possessing characteristics of magnetism. Having a north and a south pole with lines of force extending between the two.

Magnetomotive force (mmf) The force produced by a current in a coiled wire in establishing a magnetic field. The unit is ampere-turns (At).

Magnitude The value of a quantity, such as the number of amperes of current or the number of volts of voltage.

Mega A prefix designating 10^6 (one million).

Mesh An arrangement of loops in a circuit.

Micro A prefix designating 10^{-6} (one-millionth).

Milli A prefix designating 10^{-3} (one-thousandth).

Modulation The process whereby a signal containing information (such as voice) is used to modify the amplitude (AM) or the frequency (FM) of a much higher frequency sine wave (carrier).

Multimeter An instrument used to measure current, voltage, and resistance.

Mutual inductance The inductance between two separate coils, such as in a transformer.

Nano A prefix designating 10^{-9} (one thousand-millionth).

Network A circuit.

Neutron An atomic particle having no electrical charge.

Node A point or junction in a circuit where two or more components connect.

Ohm (Ω) The unit of resistance.

Ohmmeter An instrument for measuring resistance.

Open circuit A circuit in which there is not a complete current path.

Oscillator An electronic circuit that internally produces a time-varying signal without an external signal input.

Oscilloscope A measurement instrument that displays signal wave forms on a screen.

Output The voltage, current, or power produced by a circuit in response to an input or to a particular set of conditions.

Overshoot A short duration of excessive amplitude occurring on the positive-going transition of a pulse.

Parallel The relationship in electric circuits in which two or more current paths are connected between the same two points.

Peak value The maximum value of an electrical wave form, particularly in relation to sine waves.

Period The time interval of one cycle of a periodic wave form.

Periodic Characterized by a repetition at fixed intervals.

Permeability A measure of the ease with which a magnetic field can be established within a material.

Phase The relative displacement of a time-varying wave form in terms of its occurrence.

Pico A prefix designating 10^{-12} (one-billionth).

Potentiometer A three-terminal variable resistor.

Power   The rate of energy consumption. The unit is watts (W).

Power factor   The relationship between volt-amperes and average power or watts. Volt-amperes multiplied by the power factor equals average power.

Power supply   An electronic instrument that produces voltage, current, and power from the ac power line or batteries in a form suitable for use in various applications to power electronic equipment.

p/s   Pulses per second, a unit of frequency measurement for pulse wave forms.

PRF   Pulse repetition frequency.

Primary   The input winding of a transformer.

Proton   A positively charged atomic particle.

Pulse   A type of wave form that consists of two equal and opposite steps in voltage or current, separated by a time interval.

Pulse width   The time interval between the opposite steps of an ideal pulse. Also, the time between the 50% points on the leading and trailing edges of a nonideal pulse.

Qualify factor (Q)   The ratio of reactive power to average power in a coil or a resonant circuit.

Radian   A unit of angular measurement. There are 2π radians in a complete revolution. One radian equals 57.3°.

Ramp   A type of wave form characterized by a linear increase or decrease in voltage or current.

Reactance   The opposition of a capacitor or an inductor to sinusoidal current. The unit is ohms (Ω).

Reactive power   The rate at which energy is stored and alternately returned to the source by a reactive component.

Rectifier   An electronic circuit that converts ac into pulsating dc.

Reluctance   The opposition to the establishment of a magnetic field in an electro-magnetic circuit.

Resistance   Opposition to current. The unit is ohms (Ω).

Resistivity   The resistance that is characteristic of a given material.

Resistor   An electrical component possessing resistance.

Resonance   In an LC circuit, the condition when the impedance is minimum (series) or maximum (parallel).

Response   In electronic circuits, the reaction of a circuit to a given input.

Rheostat   A two-terminal, variable resistor.

Right angle   A 90-degree angle.

Ringing   An unwanted oscillation on a wave form.

Rise time   The time interval required for a pulse to change from 10% of its amplitude to 90% of its amplitude.

Rising edge   The positive-going transition of a pulse.

rms   Root mean square. The value of a sine wave that indicates its heating effect. Also known as *effective value*.

Sawtooth   A type of electrical wave form composed of ramps.

Secondary   The output winding of a transformer.

Semiconductor   A material that has a conductance value between that of a conductor and that of an insulator. Silicon and germanium are examples.

Series   In an electrical circuit, a relationship of components in which the components are connected such that they provide a single current path between two points.

Short   A zero resistance connection between two points.

Siemen (S)   The unit of conductance.

Signal   A time-varying electrical wave form.

Silicon   A semiconductor material used in transistors.

Sine wave   A type of alternating electrical wave form.

Slope   The vertical change in a line for a given horizontal change.

Source   Any device that produces energy.

Steady state   An equilibrium condition in a circuit.

Step   A voltage or current transition from one level to another.

Susceptance   The ability of a reactive component to permit current flow. The reciprocal of reactance.

Switch   An electrical or electronic device for opening and closing a current path.

Synchronous   Having a fixed time relationship.

Tangent   A trigonometric function that is the ratio of the opposite side of a right triangle to the adjacent side.

Tank   A parallel resonant circuit.

Tapered   Nonlinear, such as a tapered potentiometer.

Temperature coefficient   A constant specifying the amount of change in the value of a quantity for a given change in temperature.

Tesla (T)   The unit of flux density. Also, webers per square meter.

Tetrode   A vacuum tube that has four elements.

Thermistor   A resistor whose resistance decreases with an increase in temperature.

Tilt   A slope on the normally flat portion of a pulse.

Time constant   A fixed time interval, set by R, C, and L values, that determines the time response of a circuit.

Tolerance   The limits of variation in the value of an electrical component.

Trailing edge   The last edge to occur in a pulse.

Transformer   An electrical device that operates on the principle of electromagnetic induction. It is used for increasing or decreasing an ac voltage and for various other applications.

Transient   A temporary or passing condition in a circuit. A sudden and temporary change in circuit conditions.

Transistor   A semiconductor device used for amplification and switching applications in electronic circuits.

Triangular wave   A type of electrical wave form that consists of ramps.

Trimmer   A small, variable resistor or capacitor.

Troubleshooting   The process and technique of identifying and locating faults in an electrical or electronic circuit.

Turns ratio   The ratio of the number of secondary turns to the number of primary turns in the transformer windings.

Undershoot   The opposite of overshoot, occurring on the negative-going edge of a pulse.

Valence   Related to the outer shell or orbit of an atom.

Volt   The unit of voltage or electromotive force (emf).

Voltage   The amount of energy available to move a certain number of electrons from one point to another in an electrical circuit.

Watt (W)   The unit of power.

Wave form   The pattern of variations of a voltage or a current.

Wavelength   The length in space occupied by one cycle of an electromagnetic wave.

Weber   The unit of magnetic flux.

Winding   The loops of wire or coil in an inductor or transformer.

Wiper   The variable contact in a potentiometer or other device.

SOLUTIONS TO SELF-TESTS

Chapter 1

1. Current: amperes. Voltage: volts. Resistance: ohms. Power: watts. Energy: joules.

2. Amperes: A. Volts: V. Ohms: Ω. Watts: W. Joules: J.

3. Current: I. Voltage: V. Resistance: R. Power: P. Energy: $\mathcal{E}$.

4. (a) $100 = 1 \times 10^2$
   (b) $12,000 = 1.2 \times 10^4$
   (c) $5,600,000 = 5.6 \times 10^6$
   (d) $78,000,000 = 7.8 \times 10^7$

5. (a) $0.03 = 3 \times 10^{-2}$
   (b) $0.0005 = 5 \times 10^{-4}$
   (c) $0.00058 = 5.8 \times 10^{-4}$
   (d) $0.0000224 = 2.24 \times 10^{-5}$

6. (a) $7 \times 10^4 = 70,000$
   (b) $45 \times 10^3 = 45,000$
   (c) $100 \times 10^{-3} = 0.1$
   (d) $4 \times 10^{-1} = 0.4$

7. Convert 12×10^5 to 1.2×10^6.
   $1.2 \times 10^6 + 25 \times 10^6 = 26.2 \times 10^6$

8. $8 \times 10^{-3} - 5 \times 10^{-3} =$
   $(8 - 5) \times 10^{-3} = 3 \times 10^{-3}$

9. $(33 \times 10^3)(20 \times 10^{-4}) =$
   $(33 \times 20) \times (10^{3+(-4)}) =$
   $660 \times 10^{-1} = 66$

10. $(4 \times 10^2)/(2 \times 10^{-3}) =$
    $(4/2) \times 10^{2-(-3)} = 2 \times 10^5$

Chapter 2

1. 3

2. 6

3. $I = Q/t = 50 \text{ C}/5 \text{ s} = 10 \text{ A}$

4. $Q = It = (2 \text{ A})(10 \text{ s}) = 20 \text{ C}$

5. $V = \mathcal{E}/Q = 500 \text{ J}/100 \text{ C} = 5 \text{ V}$

6. $G = 1/R = 1/10 \ \Omega = 0.1 \text{ S}$

7. Blue: 6, 1st digit
   Green: 5, 2nd digit
   Red: 2, number of zeros
   Gold: 5%, tolerance
   $R = 6500 \ \Omega \pm 5\%$

8. $500 \ \Omega$. Center position is one-half the resistance.

9. Breakdown strength of mica = 2000 kV/cm. $V_{MAX} =$
   $(2000 \text{ kV/cm})(10 \text{ cm}) = 20,000 \text{ kV}$

10. $d = 0.003 \text{ in.} = 3 \text{ mils}$
    $A = d^2 = 9 \text{ CM}$
    $R = \rho l/A = (10.4 \text{ CM-}\Omega\text{/ft}) \cdot$
    $(10 \text{ ft})/9 \text{ CM} = 11.56 \ \Omega$

Chapter 3

1. $V = IR, I = V/R, R = V/I$

2. To find current, use $I = V/R$.
   To find voltage, use $V = IR$.
   To find resistance, use $R = V/I$.

3. $I = V/R = 10 \text{ V}/5 \ \Omega = 2 \text{ A}$

4. $I = V/R = 75 \text{ V}/10 \text{ k}\Omega = 7.5 \text{ mA}$

5. $V = IR = (2.5 \text{ A})(20 \ \Omega) = 50 \text{ V}$

6. $V = IR = (30 \text{ mA})(2.2 \text{ k}\Omega) = 66 \text{ V}$

7. $R = V/I = 12 \text{ V}/6 \text{ A} = 2 \ \Omega$

8. $R = V/I = 9 \text{ V}/100 \ \mu\text{A} = 0.09 \text{ M}\Omega$
   $= 90 \text{ k}\Omega$

9. $I = V/R = 50 \text{ kV}/2.5 \text{ k}\Omega = 20 \text{ A}$

10. $I = V/R = 15 \text{ mV}/10 \text{ k}\Omega = 1.5 \ \mu\text{A}$

11. $V = IR = (4 \ \mu\text{A})(100 \text{ k}\Omega) = 400 \text{ mV}$

12. $V = IR = (100 \text{ mA})(1 \text{ M}\Omega) = 100 \text{ kV}$

13. $R = V/I = 5 \text{ V}/2 \text{ mA} = 2.5 \text{ k}\Omega$

14. $I = V/R = 1.5 \text{ V}/5 \text{ k}\Omega = 0.3 \text{ mA}$

15. A: $I = V/R = 25 \text{ V}/10 \text{ k}\Omega$
       $= 2.5 \text{ mA}$
    B: $I = V/R = 5 \text{ V}/2 \text{ M}\Omega = 2.5 \ \mu\text{A}$
    C: $I = V/R = 1.5 \text{ kV}/15 \text{ k}\Omega$
       $= 0.1 \text{ A}$

16. A: $V = IR = (3 \text{ mA})(27 \text{ k}\Omega)$
       $= 81 \text{ V}$
    B: $V = IR = (5 \ \mu\text{A})(100 \text{ M}\Omega)$
       $= 500 \text{ V}$
    C: $V = IR = (2.5 \text{ A})(50 \ \Omega)$
       $= 125 \text{ V}$

17. A: $R = V/I = 8 \text{ V}/2 \text{ A} = 4 \ \Omega$
    B: $R = V/I = 12 \text{ V}/4 \text{ mA} = 3 \text{ k}\Omega$

C: $R = V/I = 30$ V$/150\ \mu$A
    $= 0.2$ MΩ

18. $V_2/30$ mA $= 10.$ V$/50$ mA
    $V_2 = (10$ V$)(30$ mA$)/50$ mA $= 6$ V
    (new value)
    It decreased 4 V, from 10 V.

19. The current increase is 50%, so the
    voltage increase must be the same; that
    is, $(0.5)(20$ V$) = 10$ V.
    $V_2 = 20$ V $+ (0.5)(20$ V$) = 30$ V
    (new value)

20. $R = 100$ V$/750$ mA $= 0.133$ kΩ
    $= 133\ \Omega$
    $R = 100$ V$/1$ A $= 100\ \Omega$

21. A: 2.5 mA
    B: 2.5 μA
    C: 0.1 A

22. A: 81 V
    B: 500 V
    C: 125 V

23. A: 4 Ω
    B: 3 kΩ
    C: 200 kΩ

Chapter 4

1. $P = \mathscr{E}/t = 200$ J$/10$ s $= 20$ W

2. $P = \mathscr{E}/t = 10{,}000$ J$/300$ ms
   $= 33.33$ kW

3. 50 kW $= 50 \times 10^3$ W $= 50{,}000$ W

4. 0.045 W $= 45 \times 10^{-3}$ W $= 45$ mW

5. $P_R = VI = (10$ V$)(350$ mA$)$
   $= 3500$ mW $= 3.5$ W
   $P_{BATT} = 3.5$ W

6. $P = V^2/R = (50$ V$)^2/1000\ \Omega = 2.5$ W

7. $P = I^2R = (0.5$ A$)^2(500\ \Omega)$
   $= 1250$ W

8. $P = VI = (115$ V$)(2$ A$) = 230$ W

9. $\mathscr{E} = Pt = (15$ W$)(60$ s$) = 900$ J

10. $P = 500$ W $= 0.5$ kW
    $t = 24$ h
    $\mathscr{E} = Pt = (0.5$ kW$)(24$ h$) = 12$ kWh

11. $(75$ W$)(10$ h$) = 750$ Wh

12. There are 30 days in April.
    $(750$ W$)(24$ h$) = 18000$ Wh
    $(18$ kWh$)(30$ days$) = 540$ kWh
    Reading at end of April:
    1000 kWh $+ 540$ kWh $= 1540$ kWh

13. $\dfrac{1{,}500{,}000\text{ Ws}}{(3600\text{ s/h})(1000\text{ W/kW})}$
    $= 0.4167$ kWh

14. See Figure S–1.

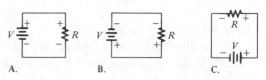

A.   B.   C.

FIGURE S–1

15. Power supply 2 has a greater load be-
    cause it must supply more current than
    power supply 1. Number 2 load has a
    smaller resistive value because it sup-
    plies more current for the same voltage.

16. 1: $P = VI = (25$ V$)(0.1$ A$) = 2.5$ W
    2: $P = VI = (25$ V$)(0.5$ A$) = 12.5$ W

17. $I_L = 12$ V$/600\ \Omega = 0.02$ A
    50 Ah$/0.02$ A $= 2500$ h

18. $(8$ A$)(2.5$ h$) = 20$ Ah

19. $P_{OUT} = P_{IN} - P_{LOST} = 2$ W $- 0.25$ W
    $= 1.75$ W

20. Efficiency $= (P_{OUT}/P_{IN}) \times 100$
    $= (0.5$ W$/0.6$ W$) \times 100$
    $= 83.33\%$

21.

I (A)	P (W)
0.01	0.15
0.02	0.6
0.03	1.35
0.04	2.4
0.05	3.75
0.06	5.4
0.07	7.35

$I(A)$	$P(W)$
0.08	9.6
0.09	12.15
0.1	15.

Chapter 5

1. See Figure S–2.

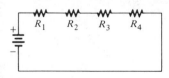

FIGURE S–2

2. Since current is the same at *all* points in a series circuit, 5 A flow out of the sixth and tenth resistors as well as all of the others.

3. See Figure S–3.

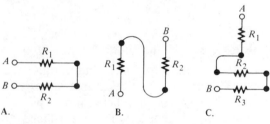

A.   B.   C.

FIGURE S–3

4. $I_{AB} = I_{BC} = I_{CD} = I_{DE} = I_{EF} = I_{FG}$
   $= I_{GH} = I_{HA} = 3$ A

5. $R_T = 75\ \Omega + 470\ \Omega = 545\ \Omega$

6. $R_T = (8)(56\ \Omega) = 448\ \Omega$

7. $R = 20$ kΩ
   $- (1\ \text{k}\Omega + 1\ \text{k}\Omega + 5\ \text{k}\Omega + 3\ \text{k}\Omega)$
   $= 10$ kΩ

8. $R_T = (3)(1\ \text{k}\Omega) = 3$ kΩ
   $I = 5\ \text{V}/3\ \text{k}\Omega = 1.67$ mA

9. A: $R_T = 100\ \Omega + 47\ \Omega + 68\ \Omega$
      $= 215\ \Omega$
      $I = 5\ \text{V}/215\ \Omega = 0.023$ A
   B: $R_T = 50\ \Omega + 175\ \Omega + 33\ \Omega$
      $= 258\ \Omega$
      $I = 8\ \text{V}/258\ \Omega = 0.031$ A
   Circuit B has more current.

10. $R_T = V/I = 12\ \text{V}/2\ \text{mA} = 6$ kΩ
   $R_{EACH} = 6\ \text{k}\Omega/6 = 1$ kΩ

11. $V_T = 8\ \text{V} + 5\ \text{V} + 1.5\ \text{V} = 14.5$ V

12. $V_T = 9\ \text{V} - 9\ \text{V} = 0$ V

13. A: $V_T = 50\ \text{V} - 30\ \text{V} - 10\ \text{V} = 10$ V
      See Figure S–4.

FIGURE S–4

   B: $V_T = 1.5\ \text{V} + 3\ \text{V} + 4.5\ \text{V} - 9\ \text{V}$
      $- 1.5\ \text{V} = -1.5$ V
      See Figure S–5.

FIGURE S–5

14. Since the resistors are all of equal value, they each drop 2 V.
   $V_T = (5)(2\ \text{V}) = 10$ V

15. $V_S = V_1 + V_2 + V_3$
   $V_3 = V_S - V_1 - V_2$
   $= 50\ \text{V} - 10\ \text{V} - 15\ \text{V} = 25$ V

16. $R_T = V/I = 30\ \text{V}/10\ \text{mA} = 3$ kΩ
   $R_T = R_1 + R_2 + R_3 + R_4 + R_5$
   $R_3 = R_T - (R_1 + R_2 + R_4 + R_5)$
   $= 3\ \text{k}\Omega - 2\ \text{k}\Omega = 1$ kΩ

17. $V_x = (R_x/R_T)(V_T)$
   $V_{50\Omega} = (50\ \Omega/200\ \Omega)(20\ \text{V}) = 5$ V

18. $V_x = (R_x/R_T)(V_T)$
   $V_1 = (30\ \Omega/100\ \Omega)(90\ \text{V}) = 27$ V
   $V_2 = (20\ \Omega/100\ \Omega)(90\ \text{V}) = 18$ V
   $V_3 = (50\ \Omega/100\ \Omega)(90\ \text{V}) = 45$ V

19. $V_x = (R_x/R_T)(V_T)$
   $3\ \text{V} = (R_x/100\ \text{k}\Omega)(9\ \text{V})$
   $R_{x1} = (3\ \text{V})(100\ \text{k}\Omega)/(9\ \text{V}) = 33.33$ kΩ
   $R_{x2} = R_T - R_{x1} = 100\ \text{k}\Omega - 33.33\ \text{k}\Omega$
   $= 66.67$ kΩ

20. $P_T = P_1 + P_2 + P_3$
$= 2.5 \text{ W} + 5 \text{ W} + 1.2 \text{ W} = 8.7 \text{ W}$

21. $V_1 = 1.5 \text{ V}$
$V_2 = 2.25 \text{ V}$
$V_3 = 3.3 \text{ V}$
$V_4 = 4.95 \text{ V}$

Chapter 6

1. See Figure S–6.

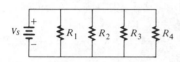

FIGURE S–6

2. $I_{BR} = I_T/2 = 5 \text{ A}/2 = 2.5 \text{ A}$

3. See Figure S–7.

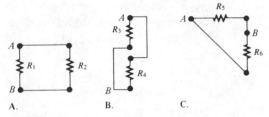

A.   B.   C.

FIGURE S–7

4. $V_{AB} = V_{AC} = V_{EC} = V_{DC} = 5 \text{ V}$
(because all of the resistors are in parallel)

5. $R_T = \dfrac{(80 \ \Omega)(150 \ \Omega)}{80 \ \Omega + 150 \ \Omega} = 52.17 \ \Omega$

6. $1/R_T = (1/1000 \ \Omega) + (1/800 \ \Omega)$
$+ (1/500 \ \Omega) + (1/200 \ \Omega)$
$+ (1/100 \ \Omega) = 0.01925 \text{ S}$
$R_T = 1/0.01925 \text{ S} = 51.95 \ \Omega$

7. $R_T = R/n = 56 \ \Omega/8 = 7 \ \Omega$

8. See Figure S–8.

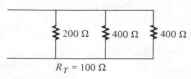

$R_T = 100 \ \Omega$

FIGURE S–8

9. $R_T = R/n = 600 \ \Omega/3 = 200 \ \Omega$
$I_T = V/R_T = 5 \text{ V}/200 \ \Omega = 0.025 \text{ A}$
$= 25 \text{ mA}$

10. A: $R_T = (50 \ \Omega)(25 \ \Omega)/75 \ \Omega$
$= 16.67 \ \Omega$
$I_T = 10 \text{ V}/16.67 \ \Omega = 0.6 \text{ A}$
$= 600 \text{ mA}$

B: $R_T = (15 \ \Omega)(100 \ \Omega)/115 \ \Omega$
$= 13.04 \ \Omega$
$I_T = 8 \text{ V}/13.04 \ \Omega = 0.613 \text{ A}$
$= 613 \text{ mA}$

Circuit B has the greater total current.

11. $R_T = V/I_T = 12 \text{ V}/3 \text{ mA} = 4 \text{ k}\Omega$
$R_T = R/n$
$4 \text{ k}\Omega = R_{EACH}/6$
$R_{EACH} = (6)(4 \text{ k}\Omega) = 24 \text{ k}\Omega$

12. $I_T = 1 \text{ A} - 2 \text{ A} + 4 \text{ A} = 3 \text{ A}$

13. $I_T = 1 \text{ A} - 1 \text{ A} = 0 \text{ A}$

14. $I_T = (5)(25 \text{ mA}) = 125 \text{ mA}$

15. $I_T = I_1 + I_2 + I_3$
$I_3 = I_T - (I_1 + I_2)$
$= 0.5 \text{ A} - (0.1 \text{ A} + 0.2 \text{ A})$
$= 0.5 \text{ A} - 0.3 \text{ A} = 0.2 \text{ A}$

16. $V_S = V_1 = V_2 = V_3$
$V_S = V_1 = I_1 R_1 = (1 \text{ A})(10 \ \Omega) = 10 \text{ V}$
$R_2 = V_s/I_2 = 10 \text{ V}/0.5 \text{ A} = 20 \ \Omega$
$I_3 = I_T - (I_1 + I_2) = 5 \text{ A} - 1.5 \text{ A}$
$= 3.5 \text{ A}$
$R_3 = V_s/I_3 = 10 \text{ V}/3.5 \text{ A} = 2.86 \ \Omega$

17. $1/R_T = (1/2000 \ \Omega) + (1/6000 \ \Omega)$
$+ (1/3000 \ \Omega) + (1/1000 \ \Omega)$
$= 0.002 \text{ S}$
$R_T = 1/0.002 \text{ S} = 500 \ \Omega$
$I_x = (R_T/R_x)(I_T)$
$I_{2k\Omega} = (500 \ \Omega/2000 \ \Omega)(1 \text{ A}) = 0.25 \text{ A}$
$I_{6k\Omega} = (500 \ \Omega/6000 \ \Omega)(1 \text{ A})$
$= 0.083 \text{ A}$
$I_{3k\Omega} = (500 \ \Omega/3000 \ \Omega)(1 \text{ A})$
$= 0.167 \text{ A}$
$I_{1k\Omega} = (500 \ \Omega/1000 \ \Omega)(1 \text{ A}) = 0.5 \text{ A}$

18. $I_1 = \left(\dfrac{R_2}{R_1 + R_2}\right) I_T = \left(\dfrac{20 \text{ k}\Omega}{25 \text{ k}\Omega}\right) 750 \text{ mA}$
$= 600 \text{ mA}$

$I_2 = \left(\dfrac{R_1}{R_1 + R_2}\right)I_T = \left(\dfrac{5 \text{ k}\Omega}{25 \text{ k}\Omega}\right)750 \text{ mA}$

$= 150 \text{ mA}$

19. $P_1 = I_1^2 R_1 = (0.6 \text{ A})^2 (5 \text{ k}\Omega) = 1800 \text{ W}$
$P_2 = I_2^2 R_2 = (0.15 \text{ A})^2 (20 \text{ k}\Omega)$
$= 450 \text{ W}$
$P_T = P_1 + P_2 = 1800 \text{ W} + 450 \text{ W}$
$= 2250 \text{ W}$

20. $I_{200\Omega} = 10 \text{ V}/200 \text{ }\Omega = 0.05 \text{ A}$
$= 50 \text{ mA}$
$I_{500\Omega} = 10 \text{ V}/500 \text{ }\Omega = 0.02 \text{ A}$
$= 20 \text{ mA}$
The total current should be 70 mA, but it is equal to the current in the 500-Ω branch only. This finding indicates that the 200-Ω branch is *open*.

21. $R_x = \dfrac{R_A R_T}{R_A - R_T} = \dfrac{(100 \text{ }\Omega)(40 \text{ }\Omega)}{100 \text{ }\Omega - 40 \text{ }\Omega}$

$= 66.67 \text{ }\Omega$

22. $I_1 = 0.012 \text{ A}$
$I_2 = 8 \text{ mA}$
$I_3 = 5.45455 \text{ mA}$
$I_4 = 3.63636 \text{ mA}$
$I_5 = 2.55319 \text{ mA}$

Chapter 7

1. R_2, R_3, and R_4 are in parallel, and this parallel combination is in series with both R_1 and R_5.

2. (a) $R_T = R_1 + R_5 + (R_2/3)$;
$R_2 = R_3 = R_4 = 30 \text{ }\Omega$.
$R_T = 10 \text{ }\Omega + 10 \text{ }\Omega + 10 \text{ }\Omega$
$= 30 \text{ }\Omega$
(b) $I_T = V_S/R_T = 30 \text{ V}/30 \text{ }\Omega = 1 \text{ A}$
(c) $I_3 = I_T/3 = 1 \text{ A}/3 = 0.333 \text{ A}$
(d) $V_4 = I_4 R_4$; $I_4 = I_3 = 0.333 \text{ A}$
$V_4 = (0.333 \text{ A})(30 \text{ }\Omega) = 10 \text{ V}$

3. (a) $R_1 \| R_2 = 2 \text{ k}\Omega/2 = 1 \text{ k}\Omega$
$R_3 \| R_4 = (300 \text{ }\Omega)(600 \text{ }\Omega)/900 \text{ }\Omega$
$= 200 \text{ }\Omega$
$R_5 \| R_6 = 1.6 \text{ k}\Omega/2 = 800 \text{ }\Omega$

$R_1 \| R_2 + R_3 \| R_4 + R_5 \| R_6$
$= 1000 \text{ }\Omega + 200 \text{ }\Omega$
$+ 800 \text{ }\Omega = 2 \text{ k}\Omega$
(b) $I_T = 10 \text{ V}/2 \text{ k}\Omega = 5 \text{ mA}$
(c) $I_1 = I_T/2 = 5 \text{ mA}/2 = 2.5 \text{ mA}$
(because I_T splits equally between R_1 and R_2)
(d) $I_6 = I_1 = 2.5 \text{ mA}$
$V_6 = (2.5 \text{ mA})(1.6 \text{ k}\Omega) = 4 \text{ V}$

4. (a) $R_4 \| R_5 = 5 \text{ k}\Omega/2 = 2.5 \text{ k}\Omega$
$R_4 \| R_5 + R_3 = 2.5 \text{ k}\Omega + 3.5 \text{ k}\Omega$
$= 6 \text{ k}\Omega$
$6 \text{ k}\Omega \| R_2 = (6 \text{ k}\Omega)(3 \text{ k}\Omega)/9 \text{ k}\Omega$
$= 2 \text{ k}\Omega$
$R_T = 2 \text{ k}\Omega \| R_1$
$= (2 \text{ k}\Omega)(10 \text{ k}\Omega)/12 \text{ k}\Omega$
$= 1.67 \text{ k}\Omega$
(b) $I_T = V_S/R_T = 6 \text{ V}/1.67 \text{ k}\Omega$
$= 3.59 \text{ mA}$
(c) The resistance to the right of AB is 2 kΩ. The current through this part of the circuit is $I = 6 \text{ V}/2 \text{ k}\Omega = 3 \text{ mA}$.

$I_3 = \left(\dfrac{R_2}{R_2 + 6 \text{ k}\Omega}\right)3 \text{ mA}$

$= \left(\dfrac{3 \text{ k}\Omega}{9 \text{ k}\Omega}\right)3 \text{ mA}$

$= 1 \text{ mA}$
$I_5 = I_3/2 = 1 \text{ mA}/2 = 0.5 \text{ mA}$
(d) $V_2 = V_S = 6 \text{ V}$

5. $V_A = 25 \text{ V}$
R from point B to gnd $= 80 \text{ }\Omega$
$V_B = (80 \text{ }\Omega/130 \text{ }\Omega)(25 \text{ V}) = 15.38 \text{ V}$
$V_C = (60 \text{ }\Omega/160 \text{ }\Omega)(15.38 \text{ V})$
$= 5.77 \text{ V}$

6. $V_A = 15 \text{ V}$
$V_B = 0 \text{ V}$
$V_C = 0 \text{ V}$

7. $V_{3k\Omega} = (1.5 \text{ k}\Omega/2.5 \text{ k}\Omega)(10 \text{ V}) = 6 \text{ V}$
The 7.5-V reading is incorrect.
$V_{2k\Omega} = (2 \text{ k}\Omega/3 \text{ k}\Omega)(6 \text{ V}) = 4 \text{ V}$
The 5-V reading is incorrect. The 3-kΩ resistor must be open to produce these incorrect voltages.

8. $V_{\text{OUT}} = (20 \text{ k}\Omega/30 \text{ k}\Omega)(30 \text{ V})$
$\phantom{V_{\text{OUT}}} = 20 \text{ V unloaded}$
$20 \text{ k}\Omega \| 200 \text{ k}\Omega = 18.18 \text{ k}\Omega$
$V_{\text{OUT}} = (18.18 \text{ k}\Omega/28.18 \text{ k}\Omega)(30 \text{ V})$
$\phantom{V_{\text{OUT}}} = 19.35 \text{ V loaded with } 200 \text{ k}\Omega$

9. $V_A = (6 \text{ k}\Omega/9 \text{ k}\Omega)(10 \text{ V})$
$ = 6.67 \text{ V switch open}$
$6 \text{ k}\Omega \| 10 \text{ k}\Omega = 3.75 \text{ k}\Omega$
$V_A = (3.75 \text{ k}\Omega/6.75 \text{ k}\Omega)(10 \text{ V})$
$ = 5.56 \text{ V switch closed}$

10. **(a)** $R_3 + R_4 = 75 \text{ }\Omega + 25 \text{ }\Omega = 100 \text{ }\Omega$
$(R_3 + R_4) \| R_2 = 100 \text{ }\Omega \| 100 \text{ }\Omega$
$ = 100 \text{ }\Omega/2 = 50 \text{ }\Omega$
$R_T = (R_3 + R_4) \| R_2 + R_1$
$ = 50 \text{ }\Omega + 50 \text{ }\Omega = 100 \text{ }\Omega$
(b) $I_T = 50 \text{ V}/100 \text{ }\Omega = 0.5 \text{ A}$

(c) $I_3 = \left(\dfrac{R_2}{R_2 + R_3 + R_4}\right) I_T$

$ = \left(\dfrac{100 \text{ }\Omega}{200 \text{ }\Omega}\right) 0.5 \text{ A} = 0.25 \text{ A}$

(d) $I_4 = I_3 = 0.25 \text{ A}$
(e) $V_A = I_3(R_3 + R_4)$
$ = (0.25 \text{ A})(100 \text{ }\Omega) = 25 \text{ V}$
(f) $V_B = I_4 R_4 = (0.25 \text{ A})(25 \text{ }\Omega)$
$ = 6.25 \text{ V}$

11. $R_{C \text{ to gnd}} = 8 \text{ k}\Omega$
$R_{B \text{ to gnd}} = 16 \text{ k}\Omega/2 = 8 \text{ k}\Omega$
$R_{A \text{ to gnd}} = 16 \text{ k}\Omega/2 = 8 \text{ k}\Omega$
$V_A = (8 \text{ k}\Omega/16 \text{ k}\Omega)(10 \text{ V}) = 5 \text{ V}$
$V_B = (8 \text{ k}\Omega/16 \text{ k}\Omega)(5 \text{ V}) = 2.5 \text{ V}$
$V_C = (8 \text{ k}\Omega/16 \text{ k}\Omega)(2.5 \text{ V}) = 1.25 \text{ V}$

12. $R_{\text{UNK}} = R_V(R_2/R_4)$
$\phantom{R_{\text{UNK}}} = (10 \text{ k}\Omega)(1 \text{ k}\Omega/8 \text{ k}\Omega) = 1.25 \text{ k}\Omega$

13. $V_A = 11.0489 \text{ V}$
$V_B = 6.86412 \text{ V}$
$V_C = 2.90853 \text{ V}$
$I_1 = 8.63406 \text{ mA}$
$I_2 = 7.36594 \text{ mA}$
$I_3 = 1.26812 \text{ mA}$
$I_4 = 0.686412 \text{ mA}$
$I_5 = 0.581705 \text{ mA}$
$I_6 = 0.581705 \text{ mA}$

Chapter 8

1. $I_S = V_S/R_S = 25 \text{ V}/5 \text{ }\Omega = 5 \text{ A}$
$R_S = 5 \text{ }\Omega$

2. $I_S = 30 \text{ V}/10 \text{ }\Omega = 3 \text{ A}; R_S = 10 \text{ }\Omega$
See Figure S–9.

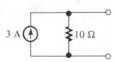

FIGURE S–9

3. $V_S = (5 \text{ mA})(25 \text{ k}\Omega) = 125 \text{ V};$
$R_S = 25 \text{ k}\Omega$
See Figure S–10.

FIGURE S–10

4. For 1-V source:
$R_T = 116.67 \text{ }\Omega$
$I_T = 8.57 \text{ mA}$

$I_3 = \left(\dfrac{R_2}{R_2 + R_3}\right) I_T = \left(\dfrac{50 \text{ }\Omega}{75 \text{ }\Omega}\right) 8.57 \text{ mA}$

$ = 5.71 \text{ mA down}$
For 1.5-V source:
$R_T = 70 \text{ }\Omega$
$I_T = 21.4 \text{ mA}$

$I_3 = \left(\dfrac{R_1}{R_1 + R_3}\right) I_T = \left(\dfrac{100 \text{ }\Omega}{125 \text{ }\Omega}\right) 21.4 \text{ mA}$

$ = 17.12 \text{ mA down}$
$I_3(\text{total}) = 5.71 \text{ mA} + 17.12 \text{ mA}$
$\phantom{I_3(\text{total})} = 22.83 \text{ mA}$

5. For 1-V source:

$I_2 = \left(\dfrac{R_3}{R_2 + R_3}\right) I_T = \left(\dfrac{25 \text{ }\Omega}{75 \text{ }\Omega}\right) 8.57 \text{ mA}$

$ = 2.86 \text{ mA down}$

For 1.5-V source:
$I_2 = I_T = 21.4$ mA up
I_2(total) $= 21.4$ mA $- 2.86$ mA
$\qquad = 18.54$ mA

6. $R_{TH} = 1$ k$\Omega/2 = 500\ \Omega$ (5-kΩ and
10-kΩ resistors are "shorted" by source)
$V_{TH} = (1$ k$\Omega/2$ k$\Omega)(12$ V$) = 6$ V
See Figure S–11.

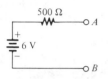

500 Ω
6 V
A
B

FIGURE S–11

7. Looking in at *open* terminals *AB*:
$R_{TH} = (20$ k$\Omega)(100$ k$\Omega)/(120$ k$\Omega)$
$\qquad = 16.67$ kΩ
$V_{TH} = (20$ k$\Omega/120$ k$\Omega)(15$ V$) = 2.5$ V
With R_L connected:
$I_L = V_{TH}/(R_{TH} + R_L)$
$\qquad = 2.5$ V$/516.67$ k$\Omega = 4.84\ \mu$A

8. $I_N = V_{TH}/R_{TH} = 2.5$ V$/16.67$ kΩ
$\qquad = 0.15$ mA
$R_N = R_{TH} = 16.67$ kΩ
See Figure S–12.

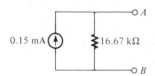

0.15 mA   16.67 kΩ A   B

FIGURE S–12

9.

$$R_{EQ} = \frac{1}{(1/20\ \Omega) + (1/10\ \Omega) + (1/50\ \Omega) + (1/10\ \Omega)}$$

$$= 3.7\ \Omega$$

$V_{EQ} =$
$$\frac{(12\ V/20\ \Omega) + (6\ V/10\ \Omega) - (10\ V/50\ \Omega) - (5\ V/10\ \Omega)}{0.27}$$
$$= 0.5/0.27 = 1.85\ V$$
See Figure S–13.

3.7 Ω
1.85 V

FIGURE S–13

10. $R_L = R_{EQ} = 3.7\ \Omega$

11. **(a)** $R_1 = (R_A R_C)/(R_A + R_B + R_C)$
$\qquad = (1$ k$\Omega)(500\ \Omega)/2.5$ kΩ
$\qquad = 200\ \Omega$
$R_2 = (R_B R_C)/(R_A + R_B + R_C)$
$\qquad = (1$ k$\Omega)(500\ \Omega)/2.5$ kΩ
$\qquad = 200\ \Omega$
$R_3 = (R_A R_B)/(R_A + R_B + R_C)$
$\qquad = (1$ k$\Omega)(1$ k$\Omega)/2.5$ kΩ
$\qquad = 400\ \Omega$
See Figure S–14.

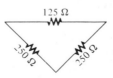

200 Ω 200 Ω
400 Ω

FIGURE S–14

(b) $R_A = \dfrac{R_1 R_2 + R_1 R_3 + R_2 R_3}{R_2}$

$= \dfrac{(50\ \Omega)(50\ \Omega) + (50\ \Omega)(100\ \Omega) + (50\ \Omega)(100\ \Omega)}{50\ \Omega}$

$\qquad = 250\ \Omega$
$R_B = 12,500/R_1 = 12,500/50$
$\qquad = 250\ \Omega$
$R_C = 12,500/R_3 = 12,500/100$
$\qquad = 125\ \Omega$
See Figure S–15.

125 Ω
250 Ω 250 Ω

FIGURE S–15

12. $R_1 = 319.149 \ \Omega$
$R_2 = 468.085 \ \Omega$
$R_3 = 702.128 \ \Omega$

Chapter 9

1. 2 loops, 5 nodes

2. The currents are shown in Figure S–16.

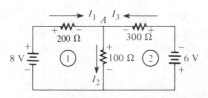

FIGURE S–16

The calculations are as follows.
Loop 1: $200I_1 + 100I_2 - 8 = 0$
Loop 2: $100I_2 + 300I_3 + 6 = 0$
Node A: $I_1 - I_2 + I_3 = 0$
Solving, we obtain
$I_1 = I_2 - I_3$
$200(I_2 - I_3) + 100I_2 = 8$
$\qquad 300I_2 - 200I_3 = 8$
$I_2 = (-6 - 300I_3)/100$

$300\left(\dfrac{-6 - 300I_3}{100}\right) - 200I_3 = 8$

$\qquad -18 - 900I_3 - 200I_3 = 8$
$1100I_3 = -26$
$\qquad I_3 = -0.024 \ \text{A}$
$100I_2 + 300(-0.024) = -6$
$\qquad I_2 = (-6 + 7.2)/100 = 0.012 \ \text{A}$
$\qquad I_1 = I_2 - I_3$
$\qquad\quad = 0.012 \ \text{A} - (-0.024 \ \text{A})$
$\qquad\quad = 0.036 \ \text{A}$

3. $I_1 = (15 - 5I_2)/2$
$6(15 - 5I_2)/2 - 8I_2 = 10$
$45 - 15I_2 - 8I_2 = 10$
$23I_2 = 35$
$\quad I_2 = 35/23 = 1.52 \ \text{A}$
Substituting, we have
$2I_1 + 5(1.52) = 15$
$\quad I_1 = (15 - 7.6)/2 = 3.7 \ \text{A}$

4. $I_1 = \dfrac{\begin{vmatrix} 25 & -12 \\ 18 & 3 \end{vmatrix}}{\begin{vmatrix} 15 & -12 \\ 9 & 3 \end{vmatrix}} = \dfrac{75 - (-216)}{45 - (-108)} = \dfrac{291}{153}$

$\qquad = 1.9 \ \text{A}$

5. The currents are shown in Figure S–17.

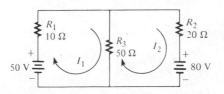

FIGURE S–17

The calculations are as follows.
$60I_1 - 50I_2 = 50$
$-50I_1 + 70I_2 = -80$

$I_1 = \dfrac{\begin{vmatrix} 50 & -50 \\ -80 & 70 \end{vmatrix}}{\begin{vmatrix} 60 & -50 \\ -50 & 70 \end{vmatrix}} = \dfrac{3500 - 4000}{4200 - 2500}$

$\quad = \dfrac{-500}{1700} = -0.294 \ \text{A}$

$I_2 = \dfrac{\begin{vmatrix} 60 & 50 \\ -50 & -80 \end{vmatrix}}{1700} = \dfrac{-4800 - (-2500)}{1700}$

$\quad = \dfrac{-2300}{1700} = -1.35 \ \text{A}$

$I_{R3} = I_2 - I_1 = 1.35 \ \text{A} - 0.294 \ \text{A}$
$\qquad = 1.056 \ \text{A down}$

6. The calculations are as follows. Also see Figure S–18.

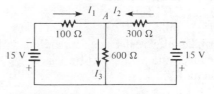

FIGURE S–18

$I_1 + I_2 = I_3$

$$\frac{-15 - V_A}{100} + \frac{-15 - V_A}{300} - \frac{V_A}{600} = 0$$

$$\frac{-V_A}{100} - \frac{V_A}{300} - \frac{V_A}{600} = \frac{15}{100} + \frac{15}{300}$$

$$\frac{-6V_A - 2V_A - V_A}{600} = \frac{45 + 15}{300}$$

$$\frac{-9V_A}{600} = \frac{60}{300}$$

$$V_A = -\frac{(600)(60)}{(9)(300)} = -13.33 \text{ V}$$

7. $V_A = 5.1$ V

Chapter 10

1. Half-scale (50 μA will produce full-scale deflection).

2. No shunt for 1-mA movement.
For 50-μA movement:
$R_{SH} = I_{MM}R_{MM}/I_{SH}$
$I_{SH} = 1 \text{ mA} - 50 \ \mu\text{A} = 950 \ \mu\text{A}$

$$R_{SH} = \frac{(50 \ \mu\text{A})(1 \text{ k}\Omega)}{950 \ \mu\text{A}} = 52.63 \ \Omega$$

3. $R_{INT} = (20{,}000 \ \Omega/\text{V})(100 \text{ V}) = 2 \text{ M}\Omega$

4. $R_M = R_{INT} - R_{MM}$
$\quad = 2 \text{ M}\Omega - 1 \text{ k}\Omega = 1.999 \text{ M}\Omega$

5. $R_{INT} = (20 \text{ k}\Omega/\text{V})(1 \text{ V}) = 20 \text{ k}\Omega$

$$R_2 \| R_{INT} = \frac{(10 \text{ k}\Omega)(20 \text{ k}\Omega)}{30 \text{ k}\Omega} = 6.67 \text{ k}\Omega$$

$V_2 = (6.67 \text{ k}\Omega/16.67 \text{ k}\Omega)1.5 \text{ V}$
$\quad = 0.6 \text{ V}$ meter reading
$V_2 = (10 \text{ k}\Omega/20 \text{ k}\Omega)1.5 \text{ V}$
$\quad = 0.75 \text{ V}$ unloaded
The internal meter resistance in parallel with R_2 causes the voltage to be less than its actual value.

6. $R = 10 \times 1 \text{ k}\Omega = 10 \text{ k}\Omega$

7. First, determine whether the meter readings are correct. The resistance (R_{T1}) of the R_2, R_3, R_4, and R_5 combination is as follows:

$$R_{T1} = R_2 + \frac{(R_2 + R_4)R_5}{R_3 + R_4 + R_5}$$

$$= 580 \ \Omega + \frac{(1970 \ \Omega)(620 \ \Omega)}{2590 \ \Omega}$$

$$= 1051.6 \ \Omega$$

The total circuit resistance normally is
$R_T = R_1 + R_{T1} = 1 \text{ k}\Omega + 1.0516 \text{ k}\Omega$
$\quad = 2051.6 \ \Omega$

$$V_A = \left(\frac{R_{T1}}{R_T}\right)V_S = \left(\frac{1051.6 \ \Omega}{2051.6 \ \Omega}\right)15 \text{ V}$$

$$= 7.69 \text{ V}$$

The voltage reading is, therefore, incorrect. If R_5 were open:

$$V_A = \left(\frac{R_2 + R_3 + R_4}{R_1 + R_2 + R_3 + R_4}\right)V_S$$

$$= \left(\frac{2550 \ \Omega}{3550 \ \Omega}\right)15 \text{ V} = 10.8 \text{ V}$$

This value corresponds to the meter reading; thus, R_5 appears to be open. To verify, calculate I_T:

$$I_T = \frac{15 \text{ V}}{3550 \ \Omega} = 4.23 \text{ mA}$$

8. $R_4 \| R_5 = \dfrac{(1 \text{ k}\Omega)(2.2 \text{ k}\Omega)}{3.2 \text{ k}\Omega} = 687.5 \ \Omega$

$$R_1 \| R_2 = \frac{(2.2 \text{ k}\Omega)(1.2 \text{ k}\Omega)}{3.4 \text{ k}\Omega} = 776.5 \ \Omega$$

$R_T = R_1 \| R_2 + R_3 + R_4 \| R_5$
$\quad = 776.5 \ \Omega + 3300 \ \Omega + 687.5 \ \Omega$
$\quad = 4764 \ \Omega$
Under normal conditions:

$$I_T = \frac{24 \text{ V}}{4764 \ \Omega} = 5 \text{ mA}$$

$$V_A = \left(\frac{687.5 \ \Omega}{4764 \ \Omega}\right)24 \text{ V} = 3.46 \text{ V}$$

Since I_T on the meter is less than the normal I_T, the total resistance of the circuit is greater than normal.

If R_4 or R_5 opened, V_A would increase; but it decreased. So, R_1 or R_2 must be open.

If R_2 opens:
$$R_T = 2.2 \text{ k}\Omega + 3.3 \text{ k}\Omega + 687.5 \text{ }\Omega$$
$$= 6.188 \text{ k}\Omega$$

$$I_T = \frac{24 \text{ V}}{6.188 \text{ k}\Omega} = 3.88 \text{ mA}$$

Thus, R_2 is open.

Chapter 11

1. $f = 1/T = 1/200 \text{ ms} = 5 \text{ Hz}$
2. $T = 1/f = 1/25 \text{ kHz} = 40 \text{ } \mu s$
3. $T = 5 \text{ } \mu s$
   $f = 1/T = 1/5 \text{ } \mu s = 0.2 \text{ MHz}$
4. $V_p = 25 \text{ V}$
   $V_{pp} = 50 \text{ V}$
   $V_{\text{rms}} = 0.707V_p = 17.68 \text{ V}$
   $V_{\text{avg}} = 0.637V_p = 15.93 \text{ V}$
5. $V_p = 1.414V_{\text{rms}} = 1.414(115 \text{ V})$
   $= 162.6 \text{ V}$
6. $I_{\text{rms}} = V_{\text{rms}}/R = 10 \text{ V}/100 \text{ }\Omega = 0.1 \text{ A}$
7. See Figure S–19.

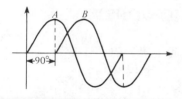

FIGURE S–19

8. $v = V_p \sin \theta$
   $v = 20 \sin 10° = 3.47 \text{ V}$
   $v = 20 \sin 25° = 8.45 \text{ V}$
   $v = 20 \sin 30° = 10 \text{ V}$
   $v = 20 \sin 90° = 20 \text{ V}$
   $v = 20 \sin 180° = 0 \text{ V}$
   $v = 20 \sin 210° = -10 \text{ V}$
   $v = 20 \sin 300° = -17.32 \text{ V}$

9. $V_p = 5 \text{ V}$
   $v = 5 \sin 30° = 2.5 \text{ V}$
   $v = 5 \sin 60° = 4.33 \text{ V}$
   $v = 5 \sin 100° = 4.92 \text{ V}$
   $v = 5 \sin 225° = -3.54 \text{ V}$

10. A: $10 \sin 30° = 5$
    B: $10 \sin 90° = 10$
    C: $10 \sin 200° = -3.42$

11. $20° \equiv -340°$
    $60° \equiv -300°$
    $135° \equiv -225°$
    $200° \equiv -160°$
    $315° \equiv -45°$
    $330° \equiv -30°$

12. $\omega = 2\pi f$
    $f = \omega/(2\pi)$
    $= (1000 \text{ rad/s})/(2\pi \text{ rad/cycle})$
    $= 159 \text{ Hz}$

13. See Figure S–20.

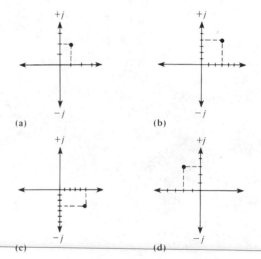

(a)   (b)

(c)   (d)

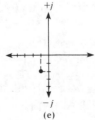

(e)

FIGURE S–20

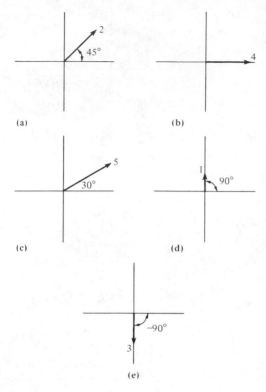

(a)

(b)

(c)

(d)

(e)

FIGURE S–21

14. See Figure S–21.

15. (a) $5 + j5 = \sqrt{5^2 + 5^2}\angle\arctan(5/5)$
$= 7.07\angle 45°$

(b) $12 + j9$
$= \sqrt{12^2 + 9^2}\angle\arctan(9/12)$
$= 15\angle 36.87°$

(c) $8 - j10$
$= \sqrt{8^2 + 10^2}\angle\arctan(-10/8)$
$= 12.8\angle -51.34°$

(d) $100 - j50 =$
$\sqrt{100^2 + 50^2}\angle\arctan(-50/100)$
$= 111.8\angle -26.57°$

16. (a) $1\angle 45° = 1\cos 45° + j1\sin 45°$
$= 0.707 + j0.707$

(b) $12\angle 60° = 12\cos 60° + j12\sin 60°$
$= 6 + j10.39$

(c) $100\angle -80° = 100\cos(-80°)$
$+ j100\sin(-80°)$
$= 17.36 - j98.48$

(d) $40\angle 125° = 40\cos 125°$
$+ j40\sin 125°$
$= -22.94 + j32.77$

17. $t_r \cong 1\ \mu s - 0.2\ \mu s = 0.8\ \mu s$
$t_f \cong 5.9\ \mu s - 5.1\ \mu s = 0.8\ \mu s$
$t_w \cong 5.6\ \mu s - 0.6\ \mu s = 5\ \mu s$
$A = 2\ V$

18. % duty cycle $= (t_w/T) \times 100$
$T = 1/\text{PRF} = 1/(2\ \text{kp/s}) = 0.5\ \text{ms}$
$= 500\ \mu s$
% duty cycle $= (1\ \mu s/500\ \mu s) \times 100$
$= 0.2\%$

19. $V_{\text{avg}} = $ baseline
$+ $ (duty cycle) (amplitude)
Duty cycle $= t_w/T = 1\ \mu s/6\ \mu s$
$= 0.167$
$V_{\text{avg}} = 5\ V + (0.167)(5\ V) = 5.835\ V$

20. $f = \text{PRF} = 1/T = 1/6\ \mu s$
$= 166.67\ \text{kHz}$

21. $f_{\text{2nd}} = 2 \times 25\ \text{kHz} = 50\ \text{kHz}$
$f_{\text{3rd}} = 3 \times 25\ \text{kHz} = 75\ \text{kHz}$

Chapter 12

1. (a) True (b) True (c) False
(d) False

2. (a) False (b) True (c) False
(d) True

3. $Q = CV = (0.006\ \mu F)(10\ V)$
$= 0.06\ \mu C$

4. $C = Q/V = (50 \times 10^{-6}\ C)/5\ V$
$= 10 \times 10^{-6}\ F = 10\ \mu F$

5. $V = Q/C = (5 \times 10^{-8}\ C)/0.01\ \mu F$
$= (5 \times 10^{-8}\ C)/(0.01 \times 10^{-6}\ F)$
$= 500 \times 10^{-2}\ V = 5\ V$

6. $(0.005\ \mu F)(10^6\ pF/\mu F) = 5000\ pF$

7. $(2000\ pF)(10^{-6}\ \mu F/pF) = 0.002\ \mu F$

8. $(1500\ \mu F)(0.000001\ F/\mu F) = 0.0015\ F$

9. $\mathscr{E} = \frac{1}{2}CV^2 = \frac{1}{2}(10 \times 10^{-6}\ F)(100\ V)^2$
$= 0.05\ J$

10. $\epsilon = \epsilon_r \epsilon_0 = (1200)(8.85 \times 10^{-12}\ F/m)$
$= 1.062 \times 10^{-8}\ F/m$

11. $C = \dfrac{A\epsilon_r(8.85 \times 10^{-12})}{d}$

$= \dfrac{(0.009 \text{ m}^2)(2)(8.85 \times 10^{-12} \text{ F/m})}{0.0015 \text{ m}}$

$= 106.2 \text{ pF}$

12. $\tau = RC = (2 \text{ k}\Omega)(0.05 \text{ }\mu\text{F})$
$= 0.1 \times 10^{-3} \text{ s} = 0.1 \text{ ms}$

13. $v_c = (0.37)(15 \text{ V}) = 5.55 \text{ V}$

14. $v = 10 \text{ V}(1 - e^{-15\text{ms}/10\text{ms}})$
$= 10 \text{ V}(0.777)$
$= 7.77 \text{ V}$

15. A: $1/C_T = (1/0.05 \text{ }\mu\text{F}) + (1/0.01 \text{ }\mu\text{F})$
$+ (1/0.02 \text{ }\mu\text{F})$
$+ (1/0.05 \text{ }\mu\text{F})$
$= 1.9 \times 10^8$
$C_T = 0.00526 \text{ }\mu\text{F}$
B: $C_T = (300 \text{ pF})(100 \text{ pF})/400 \text{ pF}$
$= 75 \text{ pF}$
C: $C_T = 12 \text{ pF}/3 = 4 \text{ pF}$

16. $1/C_T = (1/0.025 \text{ }\mu\text{F})$
$+ (1/0.04 \text{ }\mu\text{F}) + (1/0.1 \text{ }\mu\text{F})$
$= 7.5 \times 10^7$
$C_T = 0.0133 \text{ }\mu\text{F}$
$V_1 = (C_T/C_1)V_T$
$= (0.0133 \text{ }\mu\text{F}/0.025 \text{ }\mu\text{F})250 \text{ V}$
$= 133 \text{ V}$
$V_2 = (C_T/C_2)V_T$
$= (0.0133 \text{ }\mu\text{F}/0.04 \text{ }\mu\text{F})250 \text{ V}$
$= 83.13 \text{ V}$
$V_3 = (C_T/C_3)V_T$
$= (0.0133 \text{ }\mu\text{F}/0.1 \text{ }\mu\text{F})250 \text{ V}$
$= 33.25 \text{ V}$

17. A: $C_T = 100 \text{ pF} + 47 \text{ pF} + 33 \text{ pF}$
$+ 10 \text{ pF} = 190 \text{ pF}$
B: $C_T = (5)(0.008 \text{ }\mu\text{F}) = 0.04 \text{ }\mu\text{F}$

18. $X_C = \dfrac{1}{2\pi fC}$

$= \dfrac{1}{2\pi(2 \times 10^6 \text{ Hz})(1 \times 10^{-6} \text{ F})}$

$= 0.0796 \text{ }\Omega$

19. $X_C = V_{\text{rms}}/I_{\text{rms}} = 25 \text{ V}/50 \text{ mA}$
$= 0.5 \text{ k}\Omega$

20. $P_{\text{avg}} = 0$

$P_r = \dfrac{V_{\text{rms}}^2}{X_C} = \dfrac{(25 \text{ V})^2}{500 \text{ }\Omega} = 1.25 \text{ VAR}$

21. The capacitor has a very low resistance, which indicates it is faulty.

Chapter 13

1. $A = 1 \text{ mm}^2 = 1 \times 10^{-6} \text{ m}^2$
$\phi = 1200 \text{ }\mu\text{Wb}$

$B = \dfrac{\phi}{A} = \dfrac{1200 \times 10^{-6} \text{ Wb}}{1 \times 10^{-6} \text{ m}^2}$

$= 1200 \text{ tesla}$

2. $F_m = NI = (50)(100 \text{ mA})$
$= 5000 \times 10^{-3} \text{ ampere-turns}$
$= 5 \text{ ampere-turns}$

3. $v_{\text{ind}} = N(d\phi/dt) = 50(25 \text{ Wb/s})$
$= 1250 \text{ V}$

4. $1500 \text{ }\mu\text{H} = 1.5 \text{ mH},$
$20 \text{ mH} = 20,000 \text{ }\mu\text{H}$

5. $v_{\text{ind}} = L(di/dt) = (100 \text{ mH})(200 \text{ mA/s})$
$= 0.02 \text{ V}$

6. $L = (N^2 \text{ }\mu A)/l$
$N = \sqrt{Ll/\mu A}$
$= \sqrt{(30 \text{ mH})(0.05 \text{ m})/(1.2 \times 10^{-6})(10 \times 10^{-5} \text{ m}^2)}$
$= 3536 \text{ turns}$

7. $I = V_{\text{dc}}/R_W = 12 \text{ V}/12 \text{ }\Omega = 1 \text{ A}$

8. $\mathscr{E} = \frac{1}{2}LI^2 = \dfrac{(0.1 \text{ H})(1 \text{ A})^2}{2}$

$= 0.05 \text{ J}$

9. $\tau = L/R = 10 \text{ mH}/2 \text{ k}\Omega = 5 \text{ }\mu\text{s}$

10. $L_T = 75 \text{ }\mu\text{H} + 20 \text{ }\mu\text{H} = 95 \text{ }\mu\text{H}$

11. $1/L_T = (1/1 \text{ H}) + (1/1 \text{ H})$
$\quad\quad\quad + (1/0.5 \text{ H}) + (1/0.5 \text{ H})$
$\quad\quad\quad + (1/2 \text{ H}) = 6.5$
$\quad L_T = 0.154 \text{ H}$

12. $f = 2 \text{ MHz}$
$\quad X_L = 2\pi fL = 2\pi(2 \text{ MHz})(15 \text{ }\mu\text{H})$
$\quad\quad = 188.5 \text{ }\Omega$

13. $X_L = V_{rms}/I_{rms} = 25 \text{ V}/50 \text{ mA}$
$\quad\quad = 0.5 \text{ k}\Omega$
$\quad X_L = 2\pi fL$

$$f = \frac{X_L}{2\pi L} = \frac{500 \text{ }\Omega}{2\pi(8 \text{ }\mu\text{H})} = 9.95 \text{ MHz}$$

14. $L = \dfrac{X_L}{2\pi f} = \dfrac{10 \text{ M}\Omega}{2\pi(50 \text{ MHz})} = 31.83 \text{ mH}$

Chapter 14

1. Sine wave

2. $\mathbf{Z}_T = R - jX_C$
$\quad\quad = 2.5 \text{ k}\Omega - j6.37 \text{ k}\Omega$
$\quad\quad = 6.84\angle -68.6° \text{ k}\Omega$

3. $\mathbf{Z}_T = 2.5 \text{ k}\Omega - j3.19 \text{ k}\Omega$
$\quad\quad = 4.05\angle -51.9° \text{ k}\Omega$

4. $X_C = R$ at $45°$
$\quad X_C = 10 \text{ k}\Omega$
$\quad 1/(2\pi fC) = 10 \text{ k}\Omega$

$$f = \frac{1}{2\pi(10 \text{ k}\Omega)(100 \text{ pF})} = 159 \text{ kHz}$$

5. $\tan \theta = X_C/R$

$$X_C = \frac{1}{2\pi fC} = \frac{1}{2\pi(5 \text{ kHz})(0.01 \text{ }\mu\text{F})}$$
$\quad\quad = 3.18 \text{ k}\Omega$
$\quad X_C/R = \tan \theta$
$\quad R = X_C/\tan \theta = 3.18 \text{ k}\Omega/\tan 60°$
$\quad\quad = 3.18 \text{ k}\Omega/1.732$
$\quad\quad = 1.84 \text{ k}\Omega$

6. $\theta = \arctan(X_C/R)$

$$X_C = \frac{1}{2\pi fC} = \frac{1}{2\pi(100 \text{ kHz})(100 \text{ pF})}$$
$\quad\quad = 15.9 \text{ k}\Omega$
$\quad \theta = \arctan(15.9 \text{ k}\Omega/10 \text{ k}\Omega) = 57.83°$

7. $\theta = \arctan(1000 \text{ }\Omega/500 \text{ }\Omega) = 63.4°$

8. $\theta = \arctan(X_C/R) = \arctan(X_C/2X_C)$
$\quad\quad = \arctan 0.5 = 26.57°$

9. A: $\mathbf{Z} = 50 \text{ }\Omega - j30 \text{ }\Omega$
$\quad$ B: $\mathbf{Z} = 25 \text{ }\Omega - j100 \text{ }\Omega$

10. A: $\mathbf{Z} = 58.3\angle -30.96° \text{ }\Omega$

$$\mathbf{I} = \frac{\mathbf{V}_s}{\mathbf{Z}} = \frac{10\angle 0° \text{ V}}{58.3\angle -30.96° \text{ }\Omega}$$
$\quad\quad = 172\angle 30.96° \text{ mA}$
$\quad$ B: $\mathbf{Z} = 103.1\angle -75.96° \text{ }\Omega$

$$\mathbf{I} = \frac{10\angle 0° \text{ V}}{103.1\angle -75.96° \text{ }\Omega}$$
$\quad\quad = 97\angle 75.96° \text{ mA}$

11. $V = \sqrt{(6 \text{ V})^2 + (4.5 \text{ V})^2}$
$\quad\quad = 7.5 \text{ V}$

12. A: $\mathbf{Y} = G + jB_C$
$\quad\quad\quad = 4 \times 10^{-5} \text{ S} + j1 \times 10^{-4} \text{ S}$
$\quad\quad\quad = 1.08 \times 10^{-4}\angle 68.2° \text{ S}$
$\quad\quad \mathbf{Z} = 1/\mathbf{Y} = 9.26\angle -68.2° \text{ k}\Omega$
$\quad$ B: $X_C = 159 \text{ }\Omega$
$\quad\quad \mathbf{Y} = 0.01 \text{ S} + j6.28 \times 10^{-3} \text{ S}$
$\quad\quad\quad = 1.18 \times 10^{-2}\angle 32.13° \text{ S}$
$\quad\quad \mathbf{Z} = 1/\mathbf{Y} = 84.7\angle -32.13° \text{ }\Omega$

13. A: $\mathbf{I}_T = \dfrac{\mathbf{V}_s}{\mathbf{Z}} = \dfrac{5\angle 0° \text{ V}}{9.26\angle -68.2° \text{ k}\Omega}$
$\quad\quad\quad = 0.54\angle 68.2° \text{ mA}$

$$\mathbf{I}_C = \frac{\mathbf{V}_s}{\mathbf{X}_C} = \frac{5\angle 0° \text{ V}}{10\angle -90° \text{ k}\Omega}$$
$\quad\quad\quad = 0.5\angle 90° \text{ mA}$

$$\mathbf{I}_R = \frac{\mathbf{V}_s}{\mathbf{R}} = \frac{5\angle 0° \text{ V}}{25\angle 0° \text{ k}\Omega}$$
$\quad\quad\quad = 0.2\angle 0° \text{ mA}$

$\quad$ B. $\mathbf{I}_T = \dfrac{\mathbf{V}_s}{\mathbf{Z}} = \dfrac{5\angle 0° \text{ V}}{84.7\angle -32.13° \text{ }\Omega}$
$\quad\quad\quad = 59\angle 32.13° \text{ mA}$

$$\mathbf{I}_C = \frac{5\angle 0° \text{ V}}{159\angle -90° \text{ }\Omega}$$
$\quad\quad\quad = 31.45\angle 90° \text{ mA}$

$$\mathbf{I}_R = \frac{5\angle 0° \text{ V}}{100\angle 0° \ \Omega} = 50\angle 0° \text{ mA}$$

14. A: $\theta = 68.2°$ V lags I
     B: $\theta = 32.13°$ V lags I

15. $X_C = \dfrac{1}{2\pi(1 \text{ kHz})(0.005 \ \mu\text{F})}$
       $= 31.83 \text{ k}\Omega$
       $\phi = -90° + \arctan(31.83 \text{ k}\Omega/33 \text{ k}\Omega)$
       $= -90° + 43.97° = -46.03°$

16. $V_{\text{out}} = \dfrac{X_C}{\sqrt{R^2 + X_C^2}} V_{\text{in}}$
        $= \dfrac{31.83 \text{ k}\Omega}{\sqrt{(33 \text{ k}\Omega)^2 + (31.83 \text{ k}\Omega)^2}} 5 \text{ V}$
        $= 3.47 \text{ V}$

17. $X_C = \dfrac{1}{2\pi(50 \text{ kHz})(0.1 \ \mu\text{F})} = 31.83 \ \Omega$
       $\theta = \arctan(X_C/R) = \arctan(31.83/75)$
       $= 22.996°$

18. $V_{\text{out}} = \dfrac{R}{\sqrt{R^2 + X_C^2}} V_{\text{in}}$
        $= \dfrac{75 \ \Omega}{\sqrt{(75 \ \Omega)^2 + (31.83 \ \Omega)^2}} 12 \text{ V}$
        $= 11.05 \text{ V}$

19. 200 Hz. This filter is a low-pass.

20. 5 kHz. This filter is a high-pass.

21. Figure 14–67:
     $I = 5 \text{ V}/Z = 5 \text{ V}/45.85 \text{ k}\Omega$
        $= 0.109 \text{ mA}$
     $P_{\text{avg}} = I^2 R = (0.109 \text{ mA})^2 (33 \text{ k}\Omega)$
        $= 0.392 \text{ mW}$
     $P_r = I^2 X_C = (0.109 \text{ mA})^2 (31.83 \text{ k}\Omega)$
        $= 0.378 \text{ mVAR}$
     $P_a = I^2 Z = (0.109 \text{ mA})^2 (45.85 \text{ k}\Omega)$
        $= 0.545 \text{ mVA}$
     Figure 14–68:
     $I = 12 \text{ V}/Z = 12 \text{ V}/81.48 \ \Omega$
        $= 0.147 \text{ A}$
     $P_{\text{avg}} = I^2 R = (0.147 \text{ A})^2 (75 \ \Omega)$
        $= 1.62 \text{ W}$

     $P_r = I^2 X_C = (0.147 \text{ A})^2 (31.83 \ \Omega)$
        $= 0.688 \text{ VAR}$
     $P_a = I^2 Z = (0.147 \text{ A})^2 (81.48 \ \Omega)$
        $= 1.76 \text{ VA}$

22. Figure 14–67:
     $\theta = \arctan(X_C/R)$
        $= \arctan(31.83 \text{ k}\Omega/33 \text{ k}\Omega)$
        $= 43.97°$
     $PF = \cos \theta = \cos 43.97° = 0.72$
     Figure 14–68:
     $\theta = \arctan(X_C/R)$
        $= \arctan(31.83 \ \Omega/75 \ \Omega)$
        $= 23°$
     $PF = \cos \theta = \cos 23° = 0.92$

23. $v_C = V_F(1 - e^{-t/\tau}) = 5 \text{ V}(1 - e^{-5/10})$
       $= 5 \text{ V}(1 - 0.607)$
       $= 1.97 \text{ V}$

24. $\tau = RC = (1 \text{ k}\Omega)(1 \ \mu\text{F}) = 1 \text{ ms}$. See Figure S–22. The time to reach steady state is $5\tau = 5$ ms for repetitive pulses.

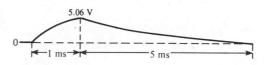

FIGURE S–22

25. $t_w = 1$ ms
     $\tau = RC = 1$ ms
     During 1st pulse:
     $v_{\text{max}} = (0.632)(8 \text{ V}) = 5.06 \text{ V}$
     Between 1st and 2nd pulses:
     $v_{\text{min}} = (0.37)(5.06 \text{ V}) = 1.87 \text{ V}$
     During 2nd pulse:
     $v_{\text{max}} = 8 \text{ V} + (1.87 \text{ V} - 8 \text{ V})(0.37)$
        $= 5.73 \text{ V}$
     See Figure S–23.

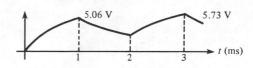

FIGURE S–23

Chapter 15

1. Sinusoidal

2. $X_L = 2\pi f L = 1.26$ kΩ
$\mathbf{Z} = R + jX_L = 3.3$ k$\Omega + j1.26$ kΩ
$= 3.53\angle 20.9°$ kΩ

3. $X_L = 2\pi f L = 2\pi(5$ kHz$)(20$ mH$)$
$= 628$ Ω
$Z = \sqrt{R^2 + X_L^2}$
$= \sqrt{(3.3 \text{ k}\Omega)^2 + (628 \ \Omega)} = 3.36$ kΩ
$\theta = \arctan(X_L/R) = 10.9°$

4. $X_L = R = 10$ kΩ at $\theta = 45°$
$f = X_L/2\pi L = 10$ k$\Omega/2\pi(100$ mH$)$
$= 15.9$ kHz

5. $X_L = 2\pi f L = 2\pi(5$ kHz$)(200$ μH$)$
$= 6.28$ Ω
$\tan 60° = 6.28 \ \Omega/R$
$R = 6.28 \ \Omega/\tan 60° = 6.28/1.73$
$= 3.63$ Ω

6. $\theta = \arctan(X_L/R)$
$= \arctan(100 \ \Omega/500 \ \Omega)$
$= 11.3°$

7. $\mathbf{Z} = 47$ k$\Omega + j20$ kΩ
$= 51.1\angle 23.05°$ kΩ

$\mathbf{I} = \dfrac{120\angle 0° \text{ V}}{51.1\angle 23.05° \text{ k}\Omega}$

$= 2.35\angle -23.05°$ mA
$\mathbf{V}_R = \mathbf{IR}$
$= (2.35\angle -23.05° \text{ mA})(47\angle 0° \text{ k}\Omega)$
$= 110.5\angle -23.05°$ V
$\mathbf{V}_L = \mathbf{I}X_L$
$= (2.35\angle -23.05 \text{ mA})(20\angle 90° \text{ k}\Omega)$
$= 47\angle 66.95°$ V

8. A: $\mathbf{Y}_T = (1/25 \text{ k}\Omega) - j(1/40 \text{ k}\Omega)$
$= 4 \times 10^{-5}$ S $- j2.5 \times 10^{-5}$ S
$= 4.72 \times 10^{-5}\angle -32°$ S
$\mathbf{Z}_T = 1/\mathbf{Y}_T = 21.2\angle 32°$ kΩ
B: $X_L = 2\pi f L = 2\pi(1$ kHz$)(2$ H$)$
$= 12.57$ kΩ

$\mathbf{Y}_T = \dfrac{1}{10 \text{ k}\Omega} - j\dfrac{1}{12.57 \text{ k}\Omega}$

$= 0.1 \times 10^{-3}$ S
$- j0.0796 \times 10^{-3}$ S

$= 10 \times 10^{-5}$ S
$-j7.96 \times 10^{-5}$ S
$= 12.78 \times 10^{-5}\angle -38.52°$ S
$\mathbf{Z}_T = 1/\mathbf{Y}_T = 7.82\angle 38.52°$ kΩ

9. A: $\mathbf{I}_T = \dfrac{5\angle 0° \text{ V}}{21.2\angle 32° \text{ k}\Omega}$

$= 0.236\angle -32°$ mA

$\mathbf{I}_R = \dfrac{5\angle 0° \text{ V}}{25\angle 0° \text{ k}\Omega} = 0.2\angle 0°$ mA

$\mathbf{I}_L = \dfrac{5\angle 0° \text{ V}}{40\angle 90° \text{ k}\Omega}$

$= 0.125\angle -90°$ mA

B: $\mathbf{I}_T = \dfrac{10\angle 0° \text{ V}}{7.82\angle 38.52° \text{ k}\Omega}$

$= 1.28\angle -38.52°$ mA

$\mathbf{I}_R = \dfrac{10\angle 0° \text{ V}}{10\angle 0° \text{ k}\Omega} = 1\angle 0°$ mA

$\mathbf{I}_L = \dfrac{10\angle 0° \text{ V}}{12.57\angle 90° \text{ k}\Omega}$

$= 0.796\angle -90°$ mA

10. $X_{L1} = 2\pi f L_1 = 3.14$ kΩ
$X_{L2} = 2\pi f L_2 = 1.26$ kΩ

$\mathbf{Y}_2 = \dfrac{1}{2.2 \text{ k}\Omega} - j\dfrac{1}{1.26 \text{ k}\Omega}$

$= 0.91 \times 10^{-3}\angle -60.2°$ S
$\mathbf{Z}_2 = 1/\mathbf{Y}_2 = 1.1\angle 60.2°$ kΩ
$= 547 \ \Omega + j955 \ \Omega$
$\mathbf{Z}_T = \mathbf{Z}_1 + \mathbf{Z}_2$
$= (3.3 \text{ k}\Omega + j3.14 \text{ k}\Omega)$
$+ (547 \ \Omega + j955 \ \Omega)$
$= 3.85$ k$\Omega + j4.1$ kΩ
$= 5.62\angle 46.8°$ kΩ

11. $X_L = 2\pi f L = 2\pi(60$ Hz$)(0.2$ H$)$
$= 75.4$ Ω
$\phi = 90° - \arctan(75.4/100) = 52.98°$

12. $V_{\text{out}} = V_L = (X_L/\sqrt{R^2 + X_L^2})V_{\text{in}}$
$= (75.4 \ \Omega/\sqrt{(100 \ \Omega)^2 + (75.4 \ \Omega)^2})5$ V
$= 3.01$ V

13. $V_{out} = V_R = (R/\sqrt{R^2 + X_L^2})V_{in}$
$= (1.2 \text{ k}\Omega/\sqrt{(1.2 \text{ k}\Omega)^2 + (628.3 \text{ }\Omega)^2})6 \text{ V}$
$= 5.32 \text{ V}$
$\phi = -\arctan(628.3 \text{ }\Omega/1.2 \text{ k}\Omega)$
$= -27.6°$

14. At 2 kHz, because X_L is greater.

15. $I = 6 \text{ V}/1354.5 \text{ }\Omega = 4.4 \text{ mA}$
 (a) $P_{avg} = I^2R = (4.4 \text{ mA})^2(1.2 \text{ k}\Omega)$
  $= 23.23 \text{ mW}$
 (b) $P_r = I^2X_L = (4.4 \text{ mA})^2(628.3 \text{ }\Omega)$
  $= 12.16 \text{ mVAR}$
 (c) $P_a = I^2Z = (4.4 \text{ mA})^2(1354.5 \text{ }\Omega)$
  $= 26.2 \text{ mVA}$
 (d) $PF = \cos \theta = \cos(27.6°) = 0.886$

16. $v_R = (0.632)(10 \text{ V}) = 6.32 \text{ V}$
$v_{L(max)} = +10 \text{ V}$
$v_{L(min)} = (10 \text{ V} - 6.32 \text{ V}) = 3.68 \text{ V}$
Note: v_L goes to -6.32 V when the
pulse falls back to zero.

17. $\tau = 10 \text{ mH}/10 \text{ }\Omega = 1 \text{ ms}$
$5\tau = 5 \text{ ms}$
$v_{out(max)} = (0.632)8 \text{ V} = 5.06 \text{ V}$
See Figure S–24

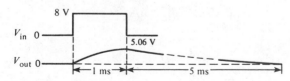

FIGURE S–24

18. $I_F = 5 \text{ V}/5 \text{ k}\Omega = 1 \text{ mA}$
$i = 1 \text{ mA}(1 - e^{-5\mu s/10\mu s}) = 393 \text{ }\mu\text{A}$

Chapter 16

1. The current leads the source voltage, because $X_C > X_L$.

2. $Z = R + j(X_L - X_C)$
$= 75 \text{ }\Omega + j(40 \text{ }\Omega - 80 \text{ }\Omega)$
$= 75 \text{ }\Omega - j40 \text{ }\Omega$
$= 85\angle-28.1° \text{ }\Omega$

3. (a) $I = \dfrac{\mathbf{V}_s}{\mathbf{Z}} = \dfrac{10\angle0° \text{ V}}{85\angle-28.1° \text{ }\Omega}$
$= 117.6\angle28.1° \text{ mA}$

 (b) $\mathbf{V}_R = \mathbf{IR}$
  $= (117.6\angle28.1 \text{ mA})(75\angle0° \text{ }\Omega)$
  $= 8.82\angle28.1° \text{ V}$
 (c) $\mathbf{V}_L = \mathbf{I}X_L$
  $= (117.6\angle28.1° \text{ mA})(40\angle90° \text{ }\Omega)$
  $= 4.7\angle118.1° \text{ V}$
 (d) $\mathbf{V}_C = \mathbf{I}X_C$
  $= (117.6\angle28.1° \text{ mA})(80\angle-90° \text{ }\Omega)$
  $= 9.4\angle-61.9° \text{ V}$
 (e) $P_{avg} = I^2R = (117.6 \text{ mA})^2(75 \text{ }\Omega)$
  $= 1.03 \text{ W}$
 (f) $P_{r(net)} = I^2(X_L - X_C)$
  $= (117.6 \text{ mA})^2(40 \text{ }\Omega)$
  $= 0.553 \text{ VAR}$

4. $X_C = X_L = 500 \text{ }\Omega$

5. $f_r = \dfrac{1}{2\pi\sqrt{LC}} = \dfrac{1}{2\pi\sqrt{(20 \text{ mH})(12 \text{ pF})}}$
$= 324.9 \text{ kHz}$
$Z_r = R = 15 \text{ }\Omega$

6. $f_r + 10 \text{ kHz} = 334.9 \text{ kHz}$
$X_L = 2\pi f L = 2\pi(334.9 \text{ kHz})(20 \text{ mH})$
$= 42.1 \text{ k}\Omega$

$X_C = \dfrac{1}{2\pi fC} = \dfrac{1}{2\pi(334.9 \text{ kHz})(12 \text{ pF})}$
$= 39.6 \text{ k}\Omega$
$Z = \sqrt{R^2 + (X_L - X_C)^2} \cong 2.5 \text{ k}\Omega$
$\theta = \arctan(2.5 \text{ k}\Omega/15) = 89.66°$

7. (a) $I = V_s/R = 1 \text{ V}/30 \text{ }\Omega = 33.3 \text{ mA}$

 (b) $f_r = \dfrac{1}{2\pi\sqrt{LC}} = 97.95 \text{ kHz}$
 (c) $V_R = V_s = 1 \text{ V}$
 (d) $X_L = 2\pi f_r L = 4.92 \text{ k}\Omega$
  $Q = X_L/R = 4.92 \text{ k}\Omega/30 \text{ }\Omega$
  $= 164$
  $V_L = QV_s = 164 \text{ V}$
 (e) $V_C = V_L = 164 \text{ V}$ (180° out-of-phase with V_L)
 (f) $\theta = 0°$ since $Z = R$

8. $\mathbf{Y} = 1/R + j(1/X_C) - j(1/X_L)$
$= 1/15 \text{ }\Omega + j(1/10 \text{ }\Omega) - j(1/20 \text{ }\Omega)$
$= 0.067 \text{ S} + j0.05 \text{ S}$
$= 0.084\angle36.7° \text{ S}$
$\mathbf{Z} = 1/\mathbf{Y} = 11.9\angle-36.7° \text{ }\Omega$

9. $\mathbf{I}_R = \mathbf{V}_s/R = 1\angle 0° \text{ V}/15\angle 0° \ \Omega$
$= 67\angle 0° \text{ mA}$
$\mathbf{I}_C = \mathbf{V}_s/\mathbf{X}_C = 1\angle 0° \text{ V}/10\angle -90° \ \Omega$
$= 100\angle 90° \text{ mA}$
$\mathbf{I}_L = \mathbf{V}_s/\mathbf{X}_L = 1\angle 0° \text{ V}/20\angle 90° \ \Omega$
$= 50\angle -90° \text{ mA}$

10. (a) $\mathbf{Z}_A = R_2 - jX_C = 22 \ \Omega - j29 \ \Omega$
$= 36.4\angle -52.8° \ \Omega$
$\mathbf{Z}_B = R_3 + jX_L = 18 \ \Omega + j43 \ \Omega$
$= 46.6\angle 67.3° \ \Omega$

$$\mathbf{Z}_C = \frac{\mathbf{Z}_A \mathbf{Z}_B}{\mathbf{Z}_A + \mathbf{Z}_B}$$

$$= \frac{(36.4\angle -52.8° \ \Omega)(46.6\angle 67.3° \ \Omega)}{42.4\angle 19.3° \ \Omega}$$

$= 40\angle -4.8° \ \Omega$
$\mathbf{Z}_T = R_1 + \mathbf{Z}_C$
$= 8 \ \Omega + 39.8 \ \Omega - j3.35$
$= 47.8 - j3.35$
$= 47.9\angle -4° \ \Omega$

(b) $\mathbf{I}_T = 12\angle 0° \text{ V}/47.9\angle -4° \ \Omega$
$= 0.25\angle 4° \text{ A}$

(c) $\mathbf{I}_L = \left(\dfrac{\mathbf{Z}_T}{\mathbf{Z}_B}\right)\mathbf{I}_T$

$= \left(\dfrac{47.9\angle -4° \ \Omega}{46.6\angle 67.3° \ \Omega}\right)0.25\angle 4° \text{ A}$

$= 0.26\angle -67.3° \text{ A}$

(d) $\mathbf{I}_C = \left(\dfrac{\mathbf{Z}_T}{\mathbf{Z}_A}\right)\mathbf{I}_T$

$= \left(\dfrac{47.9\angle -4° \ \Omega}{36.4\angle -52.8° \ \Omega}\right)0.25\angle 4° \text{ A}$

$= 0.33\angle 52.8° \text{ A}$

11. $f_r = \dfrac{\sqrt{1 - (R_w^2 C/L)}}{2\pi\sqrt{LC}}$

$= \dfrac{\sqrt{1 - [(50 \ \Omega)^2 (0.02 \ \mu\text{F})/2 \text{ H}]}}{2\pi\sqrt{(2 \text{ H})(0.02 \ \mu\text{F})}}$

$\cong 795.8 \text{ Hz}$

$Z_r = \dfrac{L}{R_w C} = \dfrac{2 \text{ H}}{(50 \ \Omega)(0.02 \ \mu\text{F})}$

$= 2 \text{ M}\Omega$

12. $I_r = 5 \text{ V}/Z_r = 5 \text{ V}/2 \text{ M}\Omega = 2.5 \ \mu\text{A}$

13. $f_r = \dfrac{\sqrt{1 - (R_w^2 C/L)}}{2\pi\sqrt{LC}} \cong 318.3 \text{ kHz}$

$X_L = 2\pi f_r L = 5 \text{ k}\Omega; Q = X_L/R$
$= 5 \text{ k}\Omega/6 \ \Omega = 833$
$\text{BW} = f_r/Q = 318.3 \text{ kHz}/833$
$= 382 \text{ Hz}$

14. $f_2 = 8 \text{ kHz} + 2 \text{ kHz} = 10 \text{ kHz}$
$f_r = (f_1 + f_2)/2 = (8 \text{ kHz} + 10 \text{ kHz})/2$
$= 9 \text{ kHz}$

Chapter 17

1. With V_2 zeroed:
$\mathbf{Z}_{V1} = \mathbf{X}_C + \mathbf{R}\|\mathbf{X}_L$
$= 100\angle -90° \ \Omega$
$\quad + 50\angle 0° \ \Omega\|25\angle 90° \ \Omega$
$= 100\angle -90° \ \Omega + 22.4\angle 63.4° \ \Omega$
$= 10 \ \Omega - j80 \ \Omega$
$= 80.6\angle -82.87° \ \Omega$

$\mathbf{I}_{V1} = \dfrac{1.5\angle 0° \text{ V}}{80.6\angle -82.87° \ \Omega}$

$= 18.6\angle 82.87° \text{ mA}$

$\mathbf{I}_{R(V1)} = \left(\dfrac{\mathbf{X}_L}{\mathbf{R} + \mathbf{X}_L}\right)\mathbf{I}_{V1}$

$= \left(\dfrac{25\angle 90° \ \Omega}{55.9\angle 26.6° \ \Omega}\right)18.6\angle 82.87° \text{ mA}$

$= 8.37\angle 146.3° \text{ mA}$
With V_1 zeroed:
$\mathbf{Z}_{V2} = \mathbf{R} + \mathbf{X}_C\|\mathbf{X}_L$
$= 50\angle 0° \ \Omega$
$\quad + 100\angle -90° \ \Omega\|25\angle 90° \ \Omega$
$= 50 \ \Omega + j33.3 \ \Omega = 60\angle 33.7° \ \Omega$

$\mathbf{I}_{R(V2)} = \mathbf{I}_{V2} = \dfrac{3\angle 0° \text{ V}}{60\angle 33.7° \ \Omega}$

$= 50\angle -33.7° \text{ mA}$
$\mathbf{I}_R = \mathbf{I}_{R(V1)} + \mathbf{I}_{R(V2)}$
$= 8.37\angle 146.3° \text{ mA}$
$\quad + 50\angle -33.7° \text{ mA}$
$= 34.64 \text{ mA} - j23.1 \text{ mA}$
$= 41.63\angle -33.7° \text{ mA}$

2. $I_{C(V1)} = I_{V1} = 18.6\angle 82.87°$ mA

$I_{C(V2)} = \left(\dfrac{X_L}{X_C + X_L}\right)I_{V2}$

$= \left(\dfrac{25\angle 90° \ \Omega}{75\angle -90° \ \Omega}\right)50\angle -33.7° \ A$

$= 16.67\angle 146.3°$ mA

$I_C = I_{C(V1)} + I_{C(V2)}$

$\quad = 30\angle -67.34°$ mA

3. $V_{th} = \left(\dfrac{X_C}{R_3 + X_C}\right)V_s$

$= \left(\dfrac{1\angle -90° \ k\Omega}{1 \ k\Omega - j1 \ k\Omega}\right)12\angle 0° \ V$

$= 8.49\angle -45°$ V

$Z_{th} = R_3 \| X_C$

$= \dfrac{(1\angle 0° \ k\Omega)(1\angle -90° \ k\Omega)}{1 \ k\Omega - j1 \ k\Omega}$

$= 707\angle -45° \ \Omega$

$= 500 \ \Omega - j500 \ \Omega$

See Figure S–25.

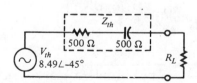

FIGURE S–25

4. $V_{th} = \left(\dfrac{X_C}{R + X_C}\right)V_s$

$= \left(\dfrac{20\angle -90° \ k\Omega}{100 \ k\Omega - j20 \ k\Omega}\right)15\angle 0° \ V$

$= \left(\dfrac{20\angle -90° \ k\Omega}{102\angle -11.3° \ k\Omega}\right)15\angle 0° \ V$

$= 2.94\angle -78.7°$ V

$Z_{th} = R \| X_C$

$= \dfrac{(100\angle 0° \ k\Omega)(20\angle -90° \ k\Omega)}{102\angle -11.3° \ k\Omega}$

$= 19.6\angle -78.7° \ k\Omega$

$I_L = \dfrac{V_{th}}{Z_{th} + R_L} = \dfrac{2.94\angle -78.7° \ V}{503.8 \ k\Omega - j19.2 \ k\Omega}$

$= \dfrac{2.94\angle -78.7° \ V}{504\angle -2.2° \ k\Omega}$

$= 5.83\angle -76.5° \ \mu A$

5. $Z_n = Z_{th} = 19.6\angle -78.7° \ k\Omega$

$\quad = 3.84 \ k\Omega - j19.22 \ k\Omega$

$I_n = \dfrac{V_{th}}{Z_{th}} = \dfrac{2.94\angle -78.7° \ V}{19.6\angle -78.7° \ k\Omega}$

$\quad = 0.15\angle 0°$ mA

See Figure S–26.

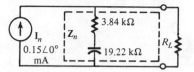

FIGURE S–26

6. $I_{eq} = \dfrac{12\angle 0° \ V}{20\angle 0° \ \Omega} + \dfrac{6\angle 0° \ V}{10\angle 0° \ \Omega}$

$\quad\quad + \dfrac{10\angle 0° \ V}{50\angle -90° \ \Omega} + \dfrac{5\angle 0° \ V}{10\angle -90° \ \Omega}$

$= 0.6 \ A + 0.6 \ A + j0.2 \ A + j0.5 \ A$

$= 1.39\angle 30.26° \ A$

$Z_{eq} = \dfrac{1}{0.05 \ S + 0.1 \ S + j0.02 \ S + j0.1 \ S}$

$= \dfrac{1}{0.15 \ S + j0.12 \ S}$

$= 5.2\angle -38.7° \ \Omega$

$= 4.06 \ \Omega - j3.25 \ \Omega$

$V_{eq} = I_{eq} Z_{eq}$

$\quad = (1.39\angle 30.26° \ A)(5.2\angle -38.7° \ \Omega)$

$\quad = 7.23\angle -8.44°$ V

See Figure S–27.

FIGURE S–27

7. $Z_L = Z_{eq} = 4.06 \ \Omega + j3.2 \ \Omega$

830

Chapter 18

1. $V_{out} = 0.707 V_{max} = (0.707)(10\ V)$
$= 7.07\ V$

2. $V_{out} = 10\ V$ dc; the sine wave frequency is shorted to ground by the capacitor.

3. V_{out} is a 15-V peak-to-peak sine wave with a *zero* average value; the dc component is blocked by the capacitor.

4. $X_C = (1/10)2.5\ k\Omega = 250\ \Omega$

$$C = \frac{1}{2\pi f X_C} = \frac{1}{2\pi(1\ kHz)(250\ \Omega)}$$
$= 0.6\ \mu F$
Use next highest standard value.

5. $V_{out} = \left(\dfrac{R}{\sqrt{R^2 + X_L^2}}\right) V_{in}$

$= \left(\dfrac{5\ \Omega}{\sqrt{(5\ \Omega)^2 + (100\ \Omega)^2}}\right) 5\ V$

$= 0.25\ V$ peak-to-peak
$V_{out} = V_{avg} = 8\ V$
The output is a 0.25-V peak-to-peak sine wave riding on an 8-V dc level.

6. See Figure S–28.

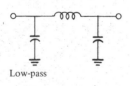

Low-pass   High-pass

FIGURE S–28

7. See Figure S–29.

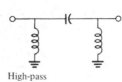

Low-pass   High-pass

FIGURE S–29

8. See Figure S–30.

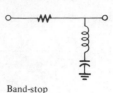

Band-pass   Band-stop

FIGURE S–30

9. See Figure S–31.

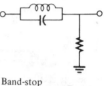

Band-pass   Band-stop

FIGURE S–31

10. $BW = f_r/Q = 5\ kHz/20 = 250\ Hz$

11. $20\ \log(4/12) = -9.54\ dB$

12. $10\ \log(5/10) = -3\ dB$

13. $f_c = \dfrac{1}{2\pi RC} = \dfrac{1}{2\pi(1\ k\Omega)(0.02\ \mu F)}$
$= 7.96\ kHz$

14. $f_c = \dfrac{1}{2\pi(L/R)} = \dfrac{1}{2\pi(0.01\ mH/5\ k\Omega)}$
$= 79.58\ MHz$

Chapter 19

1. $k = 0.9$

2. $L_M = k\sqrt{L_1 L_2} = 0.9\sqrt{(10\ \mu H)(5\ \mu H)}$
$= 6.36\ \mu H$

3. $N_s/N_p = 36/12 = 3$. It is a step-up transformer.

4. See Figure S–32 (page 831).

5. $N_s/N_p = 15/10 = 1.5$
$V_s = 1.5 V_p = 1.5(120\ V) = 180\ V$

6. $V_s = 0.5 V_p = 0.5(50\ V) = 25\ V$

7. $N_s/N_p = V_s/V_p = 60\ V/240\ V = 0.25$

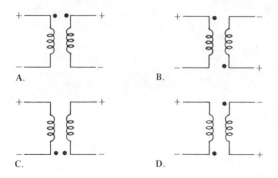

FIGURE S–32

8. $I_s = (N_p/N_s)I_p = (1/0.25)(0.25 \text{ A})$
   $= 1 \text{ A}$

9. $Z_p = (N_p/N_s)^2 Z_L = (1/5)^2 (1 \text{ k}\Omega)$
   $= 40 \; \Omega$

10. $R_s = R_{\text{reflected}}$
    $= 10 \; \Omega$ for maximum power to R_L
    $R_{\text{reflected}} = R_p = (N_p/N_s)^2 (R_L)$
    $(N_p/N_s)^2 (R_L) = 10 \; \Omega$
    $R_L = 10 \; \Omega/0.04 = 250 \; \Omega$

11. 1: $V_1 = (N_s/N_p)V_s = (100/10)120 \text{ V}$
    $= 1200 \text{ V}$
    2: $V_2 = (10/10)120 \text{ V} = 120 \text{ V}$
    3: $V_3 = (5/10)120 \text{ V} = 60 \text{ V}$

12. $1200 \text{ V}/2 = 600 \text{ V}$. The polarities are opposite.

Chapter 20

1. $v = V_{\text{max}} \sin \theta = 208 \sin 35°$
   $= 119.3 \text{ V}$

2. $f =$ (number of pole pairs)(rps)
   $= (1)(250 \text{ rps}) = 250 \text{ Hz}$

3. V_{out} increases.

4. Current through the field windings produces a magnetic field for rotor coil to cut in order to induce a voltage in the rotor windings.

5. $\mathbf{I}_L = \dfrac{30\angle 0° \text{ V}}{15\angle 22° \; \Omega} = 2\angle -22° \text{ A}$

6. $I_L = I_\theta = 12 \text{ A}$

7. $V_L = \sqrt{3} \, V_\theta = \sqrt{3} \, (180 \text{ V}) = 311.8 \text{ V}$

8. $V_L = V_\theta = 30 \text{ V}$

9. $I_L = \sqrt{3} \, I_\theta = \sqrt{3} \, (5 \text{ A}) = 8.66 \text{ A}$

10. $I_\theta = I_L = I_Z = 15 \text{ A}$

11. $V_L = \sqrt{3} \, V_\theta = \sqrt{3} \, (1.5 \text{ kV}) = 2.6 \text{ kV}$
    $V_Z = V_\theta = 1.5 \text{ kV}$

12. $V_L = V_Z = 75 \text{ V}$
    $I_L = \sqrt{3} \, I_Z = \sqrt{3} \, (3.3 \text{ A}) = 5.7 \text{ A}$

13. $V_Z = V_\theta/\sqrt{3} = 220 \text{ V}/\sqrt{3} = 127 \text{ V}$

14. $P_T = 3V_Z I_Z \cos \theta$
    $= 3(145 \text{ V})(2.5 \text{ A})(0.866)$
    $= 941.8 \text{ W}$

Chapter 21

1. Electrodynamometer.

2. A circuit that converts ac to pulsating dc.

3. Magnitude and phase.

4. (5 divisions)(0.1 V/div) $= 0.5 \text{ V}$

5. (10 divisions)(1 μs/div) $= 10 \; \mu$s

6. $T = 1/f = 1/1 \text{ kHz} = 1 \text{ ms}$
   (one cycle)
   (0.1 ms/div)(10 div) $= 1 \text{ ms}$; this allows one full cycle to be displayed.
   (1 ms/div)(10 div) $= 10 \text{ ms}$; this allows 10 full cycles to be displayed.

ANSWERS TO ODD-NUMBERED PROBLEMS

Chapter 1

1–1. (a) 3×10^3 (b) 75×10^3
     (c) 2×10^6

1–3. (a) $8.4 \times 10^3 = 0.84 \times 10^4$
                 $= 0.084 \times 10^5$
     (b) $99 \times 10^3 = 9.9 \times 10^4$
                 $= 0.99 \times 10^5$
     (c) $200 \times 10^3 = 20 \times 10^4$
                 $= 2 \times 10^5$

1–5. (a) 0.0000025   (b) 5000
     (c) 0.39

1–7. (a) 126×10^6
     (b) 5.00085×10^3
     (c) 6060×10^{-9}

1–9. (a) 20×10^8 (b) 36×10^{14}
     (c) 15.4×10^{-15}

1–11. (a) 31 mA   (b) 5.5 kV
     (c) 200 pF

1–13. (a) 2370×10^{-6}
     (b) 18.56×10^{-6}
     (c) $0.00574389 \times 10^{-6}$
     (d) $100,000,000,000 \times 10^{-6}$

1–15. (a) 5000 μA   (b) 3.2 mW
     (c) 5 MV   (d) 10,000 kW

Chapter 2

2–1. 0.2 A

2–3. 0.15 C

2–5. 20 V

2–7. (a) 10 V   (b) 2.5 V   (c) 4 V
     (d) 0.2 V

2–9. (a) 10 Ω (b) 2 Ω (c) 50 Ω
     (d) 0.333 Ω

2–11. (a) $R_{MAX} = 34.65\ \Omega$,
        $R_{MIN} = 31.35\ \Omega$
     (b) $R_{MAX} = 51,700\ \Omega$,
        $R_{MIN} = 42,300\ \Omega$

2–13. 2250 kV

2–15. 15.6 Ω

2–17. See Figure P–1.

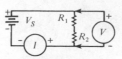

FIGURE P–1

2–19. 0.366 A

2–21. AWG #27

2–23. See Figure P–2.

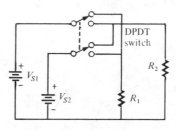

FIGURE P–2

Chapter 3

3–1. (a) 5 A   (b) 1.5 A   (c) 0.5 A
     (d) 2 mA   (e) 50 μA

3–3. 1.2 A

3–5. (a) 10 mV   (b) 1.65 V
     (c) 15 kV   (d) 3.52 V
     (e) 0.25 V   (f) 750 kV
     (g) 8.5 kV   (h) 3.75 mV

3–7. (a) 5 Ω (b) 2 Ω (c) 10 Ω
     (d) 0.55 Ω (e) 300 Ω

3–9. 3 kΩ

3–11. See Figure P–3.

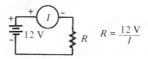

FIGURE P–3

3–13. (a) 7.5 mA   (b) 10 mA
     (c) 15 mA   (d) 20 mA
     (e) 50 mA

3–15. A: 15.15 mA (least)
B: 19.23 mA
C: 21.28 mA (most)

3–17.
```
10 CLS
20 INPUT "MINIMUM VOLTAGE (VOLTS)";VL
30 INPUT "MAXIMUM VOLTAGE (VOLTS)";VH
40 INPUT "VOLTAGE INCREMENTS";VI
50 INPUT "RESISTANCE (OHMS)";R
60 PRINT "VOLTAGE","CURRENT"
70 FOR V=VL TO VH STEP VI
80 I=V/R
90 PRINT V;"V", I;"A"
100 NEXT
```

3–19. **(a)** 59.9 mA   **(b)** 5.99 V
(c) 0.005 V

Chapter 4

4–1. 350 W

4–3. 20,000 W

4–5. **(a)** 1 MW   **(b)** 3 MW
(c) 150 MW   **(d)** 8.7 MW

4–7. **(a)** 2,000,000 μW   **(b)** 500 μW
(c) 250 μW   **(d)** 6.67 μW

4–9. 37.5 Ω

4–11. 345 W

4–13. 100 μW

4–15. 0.045 W

4–17. 0.68 W. Use 1-W resistor.

4–19. 8640 J

4–21. 216 kWh

4–23. 83.33 kWh

4–25. 100 W

4–27. 36 Ah

4–29. 13.54 mA

4–31. 4.25 W

4–33. See Program 4–33.

4–35. 20 ms

4–37. See Program 4–37.

```
10 CLS
20 PRINT "THIS PROGRAM COMPUTES THE POWER FOR EACH OF A SPECIFIED"
30 PRINT "NUMBER OF VOLTAGE VALUES AND A SPECIFIED RESISTANCE."
40 PRINT "THE RESULTS ARE DISPLAYED IN TABULAR FORM."
50 PRINT:PRINT:PRINT
60 INPUT "TO CONTINUE PRESS 'ENTER'";X
70 CLS
80 INPUT "RESISTANCE VALUE IN OHMS";R
90 INPUT "MINIMUM VOLTAGE IN VOLTS";V1
100 INPUT "MAXIMUM VOLTAGE IN VOLTS";V2
110 INPUT "VOLTAGE INCREMENTS IN VOLTS";VI
120 CLS
130 PRINT TAB(15),"VOLTAGE (VOLTS)";TAB(35),"POWER (WATTS)"
140 FOR V=V1 TO V2 STEP VI
150 P=(V*V)/R
160 PRINT TAB(15),V;TAB(35),P
170 NEXT
```
PROGRAM 4–33

```
10 CLS
20 INPUT "NUMBER OF DAYS FOR WHICH KWH ARE TO BE COMPUTED";D
30 INPUT "NUMBER OF APPLIANCES";A
40 FOR NA=1 TO A
50 PRINT "WATTAGE RATING FOR APPLIANCE";NA;"IN KW"
55 INPUT P
60 INPUT "HOURS PER DAY THE APPLIANCE IS USED";T
70 E=P*T*D+E
80 NEXT
90 CLS
100 INPUT "COST OF ENERGY PER KWH (CENTS)";C
110 TC=E*C
120 CLS
130 PRINT "THE TOTAL ENERGY CONSUMPTION FOR THE SPECIFIED PERIOD IS";E;"KWH."
140 PRINT "THE TOTAL COST IS $";TC*.01
```
PROGRAM 4–37

Chapter 5

5–1. See Figure P–4.

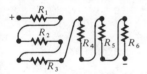

FIGURE P–4

5–3. 5 mA

5–5. **(a)** 1560 Ω   **(b)** 103 Ω
      (c) 13.7 kΩ   **(d)** 3.971 MΩ

5–7. 6 kΩ

5–9. 1144 Ω

5–11. A: 0.6875 mA
       B: 4 μA

5–13. 500 Ω

5–15. 9 V

5–17. 26 V

5–19. 20 Ω

5–21. **(a)** 4 V   **(b)** 6.67 V

5–23. $V_{AF} = 100$ V, $V_{BF} = 80$ V,
      $V_{CF} = 70$ V, $V_{DF} = 20$ V, $V_{EF} = 5$ V

5–25. 250 mW

5–27. R_2 is open; 12 V.

5–29. Add line:
      245 P(X) = V(X) ↑ 2/R(X)
      Add to end of line 250:
      ,"P"; X; "="; P(X); "WATTS"

5–31. **(a)** 19.1 mA   **(b)** 45.6 V
      (c) 342.8 Ω(⅛ W); 685.7 Ω(¼ W);
      1371.4 Ω(½ W)

5–33. See Figure P–5.

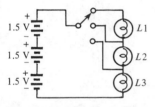

FIGURE P–5

Chapter 6

6–1. See Figure P–6.

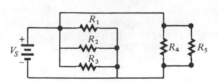

FIGURE P–6

6–3. 12 V; 5 mA

6–5. 545.45 kΩ

6–7. A: 1.43 Ω
      B: 18.22 Ω
      C: 892 Ω

6–9. 500 Ω

6–11. A: 1 A
      B: 81.25 mA

6–13. 8 kΩ

6–15. A: 3 mA
      B: 10 μA
      C: 0.5 A

6–17. 175 mA

6–19. 12.5 mA

6–21. $I_R = 4.8$ A; $I_{2R} = 2.4$ A;
      $I_{3R} = 1.6$ A; $I_{4R} = 1.2$ A

6–23. A: 66.67 μW
      B: 50 mW

6–25. 0.682 A, 3.41 A

6–27. R_1 and R_3 branches are open.

6–29.

```
10 CLS
20 INPUT "TOTAL RESISTANCE (OHMS)";RT
30 INPUT "HOW MANY RESISTORS IN PARALLEL";N
40 FOR X=1 TO N-1
50 PRINT "THE VALUE OF R";X;"IN OHMS"
60 INPUT R(X)
70 NEXT
80 RS=R(1)
90 FOR X=2 TO N-1
100 RS=1/(1/(RS)+(1/R(X)))
110 NEXT
120 CLS
130 RN=RT*RS/(RS-RT)
140 PRINT "UNKNOWN RESISTOR VALUE IS";RN;"OHMS."
```

6–31. (a) $I_2 = 50$ mA; $R_1 = 100$ Ω;
    $R_2 = 200$ Ω
(b) $P_1 = 1.25$ W; $I_2 = 74.91$ mA;
    $I_1 = 125.1$ mA; $V_S = 10$ V;
    $R_1 = 79.9$ Ω; $R_2 = 133.5$ Ω
(c) $I_2 = 0.167$ A; $I_3 = 0.1$ A;
    $R = 429.2$ Ω; $I_1 = 0.23$ A

Chapter 7

7–1. See Figure P–7.

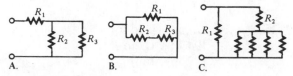

FIGURE P–7

7–3. A: R_1 and R_4 are in series with
    the parallel combination of R_2
    and R_3.
B: R_1 is in series with the parallel
    combination of R_2, R_3, and R_4.
C: The parallel combination of R_2
    and R_3 is in series with the par-
    allel combination of R_4 and R_5.
    This is all in parallel with R_1.

7–5. 2003 Ω

7–7. A: 687.5 Ω
B: 400 kΩ
C: 3.2 kΩ

7–9. A: $I_1 = 1$ mA; $I_2 = 0.453$ mA;
    $I_3 = 0.272$ mA; $I_4 = 0.181$ mA
    $V_1 = 1$ V; $V_2 = 0.453$ V;
    $V_3 = V_4 = 0.544$ V
B: $I_1 = 2$ μA; $I_2 = 0.667$ μA;
    $I_3 = 0.333$ μA; $I_4 = 2$ μA
    $V_1 = V_2 = V_3 = V_4 = 2$ V
C: $I_1 = I_2 = 0.936$ mA;
    $I_3 = 0.312$ mA; $I_4 = 0.624$ mA;
    $I_5 = 0.624$ mA;
    $I_6 = I_7 = 0.312$ mA;
    $I_8 = 0.624$ mA
    $V_1 = V_2 = 0.936$ V;
    $V_3 = V_4 = 0.312$ V;

$V_5 = 1.87$ V;
$V_6 = V_7 = 1.87$ V; $V_8 = 1.25$ V

7–11. $V_A = 3.85$ V, $V_B = 42.31$ V,
    $V_C = 50$ V, $V_D = 5$ V

7–13. No, it should be 4.39 V.

7–15. The 2-kΩ resistor is open.

7–17. 7.5 V unloaded; 7.32 V loaded.

7–19. 50-kΩ load.

7–21. $R_1 = 1000$ Ω; $R_2 = R_3 = 500$ Ω;
    $V_{5V} = 3.12$ V (loaded);
    $V_{2.5V} = 1.25$ V (loaded)

7–23. $R_T = 6$ kΩ, $V_A = 3$ V, $V_B = 1.5$ V,
    $V_C = 0.75$ V

7–25. $V_1 = 1.67$ V, $V_2 = 6.68$ V,
    $V_3 = 1.67$ V, $V_4 = 3.34$ V,
    $V_5 = 0.4175$ V, $V_6 = 2.505$ V,
    $V_7 = 0.4175$ V, $V_8 = 1.67$ V,
    $V_9 = 1.67$ V

7–27. 1800 Ω

7–29.

```
10 CLS
20 INPUT "WHAT IS THE INPUT VOLTAGE";VIN
30 PRINT:PRINT
40 FOR X=1 TO 8
50 PRINT "VALUE OF R";X;"IN OHMS"
60 INPUT R(X)
70 NEXT
80 CLS
90 RC=(R(6)*(R(7)+R(8)))/(R(6)+R(7)+R(8))
100 RB=(R(4)*(R(5)+RC))/(R(4)+R(5)+RC)
110 RA=(R(2)*(R(3)+RB))/(R(2)+R(3)+RB)
120 RT=RA+R(1)
130 IT=VIN/RT
140 VA=IT*RA
150 I(1)=IT
160 I(2)=VA/R(2)
170 I(3)=IT-I(2)
180 VB=I(3)*RB
190 I(4)=VB/R(4)
200 I(5)=I(3)-I(4)
210 VC=I(5)*RC
220 I(6)=VC/R(6)
230 I(7)=I(5)-I(6)
240 I(8)=I(7)
250 VD=I(8)*R(8)
260 PRINT "VA=";VA;"V","VB=";VB;"V","VC=";VC;"V","VD=";VD;"V"
270 PRINT
280 FOR Y=1 TO 8
290 PRINT "I";Y;"=";I(Y);"AMP"
300 NEXT
```

7–31. 109.9 Ω

7–33. $V_{OUT} = 0$ V; $I_T = 0.97$ A

Chapter 8

8–1. $I_S = 6$ A, $R_S = 50$ Ω

8–3. $V_S = 720$ V, $R_S = 1.2$ kΩ

8–5. 0.905 mA

8–7. 1.6 mA

8–9. A: $R_{TH} = 75$ Ω, $V_{TH} = 8.33$ V
 B: $R_{TH} = 71.43$ Ω, $V_{TH} = 0.86$ V
 C: $R_{TH} = 33.33$ kΩ, $V_{TH} = 1.67$ V
 D: $R_{TH} = 1.2$ kΩ, $V_{TH} = 80$ V

8–11. 11.11 V

8–13. 0.1 mA

8–15. $R_{EQ} = 0.545$ Ω, $V_{EQ} = -2.45$ V

8–17. A: 12 Ω
 B: 8 kΩ
 C: .5.67 Ω
 D: 650 Ω

8–19. A: $R_1 = 167$ kΩ, $R_2 = 500$ kΩ,
   $R_3 = 250$ kΩ
 B: $R_1 = 0.364$ Ω, $R_2 = 0.909$ Ω,
   $R_3 = 0.455$ Ω

8–21. See Program 8–21.

8–23. 32 V: 125 μA
  15 V: 236 μA

8–25. $R_{TH} = 48$ Ω; $R_4 = 229.5$ Ω

Chapter 9

9–1. 6 possible loops

9–3. $I_1 - I_2 - I_3 = 0$

9–5. $V_1 = 5.676$ V, $V_2 = 6.365$ V,
  $V_3 = 0.365$ V

9–7. $I_1 = 0.738$ A, $I_2 = -0.527$ A,
  $I_3 = -0.469$ A

9–9. $I_1 = 0$ A, $I_2 = 2$ A

9–11. $I_1 = -5.18$ mA left,
  $I_2 = -1.65$ mA up
  $I_3 = 3.53$ mA left,

9–13. $60I_1 - 10I_2 = 1.5$ V
  $-10I_1 + 40I_2 - 5I_3 = -3$ V
  $-5I_2 + 20I_3 = -1.5$ V

9–15. $I_{75Ω} = 12$ mA (L to R)
  $I_{50Ω} = 194$ mA (L to R)
  $I_{60Ω} = 182$ mA (up)

9–17. $I_1 = -6.6455$ mA,
  $I_2 = -0.0892405$ A,
  $I_3 = -0.100633$ A

9–19. $I_{R1} = 146.67$ mA; $I_{R2} = 46.67$ mA
  $I_{R3} = 100$ mA

9–21. $V_{AB} = 0.117$ V

Chapter 10

10–1. Full-scale; half-scale.

10–3. (a) 50 μA   (b) 950 μA
  (c) 9.95 mA   (d) 99.95 mA
  (e) 999.95 mA

10–5. (a) 1 kΩ   (b) 52.63 Ω
  (c) 5.025 Ω   (d) 0.5 Ω
  (e) 0.05 Ω

10–7. 0.1 V: 1 kΩ
   1 V: 18 kΩ

```
10 CLS
20 PRINT "PLEASE PROVIDE THE DELTA RESISTOR VALUES WHEN PROMPTED."
30 PRINT:PRINT
40 INPUT "RA IN KILOHMS";RA
50 INPUT "RB IN KILOHMS";RB
60 INPUT "RC IN KILOHMS";RC
70 CLS
80 RT=RA+RB+RC
90 R1=RA*RC/RT
100 R2=RB*RC/RT
110 R3=RA*RB/RT
120 PRINT "THE VALUES FOR THE WYE NETWORK ARE AS FOLLOWS:"
130 PRINT:PRINT "R1=";R1;"KOHMS"
140 PRINT "R2=";R2;"KOHMS"
150 PRINT "R3=";R3;"KOHMS"
```

PROGRAM 8–21

5 V: 80 kΩ
10 V: 100 kΩ
50 V: 800 kΩ

10–9. A: 12.5 V
B: 0.7 V

10–11. **(a)** 200 Ω   **(b)** 150 MΩ
(c) 4.5 kΩ

10–13. Actual voltage is 20.41 V. This is more than the scale setting can handle, so a 1 appears in the overflow (left-most) digit of the display, giving a reading of 10.41. Operator must increase scale range.

10–15. TP1:   14.3% low
TP2:   16.9% low
TP3:   10.1% low

10–17. 1.11 V; 11.1 μA

Chapter 11

11–1. **(a)** 1 Hz   **(b)** 5 Hz
(c) 20 Hz   **(d)** 1000 Hz
(e) 2 kHz   **(f)** 100 kHz

11–3. 2 μs

11–5. **(a)** 8.48 V   **(b)** 24 V
(c) 7.64 V

11–7. **(a)** 7.07 mA   **(b)** 6.37 mA
(c) 10 mA   **(d)** 20 mA
(e) 10 mA

11–9. **(a)** 22.5°   **(b)** 60°   **(c)** 90°
(d) 108°   **(e)** 216°   **(f)** 324°

11–11. Sine wave with peak at 75°: leading by 15°. Sine wave with peak at 100°: lagging by 10°. $\theta = 25°$

11–13. **(a)** 57.36 mA   **(b)** 99.62 mA
(c) −17.36 mA   **(d)** −57.36 mA
(e) −99.62 mA   **(f)** 0 mA

11–15.   30°:   12.99 V
45°:   14.49 V
90°:   12.99 V
180°:   −7.5 V
200°:   −11.49 V
300°:   −7.5 V

11–17. See Figure P–8.

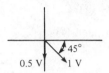

FIGURE P–8

11–19. **(a)** 9.55 Hz   **(b)** 57.3 Hz
(c) 0.318 Hz   **(d)** 199.9 Hz

11–21. 18.03

11–23. **(a)** 642.8 − j766
(b) −14.1 + j5.13
(c) −17.68 − j17.68
(d) −3 + j0

11–25. **(a)** 1.1 + j0.7   **(b)** −81 − j35
(c) 5.3 − j5.27
(d) −50.4 + j62.5

```
10 CLS
20 INPUT "THE PEAK VALUE IN AMPS";IM
30 INPUT "THE PERIOD IN SECONDS";T
40 INPUT "INSTANTANEOUS PHASE ANGLE IN DEGREES";THETA
50 CLS
60 IPP=2*IM
70 IRMS=SQR(0.5)*IM
80 IAVG=(2/3.1415927)*IM
90 F=1/T
100 I=IM*SIN(THETA/57.29578)
110 PRINT "IP =";IM;"A"
120 PRINT "O =";THETA;"DEGREES"
130 PRINT "T =";T;"SECONDS"
140 PRINT "IPP =";IPP;"A"
150 PRINT "IRMS =";IRMS;"A"
160 PRINT "IAVG =";IAVG;"A"
170 PRINT "F =";F;"HZ"
180 PRINT "INSTANTANEOUS CURRENT AT";THETA;"DEGREES =";I;"A"
```

PROGRAM 11–35

```
10 CLS
20 INPUT "REAL PART OF COMPLEX NUMBER";RP
30 INPUT "J PART OF COMPLEX NUMBER";JP
40 CLS
50 MAG=SQR(RP[2+JP[2)
60 ANG=ATN(JP/RP)*57.29578
70 PRINT "THE NUMBER IN POLAR FORM IS";MAG;"/";ANG;
```
PROGRAM 11–37

11–27. (a) $3.2\angle11°$ (b) $7\angle-101°$
  (c) $1.52\angle70.6°$
  (d) $2.79\angle-63.5°$

11–29. A: 25%
  B: 66.67%

11–31. A: 250 kHz
  B: 33.33 Hz

11–33. 25 kHz; 75 kHz; 125 kHz; 175 kHz;
  225 kHz; 275 kHz; 325 kHz

11–35. See Program 11–35 on page 839.

11–37. See Program 11–37.

11–39. 512.6 W

11–41. 3 V

Chapter 12

12–1. (a) $5\ \mu F$ (b) $1\ \mu C$ (c) 10 V

12–3. (a) 100,000 pF   (b) 2500 pF
  (c) 5,000,000 pF

12–5. (a) $0.1\ \mu F$ (b) $2200\ \mu F$
  (c) $0.0015\ \mu F$

12–7. $2\ \mu F$

12–9. 221.25 pF

12–11. 990 pF

12–13. By increasing the number of layers
  of plate material and dielectric.

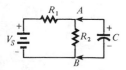

FIGURE P–9

12–15. See Figure P–9.

12–17. (a) $100\ \mu s$ (b) $500\ \mu s$
  (c) $23.5\ \mu s$ (d) 15 ms

12–19. (a) 9.48 V   (b) 12.97 V

  (c) 14.25 V   (d) 14.73 V
  (e) 14.899 V

12–21. (a) 2.72 V   (b) 5.90 V
  (c) 11.65 V

12–23. $t = -RC \ln[1 - (v_C/V_F)]$; 13.86 μs

12–25. 7.62 μs

12–27. A: 0.67 μF
  B: 68.97 pF
  C: 2.7 μF

12–29. 2 μF

12–31. A: 47 pF: 4.7×10^{-10} C
    10 pF: 100×10^{-12} C
    0.001 μF: 0.01 μC
  B: 0.1 μF: 0.5 μC
    0.01 μF: 0.05 μC
    0.001 μF: 0.005 μC
    10,000 pF: 0.05 μC

12–33. (a) 2.5 V   (b) 3.33 V   (c) 4 V

12–35. (a) 31.83 Ω (b) 111 kΩ
  (c) 49.7 Ω

12–37. 200 Ω

12–39. 0 W; 3.39 mVAR

12–41. See Program 12–41.

12–43. 7.86 kΩ

12–45. $I_{C1} = 0.1455$ mA; $V_{C1} = 7.72$ V;
  $I_{C2} = 94.6\ \mu A$; $V_{C2} = 2.28$ V;
  $I_{C3} = 50.8\ \mu A$; $V_{C3} = 1.798$ V;
  $I_{C4} = 45.5\ \mu A$; $V_{C4} = 0.482$ V;
  $I_{C5} = I_{C6} = 5.5\ \mu A$;
  $V_{C5} = 0.2892$ V; $V_{C6} = 0.1928$ V.

Chapter 13

13–1. 10 μWb

13–3. 400 turns

13–5. (a) 1000 mH   (b) 0.25 mH
  (c) 0.01 mH   (d) 0.5 mH

```
10 CLS
20 INPUT "INITIAL CAPACITOR VOLTAGE";VC
30 INPUT "CAPACITANCE IN FARADS";C
40 INPUT "RESISTANCE IN OHMS";R
50 INPUT "NUMBER OF TIME INTERVALS";TI
60 CLS
70 PRINT "TIME (SEC)","CAPACITOR VOLTAGE","% INITIAL CHARGE"
80 FOR T=0 TO 5*R*C STEP 5*R*C/TI
90 X=T/(R*C)
100 V=VC*(2.718[-X)
110 P=(V/VC)*100
120 PRINT TAB(0) T "S";TAB(20) V "V";TAB(50) P "%"
130 NEXT
```
PROGRAM 12–41

13–7. 50 mV

13–9. 89 turns

13–11. (a) 5 μs   (b) 22.73 μs
   (c) 22.73 μs

13–13. (a) 25 V   (b) 9.2 V
   (c) 1.25 V   (d) 0.46 V

13–15. (a) 17.91 V   (b) 12.84 V
   (c) 6.59 V

13–17. Positive at top of inductor; 2.4 A

13–19. 18 mH

13–21. 7.14 μH

13–23. A: 4.33 H
   B: 50 mH
   C: 0.57 μH

13–25. A: 136 kΩ
   B: 1.57 kΩ
   C: 0.0179 Ω

13–27. I_T = 10.05 A; I_{L2} = 6.7 A;
   I_{L3} = 3.35 A

13–29. 100.5 VAR

13–31. (a) Infinite resistance.
   (b) Zero resistance.
   (c) Lower R_w

13–33. See Program 13–33.

13–35. 2.86 mA

13–37. 26.1 mA

Chapter 14

14–1. 8 kHz; 8 kHz

14–3. A: 250 Ω − j100 Ω;
   269.3$\angle$−21.8° Ω
   B: 600 Ω − j1000 Ω;
   1166$\angle$−59° Ω

14–5. (a) 50 kΩ − j796 kΩ
   (b) 50 kΩ − j159 kΩ
   (c) 50 kΩ − j79.6 kΩ
   (d) 50 kΩ − j31.8 kΩ

14–7. (a) R = 30 Ω; X_C = 50 Ω
   (b) R = 271.9 Ω; X_C = 126.8 Ω
   (c) R = 697.5 Ω; X_C = 1.66 kΩ
   (d) R = 557.9 Ω; X_C = 557.9 Ω

14–9. A: 0.178$\angle$57.7° mA
   B: 0.625$\angle$38.5° mA

14–11. −17.64°

14–13. (a) 94$\angle$−57.9° Ω
   (b) 106$\angle$57.9° mA
   (c) 5.3$\angle$57.9° V
   (d) 8.47$\angle$−32.1° V

14–15. 353.4 Ω; −61.99°

```
10 CLS
20 INPUT "INDUCTANCE IN HENRIES";L
30 INPUT "WINDING RESISTANCE IN OHMS";RW
40 INPUT "DIRECT CURRENT IN AMPS";I
50 E=.5*L*I*I
60 P=I*I*RW
70 CLS
80 PRINT "ENERGY STORED BY INDUCTOR =";E;"JOULES"
90 PRINT "POWER DISSIPATED BY WINDING RESISTANCE =";P;"WATTS"
```
PROGRAM 13–33

14–17. $\mathbf{V}_C = \mathbf{V}_R = 10\angle 0°$ V;
$\mathbf{I}_T = 184.2\angle 37.1°$ mA;
$\mathbf{I}_R = 147\angle 0°$ mA;
$\mathbf{I}_C = 111.1\angle 90°$ mA

14–19. (a) $9\angle -25.5°$ Ω
(b) $10\angle 0°$ mA
(c) $4.76\angle 90°$ mA
(d) $11.1\angle 25.5°$ mA   (e) $25.5°$

14–21. 18.16-kΩ resistor in series with 190.6-pF capacitor.

14–23. Capacitive.

14–25. (a) $45.24\angle 73.7°$ mA
(b) $-73.7°$
(c) $2.26\angle 73.7°$ V
(d) $1.94\angle 73.7°$ V
(e) $1.94\angle 73.7°$ V
(f) $14.4\angle -16.3°$ V

14–27. $P_{avg} = 0.562$ W; $P_r = 0.9$ VAR

14–29. $P_{avg} = 0.19$ W; $P_r = 0.651$ VAR;
$P_a = 0.678$ VA; $PF = 0.28$

14–31. See Figure P–10.

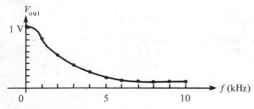

FIGURE P–10

14–33. See Figure P–11.

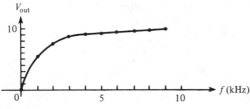

FIGURE P–11

14–35. 0.0796 μF

14–37. A: (a) 25 ms
(b) See Figure P–12.

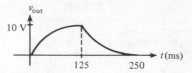

FIGURE P–12

B: (a) 0.5 μs
(b) See Figure P–13.

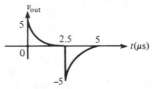

FIGURE P–13

14–39. See Figure P–14.

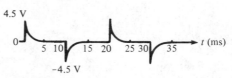

FIGURE P–14

14–41. See Figure P–15.

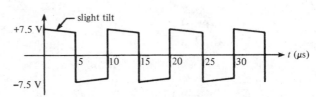

FIGURE P–15

14–43. Reduces amplitude of V_{out} to 6.19 V and shortens τ.

14–45. (a) No output pulse
(b) τ increases; C partially charges to 18.36 V.
(c) Output is 10 V, 125 ms pulse.
(d) Zero output

14–47.

```
10 CLS
20 INPUT "INPUT VOLTAGE IN VOLTS";VIN
30 INPUT "THE VALUE OF R IN OHMS";R
40 INPUT "THE VALUE OF C IN FARADS";C
50 INPUT "THE LOWEST NONZERO FREQUENCY IN HERTZ";FL
60 INPUT "THE HIGHEST FREQUENCY IN HERTZ";FH
70 INPUT "THE FREQUENCY INCREMENTS IN HERTZ";FI
80 CLS
90 PRINT "FREQUENCY(HZ)","PHASE SHIFT","VOUT", "T"
```

```
100 FOR F=FL TO FH STEP FI
110 XC=1/(2*3.1416*F*C)
120 PHI=-90+ATN(XC/R)*57.3
130 VO=VIN*XC/(SQR(R*R+XC*XC))
140 I=VIN/(SQR(R*R+XC*XC))
150 PRINT F,PHI,VO,I
160 NEXT
```

14–49. $R_x = 12 \ \Omega$; $C_x = 13.3 \ \mu F$

14–51. $0.1 \ \mu F$

Chapter 15

15–1. 15 kHz

15–3. A: $100 \ \Omega + j50 \ \Omega$;
$111.8\angle 26.6° \ \Omega$
B: $1.5 \ k\Omega + j1 \ k\Omega$;
$1.8\angle 33.7° \ k\Omega$

15–5. (a) $17.38\angle 46.4° \ \Omega$
(b) $63.97\angle 79.2° \ \Omega$
(c) $126.23\angle 84.5° \ \Omega$
(d) $251.61\angle 87.3° \ \Omega$

15–7. A: $89.4\angle -26.6° \ mA$
B: $2.78\angle -33.7° \ mA$

15–9. $37°$

15–11. See Figure P–16.

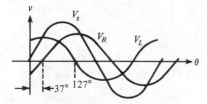

FIGURE P–16

15–13. $7.69\angle 50.3° \ \Omega$

15–15. 2387 Hz

15–17. (a) $274\angle 60.7° \ \Omega$
(b) $89.3\angle 0° \ mA$
(c) $159\angle -90° \ mA$
(d) $182.5\angle -60.7° \ mA$
(e) $60.7°$ (I_T lagging V_s)

15–19. 1.83-kΩ resistor in series with 4.21-kΩ inductive reactance.

15–21. Resistive.

15–23. (a) $16.5\angle -9.9° \ mA$
(b) $9.9°$
(c) $18\angle 0° \ V$
(d) $7\angle -56.2° \ V$
(e) $5.68\angle -91.7° \ V$
(f) $15.25\angle 22.5° \ V$
(g) $4.06\angle -1.7° \ V$

15–25. $P_{avg} = 1.28 \ W$; $P_r = 0.96 \ VAR$

15–27. $P_{avg} = 0.29 \ W$; $P_r = 0.05 \ VAR$;
$P_a = 0.297 \ VA$; PF = 0.99

15–29. See Figure P–17.

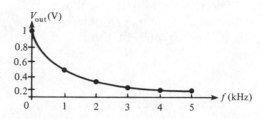

FIGURE P–17

15–31. See Figure P–18.

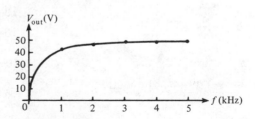

FIGURE P–18

15–33. See Figure P–19.

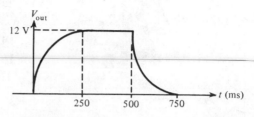

FIGURE P–19

15–35. (a) $0.2 \ \mu s$ (b) See Figure P–20.

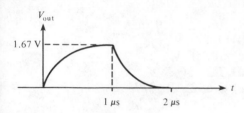

FIGURE P–20

15–37. (a) 0 V   (b) 0 V
(c) Lowers τ to 0.08 μs, thus decreasing rise and fall times of output pulse.
(d) Increases τ to 0.36 μs, thus preventing output from reaching 5 V (only 4.69 V).

15–39.

```
10 CLS
20 INPUT "RESISTANCE IN OHMS";R
30 INPUT "INDUCTANCE IN HENRIES";L
40 INPUT "LOWEST FREQUENCY IN HERTZ";FL
50 INPUT "HIGHEST FREQUENCY IN HERTZ";FH
60 INPUT "FREQUENCY INCREMENTS IN HERTZ";FI
70 CLS
80 PRINT "FREQUENCY(HZ)","PHASE SHIFT","VOUT"
90 FOR F=FL TO FH STEP FI
100 XL=2*3.1416*F*L
110 PHI=-ATN(XL/R)*57.3
120 VO=R/(SQR(R[2+XL[2))
130 PRINT F,PHI,VO
140 NEXT
```

15–41. $\phi = -128.44°$; attenuation = 0.205

Chapter 16

16–1. $479.6\angle -88.8°$ Ω; 479.5-Ω capacitive

16–3. Impedance increases.

16–5. See Figure P–21.

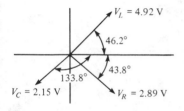

V_L = 4.92 V
46.2°
43.8°
133.8°
V_C = 2.15 V
V_R = 2.89 V

FIGURE P–21

16–7. $Z = 200\ \Omega$; $X_C = X_L = 2\ k\Omega$

16–9. 0.5 A

16–11. $99.697\angle -4.5°\ \Omega$

16–13. $I_T \cong 0.05\angle 4.5°$ A; $I_R = 0.05\angle 0°$ A
$I_L = 4.4\angle -90°$ mA;
$I_C = 8.3\angle 90°$ mA;
$V_R = V_L = V_C = 5\angle 0°$ V

16–15. A: 3.18° (V_s leads I_T)
B: 23.2° (V_s leads I_T)

16–17. 50.5-kΩ resistor in series with 1.3-H inductor.

16–19. 42.87° (I_2 leads V_s)

16–21. $I_{R1} = I_{C1} = 1\angle -36.9°$ mA
$I_{R2} = 0.707\angle -81.9°$ mA
$I_{C2} = 0.707\angle 8.1°$ mA
$I_L = 1.41\angle -171.9°$ mA
$V_{R2} = V_{C2} = V_L = 7.07\angle -81.9°$ V
$V_{R1} = 3\angle -36.9°$ V
$V_{C1} = 1\angle -126.9°$ V

16–23. 531.9 MΩ; 103.82 kHz

16–25. 62.5 Hz

16–27. 1.375 W

16–29. 1.99 mH; 0.2 μF

16–31.

```
10 CLS
20 INPUT "L IN HENRIES";L
30 INPUT "C IN FARADS";C
40 INPUT "WINDING RESISTANCE IN OHMS";RW
50 CLS
60 FR=(SQR(1-RW[2*C/L))/(2*3.1416*(SQR(L*C)))
70 Q=2*3.1416*FR*L/RW
80 BW=FR/Q
90 PRINT "FR (HZ)","Q","BW (HZ)"
100 PRINT FR,Q,BW
```

16–33. None.

16–35. $f_{r(\text{series})} = 4.11$ kHz,
$V_{\text{out}} = 9.97\angle 3.26°$ V
$f_{r(\text{parallel})} = 2.6$ kHz,
$V_{\text{out}} \cong 10\angle 0°$ V

Chapter 17

17–1. $1.3\angle 29.5°$ mA

```
10 CLS
20 INPUT "VTH (VOLTS THEN PHASE ANGLE)";VTH,ANG
30 INPUT "ZTH IN RECTANGULAR FORM (R FIRST, X SECOND)";RTH,XTH
40 INPUT "ENTER 1 IF XTH IS CAPACITIVE, 2 IF INDUCTIVE";S
50 IN=VTH/(SQR(RTH[2+XTH[2))
60 NANG=ANG-ATN(XTH/RTH)*57.3
70 CLS
75 IF S=1 THEN PRINT"ZN =";RTH;"- J";XTH ELSE PRINT "ZN =";RTH;"+ J";XTH
90 PRINT "IN=";IN;"A"
```

PROGRAM 17–15

17–3. $84.7\angle 61.6°$ mA

17–5. A: $\mathbf{V}_{th} = 15\angle -53.1°$ V;
$\quad\quad\quad \mathbf{Z}_{th} = 61\ \Omega - j48\ \Omega$
$\quad\quad$ B: $\mathbf{V}_{th} = 1.22\angle 0°$ V;
$\quad\quad\quad \mathbf{Z}_{th} = j237\ \Omega$
$\quad\quad$ C: $\mathbf{V}_{th} = 12.1\angle 11.9°$ V;
$\quad\quad\quad \mathbf{Z}_{th} = 50\ k\Omega - j20\ k\Omega$

17–7. $16.77\angle 87.9°$ V

17–9. $0.113\angle -65.6°$ mA

17–11. $\mathbf{Z}_{eq} = 3.33\angle 3.8°$ kΩ;
$\quad\quad\ \ \mathbf{V}_{eq} = 3.56\angle -8°$ V

17–13. A: $6.8\ k\Omega + j10.61\ k\Omega$
$\quad\quad$ B: $8.2\ k\Omega + j5\ k\Omega$
$\quad\quad$ C: $50\ \Omega + j301\ \Omega$

17–15. See Program 17–15.

17–17. $2.23\angle -141.5°$ A

17–19. 95.2-Ω resistor in series with
42.7-Ω inductor.

Chapter 18

18–1. 2.43 V rms

18–3. A: 9.36 V

B: 7.07 V
C: 9.95 V
D: 9.95 V

18–5. (a) 10.6 μF $\quad$ (b) 1.27 μF
$\quad\quad$ (c) 0.6 μF $\quad\quad$ (d) 0.127 μF

18–7. (a) rejected $\quad$ (b) rejected
$\quad\quad$ (c) passed $\quad\quad$ (d) passed
$\quad\quad$ (e) passed

18–9. A: 159 Hz; 7.07 V
$\quad\quad$ B: 636 Hz; 7.07 V
$\quad\quad$ C: 9.54 kHz; 7.07 V
$\quad\quad$ D: 19.9 kHz; 7.07 V

18–11. A: 1.12 kHz
$\quad\quad\ \ $ B: 2.4 kHz

18–13. A: 50.33 Hz
$\quad\quad\ \ $ B: 20.13 MHz

18–15. A: 339 kHz
$\quad\quad\ \ $ B: 10.4 kHz

18–17. A: 4.9 V
$\quad\quad\ \ $ B: 4.04 V

18–19. (a) 7.13 V $\quad$ (b) 5.67 V
$\quad\quad\ \ $ (c) 4 V $\quad\quad$ (d) 0.8 V

18–21. See Program 18–21.

```
10 CLS
20 INPUT "VALUE OF R IN OHMS";R
30 INPUT "VALUE OF L IN HENRIES";L
40 INPUT "VALUE OF WINDING RESISTANCE IN OHMS";RW
50 INPUT "INPUT VOLTAGE IN VOLTS";VIN
60 INPUT "LOWEST FREQUENCY IN HERTZ";FL
70 INPUT "HIGHEST FREQUENCY IN HERTZ";FH
80 INPUT "FREQUENCY INCREMENTS IN HERTZ";FI
90 CLS
100 PRINT "FREQUENCY (HZ)","VOUT (VOLTS)"
110 FOR F=FL TO FH STEP FI
120 XL=2*3.1416*F*L
130 V=(R/(SQR((R+RW)[2+XL[2)))*VIN
140 PRINT F,V
150 NEXT
```

PROGRAM 18–21

18–23. $C = 0.064$ μF; $L = 0.989$ mH;
$f_r = 20$ kHz

Chapter 19

19–1. 1.5 μH

19–3. 4; 0.25

19–5. A: 100 V rms; same polarity
B: 100 V rms; opposite polarity
C: 20 V rms; opposite polarity

19–7. 600 V

19–9. 0.25 (4:1)

19–11. 60 V

19–13. **(a)** 25 mA **(b)** 50 mA
(c) 15 V **(d)** 0.75 W

19–15. 1.83

19–17. 9.73 W

19–19. A: 10 V
B: 240 V

19–21. S_1: 2
S_2: 0.5
S_3: 0.25

19–23. A: 48 V
B: 25 V

19–25. 25 kVA

19–27. **(a)** 0.05 **(b)** 41.67 A
(c) 2.08 A

Chapter 20

20–1. 17.46°

20–3. $0.376\angle 41.2$ A

20–5. $1.32\angle 121°$ A

20–7. $\mathbf{I}_{La} = 8.66\angle -30°$ A;
$\mathbf{I}_{Lb} = 8.66\angle 90°$ A;
$\mathbf{I}_{Lc} = 8.66\angle -150°$ A

20–9. **(a)** $\mathbf{V}_{L(ba)} = 866\angle -30°$ V;
$\mathbf{V}_{L(ca)} = 866\angle -150°$ V;
$\mathbf{V}_{L(cb)} = 866\angle 90°$ V
(b) $\mathbf{I}_{\theta a} = 0.5\angle -32°$ A;
$\mathbf{I}_{\theta b} = 0.5\angle 88°$ A;
$\mathbf{I}_{\theta c} = 0.5\angle -152°$ A

(c) $\mathbf{I}_{La} = 0.5\angle -32°$ A;
$\mathbf{I}_{Lb} = 0.5\angle 88°$ A;
$\mathbf{I}_{Lc} = 0.5\angle -152°$ A
(d) $\mathbf{I}_{Za} = 0.5\angle -32°$ A;
$\mathbf{I}_{Zb} = 0.5\angle 88°$ A;
$\mathbf{I}_{Zc} = 0.5\angle -152°$ A
(e) $\mathbf{V}_{Za} = 500\angle 0°$ V;
$\mathbf{V}_{Zb} = 500\angle 120°$ V;
$\mathbf{V}_{Zc} = 500\angle -120°$ V

20–11. **(a)** $\mathbf{V}_{L(ba)} = 86.6\angle -30°$ V;
$\mathbf{V}_{L(ca)} = 86.6\angle -150°$ V;
$\mathbf{V}_{L(cb)} = 86.6\angle 90°$ V
(b) $\mathbf{I}_{\theta a} = 0.25\angle 110°$ A;
$\mathbf{I}_{\theta b} = 0.25\angle -130°$ A;
$\mathbf{I}_{\theta c} = 0.25\angle -10°$ A
(c) $\mathbf{I}_{La} = 0.25\angle 110°$ A;
$\mathbf{I}_{Lb} = 0.25\angle -130°$ A;
$\mathbf{I}_{Lc} = 0.25\angle -10°$ A
(d) $\mathbf{I}_{Za} = 0.144\angle 140°$ A;
$\mathbf{I}_{Zb} = 0.144\angle 20°$ A;
$\mathbf{I}_{Zc} = 0.144\angle -100°$ A
(e) $\mathbf{V}_{Za} = 86.6\angle -150°$ V;
$\mathbf{V}_{Zb} = 86.6\angle 90°$ V;
$\mathbf{V}_{Zc} = 86.6\angle -30°$ V

20–13. $\mathbf{V}_{L(ba)} = 330\angle -120°$ V;
$\mathbf{V}_{L(ca)} = 330\angle 120°$ V;
$\mathbf{V}_{L(cb)} = 330\angle 0°$ V
$\mathbf{I}_{Za} = 38.06\angle -150°$ A;
$\mathbf{I}_{Zb} = 38.06\angle -30°$ A;
$\mathbf{I}_{Zc} = 38.06\angle 90°$ A

20–15. Figure 20–52: 636 W
Figure 20–53: 149.36 W
Figure 20–54: 12.8 W
Figure 20–55: 2776.8 W
Figure 20–56: 10.9 kW

20–17. 24 W

20–19. Output phase voltages = 1 kV;
output line voltages = 1.73 kV.

Chapter 21

21–1. **(a)** one half-cycle
(b) 5 cycles
(c) 50 cycles

21–3. 0.5 V/div; 50 μs/div

INDEX